PULMONATES

PULMONATES

Volume 2A

Systematics, Evolution and Ecology

Edited by

Vera Fretter

Department of Zoology, University of Reading

and

J. Peake

British Museum (Natural History), London

1978

ACADEMIC PRESS

London New York San Francisco

A Subsidiary of Harcourt Brace Jovanovich, Publishers

ACADEMIC PRESS INC. (LONDON) LTD.
24/28 Oval Road
London NW1

United States Edition Published by
ACADEMIC PRESS INC.
111 Fifth Avenue,
New York, New York 10003

Library of Congress Catalog Card Number: 74–18501
ISBN: 0–12–267502–9

Printed in Great Britain by
Latimer Trend & Company Ltd Plymouth

Contributors to Volume 2A

W. Arthur, *Biology Building, University of Sussex, Falmer, Brighton BN1 9QG, Sussex, UK*

K. Boss, *Department of Molluscs, Museum of Comparative Zoology, Harvard University, Cambridge, Massachusetts 02138, USA*

D. S. Brown, *Department of Zoology, British Museum (Natural History), Cromwell Road, London SW7 5PD, UK*

J. Burch, *Museum of Zoology, University of Michigan, Ann Arbour, Michigan 48104, USA*

B. Clarke, *Department of Genetics, University of Nottingham, Nottingham NG7 2RD, UK*

G. Davis, *Department of Malacology, Academy of Natural Sciences, Philadelphia, Pennsylvania 19103, USA*

D. T. Horsley, *1 The Grove, Cromer Road, Holt, Norfolk, UK*

B. Hubendick, *Naturhistoriska Museet, S–400 30 Göteborg, Sweden*

P. J. Hunter, *Department of Education and Science, Elizabeth House, York Road, London SE1 7PH, UK*

A. Mead, *Department of Biological Sciences, College of Liberal Arts, University of Arizona, Tucson, Arizona 85721, USA*

J. Peake, *Department of Zoology, British Museum (Natural History), Cromwell Road, London SW7 5BD, UK*

D. T. Parkin, *Department of Genetics, University of Nottingham, Nottingham NG7 2RD, UK*

C. Patterson, *Museum of Zoology, University of Michigan, Ann Arbour, Michigan 48104, USA*

W. D. Russell-Hunter, *Marine Biological Laboratory, Woods Hole, Massachusetts 02543, USA*

A. Solem, *Field Museum of Natural History, Roosevelt Road at Lake Shore Drive, Chicago, Illinois 60605, USA*

Preface

Pulmonates are found in a wide spectrum of terrestrial and freshwater habitats, extending in some instances into the marine inter-tidal. They are not, however, the only group of gastropod molluscs to have colonized the non-marine environment, although they are frequently the most numerous. Even so, considerable variations occur in both the density of individuals and diversity of taxa. In temporary habitats or under extreme physical conditions associations of a few, or even single, species may exist in high densities; while in contrast, many closely-related forms may be found co-existing in lakes or on tropical islands, with each species being represented by comparatively few individuals. For the majority of taxa there is no detailed information concerning the niches which they occupy, although it is known that these vary from fossorial to arboreal in terrestrial habitats, or rock-clinging to benthic in fresh water.

Pulmonates have been utilized by man in many ways, as sources of food and medication or as decoration and ornaments. Such examples, however, appear trivial compared to the undoubted importance of the group as vectors of parasites of man and domesticated animals, or as pests of agricultural and horticultural crops. It is, therefore, surprising to discover the scarcity of information in readily accessible sources. There is a dearth of accounts concerning the biology of the group, and an absence of identification guides to the majority of the non-marine faunas of the world; even data concerning the more conspicuous or economically important taxa are frequently scarce. Where guides do exist there may be considerable disagreement concerning the taxonomy or the appropriate nomenclature to be applied. The reasons for this situation are not hard to elucidate; they are associated with the diffuse nature of the literature, problems in collection and preservation, the small size of many species and difficulties in assessing the importance of fine morphological variations, whether it be in the shell or the "soft parts". These difficulties have been compounded by convergence in shell form in widely divergent groups, and frequently the absence of external features to distinguish closely-related taxa.

Many of the chapters included in this volume reflect these problems. Often there is no consensus regarding the interpretation of data, and nowhere is this more apparent than in the classification of the pulmonates.

Editorial changes have therefore not been introduced to produce conformity between chapters where this does not or cannot exist, thus inconcistencies arise through the lack of information and the consequent need to speculate. Nevertheless, in order to assist the elucidation of this taxonomic and nomenclatural confusion, a collation of four of the major classifications in current use is presented in an Appendix on p. 527; this should be used with caution, and reference should be made to the original works cited.

A considerable proportion of the research undertaken on pulmonates has been restricted to a few types that represent a very biased sample of the diversity exhibited by the group. These have tended to be temperate species of economic importance or those amenable to laboratory culture, as will be readily apparent from a perusal of many physiological papers concerning the group; such reports serve to emphasize the limitations of the available information. More data is undoubtedly required on a wider range of taxa, together with observations or experiments linked to the production of more "readily testable" hypotheses. Authors in this volume have been forced to speculate, and consequently the reader is urged to read them in this light, maintaining a healthy degree of enquiry and scepticism.

The chapters in this volume are concerned with the systematics, evolution and ecology of pulmonates; consequently the subject matter is orientated towards "field" problems, while those in Volume 1 were based on laboratory work. The economic importance of the group is also considered, and is supplemented by a massive contribution on *Achatina fulica* in Volume 2b.

Unfortunately there have been delays in publication of this work and not all the chapters have been revised to take account of very recent literature. However, in nearly every instance the effect on the hypotheses and arguments which have been advanced is minimal.

September, *1978* J. PEAKE

Contents of Volume 2A

Contents of Volume 1: Functional Anatomy and Physiology

Contents of Volume 2B: Economic Malacology with Particular Reference to *Achatina fulica*

Chapter 1

Systematics and comparative morphology of the Basommatophora*

B. HUBENDICK

Naturhistoriska Museet, Göteborg, Sweden

I. Introduction

The thousands of gastropod taxa living on land or in water represent only the surviving peaks of evolution; they are the comparatively few successful units of the mass of forms that have evolved. Moreover, they illustrate only a thin cross-section through one of the main branches of molluscan evolution. Yet man, according to his nature, feels obliged to classify this variation hoping that the resultant order will represent a natural phylogenetic arrangement. However, any proposals must remain largely hypothetical for many taxa are extinct and, therefore, phylogenetic reconstructions often represent a range of possibilities rather than conclusive results.

* This manuscript was delivered in 1972. Papers after 1971 have not been included.

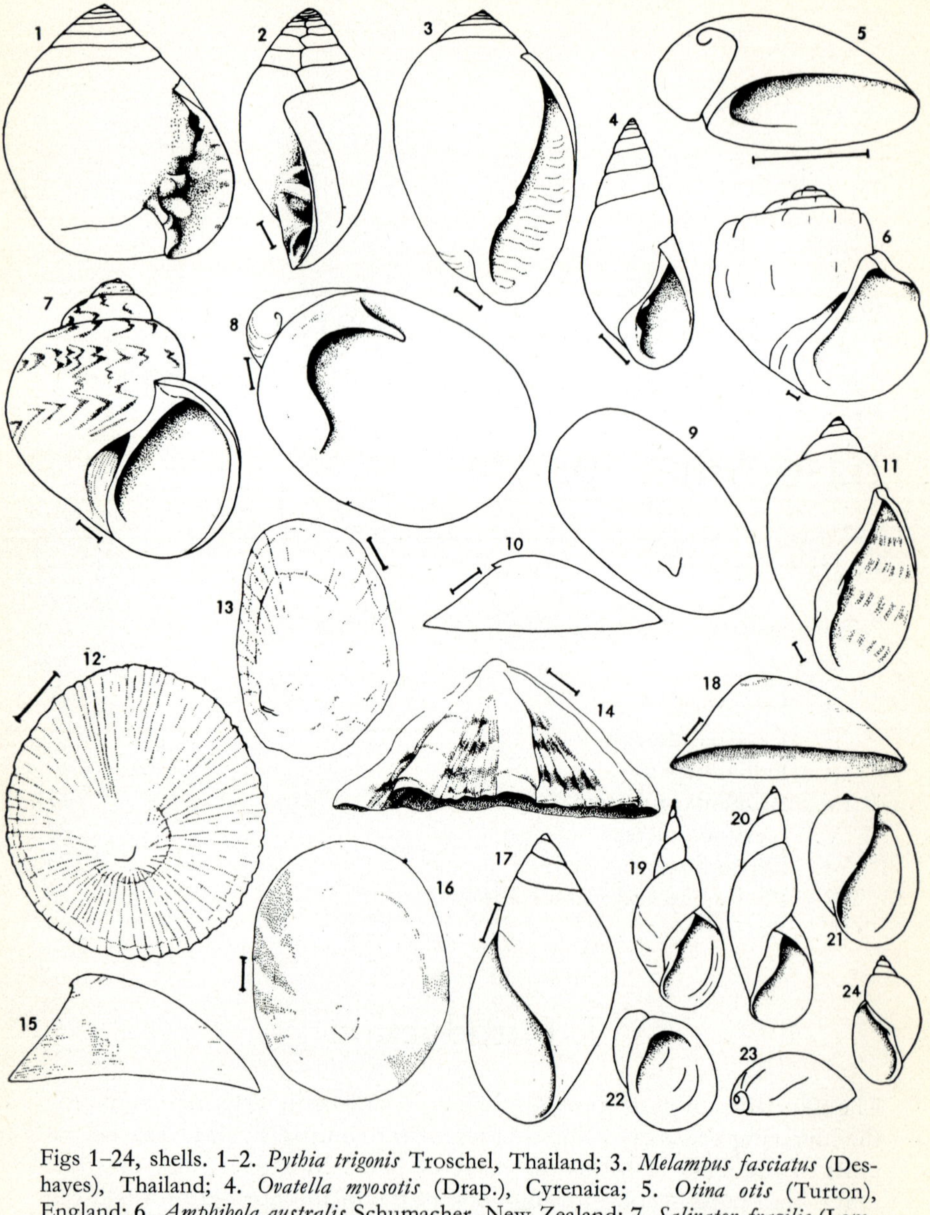

Figs 1–24, shells. 1–2. *Pythia trigonis* Troschel, Thailand; 3. *Melampus fasciatus* (Deshayes), Thailand; 4. *Ovatella myosotis* (Drap.), Cyrenaica; 5. *Otina otis* (Turton), England; 6. *Amphibola australis* Schumacher, New Zealand; 7. *Salinator fragilis* (Lamarck), Australia; 8. *Latia neritoides* Gray, New Zealand; 9–10. *Acroloxus lacustris* (L.), Sweden; 11. *Chilina pulchra* Orbigny, Chile; 12. *Trimusculus peruvianus* (Sowerby), Chile; 13. *Siphonaria lateralis* Gould, South Georgia; 14. *Siphonaria laciniosa* (L.), Celebes, Indonesia; 15–16. *Williamia vernalis* (Dall), California; 17. *Physa marmorata* Guilding, Puerto Rico; 18. *Lanx nuttalli* Haldemann, Idaho; 19. *Lymnaea stagnalis* L., Sweden; 20. *Lymnaea palustris* (Müller), Illinois; 21. *Lymnaea cumingiana* Pfeiffer, Philippines; 22–23. *Lymnaea onychia* Westerlund, Japan; 24. *Lymnaea reticulata* Gould, Hawaiian Islands. The scales correspond to 1 mm.

Within the Pulmonata the subclass Basommatophora forms a comparatively heterogeneous unit comprising a number of rather clearly defined groups. Considerable phylogenetic gaps exist between the groups, and probably it will never be possible to elucidate their relationships; it is even doubtful whether the Basommatophora represent a truly monophyletic group.

The characters distinguishing the Basommatophora from other euthyneuran snails, that is opisthobranchs and stylommatophoran pulmonates, can be considered rather minor; they include, for example, the position of the eyes near the base of a single pair of tentacles. Certainly the lack of major features reflects the doubts that have been expressed regarding the nature of the Basommatophora as a cohesive unit within the Euthyneura. However, it is not the purpose of this chapter to consider the phylogeny of the Basommatophora in detail, but rather the comparative morphology and taxonomy of the group; references to the relevant literature is presented in Table I, pp. 32–43.

II. External morphology

The form of the shell varies widely not only within the Basommatophora as a whole, but also within the individual families, it provides very little assistance in elucidating phylogenetic relationships of the higher taxonomic groups, although occasionally it is of use at lower levels. It usually reflects the external shape of the visceral hump.

A high-spired and dextral shell is the commonest form; it characterizes the families Ellobiidae (Figs 1–4), Amphibolidae (Fig. 6), Chilinidae (Fig. 11) and most Lymnaeidae (Figs 19–20). In contrast, a sinistral shell is present in the Physidae (Fig. 24) and some Planorbidae (Figs 25–26, 31–33). In most Planorbidae the spire is reduced, with all whorls being arranged in a single plane producing a discoidal shell type (Figs 39–40, 45–61). In some species exhibiting this form the inner whorls are secondarily pushed through the peripheral whorls, forming an inverted spire on the original umbilical side of the shell; such shells are termed pseudodextral or heterostrophic. They have a dextral appearance although the animal inside has a sinistral anatomy (cf. Fig. 3 in PAHO/WHO guide, 1968).

Reduction in the elevation of the spire and also the number of whorls has taken place independently on several occasions within the Basommatophora. Partial reduction of the whorls has occurred in the Otinidae (Fig. 5) and Latiidae (Fig. 8); the former having a vestigial spire on the left side of the shell, the latter being somewhat neritoid in shape. Evidence of a vestigial spire is also present in some species of patelliform Planorbidae (Figs 27–30, 41–42) and Ancylidae (Figs 62–63); moreover, the shell of the

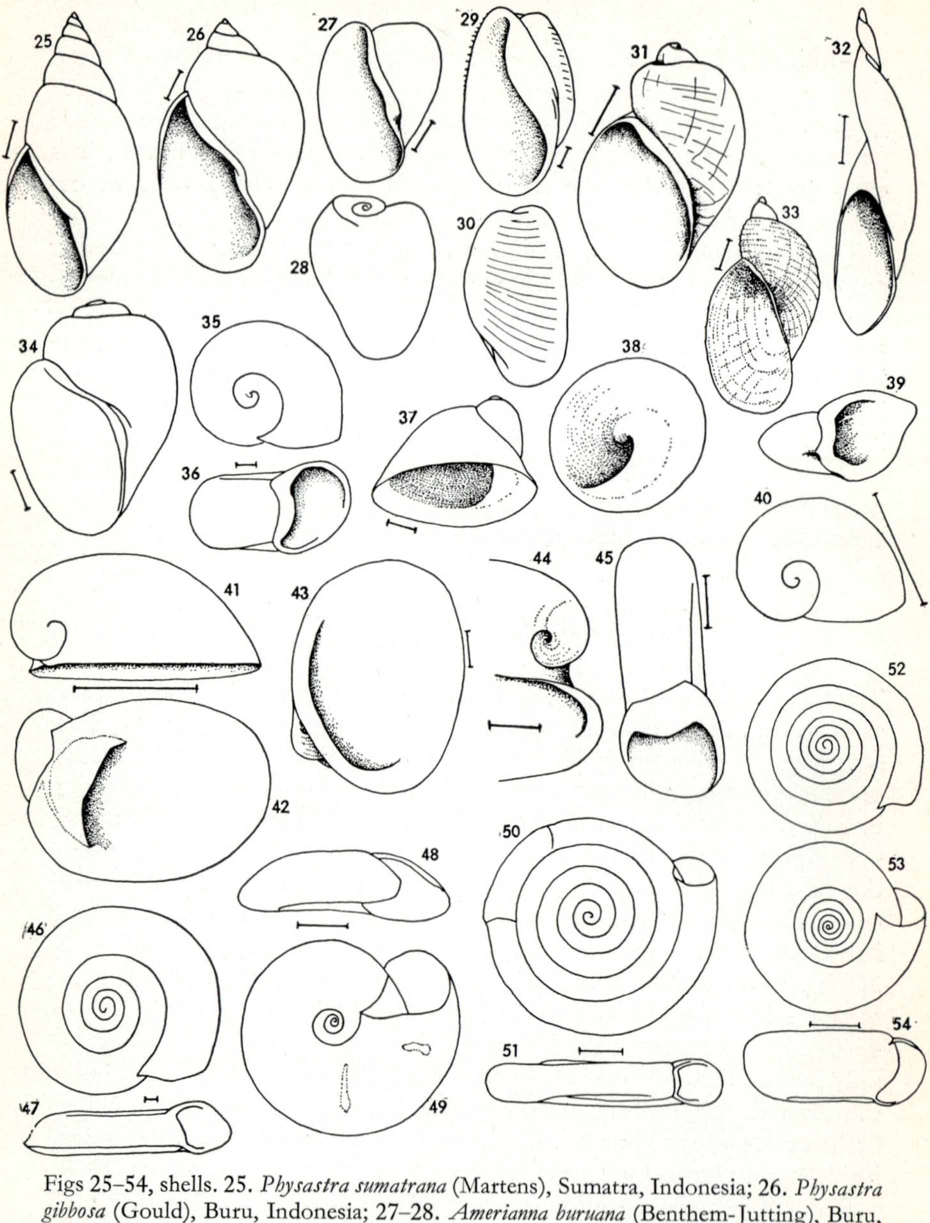

Figs 25–54, shells. 25. *Physastra sumatrana* (Martens), Sumatra, Indonesia; 26. *Physastra gibbosa* (Gould), Buru, Indonesia; 27–28. *Amerianna buruana* (Benthem-Jutting), Buru, Indonesia; 29–30. *Amerianna leopoldi* (Dupuis), New Guinea. 31. *Plesiophysa striata* (Orbigny), Puerto Rico; 32. *Camptoceras terebra* Benson, India; 33. *Camptoceras prashadi* (Clench), Japan; 34. *Bulinus truncatus* (Audouin), Egypt; 35–36. *Indoplanorbis exustus* (Deshayes), Thailand; 37–38. *Patelloplanorbis tigiensis* Hubendick, New Guinea; 39–40. *Neoplanorbis carinatus* Walker, Alabama; 41–42. *Amphigyra alabamensis* Pilsbry, Alabama; 43–44. *Ancylastrum cumingianus* (Bourguignat), Tasmania; 45. *Planorbis corneus* (Linné), Sweden; 46–47. *Planorbis planorbis* (Linné), Sweden; 48–49. *Segmentina nitida* (Müller), Sweden; 50–51. *Anisus leucostomus* (Müller), Sweden; 52–54. *Anisus contortus* (Linné), Sweden. The scales correspond to 1 mm.

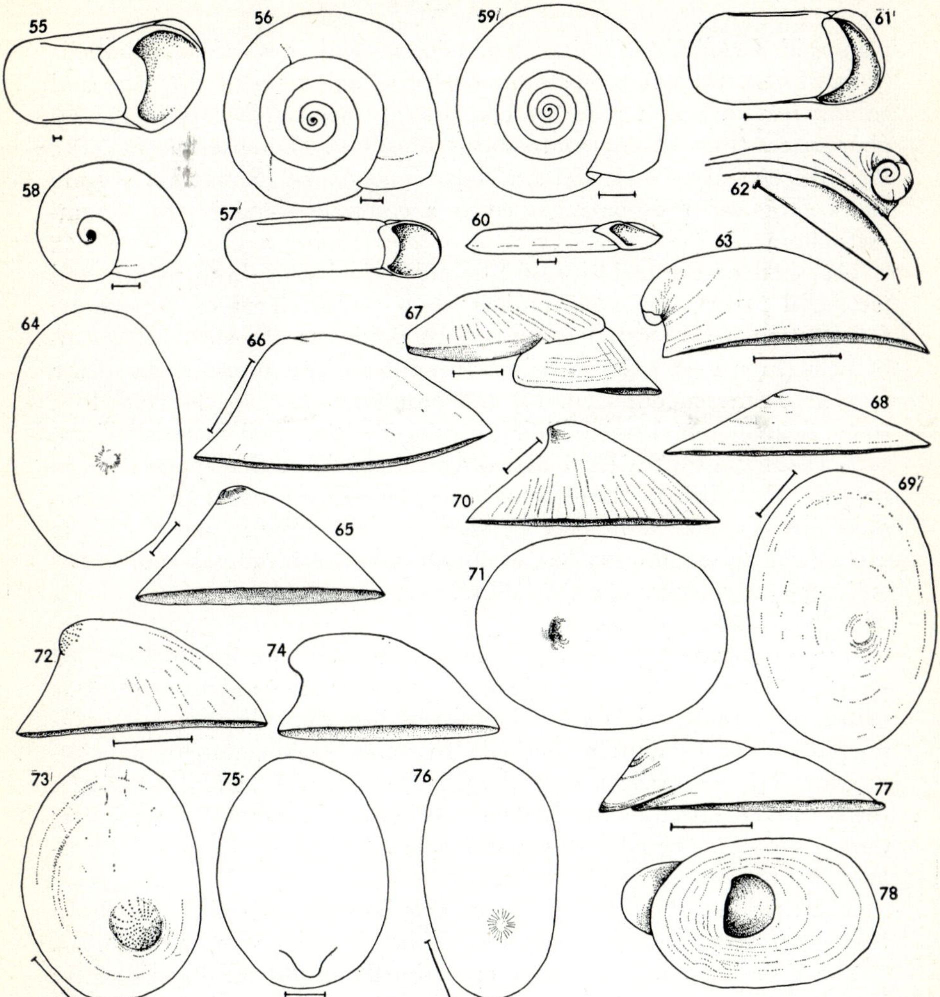

Figs 55–78, shells. 55. *Helisoma foveale* (Menke), Puerto Rico; 56–57. *Biomphalaria havanensis* (Pfeiffer), Haiti; 58. *Biomphalaria prona* (Martens), Venezuela; 59. *Drepanotrema lucida* (Pfeiffer), Puerto Rico; 60. *Drepanotrema kermatoides* (Orbigny), Guadeloupe; 61. *Drepanotrema anatina* (Orbigny), Puerto Rico; 62–63. *Brondelia drouetiana* Bourguignat, Algeria; 64–65. *Ferrissia rivularis* (Say), W. Virginia; 66–67. *Gundlachia radiata* (Guilding), Puerto Rico; 68–69. *Laevapex fuscus* (Adams), Georgia; 70–71. *Rhodacmea gwatkiniana* Walker, Alabama; 72–73. *Burnupia stuhlmanni* (Martens) Uganda; 74–75. *Ancylus fluviatilis* Müller, Pyrenees; 76. *Ferrissia* (*Pettancylus*) *neozelanica* (Suter), New Zealand; 77–78. *Ferrissia* (*Pettancylus*) *petterdi* (Johnston), Tasmania. The scales correspond to 1 mm.

planorbid *Patelloplanorbis* is calyptraeiform with a vestigial columella (Figs 37–38). Complete or almost complete reduction of the spire and coiling resulting in a more or less patelliform shell occurs in the Siphonariacea (Figs 12–16), Acroloxidae (Figs 9–10), Lymnaeidae (Fig. 18), Planorbidae (Figs 43–44) and Ancylidae (Figs 64–78). Reduction of spire and coiling can thus be regarded as a common trend in the Basommatophora.

The shell of the majority of Ellobiidae, including both marine and terrestrial forms, are comparatively thick-walled. That of the marine forms often has an external appearance similar to the shells of prosobranchs (Figs 1–4), and this is also true for the patelliform shells of the families Trimusculidae (Fig. 12) and Siphonariidae (Fig. 14) and, to a lesser degree, the Amphibolidae (Figs 6–7). The shell material of the terrestrial forms of the Ellobiidae often has an alabastrine appearance. In contrast, the shells of freshwater forms belonging to the families Chilinidae, Acroloxidae, Physidae, Lymnaeidae, Planorbiidae and Ancylidae are almost always thin-walled, although columellar calluses and similar structures may occur. The aberrant planorbid snail *Miratesta* with a thick shell represents an exception.

A pronounced shell sculpture is only associated with the thick-shelled forms, and then typically marine taxa. A much finer micro-sculpture is occasionally present in thin-shelled forms; for example, *Ferrissia* has a fine radial apical sculpture (Fig. 64). In some Ellobiidae there are well-developed palatal, parietal and columellar teeth (Fig. 1), and this family is further characterized by the gradual resorbtion of internal shell structures during growth. The ellobiid genus *Pythia* is unique by having a dorsolaterally compressed shell (Fig. 2).

A distinctive colour pattern occurs only in some Ellobiidae, nearly all Siphonariidae and, rather unexpectedly, the freshwater family Chilinidae.

The Amphibolidae are quite exceptional among all Pulmonata in having an operculum (Fig. 82). Nevertheless, an embryonic rudimentary operculum has been observed in several other groups of Basommatophora. *Otina* is also unique as the body is too large to be completely withdrawn into the shell.

In the Basommatophora, as in other Pulmonata, the edge of the mantle is fused with the body wall leaving only a comparatively small mantle opening or pneumostome, located on the right side in dextral forms (Figs 81–82) and vice versa in sinistral (Figs 86, 88–89). The lateral location of the mantle opening has arisen through true detorsion, and partially by fusion of the original opening in the anterior region of the mantle. The anal opening, renal pore and reproductive pores are closely associated with the mantle opening.

There are few external appendages, besides the eaves-like mantle-

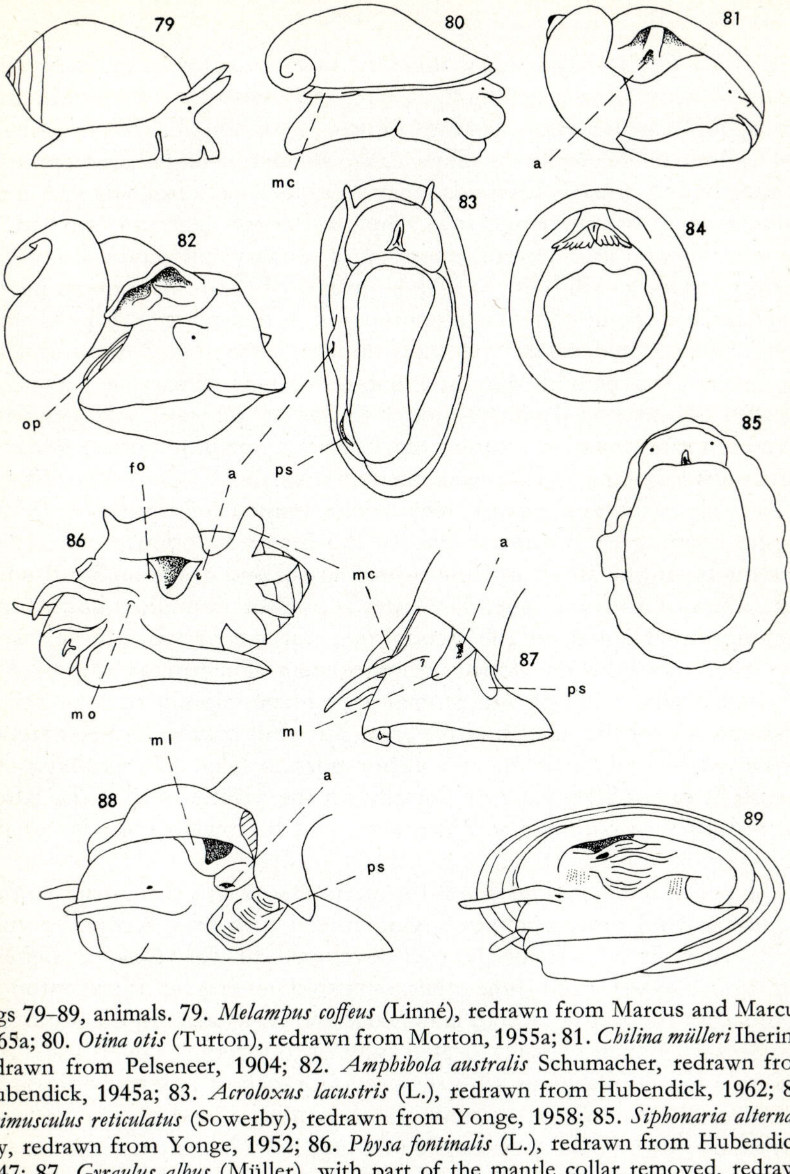

Figs 79–89, animals. 79. *Melampus coffeus* (Linné), redrawn from Marcus and Marcus, 1965a; 80. *Otina otis* (Turton), redrawn from Morton, 1955a; 81. *Chilina mülleri* Ihering, redrawn from Pelseneer, 1904; 82. *Amphibola australis* Schumacher, redrawn from Hubendick, 1945a; 83. *Acroloxus lacustris* (L.), redrawn from Hubendick, 1962; 84. *Trimusculus reticulatus* (Sowerby), redrawn from Yonge, 1958; 85. *Siphonaria alternata* Say, redrawn from Yonge, 1952; 86. *Physa fontinalis* (L.), redrawn from Hubendick, 1947; 87. *Gyraulus albus* (Müller), with part of the mantle collar removed, redrawn from Hubendick, 1947; 88. *Bulinus* sp with part of the mantle collar lifted, redrawn from Hubendick, 1948b; 89. *Gundlachia radiata* (Guilding), redrawn from Hubendick, 1964. a = anus; fo = female genital pore; mc = mantle collar; ml = mantle lobe; mo = male genital pore; op = operculum; ps = pseudobranch.

collar (Figs 80, 86–88) which is involved in shell secretion and surrounds the body below the visceral hump. One pair of tentacles is present in most forms, but these are reduced in the Otinidae, Amphibolidae, Siphonariacea and Chilinidae. Typically the tentacles are slender, cylindrical and pointed terminally, but in some Ellobiidae they are thickened terminally and in the Lymnaeidae they are triangular in shape and dorso-ventrally flattened. In the latter there is also a lateral extension of sensory epithelium posteriorly from the tentacles (Figs 86, 88). Most groups of Basommatophora have a characteristic mode of tentacle contraction and extension, for they can never be retracted as in Stylommatophora, except in some forms of Ellobiidae (Marcus and Marcus, 1965b). The pair of papillae present in certain Ellobiidae on the dorsal side of the snout, anterior to the tentacles, have been interpreted as a rudimentary second pair of tentacles. The eyes are always located at the bases of the tentacles.

Two types of appendages may occur immediately exterior to the mantle opening or pneumostome. In the lower Basommatophora they take the form of a single pneumostomal lappet and in the higher Basommatophora of a mantle lobe. The latter is possibly homologous with the pneumostomal lappet and the pseudobranch; it may vary from being well-developed to weakly developed or non-existent. The former (Figs 87–88), also called anal lobe, pneumostomal appendage, siphon or lobe palléal inférieur, is probably the primitive type. The anus may be located anterior to the pseudobranch or on its anterior portion (Fig. 89), although frequently it opens terminally or dorsally on the mantle lobe; if the latter, the rectum may continue as a dorsally open furrow to the edge of the lobe (Fig. 82).

The pseudobranch (Figs 87–89) is an accessory gill that varies from a single flap to a more complex organ, folded to a first, second or even third order. It is particularly well-developed in *Indoplanorbis*, *Bulinus*, *Miratesta*, *Physastra* and some other planorbid genera, but is present in all members of the Acroloxidae, Planorbidae and Ancylidae. In the planorbid *Plesiophysa* a similar organ is located anterior to the anus, nevertheless it is probably not homologous with the pseudobranch of other planorbids.

The mouth is located ventrally at the anterior end and is typically surrounded by a horse-shoe shaped swelling, which is often regarded as a pair of oral lappets. In *Trimusculus* these lappets are well separated by a median insertion (Fig. 84). Behind the mouth the head is ventrally demarcated from the sole by a transversal furrow. In Otinidae and certain Elobiidae the sole is divided ventrally by a transversal furrow into anterior and posterior portion (Figs 79–80).

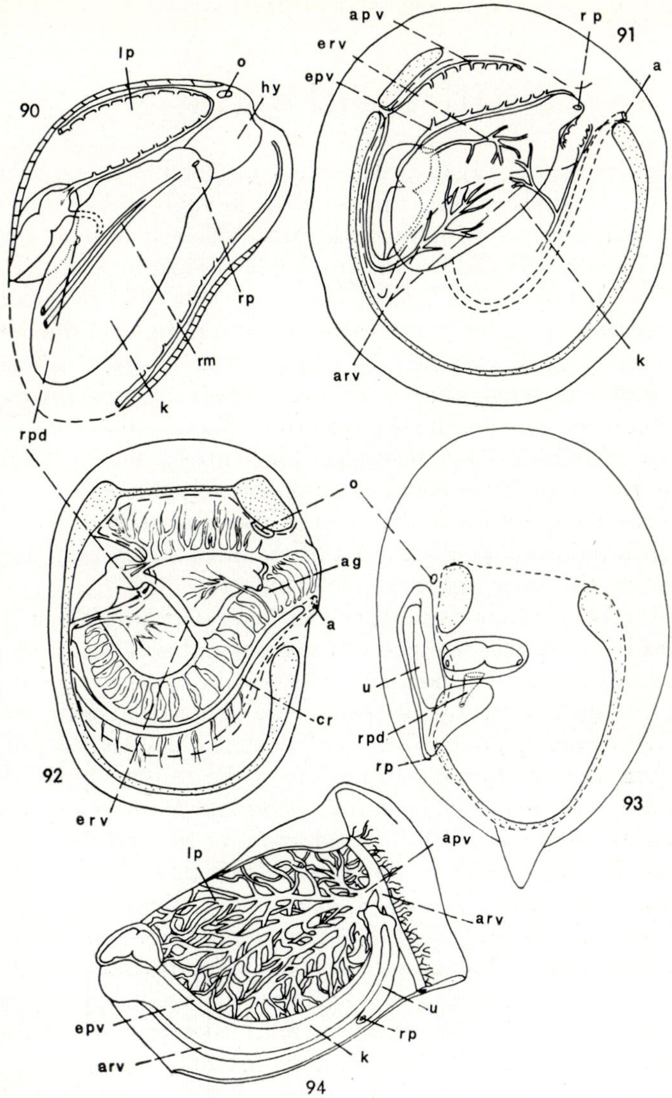

Figs 90–94, pallial organs. 90. Diagram of pallial organs, dorsal view, in *Amphibola australis* Schumacher, redrawn from Hubendick, 1945a; 91. Diagram of pallial organs, dorsal view, in *Trimusculus reticulatus* (Sowerby), redrawn from Hubendick, 1945a; 92. Diagram of pallial organs, dorsal view, in *Siphonaria sp.*, redrawn from Hubendick, 1945a; 93. Diagram of pallial organs, dorsal view, in *Ancylus tapirulus* Polinski, redrawn from Hubendick, 1964; 94. Diagram of ceiling of pallial cavity, ventral view, in *Bulinus truncatus* (Audouin), redrawn from Demian 1960. a = anus; ag = accessory gland; apv = afferent pulmonary vein; arv = afferant renal vein; cr = ciliated ridge; epv = efferent pulmonary vein; erv = efferent renal vein; hy = hypobranchial gland; k = kidney; lp = lung plexus; o = osphradium; pc = pericardium; rm = renal muscle; rp = renal pore; rpd = reno-pericardial duct; u = urethra. Dotted areas indicate shell adductor muscles; in 93 the broken line indicates the inner limit of the mantle cavity.

III. Pallial organs

Typically the mantle cavity occupies a considerable proportion of the volume of the body whorl; for example, in high spired forms it can extend to half and in certain discoidal planorbids may be even larger. Yet it is considerably reduced or almost completely lost in many patelliform snails of the family Ancylidae and *Axroloxus* (Fig. 93); although, in the patelliform Siphonariidae, it is enlarged, covering most of the dorsal side of the body (Fig. 92). As may be anticipated, the shape of the mantle cavity varies accordingly; in discoidal planorbids it forms a narrow tube which is convex–concave in cross-section, in high spired forms a somewhat triangular cavity and in certain Ancylidae a shallow impression. Typically the cavity has a concave, comparatively thin exterior wall, the mantle, and a convex floor; the latter is thin and semi-transparent and corresponds morphologically to part of the dorsal body wall. The pallial organs lie on the inner surface of the mantle.

From the pericardium a short, ciliated reno-pericardial duct (Figs 90, 92–93) leads to the proximal end of the kidney; this is homologous with the left nephridium. The kidney opens to the exterior, usually via a urethra, through a renal pore situated on the right side in dextral forms. In high-spired forms the kidney is located posterior and to the right of the pericardium, but anterior and to the left of the rectum. In the discoidal Planorbidae, with a sinistral anatomy, the kidney forms a loop along the left side of the body whorl. It is broad, more or less saccular, and dorsoventrally flattened (Fig. 91) in those patelliform Basommatophora with a wide mantle cavity. In *Siphonaria* it extends over the base of the mantle cavity in the neighbourhood of the pericardium (Fig. 92). The kidney is often divided into a proximal saccular portion and a more distal tubular portion. In most Planorbidae, discoidal as well as high-spired forms, the terminal part of the tubular portion is reflected forming a urethra (Fig. 94). In the patelliform lymnaeid *Lanx*, the urethra extends in a posterior direction. While in *Physa* the whole kidney forms a serpentine duct on the internal surface of the mantle. In both Acroloxidae and Ancylidae the urethra is elongated forming a serpentine duct running respectively in the right or left mantle collar and emptying through a tiny renal pore on the inside of the mantle collar in the posterior third of the body (Fig. 93). In several groups there is a slender muscle running along the ventral surface of the saccular part of the kidney (Fig. 90).

The heart lies within the pericardium and is divided into two chambers, the atrium and the ventricle, separated by an atrio-ventricular passage containing a valve, which in *Acroloxus* is observed to consist of two laminated muscles. The arterial system in general is incomplete and opens into a haemocoel; it has been described in *Lymnaea* (Walter, 1969), *Acroloxus* (Hubendick, 1962), *Bulinus* (Demian, 1960), *Burnupia* (Oberholzer and van Eeden, 1969) and *Ferrissia* (Boer and Lever, 1959). From the haemocoel, the blood collects into sinuses which feed into two major pallial veins, the afferent pulmonary vein and the afferent renal vein (Figs 91, 94). The latter is always well-developed, while the structure of the former is correlated with the structure of the respiratory organs. In *Lymnaea* the afferent pulmonary vein forms a diffuse plexus, but in *Bulinus* it is profusely branched (Fig. 94) and is associated with a mantle cavity adapted for both aquatic and aerial respiration. Usually the kidney and the lung are drained by the efferent pulmonary vein (Figs 91, 94). In Siphonariidae the right section of the well-developed accessory gill (Fig. 92), the mantle cavity, and the extension of the kidney (Fig. 92) are drained by one major vein; the left and middle sections of the gill and the principal portions of the kidney are drained by another. Neither of these veins can with certainty be homologized with vessels in other forms.

In some groups, for example the Siphonariidae (Fig. 92) and Planorbidae, there is a ciliated ridge, the rectal ridge running posteriorly within the mantle cavity from the mantle lobe along the rectum or, more precisely, on the dorsal wall of the rectal blood sinus. It is continuous with another, the dorsal ridge, running along the dorsal surface of the cavity to the mantle collar, yet the so called renal ridge, which in some forms runs along the free ventral surface of the kidney, is not ciliated and is thus different from the rectal and dorsal ridges.

A hypobranchial gland has been observed by Morton (1955b) in the terrestrial Ellobiid *Carychium*, and this may be homologized with "pallial gland" in *Pythia* (Morton, 1955c); the gland is expanded at its anterior end and opens to the exterior. The hypobranchial gland appears to be lacking in all aquatic Basommatophora (and in all Stylommatophora) except *Amphibola*. The "pneumostomal glands" described by Marcus and Marcus (1965a) in *Melampus* cannot be homologized with any other structures.

A sensory organ close to the anterior edge of the mantle opening is probably present in all aquatic Basommatophora, it has frequently been interpreted as an osphradium (Figs 90, 92–93), although was also called Lacazé or Lacazé-Duthier's organ by earlier workers.

IV. Alimentary system

The mouth leads into the pharynx including a buccal complex containing the jaw, the odontophore with the radula, a complicated muscular apparatus and the salivary glands. The pharynx opens into the oesophagus running to the stomach; the alimentary canal continues, receiving a duct from the digestive gland, via the intestine and rectum to the anus. The general structure of these elements within the Basommatophora is described below.

A jaw is present at the entrance of the pharynx in a dorsal or a dorsal and a lateral position, except in *Trimusculus* and *Amphibola* where it is lacking. It consists of numerous tiny bars, secreted by an epithelial cell or cell complex. The bars can be of different length and thickness and either

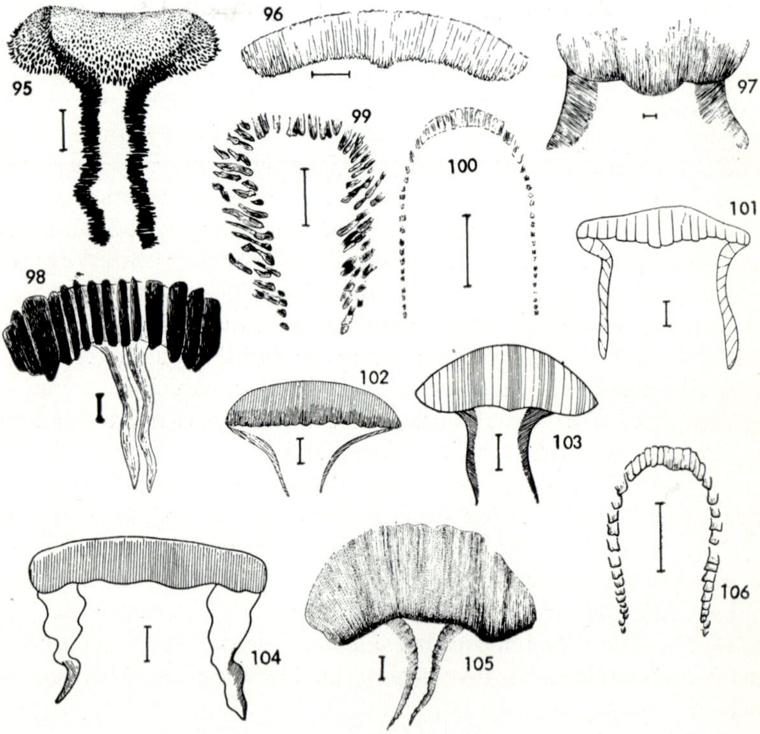

Figs 95–106, jaws. Frontal view of the jaw of various Basommatophorans. 95. *Melampus fasciatus* (Deshayes), from Knipper and Meyer, 1956; 96. *Siphonaria cochleariformis* Reeve, from Hubendick, 1945a; 97. *Lymnaea auricularia* L, redrawn from Hubendick, 1945c; 98. *Ancylatrum irvinae* Petterd, from Hubendick, 1964; 99. *Ancylus fluviatilis* Müller, redrawn from Hubendick, 1964; 100. *Gundlachia radiata* (Guilding), redrawn from Hubendick, 1964; 101. *Planorbis planorbis* (L.), from Hubendick, 1955; 102. *Bulinus africanus* (Krauss), from Hubendick, 1948a; 103. *Physastra sarasinorum* (Bollinger), from Hubendick, 1948a; 104. *Miratesta celebensis* Sarasin and Sarasin, from Hubendick, 1948a; 105. *Ancylastrum cumingianus* (Bourguignat), redrawn from Hubendick, 1964; 106. *Ferrissia rivularis* (Say), redrawn from Hubendick, 1964. The scales indicate 0·1 mm.

separate, organized into groups, or more or less fused (Figs 95–106) and even though the structure is variable between taxonomic groups (Figs 98 and 105) it appears to have little systematic value.

The buccal mass or complex (Figs 141–142) incorporates the radula (Fig. 146), functionally the principal component of this region, and structures associated with supporting and manoeuvring the organ. Direct support to the radula is provided by the complex odontophore (Fig. 143) with the so-called odontophoral cartilage (Fig. 145). These are surrounded by the complex muscular system of the bulb; this has been described in the Lymnaeacea and *Acroloxus*, thanks to the excellent works by Carriker (1947) and Demian (1962). Demian recognized a total of 43 or 44 muscles in species belonging to the families Lymnaeidae, Physidae, Planorbinae, Buliminae and Ancylidae. Nevertheless, in *Acroloxus*, there are some distinct and striking differences, even though there is a similar basic structure.

The odontophore, which protrudes into the lumen of the buccal bulb (Fig. 143), is composed largely of the so-called odontophoral cartilage. This consists not of true cartilage, but a supporting cushion of turgid connective cells and transverse muscles. Its dorso–posterial surface is concave, and in this concavity the radular sac, a branch or invagination from the pharyngeal wall, is located. In some forms, for example *Ancylus*, the radula sac extends beyond the cartilage and the buccal bulb proper. The radula is secreted from the epithelium of the inner part of the radula sac. The dorsal wall of the sac is pressed down towards or invaginated into its ventral wall, thus forming a two-layered envelope. The sides of the envelope, the edges where the dorsal and ventral epithelia of the sac merge, are rolled up towards each other forming an incomplete cylinder. This cylinder is filled with the collostyle, a translucent amorphous supporting substance. The radula is produced in the narrow lumen between the two walls of the radular sac and is thus horseshoe-shaped in cross-section. It is pushed out continuously from the sac and where the latter opens out into the pharyngeal lumen the sides of the radula (upturned in the sac) are spread out and then turned downwards over the anterior end of the odontophore (Fig. 146).

The pulmonate radula in general is of a primitive type with numerous small and comparatively simple teeth of fairly similar shape in each cross row. As each tooth is produced by a small group of cells it is reasonable to assume that a more specialized stage is represented by fewer, larger teeth with a clearly differentiated shape. In some families, for example, Ellobiidae and Planorbidae, and even within one genus, *Acroloxus*, there are representatives of both primitive and specialized radula types. Within the Basommatophora, therefore, the taxonomic value of the radula is restricted frequently to taxa of lower rank and even then the considerable

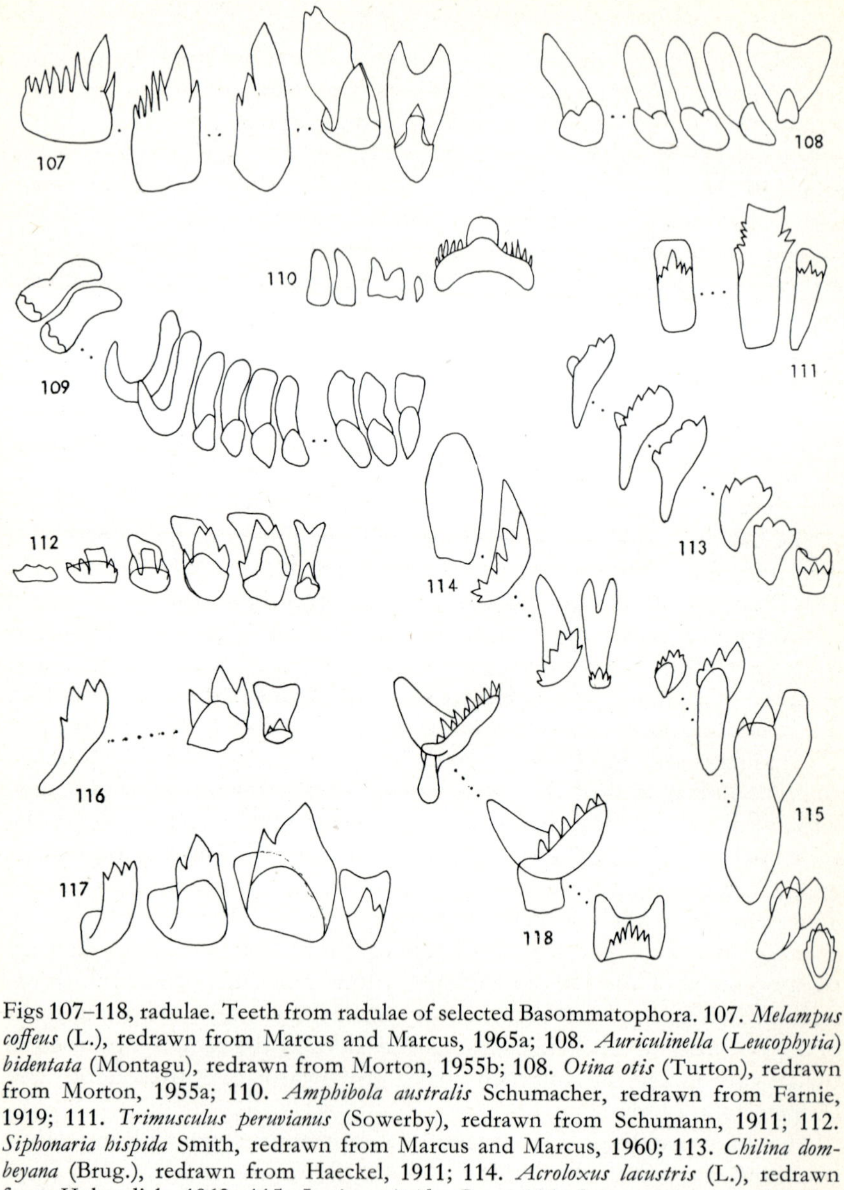

Figs 107–118, radulae. Teeth from radulae of selected Basommatophora. 107. *Melampus coffeus* (L.), redrawn from Marcus and Marcus, 1965a; 108. *Auriculinella* (*Leucophytia*) *bidentata* (Montagu), redrawn from Morton, 1955b; 108. *Otina otis* (Turton), redrawn from Morton, 1955a; 110. *Amphibola australis* Schumacher, redrawn from Farnie, 1919; 111. *Trimusculus peruvianus* (Sowerby), redrawn from Schumann, 1911; 112. *Siphonaria hispida* Smith, redrawn from Marcus and Marcus, 1960; 113. *Chilina dombeyana* (Brug.), redrawn from Haeckel, 1911; 114. *Acroloxus lacustris* (L.), redrawn from Hubendick, 1962; 115. *Latia neritoides* Gray; 116. *Lymnaea tomentosa* Pfeiffer, redrawn from Hubendick, 1951; 117. *Lanx alta* (Tryon), redrawn from Baker, 1925; 118. *Physa parkeri* Currier, redrawn from Baker, 1928.

Figs 119–140, radulae. Central, first lateral and one or two marginal teeth from the radula of selected Planorbidae and Ancylidae. 119. *Bulinus truncatus* (Audouin), redrawn from Demian, 1960; 120. *Indoplanorbis exustus* (L.), redrawn from Baker, 1945; 121. *Amerianna pesigani* Hubendick, redrawn from Hubendick, 1954; 122. *Plesiophysa striata* (Orbigny), redrawn from Hubendick, 1949; 123. *Camptoceras jiraponi* Hubendick, redrawn from Hubendick, 1967a; 124. *Biomphalaria glabrata* (Say), redrawn from Hubendick, 1961; 125. *Helisoma* sp, redrawn from Hubendick, 1961; 126. *Planorbis planorbis* (L.), redrawn from Baker, 1945; 127. *Segmentina nitida* (Müller), redrawn from Baker 1945; 128. *Anisus spirorbis* (L.), redrawn from Baker, 1945; 129. *Drepanotrema anatina* (Orbigny), redrawn from Baker, 1945; 130. *Patelloplanorbis tigiensis* Hubendick, redrawn from Hubendick, 1957; 131. *Ancylastrum cumingi* (Bourguignat), redrawn from Hubendick, 1964; 132. *Neoplanorbis carinatus* Walker, redrawn from Hubendick, 1964; 133. *Amphigyra alabamensis* Pilsbry, redrawn from Hubendick, 1964; 134. *Ancylus fluviatilis* Müller, redrawn from Hubendick, 1964; 135. *Ferrissia parallela* Haldeman, redrawn from Hubendick, 1964; 136. *Ferrissia* (*Pettancylus*) *petterdi* (Johnston), redrawn from Hubendick, 1964; 137. *Gundlachia radiata* (Guilding), redrawn from Hubendick, 1964; 138. *Burnupia caffra* (Krauss), redrawn from Hubendick, 1964; 139. *Rhodacmea cahawbensis* Walker, redrawn from Basch, 1962a; 140. *Laveapex fuscus* (Adams), redrawn from Hubendick, 1964.

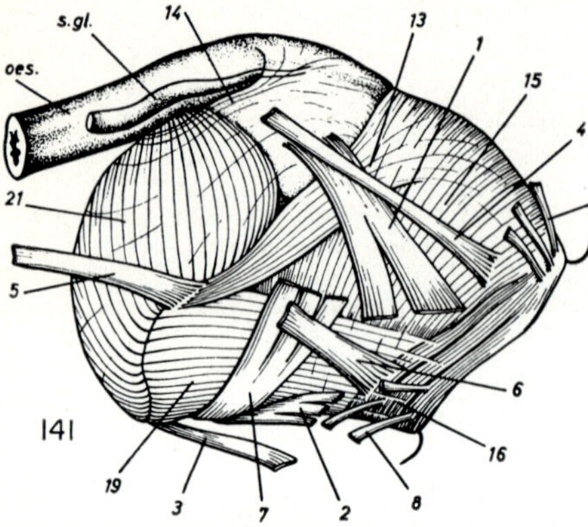

141

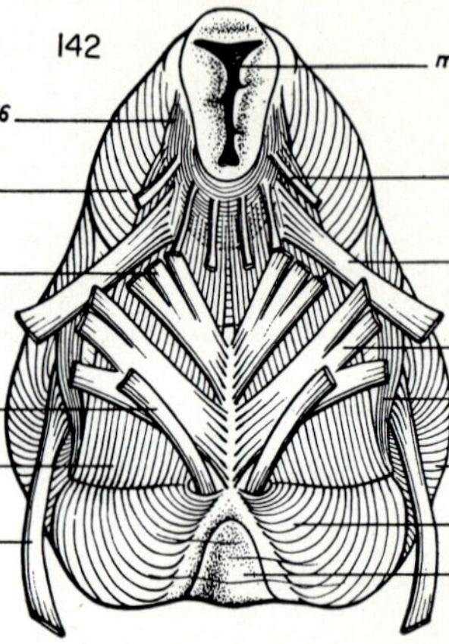

142

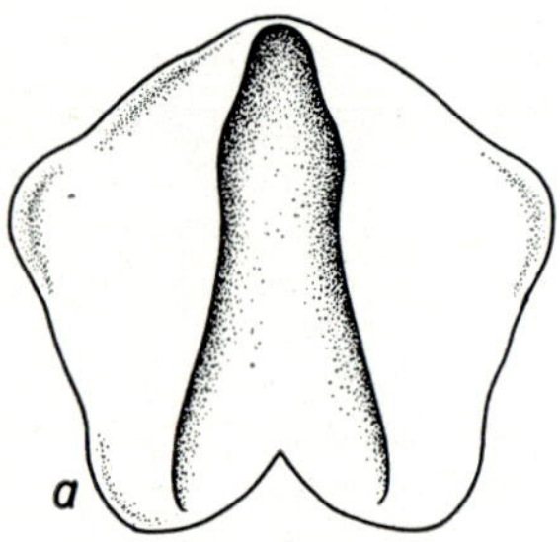

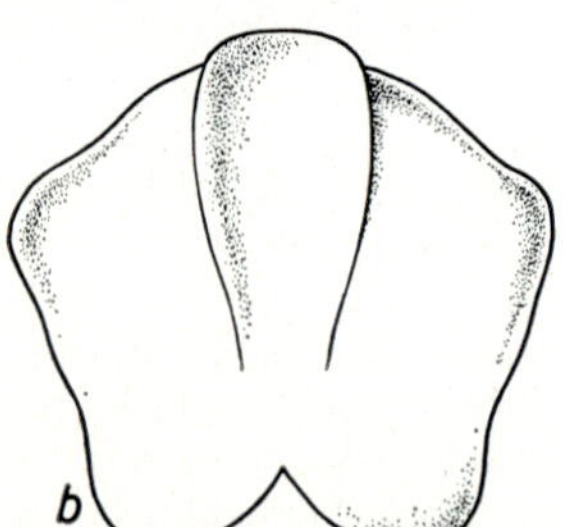

145

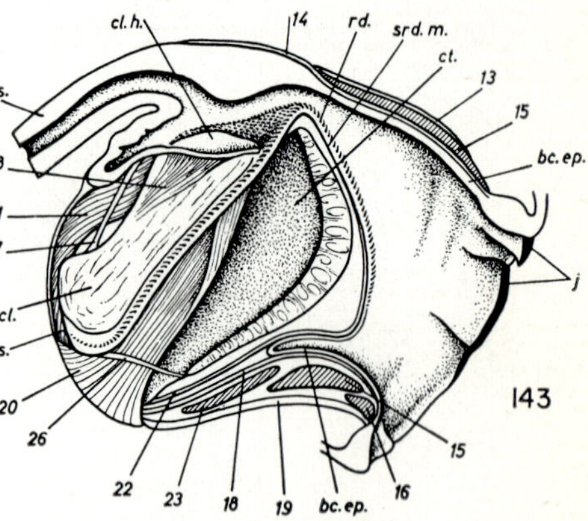

143

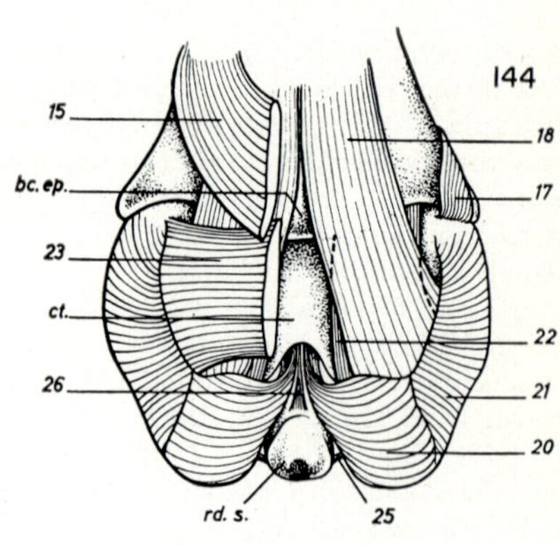

144

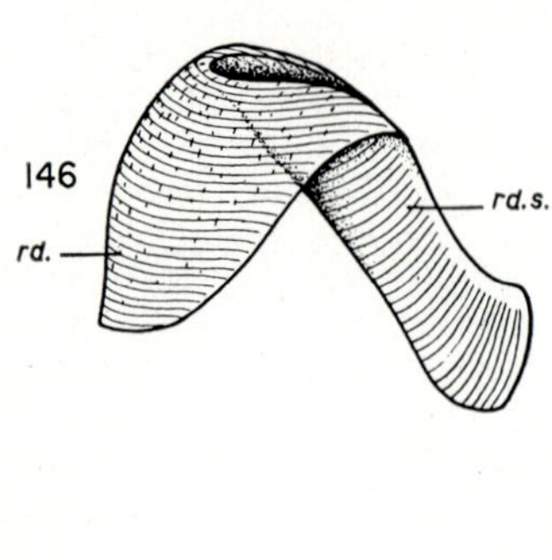

146

range of intra-specific variation has to be considered. Nevertheless, the radula has often been considered important in the classification of the Basommatophora and a selection are illustrated in Figs 107–140.

The salivary glands enter the buccal mass on each side of the oesophagus; they are somewhat elongated and more or less lobated or even tubulated, often flattened and anastomosed posteriorly. In some groups they pass through the cerebral ring.

The oesophagus leaves the pharynx dorsally and runs posterially to the stomach. The latter is a complex structure which shows considerable variation in the Basommatophora, and even within the family Ellobiidae. The portion of the gut specialized as a stomach is divided into an oesophageal or pyloric part connected with the oesophagus and an intestinal or cardiac part connected with the intestine. In the transitional area there is a caecum and, normally, the aperture of the posterior digestive gland duct. When present the anterior digestive gland duct is associated with the oesophageal section. A certain portion of the stomach may be modified to form a strongly muscular gizzard, with an internal cuticular wall; while the region between the oesophagus and the gizzard may form an oesophageal atrium or crop. A number of internal ridges, fold formations and ciliated areas assist in passing the food through the organ complex associated with the stomach.

Among the Basommatophora the most primitive type of stomach is found in *Acroloxus* (Fig. 156); in this group it is non-muscular and there is a well-developed caecum, at the base of which open two digestive gland ducts. The caecum is histologically similar to a style sac and contains a structure resembling a crystalline style.

Figs 141–146, the buccal complex of Lymnaea stagnalis (L.). 141. View from the right; 142. Ventral view; 143. Left half of buccal mass cut out medially, view from the right; 144. Ventral view, after removal of several muscles; 145. Posterior (a) and anterior (b) view of the odontophoral cartilage; 146. The radula seen from the left. All figures from Demian, 1962. bc.ep. = buccal cavity epithelium; cl.h. = collostylar hood; ct. = odontophoral cartilage; j. = jaw; m. = mouth opening; oes. = oesophagus; rd. = radula; rd.cl. = radular collostyle; rd.s. = radular sac; s.gl. = salivary gland; srd.m. = subradular membrane: 1 = dorsolateral protractor; 2 = preventral protractor; 3 = postventral protractor; 4 = dorsolateral retractor; 5 = buccal retractor; 6 = preventral levator; 7 = postventral levator; 8 = suboral dilators; 9 = dorsomandibular dilators; 9a = mandibular tensor; 13 = anterior jugalis; 14 = posterior jugalis; 15 = buccal sphincter; 16 = mandibular approximator; 17 = dorsal odontophoral flexor; 18 = ventral odontophoral protractor; 19 = infraventral odontophoral protractor; 20 = supramedian radular tensor; 21 = supralateral radular tensor; 22 = inframedian radular tensor; 23 = infralateral cartilage tensor; 24a = intercartilage contractor; 25 = lateral suspensor of the radular sac; 26 = inferior suspensor of the radular sac; 27 = superior suspensor of the radular sac; 28 = tensor of the hood. The nomenclature employed follows Carriker, 1947 and Demian, 1962; Fretter and Graham, 1962 have introduced another nomenclature which may be generally applied in the future.

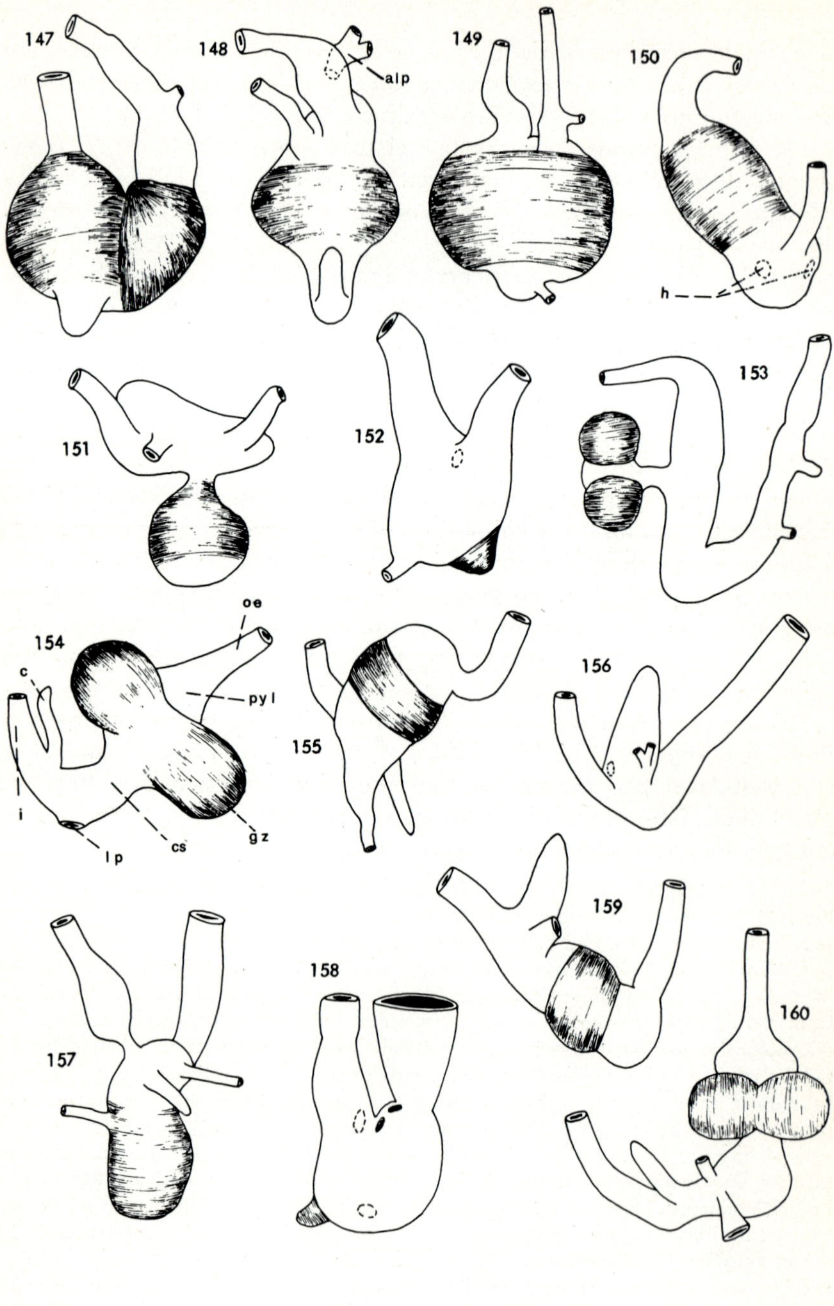
147
148
alp
149
150
h
151
152
153
154
oe
c
pyl
i
lp
cs
gz
155
156
157
158
159
160

Among other basommatophoran groups the simplest stomach is found in *Otina* (Fig. 152) and *Siphonaria* (Fig. 158), where only a small rudimentary gizzard is present; there is also a vestige of an intestinal style sac in *Otina*. The caecum is missing in both groups. While in other members of the Ellobiidae that have been examined the most primitive gastric structure occurs in *Ophicardelus*, *Ovatella* (Fig. 148), *Cassidula* and *Pythia*; the oesophageal and intestinal parts of the gizzard are, in some of these forms, still distinct and the degree of muscular development is moderate. In *Leucophytia* (Fig. 148), *Ellobium* (Fig. 149), *Pedipes*, *Leuconopsis* and *Marinula* the oesophageal and intestinal sections of the gizzard have coalesced. Posterior to the gizzard is a single posterior chamber with a caecum and, in some of the forms, the posterior digestive gland duct opening into it. The oesophagus enters and the intestine leaves a common anterior chamber in front of the gizzard. In *Melampus* (Fig. 151) the gizzard is separated from the rest of the stomach by a constriction and in the minute land snail *Carychium* (Fig. 150) the gizzard is barrel-shaped, with the oesophagus and the ducts of the anterior and posterior digestive glands entering the stomach at the opposite end to that from which the intestine emerges.

In *Trimusculus* (Fig. 157) the gizzard, with an irregularly arranged musculature, forms a sac posterior to the direct path between the oesophagus and the intestine and also just posterior to the caecum, but between the ducts of the two digestive glands. In *Latia* (Fig. 155) the region of the stomach nearest to the oesophagus forms the gizzard; the caecum plus a single digestive gland duct are joined just posterior to this organ. A similar level of organization is found in *Chilina* (Fig. 154), although there is a strongly developed and bilobed gizzard; the latter is oriented

---

Figs 147–160, stomach region of selected Basommatophora. Oesophagus to the right and intestine to the left in all figures. Gizzard region shadowed. 147. *Ovatella myosotis* (Drap.), redrawn from Morton, 1955b; 148. *Auriculinella* (*Leucophytia*) *bidentata* (Montagu), redrawn from Morton, 1955b; 149. *Ellobium pellucens* (Menke), redrawn from Marcus and Marcus, 1965b; 150. *Carychium tridentatum* Risso, redrawn from Morton 1955b; 151. *Melampus* sp, redrawn from Morton, 1955c; 152. *Otina otis* (Turton), redrawn from Morton, 1955a; 153. *Amphibola australis* Schumacher, redrawn from Farnie, 1919; 154. *Chilina fluctuosa* Gray, redrawn from Harry, 1964; 155. *Latia neritoides* Gray, redrawn from Pelseneer, 1901; 156. *Acroloxus lacustris* (L.), redrawn from Hubendick, 1962; 157. *Trimusculus peruvianus* (Sowerby), redrawn from Schumann, 1911; 158. *Siphonaria hispida* Smith, redrawn from Marcus and Marcus, 1960; 159. *Ancylus fluviatilis* Müller, redrawn from Hubendick, 1964; 160. *Lymnaea stagnalis* (L.), redrawn from Carriker, 1947. alp = anterior liver pore or anterior digestive gland duct; c = caecum; cs = cardiac or intestinal end of stomach; gz = gizzard; h = hidden entries of liver pore or digestive gland ducts; i = intestine; lp = liver pore or entry of the posterior digestive gland duct; oe = oesophagus; pyl = pyloric or oesophageal end of stomach.

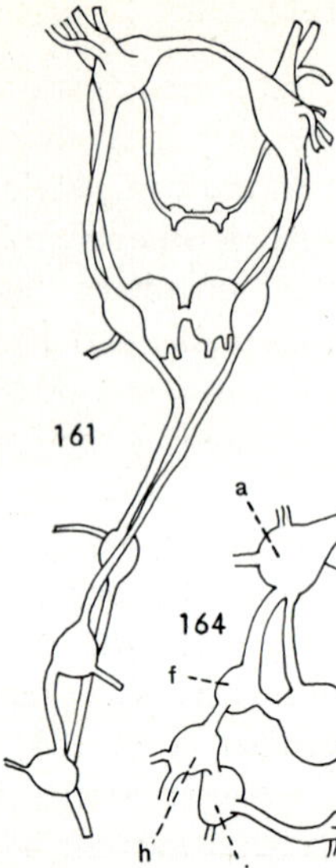
161

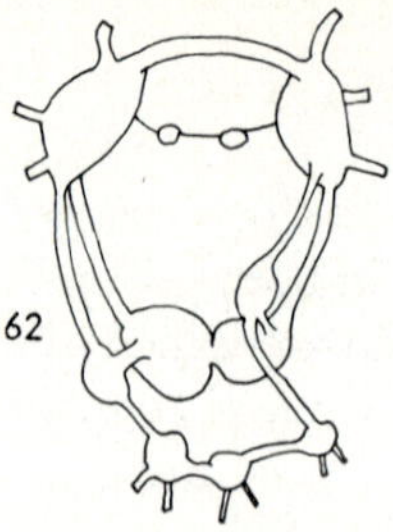
162

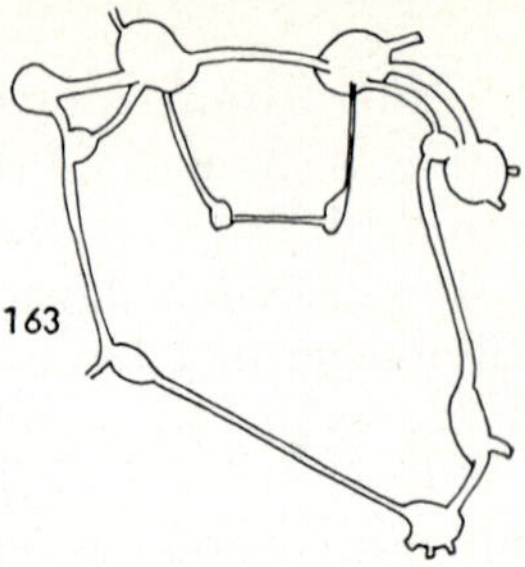
163

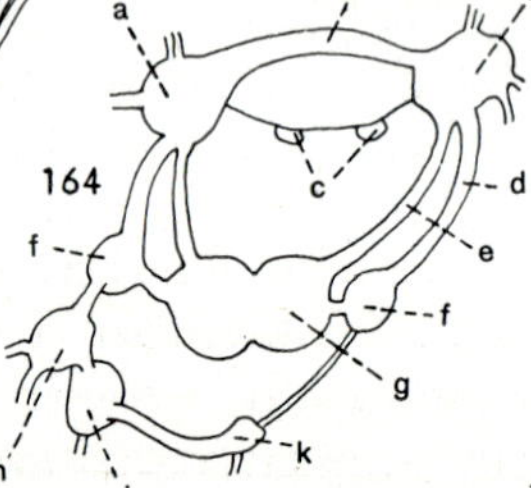

164

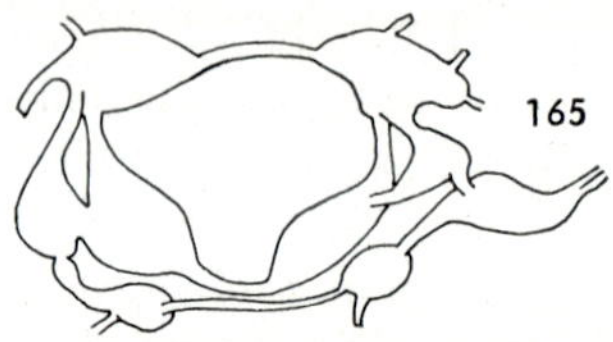
165

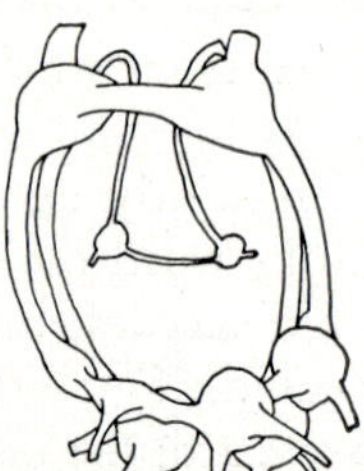
166

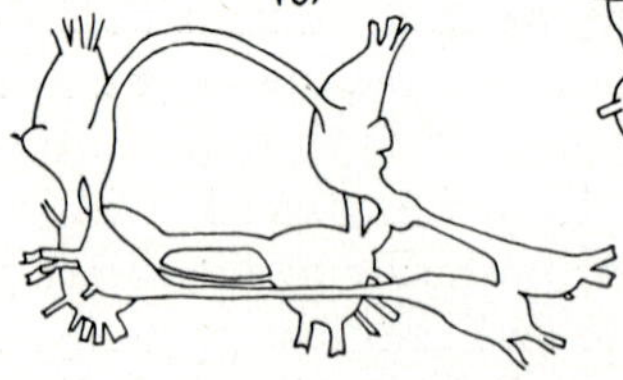
167

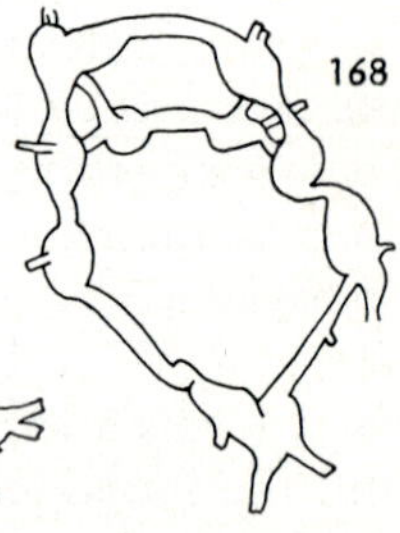
168

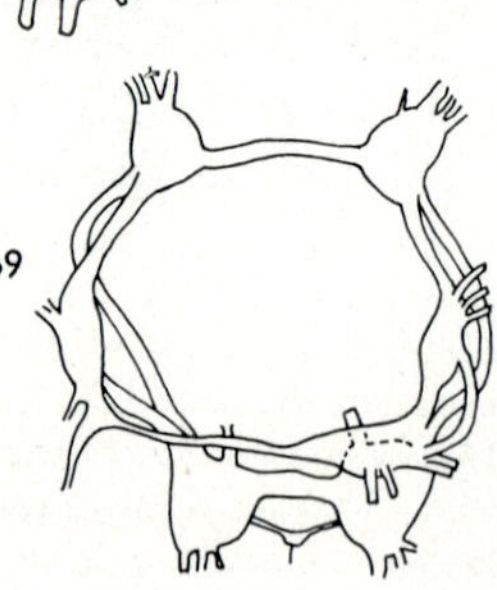
169

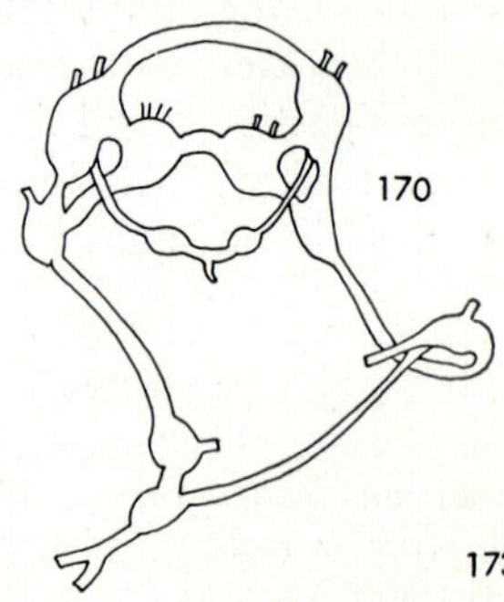
170

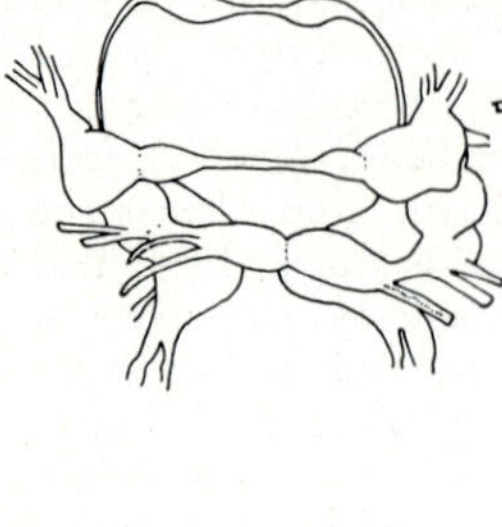
171

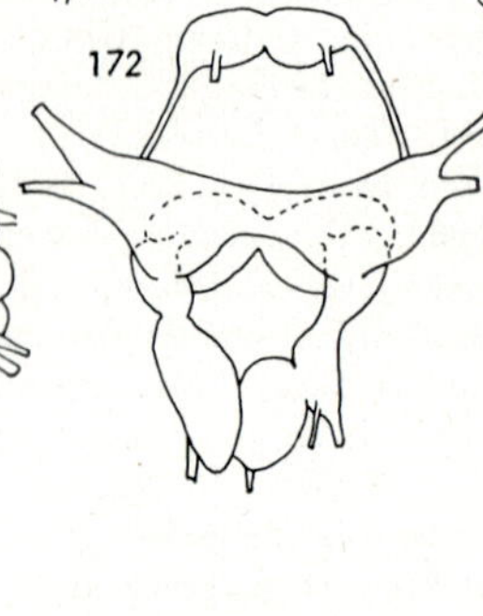
172

173

across the main axis of the gut. Posterior to the gizzard there is one single digestive gland duct and a caecum. In *Amphibola* (Fig. 153) the well-developed gizzard is located near the intestinal region of the stomach posterior to the opening of the digestive gland duct (or ducts?); it is defined by a narrow constriction of the stomach.

In the higher limnic Basommatophora, the Lymnaeidae, Physidae, Planorbidae and Ancylidae, the gastric morphology of the stomach region resembles that present in *Chilina*. Between the oesophagus and the gizzard there is a thin-walled crop. The development of the gizzard can vary from being weakly developed, as in the Physidae, to forming a cylindric muffle in the Ancylidae (Fig. 159), to consisting of two moderately or strongly developed muscle pads, as in the Planorbidae and Lymnaeidae (Fig. 160) respectively. The caecum is well-developed. When there are two, rather than a single digestive gland duct, they enter the stomach near each other.

The spacious digestive gland consists normally of two main lobes, each with one duct to the stomach. One lobe can be more or less reduced or the ducts of the two lobes fused before entering the stomach.

In the anterior part of the intestine there may be an internal ventral ridge, a typhlosole, while just posterior to this there may be a short section of the intestine with a stronger circular musculature forming a pellet compressor. Usually the intestine forms two loops, but in *Amphibola* it is remarkably prolonged to form a spiral. In *Siphonaria gigas* Sow. the posterior part of the intestine is morphologically and histologically aberrant; it is considerably wider than any other areas of the intestine and proportionately wider than in any other basommatophoran species.

Figs 161–173, the principle parts of the nervous system in selected Basommatophora. 161. *Ellobium gaziensis* (Preston), redrawn from Knipper and Meyer, 1956; 162. *Ellobium pellucens* (Menke), redrawn from Marcus and Marcus, 1965b; 163. *Auriculinella bidentata* (Montagu), redrawn from Morton, 1955b; 164. *Blauneria heteroclita* (Montagu), redrawn from Marcus and Marcus, 1965b; 165. *Marinula juanensis* Odhner, redrawn from Odhner, 1925; 166. *Melampus semisulcatus* Mousson, redrawn from Knipper and Meyer, 1956; 167. *Siphonaria hispida* Smith, redrawn from Marcus and Marcus, 1960; 168. *Amphibola australis* Schumacher, redrawn from Hubendick, 1945b; 169. *Trimusculus garnoti* Payr, redrawn from Dieuzeide, 1935; 170. *Chilina fluctuosa* Gray, redrawn from Harry, 1964; 171. *Lanx alta* (Tryon), redrawn (and slightly simplified) from Baker, 1925; 172. *Biomphalaria glabrata* (Say), redrawn from Hubendick, 1955; 173. *Laevapex fuscus* (Adams), redrawn from Basch, 1959. a = cerebral ganglion; b = cerebral commissure; c = buccal ganglia; d = cerebro-pleural connective; e = cerebro-pedal connective; f = pleural ganglion; g = pedal ganglion; h–k = visceral ganglia; h corresponds to subintestinal, i to abdominal, and k to supraintestinal ganglion in Prosobranchia.

V. Nervous system

In the Pulmonata the central nervous system tends to be concentrated into a cerebral ring around the oesophagus, with the associated commissures and connectives between the ganglia either very short, or almost non-existent. In contrast, within the Basommatophora forms occur with clearly defined and well-developed commissures and connectives, as in *Chilina* with a prominent visceral commissure. However, within the basommatophoran groups there are obvious trends towards shortening of these commissures and connectives, and a further concentration of the nervous system through fusion of ganglia.

Pulmonates are considered euthyneurous, that is, with the pleuro-visceral connectives not crossed, due to torsion, as in the Prosobranchia. However, traces of chiastoneury, where the connectives are crossed, can be recognized amongst the primitive Basommatophora. Thus the final stages of reduction in chiastoneury have taken place within the Basommatophora and various hypotheses have been proposed to account for these changes; of these, that of Naef-Merker is the most generally accepted. According to this the detorsion of the visceral commissure results from shortening of the pleuroparietal connectives, between the pleural ganglia and the supra- and sub-intestinal ganglia. Concomitant with these changes the supraintestinal ganglion, on the left side in streptoneurous forms, becomes more closely associated with the right pleural ganglion and conversely the sub-intestinal with the left pleural. It is then impossible for chiastoneury to exist.

Taxa exhibiting a remnant of chiastoneury are found in the Chilinidae and Ellobiidae. In *Chilina* (Fig. 170) the degree of chiastoneury varies between species, while in the Ellobiidae chiastoneury is found in *Ellobium* (Fig. 161), but not in *Auriculinella* (*Leucophylia*) (Fig. 163), even though the pleuroparietal connectives are only moderately shortened. In *Cassidula* and particularly *Melampus* (Fig. 166) the visceral connectives are considerably reduced, although the cerebropleural and cerebropedal connectives are of considerable length. Apparently these changes in the cerebral complex and the visceral commissures occur independently of each other, suggesting perhaps a polyphyletic origin for these taxa; compare diagrams Figs 161–173.

It is impossible to consider details of the peripheral nervous system or the central nervous system, but a guide to further information is provided in Table I.

VI. Reproductive system

All Basommatophora are hermaphroditic, with protandry common among the more primitive forms, for example, many species of Ellobiidae, and simultaneous hermaphroditism present in the more advanced taxa, including the Lymnaeacea but also in *Melampus*; an intermediary stage, a condensed protandry is found in the Physidae.

The ovotestis is located, together with the digestive gland, apically in the visceral hump. It is divided into a series of diverticula of which the number of branches tend to be proportional to the size of the animal; they are arranged in either a single row, a double row or fan-like (Figs 196–197). The ovotestis or haemophrodite duct leads from the ovotestis to the spermoviduct or to the point where the female and male ducts separate; this point is also frequently marked by the attachment of the albumen gland or its efferent duct and by the so-called fertilization pocket. It is this junction that is often called the carrefour (Fig. 188).

With few exceptions along the ovotestis duct there may be one or many morphologically discernible seminal vesicles. The structure of these can vary from a thickening of a simple duct, a thickening of a serpentine portion of the duct, numerous diverticula in two opposite rows or a mass of diverticula, one branched diverticulum (*Trimusculus* Fig. 182), to a thin-walled bladder-like appendage to the duct (*Acroloxus* Fig. 185).

In all Basommatophora there is only a single albumen gland, except for a few members of the Ellobidae where there are two. Yet the precise topographical relationships between the gland, or its efferent duct, the ovotestis duct, the fertilization pouch and the spermoviduct or oviduct and sperm duct are complicated and show considerable variation within the Basommatophora; that is between the carrefour and the associated structures. Although these relationships have been disentangled in rather few species, it is interesting to note that the Physidae are different in this respect from related families included in the Hygrophila, that is the families Lymnaeidae, Planorbidae and Ancylidae.

In the ellobiids *Pythia* (Fig. 174), *Cassidula* and *Ovatella* (Fig. 175) the genital duct is incompletely divided into male and female tracts or, rather, semi-canals. The female tract is lined with mucous glandular epithelium and the male with prostatic epithelium. There is a non-glandular genital vestibule just inside the common or functionally female pore corresponding to the vagina in other Basommatophora. From the pore in *Pythia* there is an open ciliated seminal groove beneath a narrow fold of the integument, leading between the closed vas deferens and the genital vestibule. In *Chilina* (Fig. 183) there is a vestigial external groove between the

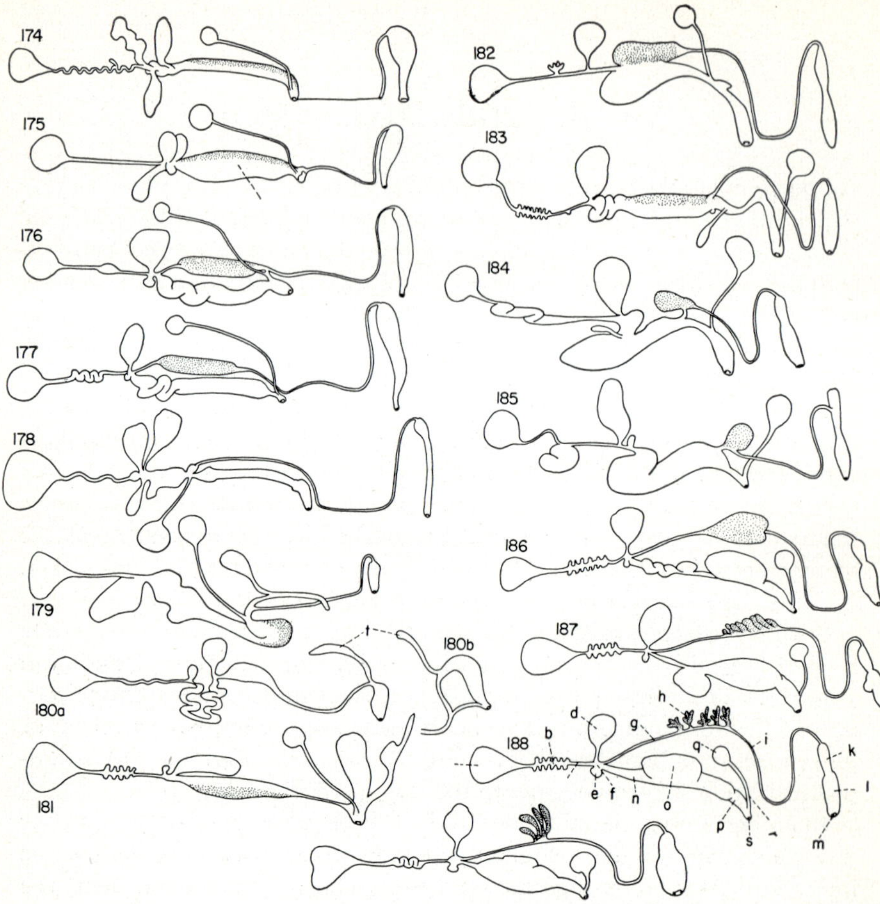

Figs 174–189, schematic diagram of the reproductive system in selected Basommatophora; 174. *Pythia scarabaeus* (L.), redrawn from Plate, 1897; 175. *Ovatella myosotis* (Drap.), redrawn from Morton, 1955b; 176. *Blauneria heteroclita* (Montagu), redrawn from Marcus and Marcus, 1965b; 177. *Auriculinella bidentata* (Montagu), redrawn from Morton, 1955b; 178. *Melampus boholensis* H. and A. Adams, redrawn from Koslowsky, 1933; 179. *Otina otis* Turton, compilated from Pelseneer, 1901, and Morton, 1955a; 180a. *Amphibola australis* Schumacher, redrawn from Farnie, 1919; 180b. *Salinator fragilis* (Lam.), redrawn from Hubendick, 1945b; 181. *Siphonaria* sp, compiled from Hubendick, 1945a; 182. *Trimusculus peruvianus* (Sowerby), redrawn from Schumann, 1911; 183. *Chilina dombeyana* d'Orb, redrawn from Haeckel, 1911; 184. *Latia neritoides* Gray, redrawn from Pelseneer, 1901; 185. *Acroloxus lacustris* (L.), redrawn from Hubendick, 1962; 186. *Lymnaea* sp, various sources; 187. *Physa* sp, various sources; 188. *Biomphalaria* sp, redrawn from PAHO/WHO guide; 189. *Burnupia caffra* (Krauss), redrawn from Hubendick, 1964; The prostate is indicated by shading; a = ovotestis; b = seminal vesicles; c = ovotestis duct; d = albumen gland; e = "fertilization pocket"; f = carrefour; g = sperm duct; h = prostate; i = vas deferens; k = penis sheath; l = preputium; m = male pore; n = oviduct; o = nidamental gland; p = "uterus"; q = spermatheca; r = spermathecal duct; s = female pore; t = flagellum; v = spermoviduct.

genital pores notwithstanding the presence of a complete vas deferens. Variations from this common pattern occur in a number of ellobiids, for example, in *Blauneria* (Fig. 176) and *Ellobium* there is a vestige of a distal connection between the female and male systems in the shape of a narrow connecting duct via the most distal portion of the spermathecal duct; in *Melampus* (Fig. 178) the hermaphrodite duct is non-glandular, the mucous gland forms a separate outgrowth, a prostate is not observed and the vagina is extensively prolongated; nevertheless, in *Auriculinella* (Fig. 177), the arrangement of the gonoducts conforms to the same pattern as in the Hygrophila.

In *Otina* (Fig. 179) the glandular spermoviduct is incompletely divided into separate tracts only in the distal portion, where part of the wall forms the prostate gland. Similarly, in *Chilina* (Fig. 183), there are separate tracts along most of the spermoviduct, but in *Trimusculus* (Fig. 182) the female and male ducts separate early and the spermoviduct is very short. In *Latia* (Fig. 184) and *Acroloxus* (Fig. 185) the spermoviduct is not divided into distinct tracts. In the Hygrophila (Figs 186–189) there is no spermoviduct, only separate female and male ducts, which commence at the carrefour.

A very different type of organization is present in the Amphibolidae (Fig. 180) and Siphonariidae (Fig. 181). The long spermoviduct has a simple lumen without separate tracts and there is no female pore in the typical position just beneath the mantle opening. Instead there is a single genital pore in the anterior position associated with the male pore.

The oviduct, like the spermoviduct, is differentiated into regions with divergent histological properties. The most distal region, the vagina, is non-glandular and the more or less globular spermatheca connects with it via the spermathecal duct. In *Melampus*, *Otina*, *Trimusculus*, and to some extent in *Latia*, the vagina is elongated and the junction between the female and spermathecal ducts is proximal to the female pore. While in the Amphibolidae the spermatheca is lacking, and in *Otina* there is a second accessory spermatheca, the duct from which enters the common duct. A similar vaginal structure in *Chilina* has been called vesicula seminalis vaginae.

In forms with a common spermoviduct the prostatic gland usually forms part of the wall of the duct. However, in *Latia* and *Acroloxus* (Figs 184–185) the prostate is separate from the spermoviduct and is associated with the proximal part of the vas deferens. In those forms without a spermoviduct, but with separate female and male ducts, the prostate is associated with the latter and forms an elongated thickened portion of the male duct. The duct between the carrefour and the prostate is termed the sperm duct or vas deferens, that between the prostate and the male copulatory organ the vas deferens. Among the Hygrophila the prostate is developed in two different ways from the male duct; in Lymnaeidae it constitutes a single (in *Lanx* two) swollen area, often with a

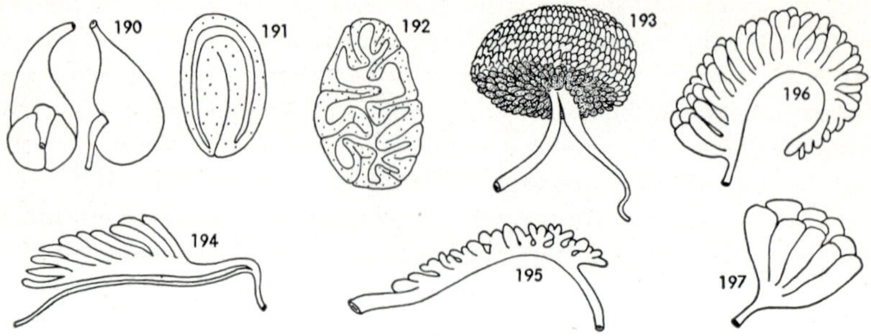

Figs 190–197, prostate and gonad. 190. Prostate of *Lymnaea auricularia* (L.), dorsal and right side, redrawn from Hubendick, 1945c; 191. Cross-section through the prostate of *Lymnaea peregra* (Müll.); 192. Cross-section through the prostate of *Lymnaea stagnalis* (L.), redrawn from Hubendick, 1951; 193. Prostate of *Physastra* sp, redrawn from Hubendick, 1955a; 194. Prostate of *Polypylis calathus* Benson, redrawn from Hubendick, 1955; 195. Prostate of *Biomphalaria amazonica* Paraense, redrawn from the PAHO/WHO guide; 196. The gonade of *Biomphalaria*, the same source; 197. The gonads of *Rhodacmea*, redrawn from Basch, 1960.

fold directed inwards from the outer surface (Figs 186, 190–192), but in the other families it is a series of invaginations from the duct (Figs 187–189, 193–195) and these may vary in number, branching and arrangement.

The male copulatory organ normally consists of a penis traversed by the vas deferens and enclosed in a penis sheath, through the terminal opening of which the penis can be everted. Frequently an extension of the penis sheath forms a preputium with internal longitudinal muscular pilasters. The penis can have a terminal or lateral pore, various surface structures, and a stylet. The latter can be formed as a condensation of the terminal tissue of the penis or as a chitinous sheath-like structure. Another form found particularly in the Bulininae (Fig. 198) where there is no true pendant penis, is a so-called ultra-penis. It is formed from the inner layer of the wall of the penial sheath, with the outer layer forming the sheath of the ultra-penis; the layers being separated by a voluminous split. When everted, the distal end of the ultra-penis at rest, becomes the proximal end.

In some forms belonging to the Ancylidae (Fig. 200) there is still a third type of penial structure. The proximal part of the copulatory organ, where the vas deferens and the flagellum meet, and the distal part of the

---

Figs 198–206, male copulatory organ. Schematic diagram of longitudinally sectioned male copulatory organ of various Ancyloplanorbidae, cut surfaces blanc, redrawn from Hubendick, 1955a and 1964; 198. *Bulinus*; 199. *Laevapex*; 200. *Gundlachia*; 201. *Miratesta*; 202. *Ferrissia*; 203. *Burnupia*; 204. *Plesiophysa*; 205. *Helisoma*; 206. *Polypylis* a = penis; b = penis sheath; c = preputium; d = ultra-penis; e = hold-fast organ; f = flagellum; g = vas deferens.

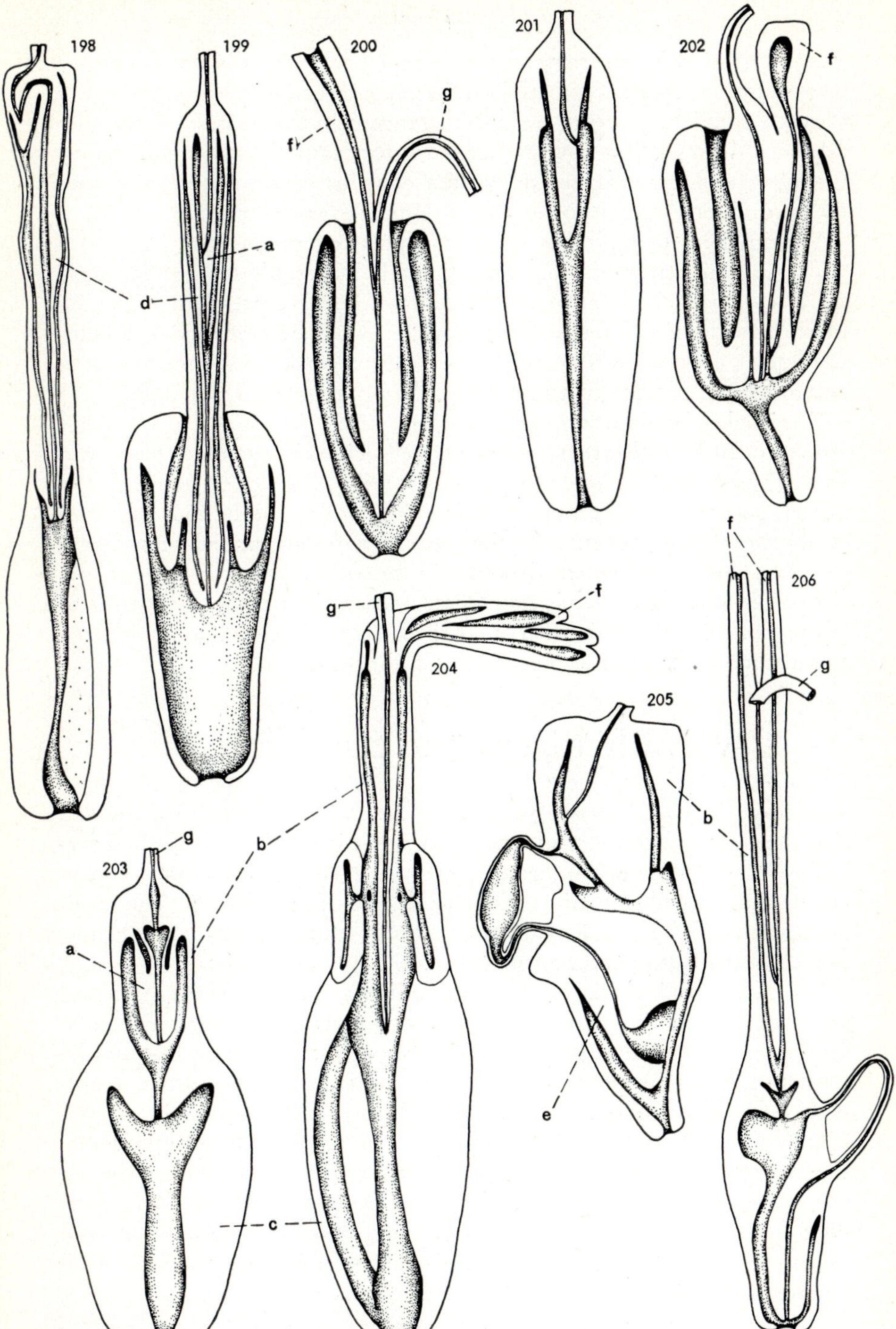
198
199
200
201
202
f
g
f
a
d
f
206
g
f
204
205
g
b
203
a
b
e
c

preputium at rest, are together invaginated into the distal part of the preputium. Further invagination in the same direction means eversion of the male copulatory organ. A similar arrangement occurs in some particularly small forms which have no flagellum, for example, the ellobiid genera *Carychium* and in *Otina*; the preputium contains no retractible penis and the intromittent organ is formed by the simple eversion of the sheath itself.

Within these major types of male copulatory organs there can be a wide range of variation; this is particularly noticeable in the Ancylidae and Planorbidae. In some of the Planorbidae there is a structure occupying most of the lumen of the preputium which, when everted, acts as a holdfast during copulation (Fig. 205). Moreover, in certain basommatophoran taxa complete absence of a male copulatory organ, aphaly, is common.

The male copulatory organ complex often comprises a glandular structure. In Physidae there is a large well-delimited preputial gland and in the Siphonariidae, where a penis is lacking, there is a so-called epiphallus gland. In many groups there is a more or less voluminous flagellum (rarely more than one) entering the copulatory organ parallel with the vas deferens; this can be either glandular or muscular.

In *Siphonaria* the sperm is transferred encased in a chitinous spermatophore, often provided with a barbed ridge, and a non-chitinized spermatophore has been observed in *Chilina* by Harry.

VII. Phylogeny and classification

As already indicated in the introduction, it cannot be taken for granted that the Basommatophora constitutes a single phylogenetic unit. Nevertheless, many of the 11 families are obviously related to each other, and all can be ranked with regard to probable degree of relationship. The result is illustrated in Fig. 207. The latter reflects studies by Duncan, Harry, Hubendick, Morton and Starobogatov.

The family Ellobiidae, although including a variety of morphological types adapted to different environments, is certainly a clearly delimited unit of monophyletic origin. In this family certain primitive features are found, such as a protandrous sexual succession, a heterostrophic embryonic shell, an embryonic operculum and, at least in one genus, a free-swimming larva with a well-developed velum. The Otinidae certainly represent a primitive group related to the Ellobiidae. However, its adaptation to withstand extreme wave exposure has led to certain morphological changes, among them small size, and this is associated with several anatomical changes. There is also at the embryonic stage reducted heterostrophy and a larval opercum, but no trace of a velum.

Among the remaining marine Basommatophora, the Amphibolidae,

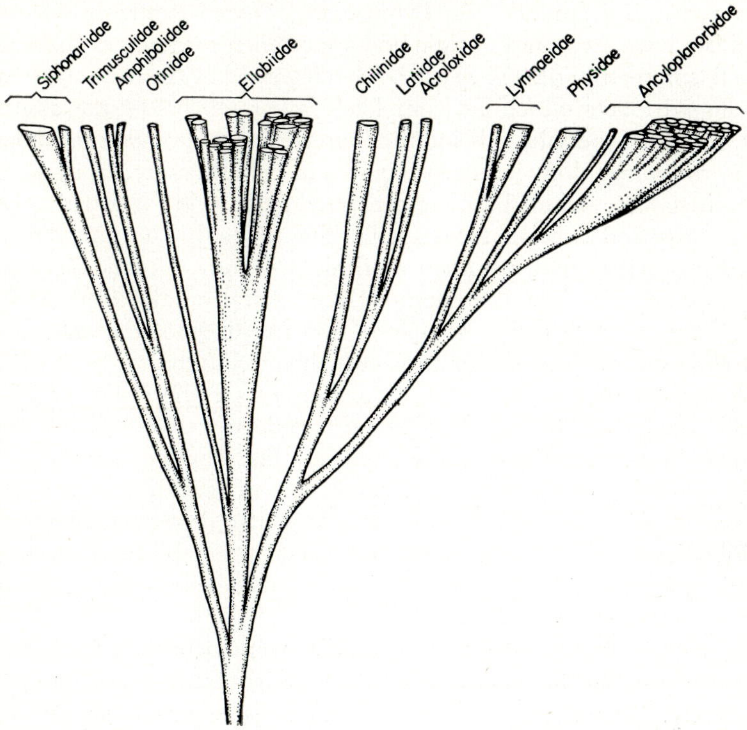

Fig. 207. Diagram of probable relationships between the Basommatophoran families.

living in estuarine mud flats, have more primitive features than the others. The shell is spiral in shape, but the operculum is retained, even in the adult. The intertidal patelliform Trimusculidae and Siphonariidae are more specialized, although this may be a consequence of a patelliform shape; a fairly well-developed mantle cavity is present in both families.

The third main branch among the Basommatophora comprises the groups often collected under the name Hygrophila. They all inhabit fresh water but have no free-swimming larval stage, nor an embryonic operculum; these are both features that characterized, at least some, members of the previously mentioned groups. Among the Hydrophila the Chilinidae are the most primitive group, as demonstrated by the comparatively long visceral commissure, with a vestige of chiastoneury, and the incompletely separated male and female tracts from the albumen gland to the female pore. The families Latiidae and Acroloxidae are patelliform derivatives from the branch including the Chilinidae; both exhibit considerable anatomical changes and reductions, some of which are paralleled in other patelliform Basommatophora. The prostate gland has a distinctive arrangement in both families.

The remaining families, the Lymnaeidae, Physidae, Planorbidae and Ancylidae, have a common origin and are often considered as the higher limnic Basommatophora. They can be characterized by the male and female reproductive ducts which are completely separated below the albumen gland. The least specialized being the Lymnaeidae, although a secondarily patelliform morphology has been acquired in the subfamily Lancinae. The Lymnaeinae have retained the typical spired gastropod form with a large mantle cavity and a simple sac-shaped kidney without a urethra. The flat, triangular tentacles may represent a primitive feature. Unique features of the Lymnaeidae include asymmetry of the central radular teeth and the form of the prostate gland, as one or two swellings of the wall of the sperm duct and with or without inwards directed folds.

Features common to the remaining families are the structure of the prostate, composed of free diverticula directed outwards from the sperm duct and the filiform shape of the tentacles. The Physidae are characterized by sinistral symmetry and a peculiar type of radula; the teeth being arranged in chevron-shaped cross rows, this may, however, represent a primitive feature. The family is characterized further by the form of the preputial gland, the serpentine canal in the kidney, the shape of the foot and a reduction of the jaw.

The families Planorbidae and Ancylidae were originally regarded as groups comprizing the planispiral and patelliform freshwater Pulmonates respectively. Subsequently it was discovered that several genera of high-spired and limpet-like forms had to be included in the Planorbidae and that the patelliform organization had obviously evolved independently on several occasions. It is hardly possible to still retain the Ancylidae as a valid phylogenetic and taxonomic unit. The spired, planispiral and patelliform groups so far referred to Planorbidae and Ancylidae form a simple complex, where characters other than the general body and shell form are of decisive phylogenetic and taxonomic importance. It is suggested that this complex should be taxonomically distinguished as the family Ancyloplanorbidae.

Certain evolutionary trends are characteristic within the Ancyloplanorbidae. The original spired body and shell form has been changed into a planospiral form on at least three independent occasions; this trend has in certain cases led to a pseudodextral form, while a change in the opposite direction has on several occasions led to a patelliform shape or limpet form. Associated with this development have been certain secondary changes, such as reduction of the mantle cavity, further development of the shell adductor muscle system, a serpentine urethra and frequently a reduction of size. These changes must have evolved in parallel several times. Other evolutionary trends which are common within the Ancyloplanorbidae are evolution of flagella in the male copulatory organ and splits in the penis and penis sheath walls leading, in certain cases, to formation of an ultra-penis structure.

The genera of the Ancyloplanorbidae can be grouped in two subfamilies, the Bulininae and the Planorbinae. In the former the gonad (cf. Fig. 197) and the prostate (Fig. 193) consist of lobate diverticula in a fanlike arrangement; all the diverticula meet in almost one point, and the radular teeth are comparatively large. In the Planorbinae the gonad consists of two lines of lobate, sometimes branched, diverticula (cf. Fig. 196); the prostate has one line of finger-shaped diverticula, emptying independently along the sperm duct (Fig. 195) or, in a few genera, in a special prostate duct (Fig. 197), while the radula is composed of comparatively small teeth, the laterals are mostly tricuspid. An exception to this division is provided by *Rhodacmea*, a genus of aberrant limpets, which might form a subfamily of its own, the Rhodacmeinae. Starobogatov (1967) distinguished Rhodacmeidae as a separate family, more related to the Physidae than to the Ancyloplanorbidae, however, its true taxonomical position must still be regarded as uncertain and the group is retained in the Ancyloplanorbidae. It is characterized by a sac-shaped prostate and a peculiar radula.

Each of these two subfamilies comprise a rich variety of forms often with peculiar specializations, particularly in the male copulatory system. I have here accepted Starobogatov's opinion that the arrangement of the gonad and prostate diverticles have a greater taxonomic bearing than the structure of the male copulatory organ. When compiling Table I, I have largely followed the taxonomy laid out by him, however, it is important to remember that there are still many weak points in the taxonomic system for the Ancyloplanorbidae presented below, which must therefore be regarded as tentative.

Table I Tabular survey of the Basommatophora

Group	Conchological type	Other morphological characteristics	Habitat	Geographical distribution	Figures in this paper	References
ORDER BASOMMATOPHORA	variable		marine, terrestrial, but predominantly freshwater	worldwide	207	Bargmann, 1930 (nervous system) Hescheler, 1900 (comparative anatomy) Hyman, 1967 (general) Simroth and Hoffmann, 1908–1928 (general) Thiele, 1931 (taxonomy) Zilch, 1959–1960 (taxonomy)
SUPERFAMILY ELLOBIACEA						
FAMILY ELLOBIIDAE (taxonomy within this family needs revision based on further comparative anatomical studies)	spired or ovoidal	protandrous hermaphroditism common				Küster 1841–1844 and Kobelt 1897–1901 in Martini and Chemnitz: *Syst. Conch.-Cab. 16* (1–2) Knipper and Meyer, 1956 (nervous system) Morton, 1955c (ecology, morphology, phylogeny) Odhner, 1925 (radulae, taxonomy)
SUBFAMILY PYTHIINAE		common hermaphrodite duct to female pore				
Genus *Pythia* Röding	dorso-ventrally compressed	open seminal grove between genital pores; albumen gland with two lobes	coastal terrestrial	Indo-Pacific	1, 2, 174	Berry *et al.*, 1967 (genital morphology) Morton, 1955c (digestive and genital system) Plate, 1897 (morphology)
Genus *Ovatella* Bivona			supratidal estuarine	Eastern Atlantic, Mediterranean	4, 147, 175	Morton, 1955b (digestive and genital systems) Pelseneer, 1894 (morphology)

Genus *Ophicardelus* Beck			estuarine, mangrove flats	Pacific		
Genus *Cassidula* Fér.			estuarine, mangrove flats	Indopacific		Berry *et al.*, 1967 (genital morphology) Knipper and Meyer, 1956 (morphology) Starmühlner, 1970 (morphology)
Genus *Cylindrotis* Moellend.				Philippines		
Genus *Allochroa* Ancey				Indopacific		
Genus *Auriculastra* Martens				Indopacific		
SUBFAMILY PEDIPEDINAE	spired or ovoidal	common hermaphrodite duct				
Genus *Pedipes* Brug.			intratidal–supratidal	tropical coasts		
Genus *Leuconopsis* Hutton			intertidal, crevice-dwelling	Australia, New Zealand		
Genus *Marinula* King			intertidal, under stones	mainly Pacific	165	Morton, 1955b (digestive and genital system) Odhner, 1925 (morphology) Robson, 1915 (morphology)
Genus *Laemodonta* Philippi (often under syn. *Plecotrema* H. and A. Adams)				Indo-Westpacific, Western Atlantic		Hubendick, 1956 (taxonomy, distribution)
Genus *Pseudomelampus* Pallary				Eastern Atlantic, Mediterranean		
Genus *Apodosis* Pilsbry and McGinty				Bahamas		
Genus *Rangitotoa* Powell			intertidal, crevice-dwelling	New Zealand		

Table I—continued

Group	Conchological type	Other morphological characteristics	Habitat	Geographical distribution	Figures in this paper	References
SUBFAMILY MELAMPODINAE	spired, ovoidal	incomplete division between male and female ducts below albumen gland				
Genus *Melampus* Montfort		simultaneous hermaphroditism occurs	supra-littoral	tropical coasts	3, 79, 95, 107, 151, 166, 178	Apley, 1970 (reproduction) Marcus and Marcus, 1965a (morphology) Morton, 1955c (digestive and genital systems) Knipper and Meyer, 1956 (morphology) Koslowsky, 1933 (morphology)
Genus *Tralia* Gray (in Turton)				tropical coasts		
SUBFAMILY ELLOBIINAE	spired, ovoidal	separate male and female ducts from albumen gland				
Genus *Ellobium* Röding		vestigial chiastoneury	mangrove flats	Indo-Westpacific	149, 161, 162	Berry *et al.*, 1967 (genital morphology) Knipper and Meyer, 1956 (morphology) Marcus and Marcus, 1965b (morphology) Morton, 1955c (digestive and genital system)
Genus *Blauneria* Schuttlew.				tropics	164, 176	Marcus and Marcus, 1965b (morphology)
Genus *Auriculinella* Tausch (often under syn. *Leucophytia* Winckw.)			intertidal, crevice-dwelling	Eastern Atlantic, Mediterranean	108, 148, 163, 177	Morton, 1955b, c (digestive and genital system)
SUBFAMILY CARYCHIINAE	spired	incomplete division of male and female ducts				

Genus *Carychium* Müll.			terrestrial	Americas, Europe, Asia including East Indies	150	Morton, 1955b (digestive and genital system)
Genus *Zospeum* Bourg.			cave-dwelling	Europe		
FAMILY OTINIDAE Genus *Otina* Gray	small, semi-limpet-like	protandrous hermaphroditic; below common duct division between male and female ducts	intertidal, crevice-dwelling or exposed	coasts of Britain and France	5, 80, 109, 152, 179	Morton, 1955a (functional morphology) Pelseneer, 1901 (morphology)
SUPERFAMILY AMPHIBOLACEA						
FAMILY AMPHIBOLIDAE	with operculum					Bouvier, 1892 (morphology) Hubendick, 1945b (taxonomy)
Genus *Amphibola* Schumacher	globose, low spire, peri-stomal notch	complete monauly	estuarine	New Zealand	6, 82, 90, 110, 153, 168, 180a	Farnie, 1919 (morphology)
Genus *Salinator* Hedley	globose, moderate spire	rudimentary diauly, but with only 1 genital pore	estuarine	Indo-Westpacific	7, 180b	Pelseneer, 1894 (morphology)
SUPERFAMILY SIPHONARIACEA (synonym Patelliformia)	patelliform					Dieuzeide, 1935 (morphology development) Hubendick, 1946 (taxonomic monograph)
FAMILY SIPHONARIIDAE		complete monauly; accessory gill in mantle cavity				Hubendick, 1945a (morphology; phylogeny)
Genus *Siphonaria* Sowerby			intertidal	almost world wide but not northern Atlantic	13, 14, 85, 92, 96, 112, 158, 167, 181	Cottrell, 1911 (morphology) Köhler, 1893 (morphology) Marcus and Marcus, 1960 (morphology) Pelseneer, 1894 (morphology) Yonge, 1952 (mantle organs)

Table I—continued

Group	Conchological type	Other morphological characteristics	Habitat	Geographical distribution	Figures in this paper	References
Genus *Williamia* Monterosato			on rocks in shallow water	Mediterranean and adjacent Atlantic; California–Panama; Galapagos; Sea of Japan	15, 16	
FAMILY TRIMUSCULIDAE (synonym Gadiniidae)		diauly but incomplete division of male and female ducts				
Genus *Trimusculus* Schmidt (often under synonym *Gardinia* Gray)				Indo-Pacific; St. Barthelmy; southern Mediterranean and adjacent Atlantic; South Africa	12, 84, 91, 111, 157, 169, 182	Schumann, 1911 (morphology) Yonge, 1958 (morphology)
SUPERFAMILY LYMNAEACEA						Bondesen, 1950 (egg capsules) Demian, 1962 (buccal mass) Duncan, 1960a, b (genital system) Harry and Hubendick 1964 (morphology) Heidermanns, 1927 (digestive system) Hubendick, 1947 (morphology, phylogeny) Lacaze-Duthiers, 1872 (nervous system)
FAMILY CHILINIDAE	spired	incomplete division of male and female ducts; vestigial chiastoneury				

Genus *Chilina* Gray			predominantly lotic	southwestern South America	11, 81, 113, 154, 170, 183	Haeckel, 1911 (morphology, phylogeny) Harry, 1964 (morphology, phylogeny) Hubendick, 1945a (morphology, phylogeny) Pelseneer, 1894 (morphology) Plate, 1894 (morphology) Plate, 1895 (nervous system)
FAMILY LATIIDAE	semi-limpet-like	incomplete division of male and female ducts; unitary globular prostate				
Genus *Latia* Gray			limnic	New Zealand	8, 115, 155, 184	Pelseneer, 1901 (morphology)
FAMILY ACROLOXIDAE	patelliform	incomplete division of male and female ducts; mantle cavity rudimentary; serpentine urethra; bladder-like seminal vesicle; unitary globular prostate				
Genus *Acroloxus* Beck (including the Bajcal *Pseudancylastrum* Lindholm etc.)			limnic	Palearctic and Rocky Mountains in the Nearctic	9, 10, 83, 114, 156, 185	Hubendick, 1962 (morphology and phylogeny) Hubendick, 1965 (comparative anatomy) Hubendick, 1969a (the Bajcal forms) Hubendick, 1969b (taxonomy)
FAMILY LYMNAEIDAE		complete diauly; prostate of thickened sperm duct wall, often with invaginated folds				

Table I—continued

Group	Conchological type	Other morphological characteristics	Habitat	Geographical distribution	Figures in this paper	References
SUBFAMILY LYMNAEINAE	high-spired to practically non-spired; exceptionally sinistral	1 prostate				Baker, 1911 (monograph of Nearctic forms) Carriker, 1947 (alimentary system) Demian, 1965 (genital system) Elo, 1938 (nervous system) Hubendick, 1945c (Radix forms)
Genus *Lymnaea* Lamarck (including *Radix* Montfort; *Stagnicola* Leach; *Galba* Schrank; *Bulimnea* Haldeman; *Acella* Haldeman; *Erinna* H. and A. Adams; *Myxas* Sow. and others which are often distinguished as genera but with doubtful taxonomic significance)			limnic, exceptionally amphibic	world wide	19–24, 97, 141–146, 116, 160, 186, 190–192	Hubendick, 1951 (taxonomic monograph) Itagaki, 1956, 1959a, b (morphology) McCraw, 1957 (morphology) Pretorius and Van Eedan, 1969 (morphology) Walter, 1969 (morphology)
SUBFAMILY LANCINAE	patelliform	2 prostates				
Genus *Lanx* Clessin			limnic	North America	18, 117, 171	Baker, 1925 (morphology)
FAMILY PHYSIDAE	spired or ovoidal, sinistral	sinistral; complete diauly; prostate of diverticula evaginated from sperm duct				Baker, 1928 Richards, 1964b
Genus *Physa* Drap.			limnic	almost world wide	17, 86, 118, 187	Duncan, 1958 (genital system)
Genus *Aplexa* Fleming (doubtful if this is truly distinct from Physa)			limnic	Holarctic, Central America		

FAMILY ANCYLOPLANORBIDAE	variable	anatomically sinistral; complete diauly; prostate of evaginations from sperm duct				Hubendick, 1955a, 1964 (comparative morphology, phylogeny) Hubendick, 1965 (comparative morphology) Starobogatov, 1958 (morphology, taxonomy) Starobogatov, 1967 (morphology, taxonomy)
SUBFAMILY RHODACMEINAE	patelliform	sac-shaped prostate; all gonad diverticula diverging from one point				
Genus *Rhodacmea* Walker			lotic	southeastern North America	70, 71, 139, 197	Basch, 1960 (morphology) Basch, 1962a (radula)
SUBFAMILY BULININAE	variable	prostate diverticula from one small area of the sperm duct; gonad diverticula arranged in a fan-like fashion				Hubendick, 1948a (morphology, taxonomy) Hubendick, 1948b (male systems)
Tribe Bulinini	variable	complete or incomplete ultra-penis present				
Genus *Bulinus* Müller (including *Physopsis* Krauss)	spired or ovoidal, sinistral	well-developed ultra-penis	limnic (1 sp. aestivates in dry soil)	Africa, Madagascar, Middle East, southern Europe	34, 88, 94, 102, 119, 198	Aardt and van Eeeden, 1969 (morphology) Brown *et al.*, 1971a, b (morphology, etc.) Demian, 1960 (morphology) Larambergue, 1939 (genital system) Mandahl-Barth, 1957b (taxonomy) Walter, 1968 (morphology)
Genus *Indoplanorbis* Annandale and Prashad	planispiral	well-developed ultra-penis	limnic	Southern Asia, East Indies	35, 36, 120	F. C. Baker, 1933 (morphology)

Table I—continued

Group	Conchological type	Other morphological characteristics	Habitat	Geographical distribution	Figures in this paper	References
Genus *Laevapex* Walker	patelliform	incomplete ultra-penis and real penis	limnic	North America	68, 69, 140, 173, 199	Basch, 1959 (morphology) Basch, 1962c (radula)
Genus *Gundlachia* Pfeiffer	patelliform; often "gundlachoid" shell growth	incomplete ultra-penis; glandular flagellum	limnic or lotic	Neotropical	66, 67, 89, 100, 137, 200	Hubendick, 1967b (taxonomy)
Tribe Physastrini						
Genus *Physastra* Tapparone-Canefri	variable spired, sinistral	true penis present muscular flagellum	limnic	East Indies, Australia, Western Pacific	25, 26, 103, 193	Hubendick, 1954a (distribution) Hubendick, 1948a, 1955a (morphology) Starmühlner, 1970 (morphology)
Genus *Amerianna* Strand	non-spired, ovoidal, sinistral	no flagellum	limnic	East Indies, Australia	27–30, 121	
Genus *Miratesta* Sarasin and Sarasin	spired, sinistral, thick-walled	no flagellum, peculiar shape of spermatheca	limnic	Celebes	104, 201	Hubendick, 1948 (morphology) Sarasin and Sarasin, 1898 (morphology)
Genus *Ancylastrum* Bourg.	patelliform with sinistral apical whorl	muscular flagellum	limnic	Tasmania	43, 44, 98, 105, 131	Hubendick, 1964 (morphology)
Genus *Protancylus* Sarasin and Sarasin	patelliform	mantle cavity anterior to left adductor muscle	limnic	Celebes		Hubendick, 1958a (morphology) Sarasin and Sarasin, 1898 (morphology)
Genus *Patelloplanorbis* Hubendick	calyptraeiform		limnic	New Guinea	37, 38, 130	Hubendick, 1957a (morphology)
Genus *Ferrissia* Walker (including *Pettancylus* Iredale)	patelliform often "gundlachoid" shell growth	aphaly not uncommon; glandular flagellum	limnic or lotic	almost world wide but missing in middle and northern Palearctic	64, 65, 76–78, 106, 135, 136, 202	Basch, 1962c (radula) Basch, 1963 (taxonomy) Boer and Lever, 1959 (circulation system) Brown, 1967 (morphology) Hoff, 1940 (morphology) Hubendick, 1967b, 1970 (taxonomy)

Genus *Burnupia* Walker	patelliform	no flagellum	limnic or lotic	Africa south of the Sahara	72, 73, 138, 189, 203	Oberholzer and van Eeden, 1969 (morphology) Walker, 1924 (taxonomy)
Tribe Camptoceratini	variable		limnic			
Genus *Camptoceras* Benson (including *Culmenella* Clench)	spired or ± ovoidal, sinistral	preputial organ; terminal penis pore	limnic	Southern and Eastern Asia	32, 33, 123	Hubendick, 1955b, 1959, 1967a (morphology, taxonomy)
Genus *Planorbarius* Duméril	planispiral	preputial organ; terminal penis pore	limnic	Palearctic	45	Hubendick, 1963 (male organ)
Genus *Helisoma* Swainson (including *Carinifex* Binney; *Parapholyx* Hanna)	spired sinistral, planispiral or pseudodextral	preputial organ with accessory duct; lateral penis pore	limnic	Nearctic	55, 125, 205	Baker, 1945 (morphology, taxonomy)
Tribe Plesiophysini						
Genus *Plesiophysa* Fischer	spired-ovoidal, sinistral	true penis; glandular flagella; penis sheath glands; pseudobranch anterior to anus	limnic	Neotropic	31, 122, 204	Hubendick, 1949 (morphology) Richards and Ferguson, 1962 (morphology)
SUBFAMILY PLANORBINAE	variable	prostate diverticula spread along part of sperm duct or separate branch from this; gonad diverticula in two lines				Baker, 1945 (morphology, taxonomy)
Tribe Planorbulini						
Genus *Planorbula* Haldeman (including *Menetus* H. and A. Adams and *Promenetus* Baker)	planispiral with tendency to pseudodextral	preputial organ with accessory duct; lateral penis pore	limnic	Nearctic		Baker, 1945 (morphology, taxonomy)
Tribe Biomphalariini		prostate diverticula from sperm duct itself				

Table I—continued

Group	Conchological type	Other morphological characteristics	Habitat	Geographical distribution	Figures in this paper	References
Genus *Biomphalaria* Preston (synonym *Australorbis* Pilsbry; *Taphius* H. and A. Adams, and others)	planispiral, sometimes with tendency to pseudo-dextral form	prostate diverticula branched, no flagellum	limnic	Africa, Madagascar, the Americas	56–58, 124, 172, 188, 195, 196	Hubendick, 1961 (morphology, taxonomy) Lever, de Vries and Jager, 1965 (nervous system) Mandah-Barth, 1957a (taxonomy of African forms) PAHO/WHO, 1968 (taxonomy of Neotropical forms) Paraense *et al.* (many papers not listed here) Richards, 1964a
Genus *Drepanotrema* Fischer and Crosse (including *Acrorbis* Odhner)	planispiral or pseudo-dextral	prostate diverticula simple; flagella present	limnic, on seepages (one sp. aestivates in dry soil)	Neotropical	59–61, 129	Harry and Hubendick, 1967 (morphology, taxonomy) Hubendick, 1961 (morphology, taxonomy) Paraense *et al.* (many papers not listed here) Richards, 1964a (genital system)
Tribe Planorbini	planispiral, sometimes pseudo-dextral or tendency to such	prostate diverticula on branch from the sperm duct; no flagellum; no preputial gland				
Genus *Planorbis* Müller	planispiral	terminal penis pore; no terminal stylet	limnic	Palearctic	46, 47, 101, 126	
Genus *Anisus* Studer (including *Bathyomphalus* Charp.)	planispiral	lateral penis pore; terminal stylet	limnic	Palearctic; probably also Oriental, possibly even West Pacific	50–54, 128	Hubendick, 1957b (male organ)

Genus *Gyraulus* Charpentier (including *Armiger* Hartmann)	planispiral or slightly pseudo-dextral	lateral penis pore; terminal stylet (or rudiment of stylet)	limnic	possibly world wide	87	Hubendick, 1958b (male organ) Radoman and Hubendick, 1959 (Ochrid species)
Genus *Choanomphalus* Gerstfeldt	pseudo-dextral	lateral penis pore; penis stylet; accessory male gland complex	limnic	Siberia		Hubendick, 1954b (morphology)
Tribe Segmentini (further anatomical investigations necessary for proper taxonomy)	planispiral, often with lens-shape	prostate diverticula on branch from sperm duct; preputial organ; multicuspid laterals				cf. Hubendick, 1955a and Zilch, 1959–60
Genus *Segmentina* Fleming	transverse lamellae inside shell	flagella and preputial organ; non-terminal penis pore	limnic	at least Palearctic	48, 49, 127	
Genus *Polypylis* Pilsbry	transverse lamellae inside shell	flagella; preputial organ with accessory duct; lateral penis pore	limnic	South and East Asia, East Indies	194, 206	
Genus *Helicorbis* Benson (including *Intha* Annandale, *Pingiella* F. C. Baker)	no transverse lamellae inside shell	flagella; preputial organ with accessory duct; non-terminal penis pore	limnic	South and East Asia		
Genus *Hippeutis* Charpentier	no transverse lamellae inside shell	flagella; terminal penis pore	limnic	Palearctic		
Genus *Segmentorbis* Mandahl-Barth (including *Acutorbis* Mandahl-Barth)	with lamellae	without or with one flagellum	limnic	Africa		
Genus *Lentorbis* Mandahl-Barth	traces of internal lamella may be present	no penial stylet	limnic	Africa		

Table I—continued

Group	Conchological type	Other morphological characteristics	Habitat	Geographical distribution	Figures in this paper	References
Plus several named groups. See Zilch 1959–1960 and Starobogatov 1967						
Tribe Ancylini						
Genus *Ancylus* Müller	patelliform	prostate diverticula unbranched and directly on sperm duct; muscular flagellum; true penis; penis sheath normally invaganated into preputium	limnic or lotic	mainly Palearctic	74, 75, 93, 99, 134, 159	Hubendick, 1960a, b (Lake Ochrid species) Hubendick, 1970 (taxonomy)
Anatomically unknown and taxonomically unsettled groups of Ancyloplanorbidae						
Genus *Amphigyra* Pilsbry	patelliform or calyptraeiform with dextral apical whorl		lotic	southeastern North America	41, 42, 133	Basch, 1962b (radula)
Genus *Neoplanorbis* Pilsbry	probably pseudo-dextral		lotic	southeastern North America	39, 40, 132	Basch, 1962b (radula)
Genus *Brondelia* Bourguignat	patelliform with sinistral apical whorl		limnic	North Africa	62, 63	Hubendick, 1970

VIII. References

Aardt, W. J. van and van Eeden, J. A. (1969). *Wetensk. Bydr. potchefstroom Univ., Reeks B: Natuurwet.* **5,** 1–72.
Apley, M. L. (1970). *Malacologia* **10,** 381–397.
Baker, F. C. (1911). *Spec. Publs Chicago Acad. Sci.* **3,** 1–539.
Baker, F. C. (1928). *Bull. Wisc. geol. nat. Hist. Surv.* **70,** 1–507.
Baker, F. C. (1933). *J. Morph.* **55,** 1–9
Baker, F. C. (1945). "The Molluscan Family Planorbidae". University of Illinois Press, Urbana.
Baker, H. B. (1925). *Proc. Ca. Acad. Sci.* **4,** 143–168.
Bargmann, H. E. (1930). *J. Linn. Soc. (Zool)* **37,** 1–59.
Basch, P. F. (1959). *Misc. Publs. Mus. Zool. Univ. Mich.* **108,** 1–56.
Basch, P. F. (1960). *Nautilus, Philad.* **73,** 88–95.
Basch, P. F. (1962a). *Nautilus, Philad.* **75,** 97–101.
Basch, P. F. (1962b). *Nautilus, Philid.* **75,** 145–149.
Basch, P. F. (1962c). *Nautilus, Philad.* **76,** 28–33.
Basch, P. F. (1963). *Bull. Mus. Comp. Zool. Harv.* **129,** 339–461.
Berry, A. J., Loong, S. C. and Thum, H. H. (1967). *Proc. malac. Soc. Lond.* **37,** 325–337.
Boer, H. H. and Lever, J. (1959). *Proc. K. ned. Akad. Wet.* **62C,** 76–83.
Bondesen, P. (1950). *Nat. Jutlandica* **3,** 1–208.
Brown, D. S. (1967). *Malacologia* **6,** 155–174.
Brown, D. S., Oberholzer, G. and van Eeden, J. A. (1971a). *Malacologia* **11,** 141–170
Brown, D. S., Oberholzer, G. and van Eeden, J. A. (1971b). *Malacologia* **11,** 171–198
Bouvier, E. L. (1892). *Bull. Soc. philomath. Paris* **4,** 146–153.
Carriker, M. R. (1947). *Trans. Wis. Acad. Sci. Arts Lett.* **38,** 1–88.
Cottrell, A. J. (1911). *Trans. Proc. N. Z. Inst.* **43,** 582–594.
Demian, E. S. (1960). *Ain Shams Sci. Bull.* **5,** 1.
Demian, E. S. (1962). *Göteborgs K. Vetensk.-O. vitterhSamh. Handl.* **9B,** 1–30.
Demian, E. S. and Ibrahim, A. M. M. (1965). *Proc. Egypt. Acad. Sci.* **19,** 21–29.
Dieuzeide, R. (1935). *Bull. Stn. Aquic. Pêche, Castiglione* **1934,** 1–196.
Duncan, C. J. (1958). *Proc. zool. Soc. Lond.* **131,** 55–84.
Duncan, C. J. (1960a). *Proc. zool. Soc. Lond.* **134,** 601–609.
Duncan, C. J. (1960b). *Proc. zool. Soc. Lond.* **135,** 339–356.
Elo, J. E. (1938). *Ann. Zool. Soc. Zool.-Bot. Fenn. Vanamo* **6,** 1–40.
Farnie, W. Ch. (1919). *Trans. Proc. N.Z. Inst.* **51,** 69–85.
Haeckel, W. (1911). *Zool. Jb. Suppl.* **13,** *Fauna Chil.* **4,** 89–136.
Harry, H. W. (1964). *Malacologia* **1,** 355–385.
Harry, H. W. and Hubendick, B. (1964). *Göteborgs K. Vetensk.-o. vitterhSamh. Handl.* **9B,** 1–79.
Heidermanns, C. (1924). *Zool. Jb. Physiol.* **41,** 335–424.

Hescheler, K. (1900). In "Lehrbuch der Vergleichenden Anatomie der Wirbellosen Tiere" (A. Lang, Ed.) Vol. 3. Gustav Fischer, Jena.
Hoff, C. C. (1940). *Trans. Am. microsc. Soc.* **59,** 224–242.
Hubendick, B. (1945a). *Zool. Bidr. Upps.* **24,** 1–126.
Hubendick, B. (1945b). *Proc. malac. Soc. Lond.* **26,** 103–110.
Hubendick, B. (1945c). *Ark. Zool.* **37A,** 1–57.
Hubendick, B. (1946). *K. Svenska VetenskAkad. Handl.* (3) **23,** 1–93.
Hubendick, B. (1947). *Zool. Bidr. Upps.* **25,** 141–164.
Hubendick, B. (1948a). *Ark. Zool.* **40A,** 1–63.
Hubendick, B. (1948b). *Proc. malac. Soc. Lond.* **27,** 186–196.
Hubendick, B. (1949). *Ark. Zool.* **42A,** 1–10.
Hubendick, B. (1951). *K. Svenska VetenskAkad. Handl.* (4) **3,** 1–223.
Hubendick, B. (1954a). *Philipp. J. Sci.* **83,** 319–323.
Hubendick, B. (1954b). *Ark. Zool.* (2) **6,** 503–599.
Hubendick, B. (1955a). *Trans. zool. Soc. Lond.* **28,** 453–542.
Hubendick, B. (1955b). *Ark. Zool.* (2) **7,** 451–457.
Hubendick, B. (1956). *Proc. malac. Soc. Lond.* **32,** 110–126.
Hubendick, B. (1957a). *Proc. K. ned. Akad. Wet.* (C) **1,** 90–95.
Hubendick, B. (1957b). *Proc. malac. Soc. Lond.* **32,** 208–212.
Hubendick, B. (1958a). *Beaufortia* **6,** 243–250.
Hubendick, B. (1958b). *Ark. Zool.* (2) **11,** 427–429.
Hubendick, B. (1959). *Venus, Kyoto* **20,** 263–274.
Hubendick, B. (1960a). *Proc. zool. Soc. Lond.* **133,** 497–529.
Hubendick, B. (1960b). *Soc. Serbe Biol. Arch. Sci. Biol.* **12,** 89.
Hubendick, B. (1961). *Göteborgs K. Vetensk.-o. vitterhSamh. Handl.* **8B,** 1–50.
Hubendick, B. (1962). *Göteborgs K. Vetensk.-o. vitterhSamh. Handl.* **9B,** 1–68.
Hubendick, B. (1963). *Proc. malac. Soc. Lond.* **35,** 64–66.
Hubendick, B. (1964). *Göteborgs K. Vetensk-o. vitterhSamh. Handl.* **9B,** 1–72.
Hubendick, B. (1965). *Proc. Eur. Malac. Congr.* 61–67.
Hubendick, B. (1967a). *Ark. Zool.* (2) **20,** 165–173.
Hubendick, B. (1967b). *Acta R. Soc. sci. litt. Gothoburg* (*Zool.*) **1,** 5–52.
Hubendick, B. (1969a). *Arch. Molluskenk.* **99,** 55–67.
Hubendick, B. (1969b). *J. Conch., Paris* **107,** 109–118.
Hubendick, B. (1970). *Acta R. Soc. sci. litt. Gothoburg* (*Zool.*) **5,** 4–52.
Hubendick, B. and Randoman, P. (1959). *Ark. Zool.* (2) **12,** 223–243.
Hyman, L. H. (1967). "The Invertebrates Vol. 6, Mollusca I." McGraw-Hill, New York.
Itagaki, H. (1956). *Venus, Kyoto* **19,** 25–38.
Itagaki, H. (1959a). *Bull. Azabu. vet. Coll.* **4,** 1.
Itagaki, H. (1959b). *Venus, Kyoto* **20,** 274–288.
Knipper, H. and Meyer, K. O. (1956). *Zool. Jb. Syst.* **84,** 99–112.
Köhler, A. (1894). *Zool. Jb. Anat.* **7,** 1–92.
Koslowsky, F. (1933). *Jena. Z. Naturw.* **68,** 117–192.
Lacaze Duthiers, H. de (1872). *Arch. Zool. exp. gén.* **1,** 437–501.
Larambergue, M. de (1939). *Bull. biol. Fr. Belg.* **73,** 19–231.
Lever, J., de Vries, C. M. and Jager, J. C. (1965). *Malacologia* **2,** 219–230.
McCraw, B. M. (1957). *Can. J. Zool.* **35,** 751–768.

Mandahl-Barth, G. (1957a). *Bull. Wld Hlth Org.* **16,** 1103–1164.
Mandahl-Barth, G. (1957b). *Bull. Wld Hlth Org.* **17,** 1–38.
Marcus, Ev. and Marcus, Er. (1960). *Bolm. Fac. Filos. Ciênc. Univ. S. Paulo Zool.* **260, 23,** 107–139.
Marcus, Ev. and Marcus, Er. (1965a). *Bolm. Fac. Filos. Ciênc. Univ. S. Paulo Zool.* **287, 25,** 19–82.
Marcus, Ev. and Marcus, Er. (1965b). *Bolm. Fac. Filos. Ciênc. Univ. S. Paulo Zool.* **287, 25,** 425–453.
Märkel, K. (1957). *Z. wiss. Zool.* **160,** 213–289.
Merker, E. (1912). *Zool. Anz.* **41,** 337–354.
Morton, J. E. (1955a). *J. mar. biol. Ass. U.K.* **34,** 113–150.
Morton, J. E. (1955b). *Phil. Trans. R. Soc. Lond.* **239B,** 89–160.
Morton, J. E. (1955c). *Proc. zool. Soc. Lond.* **125,** 127–168.
Oberholzer, G. and van Eeden, J. A. (1969). *Wetensk. Bydr. Potchefstroom Univ.* **7,** 1–69.
Odhner, N. H. (1925). *Ark. Zool.* **17A,** 1–15.
PAHO/WHO Working Group for the development of guidance for identification of American Planorbidae (1968). *PAHO/WHO Sci. Publ.* 168. Washington.
Pelseneer, P. (1894). *Mém. Acad. r. Sci. Lett. Belg.* **53,** 1–157.
Pelseneer, P. (1901). *Mém. Acad. r. Sci. Lett. Belg.* **54,** 1–76.
Plate, L. H. (1894). *SB. König. Preus. Akad. Wiss. Berlin* **72,** 1267–1276.
Plate, L. H. (1895). *Zool. Jb. Anat.* **9,** 162–206.
Plate, L. (1897) *Verh. dt. zool. Ges.* **7,** 119–140.
Pretorius, S. J. and van Eeden, J. A. (1969). *Wetensk. Bydr. Potchefstroom Univ.* **6,** 1–109.
Richards, C. S. (1964a). *Malacologia* **2,** 105–129.
Richards, C. S. (1964b). *Publ. Health Rep.* **79,** 1025.
Richards, C. S. and Ferguson, F. F. (1962). *Trans. Am. micros. Soc.* **81,** 251–256.
Robson, G. C. (1915). *Ann. S. Afr. Mus.* **13,** 109–112.
Sarasin, P. and Sarasin, F. (1898). "Die Süsswasser-Mollusken von Celebes". Wiesbaden.
Schumann, W. (1911). *Zool. Jb. Suppl.* **13,** *Fauna Chil.* **4,** 1–88.
Simroth, H. and Hoffmann, H. (1908–1928). *In* "Bronns Klassen und Ordnungen des Tier-Reichs" Vol. 3 (2) 2. Leipzig.
Starmühlner, F. (1970). *Cah. Off. Rech. Sci. Tech. Outre-Mer* (Hydrobiol.) **4,** 3–127.
Starobogatov, Y. I. (1958). *Bull. Invest. Nat. Sci. Soc. Moscou* **63,** 37.
Starobogatov, Y. I. (1967). *Trudy zool. Inst. Leningr.* **42,** 280–304.
Thiele, J. (1931). "Handbuch der systematischen Weichtierkunde." Fischer, Jena.
Walker, B. (1924). "The Ancylidae of South Africa". Printed for the author.
Walter, H. J. (1968). *Malac. Rev.* **1,** 35–89.
Walter, H. J. (1969). *Malac. Rev.* **2,** 1–102.
Yonge, C. M. (1952). *Proc. malac. Soc. Lond.* **29,** 190–199.
Yonge, C. M. (1958). *Proc. malac. Soc. Lond.* **33,** 31–37.
Zilch, A. (1959–1960). Gastropoda. Teilz Euthyneura Handbuch der Paläozoologie **6** (2), 1–834.

Chapter 2

Classification of the land mollusca*

A. SOLEM

Field Museum of Natural History, Chicago, Illinois, USA

* This manuscript was submitted 15th February, 1972 and first returned for minor corrections on 17th October, 1977. Time for complete up-dating was not available and only a very few post-1971 references have been inserted. Little fundamental phylogenetic work on land snails has been published in the interim so this review is not essentially out-dated despite the long delay in publication.

I. Introduction

Detailed knowledge of the structure and physiology of land molluscs is still restricted to a few of the most advanced taxa. Information needed to establish phylogenies of many characters and ultimately to elucidate the points of origin for pulmonate groups from the prosobranch–opisthobranch–pulmonate interface has not advanced since the studies of Morton (1955, p. 164).

The taxonomic restriction of functional and physiologic data is well illustrated by the systematic index in Fretter and Peake (1975). For the pulmonate land snails and slugs, only 25 of 60 families are even mentioned; only 53 (4·9%) of the 1091 genera listed in Zilch (1959–1960) are named; and a mere 47 (0·2%) of the estimated 20,500 species of land pulmonates are mentioned even once. Frequency of citation is even more indicative. Of 179 index lines devoted to land snails and slugs, 79 (43·6%) are for *Agriolimax* (= *Deroceras*) (16 lines), *Arion* (16), *Helix* (30), *Limax* (9), and *Otala* (7). Geographic restriction is equally obvious. There are 47 European, 8 North American, 6 African, 5 Oriental, 3 South American, 2 Australian, and 1 Pacific Island taxa mentioned.

Far from being well known, the land snails remain one of the less known groups of terrestrial organisms. The main function this survey can perform is to outline the progressive evolutionary trends in land snails, to propose a revised ordinal classification that reflects significant evolutionary trends, and to point out areas for investigation that can yield the data needed for conventional phylogenetic analysis.

In order to accomplish this, it is necessary to review and assess the accumulated knowledge concerning land snail diversity, to outline the historical development of and the bases used for current classifications, and then to present a critique of the criteria used to interpret the evidence. Following this, a discussion outlining the levels of land snail evolution precedes an analysis of the Succineidae, Veronicellidae and Onchidiidae, groups whose position in the Gastropoda has been controversial for a long time. Finally, an outline classification to the family level is presented that reduces the ordinal groupings of the Stylommatophora to 3 instead of the customary 4 or 5.

II. Patterns of diversity

A somewhat artificial, but highly stimulating mode of analysing the course of evolution involves recognizing patterns of diversification where a taxon has radiated into local niches or spread over a geographic area, in contrast to progressive changes that involve crossing an adaptive threshold. Usually the latter is followed by diversification, but a distinction between the two phenomena focuses attention on major shifts as opposed to repetitive specializations.

For a variety of reasons, study of non-marine molluscs has lagged behind that of marine molluscs. Most species are smaller and are far less conspicuous, besides being much more difficult and usually uncomfortable to collect. Once found, they are very difficult to identify since the literature is scattered and available only in the largest libraries. In common with all groups of organisms, knowledge concerning the land snails has accumulated in an irregular pattern. First there was a simple amassing of species descriptions. Preliminary attempts to cluster these into genera, subgenera and sections preceded formulation of major taxonomic groupings. Subsequently, monographic treatments of faunistic and family units began to introduce more basic biologic queries concerning the meaning of observed structural variations. From this level of knowledge, detailed comparative and analytic studies led to a phyletic treatment of classification, which in turn provided feedback to faunistic and revisionary studies, finally allowing yet a higher level of understanding.

Because comparatively few people have worked on the phyletic and systematic aspects of land snails in this century, our knowledge concerning higher categories is far less satisfactory than our understanding of generic and specific level diversity. Problems concerning phyletic aspects occupy the bulk of this contribution, but a brief review of specific to family level diversity provides essential background data, as well as emphasizing our fragmentary knowledge of this group.

A. Generic and specific level

Table I surveys the historical pattern of increasing knowledge. The sharp rise in number of genera utilized, from 3 in Müller (1774) to 1091 in Zilch (1959–1960), actually understates the extent of supraspecific clustering. A peculiarity of terrestrial malacology, when compared with marine studies and much less reports on vertebrate or insect taxa, is the very extensive use of subgeneric and sectional groupings. These cannot be tallied easily and their omission sharply reduces the actual degree of cluster recognition. Table II summarizes the periods in which stylommatophoran genera accepted by Zilch (1959–1960) were described. Some 58·4% of the genera were proposed between 1870 and 1929. The sharp drop off since 1930 is indicative of the comparative neglect of land snail taxonomy during this period. Changes in family level units were particularly great between the reviews of Tryon (1884) and Thiele (1931).

Table I. Historical pattern of knowledge concerning Stylommatophoran diversity.

Source	Families	Genera	Named sections	Species
Müller (1774)	—	3	—	121
Lamarck (1822)	—	12	—	231
Beck (1837)	4*	12	126	1477
Albers (1850)	—	23	168	1760
Martens (1860)	5	33	267	3549
Clessin (1881)	14	80	390	8092
Fischer (1883)	12	83	not counted	—
Tryon (1884)	16	96	not counted	—
Thiele (1931)	46	663	not counted	—
Zilch (1959–1960)	57	1091	not counted	—

* Tribes used in the sense of families.

Table II. Dates of description for accepted Stylommatophoran genera.

Years	Number of genera
1758–1779	3
1780–1809	10
1810–1839	120
1840–1869	203
1870–1899	326
1900–1929	311
1930–1959	118
TOTAL	1091

During this period the anatomical investigations of H. A. Pilsbry, H. Watson, C. M. Steenberg, H. B. Baker, and H. von Ihering were focused on patterns of differentiation between groups of genera. This resulted in multiplication of family units.

Data concerning the level of specific differentiation is more difficult to obtain or to assess. Although reasonable lists and summaries of what were then considered to be valid species can be obtained from works through the mid-1800s, the rate of discoveries in the 1880s and 1890s overwhelmed bibliographers. The efforts of Pilsbry, Möllendorff, Kobelt, Böttger, Fulton, Godwin-Austen, Dautzenberg, Pallary and Gude in enumerating faunas of previously unknown areas produced a flood of systematic papers. Pilsbry's "Manual of Conchology, Second Series, Pulmonata" attempted to deal with the Stylommatophora. After 28 volumes concerning more than 11,000 species, issued over almost 50 years, it was still far from completed. The last individual who attempted a bibliographic listing of all stylommatophoran species was Clessin (1881). This was based upon the annotated catalogues of Pfeiffer (1847–1876). The 8092 species tallied by Clessin (1881) represented a dramatic increase over those recorded by earlier workers (see Table I).

The only recent attempt to estimate the species level diversity of stylommatophoran molluscs on a family by family basis is that of Boss (1971). While a useful first approximation, it suffers from infatuation with a ratio and an uncritical evaluation of the literature. Boss focused on the total number of names compared with the number of species listed in modern monographs. He proposed a "synonymy ratio", or "SR", as an index to diversity, with the number of names divided by the number of species. By citing the most notorious systematic messes of the worst workers, and by counting each varietal, colour form or aberrational name in a group as equivalent to a "species name", he documented an impressive sounding case against malacologists and snail diversity. It becomes much less impressive when examined more closely. For example, Boss (1971, p. 92) stated that Laidlaw and Solem (1961) "recognize 74 species for 309 named *Amphidromus*", implying a 4·18 synonymy ratio. This overlooks the fact that 111 of these names had been proposed merely as "varietal" appellations and an additional 53 as geographic races. Only 145 names ever had been proposed as full species, producing a synonymy ratio of 1:96 at most. Since a majority of those names had been synonymized before 1900, a better picture of knowledge concerning species diversity is indicated by the facts that Fulton (1896) recognized 64 species, and Pilsbry (1900c) listed 81 species units (indicating that 7 were probably varieties and 8 were not seen and thus could not be evaluated). Laidlaw and Solem (1961) recognized 74 species by name, 1 as probably new but unnamed, placed 7 "species" into *incertae sedis*, and sunk only 3 species recognized as valid by Pilsbry. Instead of a tremendous reduction from the level of diversity understood in the late nineteenth century, their estimate was retained virtually intact and a modest new number of species added. Since *Amphidromus* contains large and colourful snails that have been avidly sought by collectors, the naming of colour varieties in this group was greater than most, and pre-1900 collecting was far more extensive than for most land snail genera.

The South American camaenid genera *Labyrinthus* and *Isomeria* (reviewed by Solem, 1966b) are cited by Boss (1971, p. 92) as having "a (synonymy) ratio of nearly 2:1". This can be achieved only if every misidentification, every proposal of a new name for one that was preoccupied, and every varietal term is counted as a separate "name". For 70 species and subspecies of these genera, there are 108 specific, subspecific and varietal names, of which 15 were varietal or form names. Hence 93 names for 73 taxa gives a synonymy ratio of only 1:27, which is quite different from the statement of Boss. Furthermore, Pilsbry (1893–1894) listed 52 species in these 2 genera, plus a number of varieties and geographic races, while Solem (1966b) listed 58 species and 3 *nomina dubia*.

These examples involve large and conspicuous snails, while the small and minute species were virtually ignored until after the 1870s. Many other examples could be cited, but it cannot be implied or shown that Pilsbry was overstating specific level diversity in his monographic studies. While papers by such workers as Bourguignat, Servain, Bartsch, Pallary and Germain have been shown to include horrendously oversplit specific level taxa, no such stigma attaches to the publications of Pilsbry, Pfeiffer, Möllendorff, Kobelt, Hedley, Fulton, H. B. Baker or von Martens. Rather than using a "synonymy ratio" (SR) to evaluate specific level diversity, it would make more sense to use a "confidence index" (CI) based on the subsequent evaluation by modern workers of early monographic studies.

An alternative method used by Boss (1971) consisted of assigning arbitrary numerical values to notations in Zilch (1959–1960) concerning whether there were one, a few, many, numerous or large numbers of species in particular genera. This is the same procedure followed by Schilder (1947) in estimating the number of prosobranch species. As has been shown previously (Solem, 1961b), this also leads to gross underestimates. Schilder's estimate of 1562 species of cyclophoraceans is almost 48% under a census of standard references to this taxon. To cite several stylommatophoran examples, estimates given by Boss (1971, pp. 108–109) are contrasted in Table III with species counts from standard monographs supplemented by quick counts from the Zoological Record. The latter covers species described from the time of monograph publication through 1954. Most of these counts were made a number of years

Table III. Estimated specific diversity in several Stylommatophoran families.

Family	Estimate by Boss (1971)	Census from monographs and Zoological Record
Amastridae	198	331
Pupillidae, *s.l.*	535	788
Partulidae	78	121
Endodontoid groups	502	1112
Urocoptidae	390	593
Bulimulidae	559	1294
Corillidae	60	91
Total	2322	4330

ago and have not been updated. This provides a margin for new synonymy and does not materially affect the numbers. The estimate of endodontoid taxa is based on a recently completed series of monographs. On the average, the estimates by Boss (1971) are only 53·7% of those obtained from monographs and subsequent publications by workers with a high "CI". If this ratio holds for all groups of stylommatophorans, then Boss's estimate of 10,719 stylommatophoran species should be upped to 20,000. No ready means of estimating the true diversity in such large and widely distributed families as the Succineidae, Clausiliidae, Helicarionidae, Zonitidae, Camaenidae, Bradybaenidae and Helicidae are available, but where tentative censuses could be made, Boss's estimates are shown to be far too low.

Much more attention has been paid to stylommatophoran faunistics than to revisionary studies in the last 40 years. Yet another approach to species diversity involves analysis of these faunistic reports. Table IV lists

Table IV. Species diversity in selected land snail faunas.

Source	Area or taxon	Total species now recognized	Species named after 1880
Connolly (1939)	South Africa	756	485 (64·2%)
Yen (1939)	China	497	419 (84·3%)
Pilsbry (1940)	United States (Polygyridae)	154	78 (50·6%)
Iredale (1937a, b, 1938)	Australia	598	233 (39·0%)
Powell (1957)	New Zealand	270	217 (80·4%)
Solem (1961a)	New Caledonia	154	37 (24·0%)
TOTAL		2429	1469 (60·48%)

several non-overlapping stylommatophoran faunas, with a breakdown into species currently regarded as valid that were described prior to 1880 and those named subsequently. A variety of situations has been chosen. New Caledonia is an area where the major descriptive and faunistic work was done in the 1870s and hence would have a low post-1880 percentage. New Zealand and China have snail faunas that were virtually unknown prior to 1880. South Africa and the United States present situations where considerable work has been done in both periods. Australia has much of the continent as yet unexplored, but considerable descriptive work on coastal area taxa was accomplished prior to 1880. If these numbers are added up, on the average some 60·48% of the species now accepted as valid were named after 1880.

Since the last catalogue of stylommatophoran species (Clessin, 1881), was based on studies by a worker (Pfeiffer) with a high confidence index and included 8092 species, then a simple calculation becomes possible. If 8092 species represent 39·52% of those species now accepted as valid, then the number of valid species should be 20,474. This figure is remarkably close to the 20,000 estimate arrived at by tallying actual species numbers in several families and then increasing Boss's overall estimates by the same percentage.

Both procedures are open to the criticism that a portion of the species described since the major monographs will be synonyms and that a portion of the species accepted as valid by Clessin (1881) have been shown to be synonyms. Balanced against this must be the fact that large portions of South America, Australia, Africa, east Asia, Indonesia, New Guinea and the Pacific Islands remain *terra incognita* in terms of land snail diversity. As examples, the Family Endodontidae (Solem, 1976) now contains 185 named species level taxa, of which 101 were described in that monograph, and an additional estimated 290 undescribed Hawaiian species level taxa are preserved in museum collections. Field work from 1974 through 1977 in Western Australia and Central Australia has increased the number of camaenid species from 40 to about 125. Any reduction in the 20,000–20,500 estimates because of over splitting will be more than compensated for by new discoveries.

I estimate that the diversity of stylommatophoran land snails is, conservatively speaking, in the neighbourhood of 20,500 species. With 1091 genera, this would be an average of 18·8 species per genus. Although far higher than in most vertebrate taxa, the practice by malacologists of recognizing many subgenera and sections helps explain the apparent gap.

Applying the same techniques to estimating the numbers of land prosobranchs, preliminary figures of 2300 cyclophoraceans, 600 helicinids, 600 pomatiasids and chondropomids, plus some 150 hydrocenids, acmeids, assimineids, and truly terrestrial truncatellids suggest approximately 3650 species. Hence the total land snail fauna would be close to 24,000 species. This estimate contrasts with the "fewer than 15,000 species" indicated by Boss (1971, p. 84).

B. History of classification

Keferstein (1862–1866, pp. 1235–1244) and Pilsbry (1893–1894, pp. xvii–xxvi) have given detailed sketches of pre-1890 stylommatophoran classification, but a brief review is presented here to highlight the more important contributions. During the initial descriptive period of terrestrial malacology, land snails were placed in either *Helix* or *Bulla*. The former genus occupied 8 pages in the 10th edition of Linné's *Systema Naturae* (1758) and was expanded to 11 pages in the 12th edition published 9 years later. Müller (1774) included 119 species in *Helix*, 11 slugs under *Limax*, and proposed new genera for single species of *Vertigo* and *Carychium*. Draparnaud (1805, pp. 23–30) listed 10 genera of terrestrial and 6 genera of freshwater taxa.

The gradual increase in generic units continued and a more complex hierarchy of classification developed. The anatomical studies of Cuvier were instrumental in defining molluscs as a group, while varying subphylum groupings were proposed by several authors. A review of these is given by Hyman (1967, pp. 1–5) together with references to the early literature. Although containing no novelties in classification, the monograph of Anton (1839) with 5 levels, roughly equivalent to class, order, family, genus, and subgenus, is a good example from this period. He separated all molluscs into 2 classes, Cephalophora and Acephala, then divided the gastropods into 24 families.

Major advances during this era involved generic clustering. Beck (1837) showed an intuitive grasp of interrelationships that was not matched until the studies by Pilsbry in the 1890s and early 1900s. Even today, most of the generic units proposed by Beck are utilized in close to their original concept. Albers (1850) and the 2nd edition prepared by Martens (1860) marked the acme of classification by shell characters. In the words of Pilsbry (1893–1894, p. xxii), it is their "... accurate feeling for subtle affinities for which no good reason can be given in words—that has rendered this work the basis of classification for three and a half decades". Simultaneously Pfeiffer was beginning his multivolume catalogue of pulmonate species (Pfeiffer, 1847–1876) that provided clear descriptions, careful synonymy, and logical order for the masses of species being sent back to European students.

A few people began looking at the animal inside the shell. Joseph Leidy (1851) discussed anatomical structures in a variety of North American species, illustrating his observations with perhaps the most exquisite drawings of snail anatomy ever published. Moquin-Tandon (1855) incorporated anatomical notes in his monograph of French species. Schmidt (1855) reported on the anatomy of many European species, recognized the distinction between Basommatophora and Stylommatophora, and presented diagramatic illustrations of the genitalia for numerous species. Although lacking the esthetic quality and fine detail of Leidy's illustrations, Schmidt's drawings carefully indicated the origin and insertion of each tube, and showed relative sizes very clearly. Far too many authors today fail to present these details, thus greatly reducing the usefulness of their illustrations and descriptions.

The Danish zoologist, O. A. L. Mörch (1860, 1865) proposed a hierarchy based upon jaw structure, and this still influences some modern classifications. Expanding on the work of Leidy, Moquin-Tandon,

Schmidt and Mörch, Carl Semper (1870–1880) dissected numerous taxa from non-European regions. The jaw was considered a subordinate character, with features of other organ systems given greater weight in determining affinities. In considering a variety of weighted characters from several organ systems, Semper was the first truly evolutionary malacologist and laid the groundwork for all subsequent classifications. If he had survived longer, the direct effects of his work on pulmonate classification would have been even greater. In his last year he was working on the pallial structures of pulmonates and a posthumous paper (Semper and Simroth, 1894) illustrated the basic morphologic types that were later (Pilsbry, 1900a) cited as key ordinal characters. Simroth's résumé of Semper's data did not exploit their importance, and hence the credit for recognizing and utilizing the data belongs to Pilsbry. Author of more than 5680 new taxa in some 1151 articles published over a period of 75 years, Pilsbry was blessed with a huge capacity for work, an encyclopaedic mind, and an intuitive sense of relationships that exceeded the genius shown by Beck a half century earlier. His first major work involved bringing order to the chaos of the Helicoid taxa, effectively dismembering the Linnean *Helix* into a number of family units (Pilsbry, 1893, 1893–1894). After a brief look at the zonitoid and endodontoid taxa (Pilsbry, 1896), he proceeded to study various Pacific Island forms and proposed (Pilsbry, 1900a, 1900b) a division of the Stylommatophora into several subordinate groups; his thoughts on the latter were summarized in two diagrams (Pilsbry, 1900a, p. 564, 1900b, p. 570). It is unfortunate that he did not subsequently define all these taxa in full and augment his preliminary sketch (Pilsbry, 1893–1894, pp. xxvi–xxviii, of criteria and methods for classifying and unravelling phyletic relationships. A subsequent chart (Pilsbry, 1919, p. 30) contained the significant footnote statement that "If the orders were elevated to subclasses and the suborders to orders, the values would be more closely comparable to the subclasses and orders of vertebrates". Most of Pilsbry's later work was monographic and faunistic in scope. While clarifying the position of many genera and revising numerous families, he did not propose any major alterations in classification.

Little attention was paid to Pilsbry's work by continental European workers. The classification of Pelseneer (1906, pp. 186–190) was based on the jaw structure divisions proposed by Mörch and the separation into Monotremata and Ditremata utilized by Fischer (1883). Thiele (1931) divided the Stylommatophora into 14 co-ordinate superfamilies.

In contrast, Watson accepted the importance of excretory organs in recognizing major suprafamilial units. His work on the evolution and classification of carnivorous snails (Watson, 1915) stands as a landmark for its grasp of evolutionary concepts and rigorous analysis of character variation. Subsequently he investigated the affinities of minute British snails (Watson, 1920), laying the groundwork for the elegant work of Steenberg (1925) concerning the affinities of orthurethran genera and families.

A protégé and friend of Pilsbry's, H. B. Baker, during the 1920s and 1930s investigated the anatomy of many Neotropical and North American taxa, focusing on the zonitoids and "oleacinids", with excursions into carnivorous, bulimoid, sagdid, veronicellid, neritid, and helicinid taxa. His thoughts on land snail classification were presented in a series of reports (Baker, 1955, 1956a, b, 1961, 1962a, b, 1963) that stylistically stand in utter contrast to Watson's rather discursive papers. Baker's often cryptic comments require extensive delving into his previous publications for understanding. His descriptions of the "suborder Mesurethra" (Baker, 1955, p. 109) and "infraorder Holopodopes" (Baker, 1962a, p. 116) approach Iredalean standards of brevity, but fortunately with considerably more information content.

Recent classifications of the molluscs by Zilch (1959–1960), Taylor and Sohl (1962) and Franc (1968) have been adaptations of the Pilsbry–Baker outline. In addition to the perennial queries concerning the Streptoneura–Euthyneura versus Pulmonata–Prosobranchia–Opisthobranchia division of the Gastropoda, recent work on gastropod classification and possible phylogeny has centred on whether the Onchidiidae, Veronicellidae and Rathouisiidae should be considered Opisthobranchia or Pulmonata, and how the Succineidae should be classified. Both of these subjects are discussed later.

C. Characters used in devising classifications

Systematics has emerged from a decade in which biologists met and mingled with computers. The resulting love–hate relationship is affecting the methodology of some systematic research, but has had only passing influence upon the underlying principles. An initial flurry of purely phenetic studies was based on Adansonian concepts with all characters being equal in the eye of the beholder. This fad is now history and we have returned to questions concerning phylogenetic interpretations. Computers are used routinely as aids in handling massive quantities of data, while traditional systematists have been forced to evaluate their methods of selecting and weighting characters. The honeymoon of computer systematics is over, but the split between on one hand the cladistic school of Hennig and Brundin and on the other the phyletic school of Simpson and Mayr still remains. Current arguments concern how to sort out primitive from derived character states, the merits of different criteria for character weighting, and how to proceed from phylogenetic reconstructions to classifications.

It is indicative of molluscan systematics that the decade of ferment had passed before a computer study of molluscan classification appeared. It probably is predictive of the 1970s that this work on lucinid phylogeny (Bretsky, 1970, 1971) is a strong statement for the phylogenetic method.

Here I wish to review briefly the character complexes of pulmonate snails used most frequently to derive classifications, and to suggest neglected areas of study. Part of this was done by Pilsbry (1893–1894, pp. vii–xvii, xxvi–xxviii) in language as applicable today as when it was first written. His suggested use of uncorrelated characters, clear recognition of convergent evolution, concise definition of character weighting precepts, and conclusion (Pilsbry, 1893–1893, p. xxvii) that "a natural (=phylogenetic) classification of Pulmonates should be based upon: organs of protection (shell, mantle, integument of body); organs of locomotion (foot with pedal-grooves, tail gland, etc.); organs of reproduction (genitalia, comparative size of eggs, etc.); organs of nutrition (jaw and teeth, intestinal tract, kidney); nervous system (including sense-organs such as tentacles, etc.); muscle system" contrasts with many recent papers that attempt sweeping conclusions on single character comparisons.

The order followed here represents roughly the time sequence of initial use of the character complexes in classification and does not suggest the order of phyletic import.

1. *Shell form and structure*

An initial dependence on gross shape differences, *Helix* versus *Bulimus* for example, led to the efforts of Ferussac (1821–1822) to use peripheral shape, umbilical contours, apertural barriers, and surface sculpture to cluster species into groups. Beck (1837), Albers (1850) and Martens (1860) progressively refined and modified the selection and weight given shell characters. A few mid-twentieth century workers such as Iredale (1944, p. 323, 1945, pp. 58–59) still insisted that shell characters were primary in establishing classification, but the vast majority recognize that convergence can arise in shell structures as well as other features. There is no doubt that ". . . . positive shell characters (e.g. definite sculpture, columellar folds or teeth) are very useful in classification, but that negative ones (i.e. the convergence exhibited by smooth, featureless or even vestigial shells) have been repeatedly produced in widely divergent groups and may mean practically nothing" (Baker, 1938, p. 6). Even complex shell structures, such as the barriers in the apertures of land snails, have evolved several times in different lineages (Solem, 1972c). While optical examination has failed frequently to differentiate such convergences, analysis by the scanning electron microscope has demonstrated family level differences in microarmature on the apertural barriers (Solem, 1973b).

Using the tool of scanning electron microscopy, there is an untapped world of observations concerning the crystalline shell layers, structure of the periostracum, and methods of bonding the two together. From both fracture and etched sections considerable data can be obtained. Use of microprobe and histochemical techniques will permit the chemical composition of the periostracum to be determined. The report by Meenakshi *et al.* (1969) that sigmurethrous stylommatophoran snails have 1 periostracal layer compared with 2 in freshwater prosobranchs is only the first example.

On a somewhat broader level, the presence of "macro" and "micro" sculpture on the surface of litter dwelling snail shells is a common, but little understood phenomenon. How is this sculpture formed? What is its significance? What are the relative contributions of periostracal and calcareous elements? Can the spacing pattern be correlated with environmental conditions? In the Pacific Island endodontoid taxa, for example, average patterns of major sculpture features are identical under optical study, although in one group (Charopidae) the sculpture is virtually completely periostracal, while in the other (Endodontidae) the periostracum provides a template with the calcareous layers underlying even the microridges (Solem, 1976).

2. *Jaw*

Hooke (1665) called the jaw of *Helix aspersa* its teeth, but not until the proposals of Mörch (1860, 1865) was major use made of the jaw in classifying genera. Absent in most carnivores and virtually undetectable in many orthurethrans (see Cooke and Kondo, 1960, p. 39), the variation in size, strength and ribbing is so great that no systematic importance can be given to its gross features. No significant studies have been made of the chemical structure, mode of formation, and function of the jaw. Quite possibly major differences will be discovered.

3. *Radula*

Traditionally the optical examination of radulae was restricted to vertical views of cusp form and anterior basal plate edges. For minute species, only the sketchiest information was obtainable. Scanning electron microscope studies by Runham and Thornton (1967), Runham (1969), Runham *et al.* (1969) and Solem (1972a, 1973a, 1974a, b) demonstrate a wealth of detail. Herbivorous species are now known to have a complex system of interrow support for the lateral teeth during the stress of cutting (Solem, 1972a, 1973a, 1974a). While the details of this support can differ radically in confamilial sympatric species (Solem, 1972a, Figs 21–29), in stylommatophorans studied to date the support pattern of the lateral teeth depends on the tooth directly in front of it (Fig. 1). In the Ellobiids *Pythia*

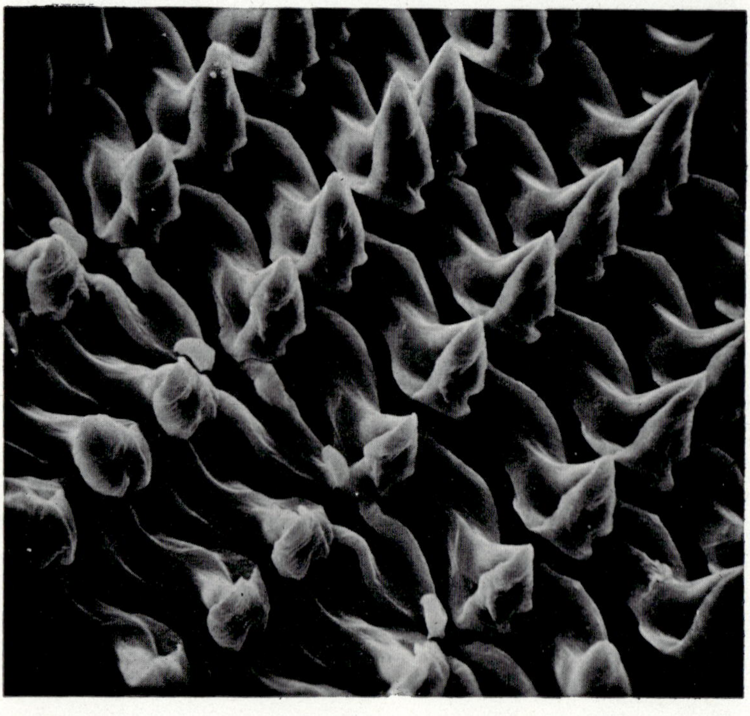

(Fig. 2) and *Melampus* the support system is different in that the first lateral teeth find support from the basal plate of the central tooth in the next anterior row, the second teeth would lie against the first lateral basal plates, etc. Few marginal teeth in any taxa have interrow support systems. Use of microprobe techniques to survey chemical structure and combining freeze-dry or critical point drying techniques with scanning electron microscope observations will revolutionize our knowledge of pulmonate feeding mechanisms. Although mainly useful at the generic and specific level of classification, in some groups such as the Arionacea (see Solem, 1976) radular structure gives the quickest indication of family level grouping. Basic differences in the support for the cutting stroke for carnivorous taxa (Solem, 1974b) reinforce the conclusion that the carnivorous groups have a polyphyletic origin.

Data accumulated from scanning electron microscope surveys of pulmonate radulae (Solem, unpublished) show many exceptions to the "law of mesometamorphosis" (see Pilsbry, 1893–1894, p. xiii), but the number of species examined are too few in number at this time to formulate new generalizations. Convergence in radular structure for algal scraping is frequent, but easily detected by support structure changes (Solem, 1973a, Figs. 7–16, 18–21).

From 1894 to the commercial production of the scanning electron microscope, virtually no advances in knowledge of radular evolution were made. A world of unknown detail is now accessible for observation.

4. *Tentacle structure and retraction*

Schmidt (1855) first pointed out the significance of different eye-spot position and structure of the basommatophoran and stylommatophoran ommatophores. Subsequent to this came the discoveries that some snails have only two tentacles and accessory "labial palps" develop in some carnivorous forms. Differences in eye spot structure and complexity are known in only the sketchiest detail for mainly European species; most detailed examination of eye structure, for example, has been based on

---

Fig. 1. Interrow basal plate support mechanism on the radula of the stylommatophoran succineid land snail, *Oxyloma hawkinsi* (Baird), from Northwest Territories, Canada. Note that the support is provided by the tooth directly in front (left) of the elevated cusp, × 2000.

Fig. 2. Interrow basal plate support system on the radula of the basommatophoran land snail *Pythia pachyodon* Pilsbry and Hirase from Lan Yi Island off Formosa. Note that the support is provided by the central teeth (C) to the lateral on each side of it in the next anterior row, and that succeeding lateral teeth receive support from the basal plate of the tooth in the next row forward (right) that is 1 tooth nearer the central tooth, × 680.

European helicid and zonitid taxa, with no information recorded concerning the more primitive Southern Hemisphere taxa. Quite possibly this is a significant gap in our knowledge.

Among the characters suggested recently as having possible phyletic significance, the pattern of tentacle retraction has least value. Burch (1968) initially concluded that the partly retractile tentacles in the Athoracophoridae indicated that this family was basically different from other Stylommatophora. Subsequently Burch and Patterson (1969) reported partly invaginable tentacular retraction in the Succineidae, which they concluded was intermediate between the normal stylommatophoran pattern of strictly invaginable tentacles and the basically retractile athoracophorid tentacle. They stated that ". . . the Athoracophoridae are related to the Succineidae . . ." and ". . . the Succineidae would seem to be an ideal ancestral type to both groups (Athoracophoridae and typical Stylommatophora)" on the basis of their observations. This is not supportable.

Invaginable tentacles reduce the body surface area exposed to evaporation and provide a mechanism for quick withdrawal of a sense organ from possible harm. Effecting this invagination requires a retractor muscle that is attached to the eye-spot at the tentacular tip being pulled back into the body cavity. There must be sufficient space available within the body cavity for the tentacle to be "turned outside in". If a tentacle is 2 mm long, then the muscle must contract essentially 4 mm before the tentacular base is fully withdrawn. In a typical snail this is readily accomplished. If the head and anterior foot of the snail are radically shortened, as in the Succineidae, there is a space shortage. By fusing or attaching a portion of the tentacular walls to the retractor muscle, a significant reduction in contraction space can be achieved. If 1 mm of the same tentacle was fused to the muscle, then a 3 mm muscle contraction would complete tentacle withdrawal. In the Succineidae there is loose attachment between the retractor muscle and the proximal tentacular walls, hence withdrawal of the tentacular base can proceed before or simultaneously with invagination of the eye tip. This is a simple mechanical adjustment to a space problem. Far from being a character state possibly ancestral to both the athoracophorid and typical stylommatophoran patterns as concluded by Burch and Patterson (1969), this is a secondarily derived solution to a problem resulting from body shortening.

In the Athoracophoridae, the free muscle system has virtually disappeared. Very short tentacular retractors (Plate, 1898) are attached to the body wall lateral to the anterior portion of the buccal mass. Their total length is only slightly greater than that of the tentacles themselves. Here the distal section of the tentacle is fused to the muscle and the purely retractile pattern of tentacle withdrawal described by Burch (1968) is a functional necessity.

A similar space problem has been solved in different ways. The Succineidae use proximal tentacle wall to muscle fusion; the Athoracophoridae use distal tentacle wall to muscle fusion. This is parallel evolution, not part of a phylogenetic series.

5. *Genitalia*

Although Jan Swammerdam's figures were not published until almost 100 years after his observations had been made (Swammerdam, 1752, Plates 4–6), these first illustrations of snail genitalia and anatomy were not equalled in the power of their detail until the work of Leidy (1851). Quite early Fischer (1883, p. 447) used the number of external genital openings as a basis for classification, separating the veronicellid and onchidiid slugs as the Ditremata. The early anatomists focused attention on the total genital system, but it soon became obvious that there were easily detectable differences in the structure of the terminal genitalia, notably the penial complex. This is the region involved in species recognition by the living, lusting snail. Particularly in situations involving congeneric sympatric species, extensive character displacement occurs on the functioning surfaces of the terminal male and female systems (Solem, 1976, pp. 80–82).

The classical German and French histologists amassed detailed information on a few species, but most effort had been devoted to cataloguing the variety of accessory structures and proposing coherent systems of terminology for the organs. Currently an English school focusing from the helicid and zonitid taxa down the phylogenetic "tree" uses one set of terms (e.g. Duncan, 1960; Rigby, 1963, 1965), while North American workers, focusing from the Orthurethra upwards prefer a different nomenclature (e.g. Solem, 1966a). The choice of terms reflects different interpretations of phylogeny. Resolution of these differences is far beyond the scope of this report. The terminology I use is that developed by Pilsbry and Baker and hence may differ from definitions given elsewhere in these volumes.

The basic "systematically useful" points to observe in dissection of a pulmonate were outlined by Pilsbry (1893) as including:

> Shape of the penis and presence or absence of internal papilla and external appendix; presence or absence of flagellum or epiphallus; point of insertion of the retractor muscle and of the vas deferens. Upon the female system should be noticed the absence or presence and form of dart sacs, darts, mucous glands or appendicula; the length of the spermathecal duct; the form of the caeca of the ovotestis and whether they are embedded in the liver or free; and finally whether the right eye-peduncle is retracted between the branches of the genitalia or to the left side.

Most systematic monographs will contain sections delineating variations in these "systematically important" characters within the taxon being revised. Baker (1938, pp. 6–10), Wurtz (1955, pp. 102–113), Cooke and Kondo (1960, pp. 29–39) and Van Mol (1970, pp. 15–27) present good examples of this approach. Thus a great amount of detailed observations on the more variable structures of the genitalia has accumulated as a by-product of systematic revisions.

Unfortunately, comparatively little attention has been paid to interpreting major trends in genital structures. Since the original suggestion by Pelseneer (1894, p. 19) that the hermaphroditic condition in molluscs was achieved secondarily by grafting a male system upon a female, it has been axiomatic among malacologists that a common pallial "spermoviduct" is primitive, and that functional division into separate male and female pallial tracts is a derived state. Elsewhere (Solem, 1972b) I have reviewed evidence suggesting that totally separated pallial gonoducts are the primitive condition, with partial to complete fusion having evolved several times in different lineages. Considerable study of histology and histochemistry of structures in the pallial genitalia will be required to interpret the varying functions and structures in this region. Van Mol (1972, pp. 202–207) discusses variation in the Bulimulidae and lists a few earlier records, but this is only a first step. Equal attention should be paid to the fine structure and function of the talon-carrefour area (receptaculum complex). Preliminary studies by Rigby (1963, 1965), Flasar (1967), and Van Mol (1972, pp. 199–202) show that there are major differences in this region of the genitalia. The pattern ranges from virtually no differentiation in such taxa as the Achatinellidae (Solem, 1972b), and primitive Camaenidae (Solem, 1973c) to highly complex and elaborate tubules in the Bulimulidae (Van Mol, 1972).

Many structures are known to be independently derived. The various mucous glands and dart apparatuses in helicoid and zonitoid taxa, the elaborate appendices and flagelli found on various organs, specialization of the lower vas deferens into an area that produces a spermatophore (=epiphallus), and stimulatory armature on the functioning surfaces of the penis or vaginal region are all features that are found in species scattered through almost every phyletic line. While formerly often cited as evidence for phyletic affinity, more critical work has shown that they differ in detail of structure and are convergent or parallel in evolution. Below the family level of classification, such structures may provide significant data concerning phylogeny, but even here the phenomenon of secondary reduction or loss must be taken into account when interpreting trends.

While knowledge of gross structural patterns is rather extensive, details on the histological level have been ignored in all but a few studies, functional aspects of the reproductive cycle have been almost totally neglected, and major sections of the reproductive tract remain to be analysed.

It is possible to make a few generalizations. In such taxa as the Veronicellidae (Oberzeller, 1970, p. 330, Fig. 6), Achatinellidae (Solem, 1972b, Fig. 3b), Endodontidae (Solem, 1976), primitive Camaenidae (Solem, 1973c), and Microcystinae (Baker, 1938, Plate 1, Fig. 4) the pallial gonoducts are totally separated between the carrefour and atrium. In many Orthurethra (see Steenberg, 1925), including the Partulidae (Solem, unpublished), and the Succineidae (Rigby, 1965) there is partial fusing of the 2 tracts, while in the higher taxa such as the Charopidae (Solem, 1976), most zonitoids (Rigby, 1963), and helicoid genera there is virtually complete fusion of the pallial gonoducts, usually resulting in a common lumen with physiological tract differentiation. The change from separate to fused tracts has occurred several times independently.

Spermatophore formation is indicative of a higher phyletic level than simple transfer of unorganized sperm masses. Together with spermatophore formation, the differentiation of an epiphallus is a progressive characteristic within a given lineage. Elaboration of stimulatory and glandular or dart structures occurs where the need for species recognition signs are the greatest, for example, under conditions of sympatry in areas of historically fluctuating climate and vegetational cover or in island situations involving explosive speciation. All of the above changes are of multiple origin and the presence of particular states in 2 families cannot be used to "prove" phyletic affinity between them.

6. *Foot and pedal grooves*

About 1890 there was a flurry of speculation concerning the significance of a caudal mucous gland, but Pilsbry (1896) showed this was a trivial character. He proposed that the presence of pedal grooves above the angle of the foot margin, which results in a portion of the foot sole extending laterally upwards, delineates the Aulacopoda as opposed to the Holopoda where the pedal groove is at or close to the foot margin. Watson (1920, p. 7, Fig. 1), Wächtler (1935), and Baker (1955) provide further discussion. While a few genera (see Pilsbry, 1946, pp. 231–232) depart from the normal pattern for their group, the basic distinction seems a real one. The functional significance of the change is uncertain.

7. *Pallial complex*

Semper and Simroth (1894), Pilsbry (1900a, b), Watson (1920), Wächtler (1934), and Baker (1955) presented the most significant data concerning the configurations of the pallial organs. Summaries of the basic patterns are given by Solem (1959, pp. 32–36) and Minichev and Slavoshevskaia

(1971). Variables that are cited include the shape and position of the kidney, its relation to the heart, the presence or absence of urinary tubes, the extent and position of lung venation and the pattern of pneumostomal openings. Pilsbry (1900a) and Baker (1955) based their ordinal groupings on the kidney–ureter configurations, while Wächtler (1934) criticized this on the basis that intermediate configurations are known.

The significance of these arrangements are tied up with factors of slug evolution and are considered later (pp. 76–82). Here it is sufficient to note that there are very few studies on the fine structure of pallial organs in other than helicids. Bouillon (1960) summarized previous literature and discussed the ultrastructure of the kidney in *Helix*. Later (Bouillon and Delhaye, 1970) he studied the kidneys of *Succinea* and some basommatophorans, concluding that the kidneys of *Helix* and *Succinea* were the same in basic structure and quite different from those found in the Basommatophora. Recent work in the same laboratory suggests that the kidney structure in the Orthurethra is quite different from the Sigmurethra (Van Mol, personal communication). As in the previous organ systems, gross anatomical data need to be supplemented by ultrastructural investigations using sophisticated techniques.

8. *Mantle lobes and shell laps*

Extensions of the mantle collar forward or laterally are called lobes, while reflected sheets of tissue that partly to completely cover the shell or shell remnant are called shell laps or pseudopallium. These are developed to a greater or lesser extent in such taxa as the Helicarionidae (see Solem, 1966a, pp. 51, 59, 69, 80, 87), Urocyclidae (Van Mol, 1970), some Charopidae such as *Ranfurlya* and *Otoconcha*, and problematic taxa such as *Aillya* (Odhner, 1927). They reach their greatest development in slugs such as *Muangnua* (Solem, 1966a), Arionidae, Parmacellidae and Limacidae. In these taxa the shell remnant is completely or virtually completely enclosed by the flaps. Finally, in the Philomycidae, there is no shell remaining.

Although there has been little detailed investigation of the shell lap and mantle lobe structure, their main function probably is in providing adequate gas exchange surface. When the pallial roof is reduced in area, venation in the Helicarionidae will become concentrated near the mantle collar edge with prominent veins entering the collar and extending into the shell laps (see Solem, 1966a, p. 53, Fig. 8 and p. 81, Fig. 18). Exactly the same result is seen in the Urocyclidae (Van Mol, 1970, p. 86, Fig. 47 and p. 105, Fig. 63) and *Aillya* (Odhner, 1927, p. 5). This is one effective method of coping with the respiration demands resulting from visceral hump reduction. The proposal by Minichev and Slavoshevskaia (1971, p. 359) of a new order, Aillyida, based upon "dermal respiration" in the shell laps of *Aillya*, cannot be accepted. The arrangement is not restricted to *Aillya*, but has evolved in many groups that have undergone visceral hump reduction.

In addition to the gas exchange function, these laps and lobes probably are involved in water balance (Peake, 1968, pp. 337–338).

9. *Nervous system*

The classic study by Van Mol (1967) on the cerebral ganglion provided a quantum jump over our previous level of knowledge. He proposed 4 basic groupings of the pulmonates—Archaeopulmonata for the Ellobiidae and Otinidae, Basommatophora for the rest of the normally included families, Soleolifera for the Onchidiidae, Veronicellidae and Rathouisiidae, and Sigmurethra for the remaining families. Equally detailed studies are needed on other parts of the nervous system. Understanding functional correlations with habitat differences will be required before the details of his phylogenetic conclusions can be confirmed or modified. Previous studies on the gross morphology of the nervous system such as by Bargmann (1930) provided much useful data, but lacked functional interpretations. At lower levels of classification, data on penial enervation have proven useful criteria.

Even less data are available concerning statocyst structure and chemoreceptors. Again we have a neglected field of study.

10. *Other organs*

Data on the free muscle system, circulatory system, and epidermal structure have not yet been significant factors in developing classifications. While often great attention is paid to the orientation of the heart, this is basically determined in taxa with reduced visceral humps by chance factors resulting from space compaction and probably has minimal phyletic significance.

11. *Cytotaxonomy and biochemistry*

Other contributions will discuss these subjects in some depth. The basic point to remember is that these disciplines provide additional sets of data for interpretation. Chromosome numbers are subject to variation through selective pressures just as any other character complex (see discussion of Succineidae on pp. 84–87). The potentially much more significant and difficult task of karyotype analysis has been applied to only a few taxa (see Butot and Kiauta, 1967; Babrakzai and Miller, 1975a, 1975b, 1976). While millions of years ago there may have been a firm correlation between chromosome number and phyletic level, sufficient time has past for aneuploid or polyploid phenomena to mask such simplistic relationships.

Currently much attention is focused on biochemical phenetics in the sense of Throckmorton (1968, p. 108). Comparisons of protein samples from different species yield an index of similarity, but no information as to whether this similarity results from divergence or convergence. It is naïve to assume otherwise. Whether one studies changes in DNA base sequence or amino acid sequence, this is still testing phenetic similarity (Throckmorton, 1968, p. 109). Both cytology and biochemistry can provide major sets of data concerning degrees of similarity. Some of these data may prove to be of great importance in recognizing phyletic lines, while other sets of data will be shown to be of as little importance as is currently given jaw form, presence or absence of a caudal mucous gland, or basic shell shape. Recognition as to which is significant and which is trivial will require far more data than are now available. Many additional studies extending the preliminary reports of Ghiselin *et al.* (1967) and various electrophoresis and serology reports will be needed before these kinds of data can be evaluated.

12. *Summary of characters*

Data from shell structure, radula composition and functioning, genital cytology and functioning, excretory and respiratory system details, neuroanatomy, karyotype analysis and biochemical studies can all be expected to provide information that will supplement and modify conclusions reached from study of the traditional morphological features of shell form and sculpture, jaw and radular cusp form, terminal genital variations, pallial region configurations and external body features. The latter suite of character complexes have been used to create the present classification schemes.

III. Character weighting

Much of the recent literature on systematic theory has been concerned with problems of weighting characters for use in phylogenetic analysis. Data concerning the direction of character change for a multitude of structures are assembled in various fashions, then correlated by one of several mathematical techniques to arrive at either phenetic clusters based on overall similarity or hypothetical phyletic sequences based on applying the "law of parsimony". The method used to select characters for analysis is crucial, with no agreement existing as to procedures (see Voris, 1971, pp. 451–453, 458, for a convenient summary and references).

This approach to systematics has been developed through studies on arthropods and vertebrates, organisms with a wealth of structures that can be quantified, and where the functional significance of many such structures is uncertain or unknown. These methods represent a significant advance over the "single character" approach to classifications endemic to the early 1900s. Unfortunately, the latter occasionally lingers on today in molluscan sytematic literature. In a sense, this approach is not new but merely quantifies the basic analytic procedures of the better "phylogenetic systematists". They traditionally reviewed, analysed subjectively and then organized reams of phenotypic data to produce phylogenies and classifications. While often a few "key characters" would be cited as identification aids, the analysis would be based on a substantially greater number of features.

For most systematic study, this procedure, whether it involves sophisticated mathematical analysis or subjective evaluation of characters, will remain routine and useful. Evolution of characters and degree of character correlation are the level of study. But it leaves unsolved a key problem: what relative weight should be assigned to each character? As each organ or organ system in the snail was analysed, for example, similarities and differences between species and families were recognized. The initial tendency was to equate similarity of phenotype with common descent and to propose "one character" groupings. At one time the shape of the snail jaw, the presence or absence of a caudal mucous pore, and whether the shell form was helicoid or bulimoid each was considered pre-eminent in sorting out family and ordinal categories, yet today they are recognized as varying too much to be very useful at anything above the generic level.

Is there any way to separate "important" from "trivial" characters in deciding what features of an organism today are indicative of major changes that occurred in the past and hence may serve as guidelines to delineating higher categories? Evolution on the molecular, biochemical, cytological, organismal and population levels has proceeded at differing rates at different times. There have been divergences and convergences at all levels of structure.

Can we distinguish recent convergence between lineages that separated a long time ago from weak divergence in recently branched taxa? The "classical pheneticists" claim that we cannot. I think that a reasonable approximation can be made once a certain level of knowledge concerning a group of organisms has been achieved. Making this approximation involves understanding the functional significance of observed major variations and combining this knowledge with decisions as to their relative ecological value. The basic premise is that progressive evolution requires ecologic shifts and should be reflected in higher level categories. In the absence of a meaningful fossil record this necessitates determining what would be the primitive state of both structure and ecology from which the observed variants could have been derived (=outgroup comparisons). Given these data, then decisions as to direction and significance of evolutionary changes become less arbitrary and mysterious.

I suggest that there are 3 levels of change that may be delineated for use in phyletic analysis. First, changes that permit a major shift in adaptive zone. For molluscs this would be initially the migration from water to land life and then the improvement of basic adaptive mechanisms for life on land. Second, there are changes that reflect patterns of diversification. Radiation within broad adaptive zones, small to medium shifts in adaptive zones themselves, and finally fine grain division of adaptive zones operate on this level of evolution. The changes from terrestrial to arboreal, from snail to slug, from herbivore to carnivore, from moist forest to desert life are examples of land snail evolution at this level. Experiments in niche exploitation conceivably can occur without involving major aspects of competition, but most frequently would involve, at least initially, aspects of the third and last level, the level of changes reflecting sympatric species interactions. This can involve simple character displacement in genital structure to enable species recognition, minor specialization in feeding niche to avoid direct competition, or selection for characteristics that would reduce predation or lessen death from local physical factors.

There is, in principle, no clear division between the 3 levels. When molluscs invaded the land, the first slithering snails would have shown only minor variations from their water-bound relatives. Selected improvements over the eons mean that retrospective appraisal of modern land snail structure enables recognizing that there were several different strategies used by snails to adjust to the demands of terrestrial life. We can say with great confidence that the threshold from water to land was crossed several times by members of different lineages. Looking then at patterns of change within each of these lineages, we can correlate certain ecologically adaptive shifts with patterns of structural variation. Finally, within progressively tighter clusters of taxa we can recognize minor structural changes indicative of specific level interactions and adaptive alterations.

Application of this method requires knowledge of structure, function and ecology. It demands recognition that an organism is an integrated unit, so that second and third level adaptations may mask or alter first level structures. Logically, this is a very inelegant procedure. It demands establishing a hierarchy of character variations, saying these sets are involved in establishing the snail on land, these sets are involved in environmental specializations on land, and the remaining sets involve species recognition and local survival tactics. Such sets are then used to produce clusters at different taxonomic levels, resulting in categories that reflect the initial judgement decisions. The circularity inherent is obvious, but I consider it inevitable and unavoidable. It is based on the premise that progressive evolution involves major ecological shifts and should be reflected in classification at higher levels. Diversification, the filling and subdividing of major adaptive zones, results in lesser changes and should be given lower taxonomic level recognition, generally familial and generic in nature. Finally, fine grain habitat shifts and competitive interactions result in speciation. Thus progressively higher taxonomic categories should reflect more significant shifts, and categories of the same rank should indicate different lineages and ways of making the particular transition. A stable old lineage with major diversification into genera and species would involve lower taxonomic levels than a more recent shift that enabled a group to enter a new adaptive zone. While speciation might be minimal, the potential for a new radiation is there and an opportunity threshold has been crossed. Hence I do not follow the cladistic school of hierarchy recognition.

This approach stands in contrast to much of the current literature on character weighting. My emphasis is on functional and ecological analysis of structural importance, rather than on attempts to correlate many characters and to determine the direction of evolutionary change for in-individual characters. Partly this reflects the nature of the group being studied. Compared with the mammalian skeleton and the insect exoskeleton, the snail body is a blob of unformed protoplasm, capable of infinite individual distorsion. Partly this reflects my own aversion to complicated mathematical procedures.

Variation patterns have been evaluated using this methodology. Our knowledge is extensive enough to outline the major directions of land snail evolution, but it is inadequate to sort out most trends between the family and ordinal levels. Study of function and ultrastructure in the areas outlined above (pp. 59–70) will be required before the evolutionary significance of many changes and the reality of many superfamily and subordinal groupings can be determined. Hopefully this review will stimulate others to begin such studies, since several lifetimes will be required to explore the obvious problems using just currently available techniques.

IV. Levels of land snail evolution

The total pattern of land snail evolution is one of cross currents. It is possible to segregate certain basic progressive changes that enabled effective radiation on land, to recognize habitat shifts and evolutionary trends that were released by attaining a particular evolutionary level, and to identify structures regularly altered through interspecific reactions. There are characters correlated with habitat and diet, regardless of phyletic line, but there also are some identifiable trends that seem to have quietly progressed in several lineages without regard to the major changes. One of these is the tendency towards fusion of the pallial gonoducts described on

p. 67. Another is the progression from a loose sperm mass to a soft sperm packet, and finally to a horny spermatophore such as is typical of the Urocyclidae and Helicarionidae. Space does not permit giving further consideration to these patterns or various physiological trends.

Major emphasis here is first on the basic adaptations to life on land, secondly on the selective forces responsible for and patterns of morphologic change resulting from evolution towards a slug-like form, thirdly a very brief review of habitat and dietary shifts, and fourthly a quick survey of changes at the level of interspecific reactions.

A. Basic adaptations to life on land

The key factor to be considered concerning snails on land is the problem of water retention. Although most authors (e.g. Hunter, 1964, pp. 103–113 and Hyman, 1967, pp. 626–633) emphasize the behavioural aspects of seeking out moist microhabitats and their ability to aestivate during dry periods, there are basic structural adaptations involved.

Land prosobranchs retain the open mantle cavity found in marine forms and depend upon retreating into the shell with the operculum sealing the opening. The effectiveness of this seal is indicated by the apparent necessity for many monsoon climate taxa to have accessory breathing tubes (Rees, 1964). This combination is a superb device for lasting over dry spells, but the open mantle cavity with its configuration of openings basically unchanged from marine ancestors is highly inefficient. There is considerable area exposed to evaporation, and the discharge of metabolic wastes requires some water loss to flush out the anterior cavity. Activity periods must be restricted to times of quite high humidity if excess water loss is to be avoided.

Pulmonates have the mantle edge fused to the foot and neck of the body, usually with only a single opening to the exterior. In addition to reducing the surface area exposed to direct evaporation, the closing of the mantle produces a large internal cavity behind the pneumostome in which a supply of water can be carried. This "pallial water" (Blinn, 1964) provides an ambulatory snail with its own water supply that can be drawn upon as needed. This permits greater freedom of activity for the land pulmonate than the land prosobranch as activity can continue in times of less than optimum humidity. Undoubtedly the almost complete restriction of land prosobranchs to the wet tropics reflects this difference. The helicinid *Hendersonia* in North America plus the modestly diverse pomatiasid genus *Pomatias*, Acmeidae, and the diplommatinid *Cochlostoma* in Europe are the only Northern land prosobranchs.

The pulmonate lacks the opercular seal and must depend on other devices during dry spells. Most species build replaceable closing sheets of mucus or calcified mucus called epiphragms to retard water loss when the snail is retracted. In such species as *Helix aperta* Born, the epiphragm is so heavily calcified as to approach an operculum in effectiveness. Frequently several epiphragms will be secreted at intervals. Other species, particularly arboreal taxa such as *Liguus* or *Achatinella*, or desert taxa such as *Eremina*, *Sonorella* and various bulimulids, will seal the shell aperture to a leaf or rock surface in addition to epiphragm production.

For either the prosobranch or the pulmonate strategy to have been effective, the basic preadaptive value of the snail's pallial cavity cannot be overemphasized. It has a simple empty space function, providing a volume capacity into which the head and foot can be withdrawn. In marine taxa this space provides the head and foot some protection from predation. For upper level tidal zone snails, this space allows the snail to reduce the surface area exposed to desiccation. The shell provides permanent protection for the visceral hump, but the extended foot and head have a high surface area–volume ratio. Retraction into the pallial cavity (and hence the shell) reduced the evaporative surface to the cross-sectional area of the shell aperture, a small percentage of the exposed body area in a crawling snail.

In moving from water to land, snails had a movable retreat against drying out (the shell), equipped with a built-in garage (pallial cavity) for the ambulatory apparatus (foot) and feeding-sensory areas (head). The "garage" also can serve as a water supply "savings bank". The body of a "... land snail is tantamount to a leaky bag of water, its existence threatened by water depletion and by its polar opposite—the danger of waterlogging through excess osmotic uptake" (Blinn, 1964, p. 336). The ability to tolerate wide swings in water content is well documented, particularly for higher helicids and slugs. As summarized by Hyman (1967, p. 627), *Helix*, *Arion* and *Limax* can survive body weight water loss of, respectively, 50%, 60–66%, and 75–80%. As usual, practically no data are available concerning less advanced taxa, the data of Warburg (1965) on an Australian pupillid and two camaenids providing a notable exception.

Given the basic preadaptive equipment of shell and pallial cavity retreat area, what improvements would increase the efficiency of water conservation? The development of slime that retards evaporation of water is an obvious change. But data on the properties of pulmonate slime are minimal.

A second method is to provide for water resorption from excretion products and to minimize any external venting of pallial water. In the marine prosobranchs, the opening of the kidney is into the pallial cavity and water currents passing through the pallial cavity carry off wastes. Little is known concerning the physiology of excretion in the pulmonates, with short summaries by Martin and Harrison (1966, pp. 371–372),

Hyman (1967, pp. 588–591) and Runham and Hunter (1970, pp. 77–82) providing entry to the literature. What is known in general outline for the pulmonates is the pattern of development for a new organ, a ureteric groove or tube. From the extent of vascularization associated with this, water resorption is assumed, but direct evidence is available for only a handful of species.

In many pulmonates, the kidney opens inside the pallial cavity at a greater or lesser distance from the pneumostome. Excretion products must then be flushed out of the cavity to the exterior. In the Sigmurethra there are various combinations of ciliated grooves and/or closed tubes that arise from the kidney, pass posteriorly to the hindgut, then reflect forward and continue partway to or reach the pneumostomal opening. Having the distinct groove or tube permits water conservation. Having it exit directly to the anterior means that water need not be "wasted" in flushing out the cavity.

In addition to the basic behavioural patterns and use of a operculum or epiphragm that is common to all land snails, the pulmonate pattern of a "closed" pallial cavity with reserve water supply is a major progressive step. With the addition of a water resorption tube that opens to the exterior from the excretory system, another major increase in efficiency resulted. The evolutionary progression is open pallial cavity, closed pallial cavity with excretory product flushing, closed pallial cavity with ureter opening externally.

B. Evolution of slugs

There are situations where the problem of water supply for land snails essentially does not exist. In the true rain forest regions and on windward slopes of high mountains, moisture supply is abundant through the year. The value of the pallial cavity as a retreat to avoid dessication is greatly lessened if not completely lost. At the same time, snails are faced with a great physiological burden in extracting sufficient calcium from the environment to build a shell. The shell also serves a protective function against predation, but this requires a heavy calcium investment to make a thick shell. If the "space retreat" function is lessened because of the adequate moisture and the calcium extraction burden is great, then there would be a dual selective pressure for shell reduction. To reduce the burden of calcium extraction leads first to a thinner shell, since correlations between shell thickness and calcium availability were demonstrated nearly 50 years ago (Oldham, 1929). Mutations that reduce the size of the shell would further the process. The initial stage in such shell size reduction would involve compacting the pallial cavity space and thus shortening the visceral hump. This would not have a deleterious effect on snails in a constantly moist environment, since the "space retreat" function is unnecessary.

The logical end result of a shell and visceral hump reduction is a slug. If we look at areas of the world where slug diversity is greatest, we find such regions as the southern slopes of the Himalayas, the Salmon River and coastal mountains of Western North America, mountains of Central Africa, the Alps, the Caucasus, and Appalachian chain of Eastern North America. These are, for the most part, basically volcanic mountains. There is low calcium availability and a constant moisture supply. I submit that the basic selective pressures that have led to slug evolution are steady availability of moisture and scarce calcium resources. The fact that slugs now are associates of human habitation merely indicates that man provides microenvironments with adequate water supply, but the basic areas of slug diversity are in mountain regions with little calcium and high rainfall.

It is equally noteworthy that the evolution of slugs is not random among the Pulmonata. There are no slugs derived from either the Orthurethra or the Mesurethra. This has been pointed out by Runham and Hunter (1970, p. 22) among others, but no explanation has been offered previously for these facts. Reflection on basic structure shows that in both of these groups the excretion products are released well within the pallial cavity and are subsequently flushed out. Both groups lack a water-resorbing ureter to conserve water. They need the "pallial space" for maintaining a water reserve. Possession of a water resorbing ureter was preadaptive to slug evolution among the Stylommatophora. Problems connected with the systellommatophoran slugs will be discussed later.

Slug evolution is a gradual process, and in several areas of the world extant species show virtually every intermediate stage from fully shelled snail to naked slug. The pallial cavity of a stylommatophoran snail involves more than just space. Its roof is vascularized for gas exchange, the hindgut runs along the parietal-palatal margin to the pneumostome, the heart plus kidney nestle at the posterior margin, while in the Sigmurethra the ureter runs to the posterior and then reflects forward along the hindgut to the pneumostome. If the pallial area is compacted, adequate provision must be made for respiratory needs while the pallial roof organs must be jammed and distorted into more limited space.

Initially the lung roof can become much more heavily vascularized to make up for reduced area as in amphibulimine Bulimulidae (see Van Mol, 1972, pp. 188–193, Figs 2–5), but frequently the compaction continues to a point that lung roof area is inadequate. As mentioned above (p. 68) the vascularization can invade mantle lobe or shell lap, it can become concentrated in pouch-like lobes that hang pendulously from the remaining roof margin in such completely unrelated taxa as *Omalonyx*, *Cystopelta* and

Parmacella (Solem, unpublished), or in the final stage of compaction it can become a mass of blind tubules radiating from the pneumostome as in the Athoracophoridae (Plate, 1898, Plate 14, Figs 25, 26). These variations in structure are all suitable solutions to a space problem. Retention of adequate gas exchange area has been solved by extending the mantle edges back over the shell, clumping the vascularized area in hanging pouches, and, in the most extreme form, concentrating the gas exchange area into the "pseudotracheal" system of the Athoracophoridae. These changes are compensations for space alterations in the process of slug evolution. They should be viewed as secondary level phenomena.

Alterations in the kidney–heart–hindgut relationships are equally varied. In the Thailand Helicarionidae, for example, there is a clear sequence. *Durgella libas* Solem (1966a, p. 53, Fig. 8) has a perfectly normal pallial cavity in a shell of 3·5–4 whorls; *Megaustenia siamensis* (Haines) (loc. cit., p. 81, Fig. 18) has the pallial cavity shortened with the shell reduced to 2·25 whorls and the kidney bent downwards by the intruding intestinal loop; *Cryptaustenia gadinodromica* Solem (loc. cit., p. 60, Fig. 11) has the kidney not only bent downwards but actually reflected forwards into a partial "U" pattern; while in both the "near-slug" *Austenia doisutepensis* Solem (loc. cit., p. 89, Fig. 21) and the slug *Muangnua limax* Solem (loc. cit., p. 71, Fig. 15) the kidney is completely "U"-shaped with the reflected lobe longer than the pallial roof section. The latter two differ in their degree of compaction. In *Austenia* the pallial length has been shortened to the point the heart lies virtually perpendicular to the body axis, while in *Muangnua* the heart and kidney lie at about a 45° angle. In no sense are these genera in linear evolutionary relationship. They do show successive stages in compaction resulting from simple bending of the kidney as the pallial cavity shortened.

African Urocyclidae (Van Mol, 1970) show first shortening, then slight bending of the kidney downward. This is followed by a lateral extension of the bent portion towards the hindgut (Van Mol, 1970, p. 143, Fig. 89) and eventually to a transverse kidney with anteriorly located transverse heart (loc. cit., p. 12, Fig. 2).

In Pacific area Charopidae (Figs 3 and 4), the typical pattern is for a "U"-shaped kidney, with one lobe next to the heart and the other lobe of almost equal length appressed to the hindgut (Fig. 3c). A narrow strip of lung roof is visible between the primary and secondary ureters. When the pallial cavity is shortened, the kidney becomes progressively shorter, broader, and the angle between the primary and secondary ureter becomes greater (Figs 3f, 4c, f). This correlates quite well with the gross reduction in whorl count, from 5·38 in "*Charopa*" *complementaria* (Mousson), to 4 in *Pilsbrycharopa gressitti* (Solem), only 3·13 in *Paryphantopsis dauloensis* (Solem), and 3·25 in *Flammulina zebra* (Le Guillou) (see Figs 3

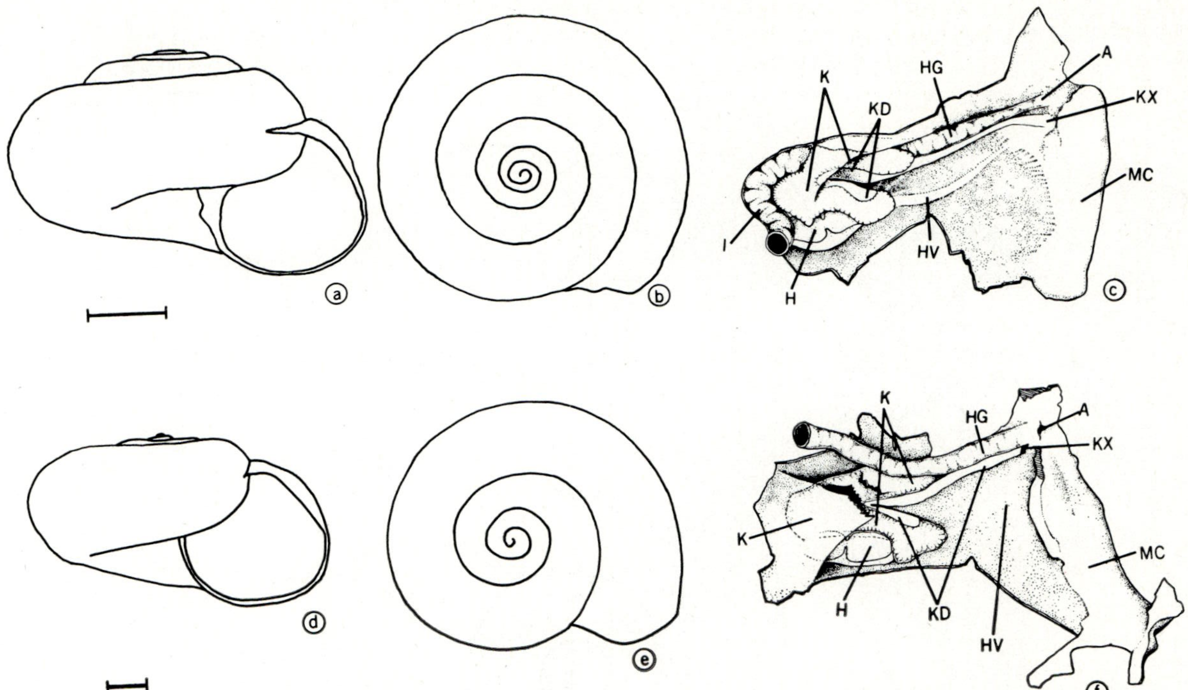

Fig. 3. Change in shell form and pallial organ configuration in "*Charopa*" *complementaria* (Mousson) from Western Samoa (a–c) and *Pilsbrycharopa gressitti* (Solem) from New Guinea (d–f). Scale lines equal 1 mm. Explanation of anatomical symbols: A=anus; H=heart; HG=hindgut; HV=principal pulmonary vein; I=intestine; K=kidney; KD=ureter; KX=external ureteric pore; MC=mantle collar.

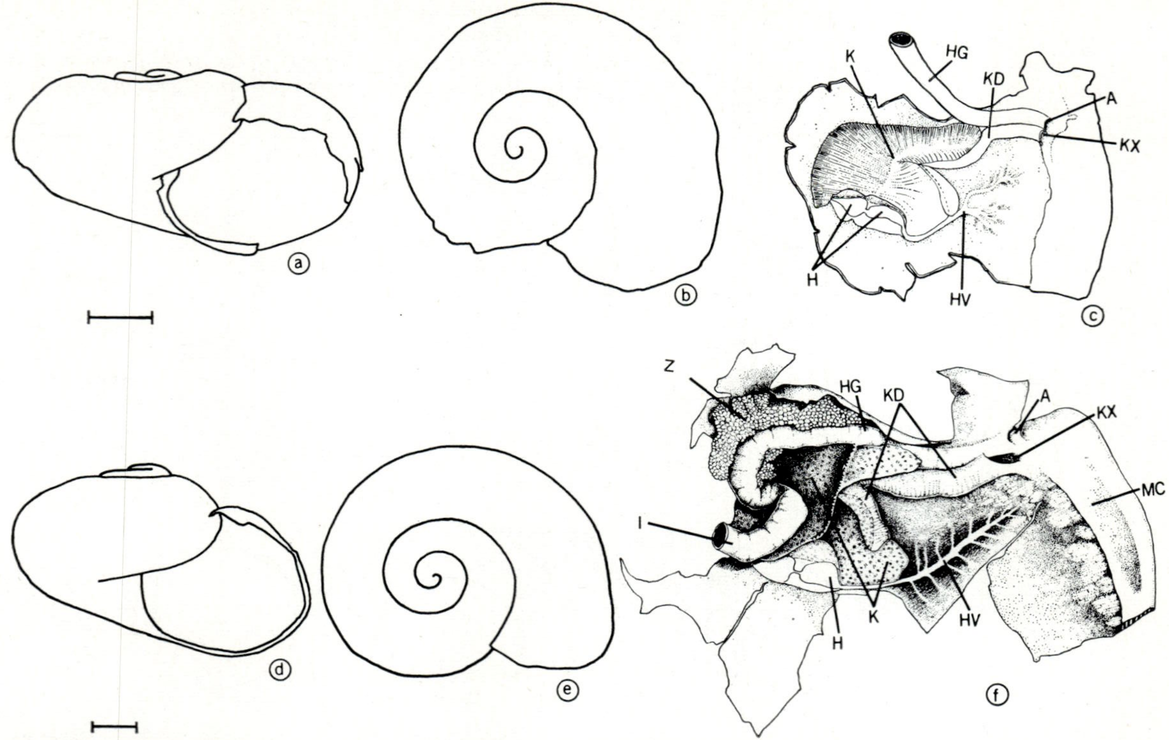

Fig. 4. Change in shell form and pallial organ configuration in *Paryphantopsis dauloensis* (Solem) from New Guinea (a–c) and *Flammulina zebra* (Le Guillou) from North Island, New Zealand. Scale lines equal 1 mm. For explanation of anatomical symbols, see Fig. 3.

and 4). There is progressive rotation of the rectal kidney arm away from the path of the hindgut along the parietal-palatal margin in the New Guinea *Pilsbrycharopa* and *Paryphantopsis*, but in the New Zealand *Flammulina* there is no separation of kidney and hindgut. Note how the whorl count reduction correlates with proportional apertural cross-sectional area. All of these taxa retain a helicoid shell. If, however, the spire of *Flammulina* or *Paryphantopsis* was elongated, then the shell shape would be that of *Succinea*. It is noteworthy that only a slight continuation of the trend seen in these charopids would result in a completely elongated, transversely oriented kidney and primary ureter. Slight flattening of the kidney would then exactly duplicate the "heterurethrous" kidney of the Succineidae. With his great taxonomic intuition, Pilsbry (1948, p. 771) indicated that "the short, wide shape of the kidney may be related to the shortening of the whole pallial cavity in this group."

Division of the Bulimulidae into several families has been proposed by Zilch (1959–1960, pp. 473–519). In a recent paper Van Mol (1972, p. 223) reviewed structures in several organ systems and concluded that reduction of the free muscle system and rotational compaction of the pallial organs were the sole characteristic by which the Amphibuliminae could be separated from the Bulimulinae. The closure of the secondary ureter, altered orientation of the heart, and concentration of lung venation (Van Mol, 1972, pp. 188–194) in the pallial region also duplicates the heterurethrous pattern of the Succineidae. His study provides another example of evolution towards a slug in a sigmurethrous group.

While the details of change in the above taxa obviously differ, the basic alteration is that of adjusting organ position and form to reduced space. This has happened in many ways. Hence observations that in the Arionidae and Philomycidae the kidney surrounds the pericardium (Pilsbry, 1948, pp. 661, 749), in the South African *Apera* the kidney and ureter are crammed above the tip of the tail (Watson, 1915, Plate IX), and in the Athoracophoridae the kidney and ureter undergo incredible convolutions (Plate, 1898, Plate 14, Figs 23, 25) must be interpreted as representing secondary phenomena, different means of coping with the problems resulting from reduced pallial space.

Reduction in pallial space also leads to a host of shifts in digestive and reproductive organs. The stomach moves into the foot area, the p state uterine areas of the genitalia become greatly shortened or coiled, and the ovotestis may become displaced. Excellent diagrams illustrating one coherent pattern of change are given by Van Mol (1970, p. 14, Fig. 3). Different patterns are shown by the Thailand genera *Muangnua* and *Austenia* (Solem, 1966a, pp. 74–76). Each slug taxon has its own peculiarities (see also Van Mol, 1972, pp. 188–194).

Finally, the shortening leads to minor alterations in the pneumostomal region. There is a tendency for a "bladder" to be formed lateral to the external ureteric pore (Runham and Hunter, 1970, p. 78, Fig. 32) and the pneumostome may be divided into two or three separate pores (compare Solem, 1966a, Figs 8, 11b, 15, 17c). Quite frequently the free muscle system is drastically reduced, as in athoracophorids, limacids, helicarionids (Solem, 1966a, pp. 73, Fig. 16a, p. 91, Fig. 22b) and the amphibulimine bulimulids (see Van Mol, 1972, pp. 194–196).

In summary, the initial impetus for slug evolution is ecological, with scarcity of calcium and continued availability of moisture providing dual selection pressure for shell and hence visceral hump reduction. Possession of a water resorbing ureter is an apparent preadaptation. Modifications in respiratory surfaces, pallial organ configuration, and digestive–reproductive systems should be viewed as space adjustments. The bewildering variety of these changes only indicates that the trend towards a slug form has been popular among sigmurethrous land snails.

C. Trends in diet and habitat shifts

Land snails are basically scavengers and detritus feeders in the litter zone, utilizing bits of decaying vegetation and fungal hyphae for food. Several divergent dietary trends can be recognized. The rastriform teeth of achatinellids (Solem, 1972a, Figs 18–20), punctids (Solem, 1973b), and some arboreal helicarionids such as *Durgella* (Solem, 1966a, p. 64, Fig. 13b) probably represent convergent modifications for piercing individual cell walls of algae and fungi. The broad, spade-like cusps of such diverse taxa as *Papuina* (Solem, 1972a, Figs 12–16), partulids (Solem, 1973a, Figs 13–16), enids (Solem, loc. cit., Figs 9–12), and various bulimulids are typical of arboreal taxa. They function in scraping loose sheets of algal or lichen tissue. Data concerning the patterns of morphological change involved in carnivorous snails have been presented by Watson (1915), while data on radular functioning are given in Solem (1974b). Each of these shifts from the basic detritus feeding has occurred independently in several groups. At present, recognition of these trends is based mostly on cusp outline optical studies, but data accumulating from scanning electron microscope analysis are adding significant new dimensions. Divergent patterns of musculature, radular position, and feeding strokes need to be studied for full data analysis.

Arboreal snails, in addition to the blunt spade-like radular cusps (Pilsbry, 1893–1894, p. xiv), tend to have brightly coloured shells with variegated patterns (*Liguus*, *Achatinella*, *Partula*, *Papuina*, *Amphidromus*), a well-developed caudal mucous pore (many helicarionids and some charopids), and, on balance, a greater tendency towards ovoviviparity.

Microsnails in the litter zone tend to have elaborate sculpture on the shell surface as a means of lessening adherence of particles to the shell, although the higher zonitoids seem to have adopted the strategy of a glistening, almost waxy periostracum. Microbarriers in the shell aperture arose in numerous families, probably from small irregularities that aided extension of the snail's body (Solem, 1972c). Large barriers in the shell aperture serve to reduce arthropod predation (Solem, loc. cit.).

Dry area and desert snails tend to have thick white shells with heavy sculpture on the shell surface if the species aestivates in the open or on bushes. More secretive taxa, such as *Thysanophora*, *Holospira* and *Sonorella*, stay hidden in rock piles and retain the colouration of wet area relatives. Behavioural and physiological adaptations to desert life involve the ability to aestivate for periods of up to several years, reduction in both the frequency and number of eggs produced (Walton, 1963), and buffering waste products by resorbing part of the shell itself. Structural changes can involve drastic reduction in the size of the genital organs in such taxa as *Sonorella*, where the penial complex length may be almost twice the shell diameter in wet area forms such as *S. virilis* Pilsbry to only one-fifth the shell diameter in arid zone species such as *S. delicata* Pilsbry and Ferriss (see Pilsbry, 1939). Other taxa show actual loss or drastic size reduction in accessory genital apparatus such as the dart sacs and mucus glands (Gregg, 1960).

Few people have attempted comparative studies of ecological equivalents from different biogeographic regions. A world of opportunity exists to expand the brief comments above into a full analysis concerning the patterns of structural changes associated with changes in diet and habitat.

D. Specific level interactions

Sympatric species must be able to discriminate between reproductively compatible mating partners and those that are genetically incompatible. The mechanism for this is partly behavioural as determined by numerous studies of courtship patterns, but mainly structural through elaboration of stimulators and glandular pads on the functioning surfaces of male and female terminal genitalia. When a group such as the Endodontidae dispersed across the Pacific by island hopping, the pressure for change in genital structure was minimal, as long as only one species lived in an area. Whenever 2 congeneric species are sympatric, one or both will show strong character displacement in terms of functioning penial surfaces (Solem, 1976). In continental areas subject to the drastic climatic changes of the Pleistocene, the ebb and flow of taxa with shifting vegetation zones caused innumerable contacts between congeneric taxa. The wild diversity of pulmonate genitalia in North American and European species probably directly stems from this mixing of species during climatic fluctuations. On islands where one family has diversified extensively, such as the Charopidae in New Zealand and New Caledonia, the variety produced by such interactions is almost unbelievable. The phenomenon is not restricted to pulmonates, with members of the poteriid prosobranch genus *Ostodes* showing equivalent changes in genitalia of sympatric species (Girardi, 1978).

Most of the "systematically useful" variations in the terminal genitalia of pulmonates as outlined by Pilsbry (1893, p. 389) are in the morphological zone affected by sympatric phenomena. Interpretation of penial and vaginal structures in any pulmonate must take into account the possibility of sympatric character displacement. Variations in spermatophore structure also lie at this level, while epiphallic structures, for example, would be more indicative of second level evolution.

Besides the variations in the genitalia, the structure of the radular cusps and inter-row support mechanism can vary rather drastically in closely related sympatric species from a West Australian desert area (Solem, 1972a, Figs 21–19). This suggests food specialization within one microgeographic area as a means of lessening competition in a region where plant cover is sparse and slow growing. Preliminary radular observations on tropical taxa suggest that similar feeding niche specializations may exist in rain forest species.

Both genetic and phenotypic variation in shell form, size, colour, thickness and growth rates are involved in adjustments of populations to local conditions. From the well-documented colour selection by predators in *Cepaea* to the variation in populations of the North African desert *Eremina*, local selective pressures have resulted in great variation that must be interpreted in light of local ecological factors before being assigned major systematic significance.

E. Position of the Succineidae and Aillyidae

Since the initial description of the group Heterurethra (Pilsbry, 1900a, p. 564), the Succineidae have been a prime focus of attention from people attempting to discuss pulmonate phylogeny. The basic criterion used to establish this group was the elongated, transverse kidney position. Systematic criteria for genera and subfamilies were developed by Quick (1933) and Odhner (1950). Baker (1955, p. 111) concluded that ". . . the ancestors of the Sigmurethra probably had an aulacopod foot and a heterurethrous lung, which perhaps was rather short." Rigby (1965, p. 483) suggested ". . . that the Succineidae are more appropriately placed among the opisthobranchs than among the Stylommatophora" based upon many misinterpretations of structural significance. Burch (1965, p. 269) reported extremely low chromosome numbers in several catinelline succineids and speculated the low chromosome numbers "... may be a very primitive number from which higher numbers resulted ..." after having earlier on the same page cautioned that "these extremely low numbers perhaps shed some doubt on the usefulness of chromosome numbers in assessing the degree of specialization within and among systematic groups." Subsequently, Butot and Kiauta (1967, p. 163) and Butot (1967) proposed full ordinal status for the Succineidae within the Stylommatophora. Additional chromosomal studies by Patterson (1968, 1970a), Kiauta and Butot (1968) focused on, respectively, number counts and attempted karyotype analysis. Solem (1969) suggested that the Succineidae be classified as a superfamily of the Aulacopoda. In contrast, Minichev and Slavoshevskaia (1971) proposed separate ordinal status for both the Succineidae and Aillyidae.

In view of this wide divergence of opinions, special attention must be paid to the morphological patterns exhibited by the group. The stomach has a posterior excretory caecum in Succineidae that is convergent with similar structures in Basommatophora, other Stylommatophora, and browsing marine opisthobranchs (Rigby, 1965, p. 474). All these groups can feed on algal materials and this convergent food specialization should not be used to imply phyletic affinity. The rectal caecum found in many limacid and athoracophorid slugs is another example of this structure (Pilsbry, 1948, p. 531, Fig. 286B). The position in *Limax* is different than in *Succinea*, identifying the convergent nature of these caeca.

Reproductive tract characters cited by Rigby (1965, pp. 475–481) as linking the Succineidae with the opisthobranchs are the presence of a discrete prostate gland on the vas deferens rather than a fused spermoviduct, lack of a spermatophore, triaulic pallial reproductive tract, and eggs bound into a mass of spawn. Her comparisons were made with an advanced zonitid, *Oxychilus*, and ignore the vast literature on less specialized stylommatophorans. Discrete prostate glands, absence of a hard spermatophore, and separated diaulic or triaulic pallial gonoducts are more common in the Stylommatophora than are the specialized conditions described for *Oxychilus* and *Helix* (see Solem, 1972b). Rather than differentiating them from Sigmurethra, these features emphasize that succineids are generalized taxa within the Sigmurethra. Laying an egg mass simply correlates with the wetter habitat of the group and probably represents a simple reversal of the trend to a shelled egg.

Based on the reports of very low chromosome numbers in succineids belonging to the Catinellinae (see Patterson, 1968 for the scattered references), Rigby (1965, p. 481) suggested that the Succineidae could be placed in the Opisthobranchia. The chromosome numbers 5 and 6 reported in several Hawaiian, North American and European species are the lowest known in the Stylommatophora, but low number does not necessarily indicate primitiveness. Stebbins (1966, p. 1466) reviewed the significance of reduction in chromosome numbers in terms of ecological fitness. Aneuploid reduction in plants has occurred in ". . . species groups in which outcrossing prevails, so that the plants are heterozygous at a large portion of their gene loci, and which occupy pioneer habitats." The desirability of colonizing and populating a pioneer, ephemeral habitat with individuals having minimally varied genotypes has high selective value. Reduction in chromosome numbers reduces both recombination and cross-over possibilities and hence variation. The habitat of the North American and European succineids with low chromosome numbers correlates perfectly with their behaviour as pioneer and ephemeral species. According to Pilsbry (1948, p. 843), "survival of the few widely scattered Quickellae (=Catinellinae) may have been favoured by their occupation of an ecologic niche . . . small thickets in the sandy shore zone." The European *Quickella arenaria* (Bouchard-Chantereaux) lives in damp depressions among shifting sand dunes (Baker, 1965). This qualifies as a pioneer, short term habitat under any criteria. The Hawaiian occurrence might seem anomalous, since Hawaiian succineids occur in high, wet forest. But the Hawaiian fauna is derived through overseas dispersal. A shore sand dune succineid accidentally stuck to the foot or feathers of a shore bird is more likely to have represented the first introduction than is an inland, wet marsh succineid and bird carrier. The latter birds are far less likely to fly to Hawaii from mainland areas than are shore birds, hence the low chromosome Hawaiian succineids in wet forest can represent secondary invasions from introduced sand dune shore forms.

As indicated earlier, the method of tentacle retraction (p. 64) and pattern of kidney–ureter arrangement (p. 68) in succineids are space adjustment changes. The kidney structure itself (Bouillon and Delhaye 1970) is like that of other Sigmurethra and does not indicate a primitive position for the family. The nervous system of the Succineidae differs from the typical Stylommatophora in several features that Van Mol (1967, pp. 131, 137–140) interpreted as being primitive. It is not known if these structures have functional significance related to the semiaquatic habitat of the Succineidae, but this possibility cannot be overlooked. They could be relict structures with the Succineidae representing an early offshoot from the Sigmurethra, or they could be secondarily derived.

None of the characters cited as indicating opisthobranch or very primitive characteristics for the Succineidae have major phyletic significance. The aulacopod foot, sigmurethrous kidney, only partly fused pallial genital tracts, radular structure (Solem, unpublished), and pallial complex are most consistent with the aulacopod Sigmurethra. The Succineidae are distinctive in their partly amphibious habitat, having undergone shell reduction without mantle lap or lobe development, and in their capacity for self-fertilization (see Patterson, 1970b) which is shared among the Stylommatophora mainly by aulacopod taxa—hence my assigning the Succineidae to the Aulacopoda as a superfamily most closely related to the Arionacea and less advanced than the Limacacea. The Athoracophoridae are placed in the same superfamily, but with far less confidence. They are so modified in pallial structure that relating them to a shelled family will require detailed analyses of other organ systems. Their possession of a multifolded sigmurethrous ureter and high chromosome count of 44 (Burch and Patterson, 1971) suggests a highly aberrant group. Where they eventually will be classified is uncertain, but continued questionable association with the Succineidae is preferable to a new outright guess.

The Aillyidae has been variously referred to the Bulimulidae, zonitoid complex, or placed with the Succineidae in the Heterurethra (see Solem, 1959, pp. 33–35 for references). Most recently it was placed in a separate order, the Aillyida, by Minichev and Slavoshevskaia (1971) on the basis of having "dermal respiration". The latter characteristic is a result of visceral hump reduction (see p. 68). Ordinal status cannot be maintained. Most of the features cited by Odhner (1927) as representing bulimulid or succineid affinities are correlatives of visceral hump reduction. Without further study of its anatomy, firm assignment is impossible, but the statement of Pilsbry (1946, p. 1) that it is an aberrant zonitoid snail seems most reasonable and I tentatively place the Aillyidae next to the Urocyclidae.

F. Position of the Onchidiidae, Veronicellidae and Rathouisiidae

Whether these families are classified in the Pulmonata (Baker, 1955; Ghiselin, 1966) or the Opisthobranchia (Fretter, 1943; Böttger, 1954) has depended essentially upon how the person defined the categories and whether they were looking at the marine Onchidiidae (Fretter, 1943) or the two terrestrial families (Baker, 1955; Solem, 1959). Baker (1955, p. 110) has pointed out that peculiarities of their kidney and mantle organization can result from rotational trends seen in numerous other slugs, while Ghiselin (1966, pp. 367–368), after a comprehensive analysis of the opisthobranch genital system, placed the Onchidiidae in the Pulmonata. Stringer (1963) reported a spiral spermatozoa of the pulmonate type in an onchidiid. Salvini-Plawen (1970) supports opisthobranch affinities, while Minichev and Slavoshevskaia (1971) separate it from the Stylommatophora as a distinct subclass, the Opisthopneumona. This, in essence, follows Pilsbry (1948, p. 1062) who used the name Systellommatophora and treated it as a group equal in rank to the Basommatophora and the Stylommatophora.

There is no question but that these 3 families show very unusual features. The posterior lung, kidney and anus; widely separated genital pores; and peculiar dorsal integument are very different from the structure in any Stylommatophora. Nevertheless, study of the digestive and reproductive systems of the terrestrial forms (Peterellis and Dundee, 1969; Oberzeller, 1970) shows many more similarities to stylommatophoran pulmonate structures than to any opisthobranch configurations. Van Mol (1967, pp. 143–150) indicated their nervous system is pulmonate in its microanatomy, while Oberzeller (1970) compared veroncellid and athoracophorid ganglionic configurations.

Until detailed study of excretory and reproductive systems in these families have been completed, the exact relationship to the Stylommatophora and Basommatophora will remain uncertain. The facts of their basically pulmonate reproductive, nervous and digestive systems, their moderately successful invasion of the land, their are breathing status, and the compatibility of their pallial complex with rotational trends in pulmonate snails, combine to suggest that they have reached the pulmonate grade of structure. They are placed here as a pulmonate taxon, Systellommatophora, that is equivalent in rank to the Basommatophora and Stylommatophora.

V. Outline classification of land molluscs

Retention of the Prosobranchia–Opisthobranchia–Pulmonata division of the Gastropoda reflects major ecological shifts and paths of evolutionary change far better than combining the latter two into the Euthyneura. As pointed out by Fretter and Graham (1962, p. 612), euthyneury seems to have been derived independently by "... real detorsion in opisthobranchs, but by internal rearrangement of the nervous system in the pulmonates ..."; hence Euthyneura is a diphyletic taxon without ecological significance.

It is quite probable that Pulmonata is a grade or level of organization and polyphyletic in exactly the same sense as the monotreme, marsupials and eutherians comprising the vertebrate class Mammalia. The latter groups seem to have originated from slightly different stocks of "true reptiles" and hence emerged independently to a level defined as mammalian (see Reed, 1960 for references). I consider that the Basommatophora, Systellommatophora and Stylommatophora represent emergences from the prosobranch–opisthobranch–pulmonate interface that are independent in the same sense as the mammalian stocks. By pushing the definition of "Pulmonata" back to the same extent that Reed (loc. cit.) extended the Mammalia, then "Pulmonata" would be a monophyletic taxon. If narrowly defined to include just the extant taxa, then it would be a polyphyletic grade. Unfortunately (or fortunately?) we lack fossil evidence of protopulmonates so that the practical problem of taxon definition does not arise.

Subdivision of the Gastropoda traditionally has been very conservatively handled. The degree of evolutionary divergence is greater than current levels of classification suggest. Hence I am carrying the proposal of Pilsbry (1919, p. 30) one step further by raising in rank the categories between superfamily and class. This brings the classification of the pulmonates closer in line to that existing in other phyla. Basic differences in structure and functioning have been obscured by the similar and plastic body form present in the group.

Land prosobranch classification follows Fretter and Graham (1962), except in recognizing several families in the Cyclophoracea. Division of the Pulmonata into the superorders Systellommatophora, Basommatophora and Stylommatophora follows the arguments developed above that these are separate but equal air breathers, differing significantly from opisthobranch taxa in structure, functioning and ecology. I prefer the well-known and clearly defined Basommatophora–Stylommatophora names, with which Systellommatophora is in etymological harmony, rather than a return to the greatly altered idea of Geophila–Gymnophila as resurrected by Baker (1955, 1963). Fortunately priority does not apply to ordinal names.

Division of the Stylommatophora normally involves recognizing 5 major divisions: Heterurethra, Tracheopulmonata, Orthurethra, Mesurethra and Sigmurethra. Based on the evidence presented above, I reduce this to 3 orders: Orthurethra, Mesurethra and Sigmurethra. I consider the heterurethrous Succineidae to be Sigmurethra with a peculiar habitat, while the abberrant tracheopulmonate slugs of the Athoracophoridae should be viewed as Sigmurethra who went weirdly sluggish. It is not possible to relate them to a shelled family at this time.

Ordinal divisions of the Stylommatophora are based upon patterns of excretion and water conservation. In the Orthurethra, the kidney extends anteriorly to open near the pneumostome, the renal pore frequently being bordered by a short reflected ridge. Except on the Pacific Islands and for the Enidae, orthurethrans are small in size. I think it highly significant that a pseudosigmurethrous kidney is known in several enids (Solem, 1964), probably indicating a separate experiment in water retention from that which resulted in the typical Sigmurethra. The Mesurethra have a kidney opening into the posterior of the pallial cavity, are large in size, and relict in distribution except for the very successful Clausiliidae. Available data are not sufficient to evaluate the relationships between the Mesurethra and Sigmurethra, or to define adequately subdivisions of the

Sigmurethra. While the Aulacopoda and Holopoda seem coherent assemblages, the position of the Holopodopes is very uncertain. Most of these taxa show a similarly elongated shell form and the pallial features cited by Baker (1955, pp. 110–111) when he described the group (Baker, 1962a, p. 116) may be convergent. Every stage from a short secondary ureter opening at the posterior section of the pallial cavity (Endodontidae), to secondary ureters with closed tubes (Limacacea and most "helicoid" taxa) exist in the Sigmurethra. This may be another grade of structure, but without detailed analyses of organ systems in a series of families, resolution of this problem will not be possible.

The subdivisions do represent collections of similar morphotypes. Much additional work will be required to interpret the functional and ecological significance of such systematically useful features as the pedal grooves, basic radular differences between the Arionacea and Limacacea, and the many partly closed ureters in sigmurethrous taxa.

Assignment of family names to particular groups generally follows Baker's arrangement, although the Corillidae have been shown to be Polygyracea (Solem, 1966a), the Aillyidae (see above) are transferred to the Limacacea, and some of Baker's later shifts of family groups (Baker, 1963) are not accepted, since his reasons for making the changes were never published. Classification of the Arionacea follows Solem (1976). A question mark precedes those names of uncertain validity and/or position.

This first approximation of a phyletic classification undoubtedly will be altered when some of the investigations outlined above have been completed. At the present time it can serve as a tentative working arrangement.

The list of names includes commonly encountered alternate family names, plus an indication of where family units listed in Zilch (1959–1960), but not accepted here, have been synonymized. Following Baker (1956a, 1956b) I consider that accepting strict priority for family names would cause chaos and have continued traditional usage below.

The classification is:

Class Gastropoda
- Subclass Prosobranchia
  - Order Diotocardia (=Archaeogastropoda)
    - Superfamily Neritacea
      - Family Hydrocenidae
      - Family Helicinidae
  - Order Taenioglossa (=Mesogastropoda)
    - Superfamily Cyclophoracea (=Architaenioglossa)
      - Family Poteriidae
      - Family Maizaniidae
      - Family Pupinidae
      - Family Diplommatinidae
      - Family Cyclophoridae

Superfamily Littorinacea
Family Pomatiasidae
Family Chondropomidae
Superfamily Rissoacea
Family Truncatellidae (partly)
Family Assimineidae (partly)
Family Acmeidae (=Aciculidae)
Subclass Pulmonata
Superorder Systellommatophora
Order Onchidiacea
Family Onchidiidae
Order Soleolifera
Family Veronicellidae (=Vaginulidae)
Family Rathouisiidae
Superorder Basommatophora (see Hubendick, this volume)
Superorder Stylommatophora
Order Orthurethra
Superfamily Achatinellacea
Family Achatinellidae (+ Tornatellinidae)
Superfamily Cionellacea
Family Amastridae
Family Cionellidae (=Cochlicopidae)
Superfamily Pupillacea
Family Pupillidae (+Vertiginidae, Orculidae, Chondrinidae)
Family Pleurodiscidae (+Pyramidulidae)
Family Valloniidae
Family Strobilopsidae
Superfamily Partulacea
Family Partulidae
Family Enidae (=Buliminidae)
Order Mesurethra
Superfamily Clausiliacea
Family Clausiliidae
Family Cerionidae
Superfamily Strophocheilacea
Family Dorcasiidae
Family Strophocheilidae
Order Sigmurethra
Suborder Holopodopes
Superfamily Achatinacea
Family Ferussaciidae
Family Subulinidae
Family Spiraxidae
?Family Megaspiridae
Family Achatinidae
Superfamily Streptaxacea
Family Streptaxidae

Superfamily Rhytidacea
Family Haplotrematidae
Family Systrophiidae
Family Rhytididae (=Paryphantidae)
Family Aperidae (=Chlamydephoridae)
?Family Macrocyclidae
?Superfamily Acavacea
Family Acavidae (+Clavatoridae)
Family Caryodidae
Superfamily Bulimulacea
Family Urocoptidae
Family Bulimulidae (+Orthalicidae, Amphibulimidae, Odontostomidae)
Suborder Aulacopoda
Superfamily Arionacea
Family Punctidae
Family Endodontidae
Family Charopidae (+Otoconchidae)
Family Helicodiscidae
Family Discidae
Family Arionidae
Family Philomycidae
Superfamily Succineacea
Family Succineidae
?Family Athoracophoridae
Superfamily Limacacea
Group "A"
Family Helicarionidae (+Euconulidae, Ariophantidae)
Family Urocyclidae
Family Aillyidae
Group "B"
Family Zonitidae (+Trochomorphidae, Vitrinidae)
?Family Thyrophorellidae
Family Parmacellidae
Family Limacidae (+Milacidae)
Family Trigonochlamydidae
?Family Testacellidae
Suborder Holopoda
Superfamily Polygyracea
Family Polygyridae
Family Sagdidae
Family Corillidae (=Plectopylidae)
?Superfamily Oleacinacea
Family Oleacinidae
Superfamily Camaenidae
Family Camaenidae
Family Ammonitellidae

Family Oreohelicidae
Superfamily Helicacea
Family Bradybaenidae
Family Helminthoglyptidae
Family Helicidae

VI. References

Albers, J. C. (1850). "Die Heliceen nach natürlicher Verwandtschaft systematisch geordnet". Enslin, Berlin.

Anton, H. E. (1839). "Verzeichniss der Conchylien". Anton, Halle.

Babrakzai, N. and Miller, W. B. (1975a). *J. Ariz. Acad. Sci.* **10,** 126–127.

Babrakzai, N. and Miller, W. B. (1975b). *J. Ariz. Acad. Sci.* **10,** 128–130.

Babrakzai, N. and Miller, W. B. (1972). *AMU-WSM Bull.* **1975,** 72.

Baker, H. B. (1938). *Bull. Bernice P. Bishop Mus.* **158,** 1–102.

Baker, H. B. (1955). *Nautilus, Philad.* **68,** 109–112.

Baker, H. B. (1956a). *Nautilus, Philad.* **69,** 128–139.

Baker, H. B. (1956b). *Nautilus, Philad.* **70,** 28–31.

Baker, H. B. (1961). *Nautilus, Philad.* **75,** 33–37.

Baker, H. B. (1962a). *Nautilus, Philad.* **75,** 116–121.

Baker, H. B. (1962b). *Nautilus, Philad.* **75,** 142–145.

Baker, H. B. (1963). *Proc. Acad. nat. Sci. Philad.* **115,** 191–259.

Baker, R. E. (1965). *Proc. malac. Soc. Lond.* **36,** 259–265.

Bargmann, H. E. (1930). *J. Linn. Soc. (Zool.)* **37,** 1–92.

Beck, H. (1837). "Index Molluscorum Praesentis aevi Musei Principis Augustissimi Christiani Frederici. Fasciculus Primus et Secundus. Mollusca Gastraeopoda Pulmonata". Copenhagen.

Blinn, W. C. (1964). *Physiol. Zool.* **37,** 329–337.

Boss, K. J. (1971). *Occ. Pap. Mollusks, Harv.* **3,** 81–135.

Böttger, C. R. (1954). *Verhl. dt. zool. Ges.* **1954,** 253–280.

Bouillon, J. (1960). *Annls Sci. nat. (Zool. Biol. Anim.)* (12) **2,** 719–749.

Bouillon, J. and Delhaye, W. (1970). *Annls Sci. nat. (Zool. Biol. Anim.)* (12) **12,** 1–26.

Bretsky, S. S. (1970). *Bull. geol. Inst. Univ. Upsala* **2,** 5–23.

Bretsky, S. S. (1971). *Syst. Zool.* **20,** 204–222.

Burch, J. B. (1965). *Proc. First Eur. malac. Congr.* **1962,** 215–241.

Burch, J. B. (1968). *J. malac. Soc. Aust.* **11,** 62–67.

Burch, J. B. and Patterson, C. M. (1969). *Malacologia* **9,** 259–260.

Burch, J. B. and Patterson, C. M. (1971). *Malac. Rev.* **4,** 25–26.

Butot, L. J. M. (1967). *Genetics, Princeton* **38,** 398–400.

Butot, L. J. M. and Kiauta, B. (1967). *Beaufortia* **14,** 157–164.

Clessin, S. (1881). "Nomenclator Heliceorum viventium". Fischer, Kassel.

Connolly, M. (1939). *Ann. S. Afr. Mus.* **33,** 1–660.

Cooke, C. M. and Kondo, Y. Y. (1960). *Bull. Bernice P. Bishop Mus.* **221,** 1–303.

Draparnaud, J. P. R. (1805). "Histoire Naturelle des Mollusques". Levrault, Schoell, Paris.

Duncan, C. J. (1960). *Proc. zool. Soc. Lond.* **134,** 601–609.
Férussac, A. E. J. (1821–1822). "Tableaux Systématiques des Animaux Mollusques Classés en Familles Naturelles. Tableau Systématique de la Famille des Limafons, Cochleae". Bertrand, Paris.
Fischer, P. (1883). *Manuel de Conchyliologie et de Paléontologie Conchyliologique* **5,** 417–512.
Flasar, I. (1967). *Vest. csl. Spol. zool.* **31,** 150–158.
Franc, A. (1968). *In* "Traité de Zoologie: Anatomie, Systématique, Biologie. **V**. Mollusques Gastéropodes et Scaphopodes" (P.-P. Grassé, Ed.) Vol. 3, pp. 325–607. Maison et Cie, Paris.
Fretter, V. (1943). *J. mar. biol. Ass. U.K.* **25,** 685–720.
Fretter, V. and Graham, A. (1962). "British Prosobranch Molluscs". Ray Society, London.
Fretter, V. and Peake. J. F. (1975). "Pulmonates. Vol. 1. Functional Anatomy and Physiology". Academic Press, London and New York.
Fulton, H. (1896). *Ann. Mag. nat. Hist.* (6) **17,** 66–94.
Ghiselin, M. T. (1966). *Malacologia* **3,** 327–378.
Ghiselin, M. T., Degens, E. T., Spencer, D. W. and Parker, R. H. (1967). *Breviora* **262,** 1–35.
Girardi, E.-L. (1978). *Veliger* **20,** 191–250.
Gregg, W. O. (1960). *Am. Malacol. Union Ann. Rep.* **1959,** 45–46.
Hooke, R. (1665). "Micrographia: or Some Physiological Descriptions of Minute Bodies Made by Magnifying Glasses with Observations and Inquiries Thereupon". Mattyn and Allestry, London.
Hunter, W. R. (1964). *In* "Physiology of Mollusca" (K. M. Wilbur and C. M. Yonge, Eds) Vol. 1, 83–115. Academic Press, New York and London.
Hyman, L. H. (1967). "The Invertebrates Vol. 4. Mollusca I". McGraw-Hill, New York.
Iredale, T. (1937a). *Aust. Zool.* **8,** 287–333.
Iredale, T. (1937b). *Aust. Zool.* **9,** 1–39.
Iredale, T. (1938). *Aust. Zool.* **9,** 83–124.
Iredale, T. (1944). *Aust. Zool.* **10,** 299–334.
Iredale, T. (1945). *Aust. Zool.* **11,** 46–71.
Keferstein, W. (1862–1866). *Klassen und Ordnungen der Weichthiere* (*Malaczoa*) **3,** 1–1500.
Kiauta, B. and Butot, L. J. M. (1968). *Beaufortia* **16** (210), 1–11.
Laidlaw, F. F. and Solem, A. (1961). *Fieldiana Zool.* **41,** 509–673.
Lamarck, J. P. B. A. de (1822). *Histoire Naturelle des Animaux sans Vertebres* **2,** 1–252.
Leidy, J. (1851). *In* "The Terrestrial Air-breathing Mollusks of the United States, and the Adjacent Territories of North America". (A. Binney, Ed.) pp. 198–260. Little and Brown, Boston.
Linné, K. von (1758). *Systema Naturae* (10) **1.**
Linné, K. von (1767). *Systema Naturae* (12) **1,** part 2.
Martens, E. von (1860). "Die Heliceenn ach natürlicher Verwandtschaft systematisch geordnet". Zweite Ausgabe, Engelmann, Leipzig.
Martin, A. W. and Harrison, I. M. (1966). *In* "Physiology of Mollusca" (K. M.

Wilbur and C. M. Yonge, Eds) Vol. 2, pp. 353–386. Academic Press, New York and London.

Meenakshi, V. R., Hare, P. E., Watabe, N. and Wilbur, K. M. (1969). *Comp. Biochem. Physiol.* **29,** 611–620.

Minichev, I. S. and Slavoshevskaia, L. V. (1971). *Zool. Zh.* **50,** 350–360.

Moquin-Tandon, A. (1855). "Histoire Naturelle des Mollusques Terrestres et Fluviatiles de France" Vol. 1–2. Bailliere, Paris.

Mörch, O. A. L. (1860). *Malakozool. Bl.* **6,** 102–126.

Mörch, O. A. L. (1865). *J. Conch., Paris* **13,** 265–283.

Morton, J. E. (1955). *Proc. zool. Soc. Lond.* **125,**, 127–168.

Müller, O. F. (1774). *Vermium Terrestrium et Fluviatilium* **2,** 1–214.

Oberzeller, E. (1970). *Annls Naturh. Mus. Wien* **74,** 325–341.

Odhner, N. H. (1927). *Ark. Zool.* **19A** (20), 1–16.

Odhner, N. H. (1950). *Proc. malac. Soc. Lond.* **28,** 200–210.

Oldham, C. (1929). *Proc. malac. Soc. Lond.* **18,** 143–144.

Patterson, C. M. (1968). *Malac. Rev.* **1,** 1–13.

Patterson, C. M. (1970a). *Malac. Rev.* **3,** 25–36.

Patterson, C. M. (1970b). *J. Conch. Paris* **108,** 61–85.

Peake, J. F. (1968). *Symp. Zool. Soc. London.* **22,** 319–346.

Pelseneer, P. (1894). *Q. J. microsc. Sci.* **37,** 19–46.

Pelseneer, P. (1906). *In* "Treatise on Zoology". (R. Lankester, Ed.) Part V. Adam and Charles Black, London.

Peterellis, L. S. and Dundee, D. S. (1969). *Trans. Am. microsc. Soc.* **88,** 547–558.

Pfeiffer, L. (1847–1876). "Monographia Heliceorum Viventium". Vols I–VIII. Brockhaus, Leipzig.

Pilsbry, H. A. (1893). *Proc. Acad. nat. Sci. Philad.* **1892,** 387–404.

Pilsbry, H. A. (1893–1894). "Guide to the Study of Helices" *Manual Conch.* (2) **9,** 1–366.

Pilsbry, H. A. (1896). *Nautilus, Philad.* **9,** 109–111.

Pilsbry, H. A. (1900a). *Proc. Acad. nat. Sci. Philad.* **1900,** 561–567.

Pilsbry, H. A. (1900b). *Proc. Acad. nat. Sci. Philad.* **1900,** 568–581.

Pilsbry, H. A. (1900c). *Manual Conch.* (2) **13,** 1–253.

Pilsbry, H. A. (1919). *Bull. Am. Mus. nat. Hist.* **40,** 1–370.

Pilsbry, H. A. (1939). "The Land Mollusca of North America (North of Mexico)" **1,** 1–573.

Pilsbry, H. A. (1940). "The Land Mollusca of North America (North of Mexico)" **1,** 575–994.

Pilsbry, H. A. (1946). "The Land Mollusca of North America (North of Mexico)" **2,** 1–520.

Pilsbry' H. A. (1948). "The Land Mollusca of North America (North of Mexico)" **2,** 521–1102.

Plate, L. H. (1898). *Zool. Jb. Anat.* **11,** 193–280.

Powell, A. W. B. (1957). "Shells of New Zealand" 3rd edition. Whitcombe, Christchurch.

Quick, H. E. (1933). *Proc. malac. Soc. Lond.* **20,** 295–318.

Reed, C. A. (1960). *Evolution, Lancaster, Pa.* **14,** 314–322.

Rees, W. J. (1964). *Proc. malac. Soc. Lond.* **36,** 55–67.

Rigby, J. E. (1963). *Proc. zool. Soc. Lond.* **141,** 311–359.
Rigby, J. E. (1965). *Proc. zool. Soc. Lond.* **144,** 445–486.
Runham, N. W. (1969). *Malacologia* **9,** 179–185.
Runham, N. W. and Hunter, P. J. (1970). "Terrestrial Slugs". Hutchinson University Library, London.
Runham, N. W. and Thornton, P. R. (1967). *J. Zool., Lond.* **153,** 445–452.
Runham, N., Thornton, P. R., Shaw, D. A. and Wayte, R. C. (1969). *Zeitschr. Zellforsch.* **99,** 608–626.
Salvini-Plawen, L. V. (1970). *Zool. Jb. Syst.* **97,** 285–299.
Schilder, F. A. (1947). *Arch. Molluskenk.* **76,** 37–44.
Schmidt, A. (1855). *Abh. naturw. Ver. Halle* **1,** 1–52.
Semper, C. (1870–1880). "Reisen im Archipel der Philippinen". Zweiter Theil, Wissenschaftliche Result. Dritter Band, Landmollusken: 1–264.
Semper, C. and Simroth, H. (1894). "Reisen im Archipel der Philippinen". Zweiter Theil, Wissenschaftliche Result. Dritter Band, Landmollusken, 45–91.
Solem, A. (1959). *Fieldiana Zool.* **48,** 1–359.
Solem, A. (1961a). *Fieldiana Zool.* **41,** 419–498.
Solem, A. (1961b). *Am. Malacol. Union Ann. Rep.* **1961,** 16–17.
Solem, A. (1964). *Veliger* **6,** 115–119.
Solem, A. (1966a). *Spolia zool. Mus. haun.* **24,** 1–108.
Solem, A. (1966b). *Fieldiana Zool.* **50,** 1–214.
Solem, A. (1969). *Malacologica* **9,** 289.
Solem, A. (1972a). *Veliger* **10,** 327–336.
Solem, A. (1972b). *Proc. malac. Soc. Lond.* **40,** 93–114.
Solem, A. (1972c). *Veliger* **15,** 81–87.
Solem, A. (1973a). *Veliger* **15,** 165–171.
Solem, A. (1973b). *Veliger* **15,** 300–306.
Solem, A. (1973c). *J. Malac. Soc. Aust.* **2,** 377–385.
Solem, A. (1974a). "The Shell Makers: Introducing Mollusks". John Wiley, New York.
Solem, A. (1974b). *Veliger* **17,** 81–88.
Solem, A. (1976). "Endodontoid Land Snails from Pacific Islands, Part 1. Family Endodontidae". Field Museum of Natural History, Chicago.
Stebbins, G. L. (1966). *Science, NY* **152,** 1463–1469.
Steenberg, C. M. (1925). *Vidensk. Meddr. dansk. naturh. Foren.* **80,** 1–211.
Stringer, B. L. (1963). *Nature, Lond.* **197,** 621–622.
Swammerdam, J. (1752). "Bibel der Natur". Leipzig.
Taylor, D. W. and Sohl, N. F. (1962). *Malacologia* **1,** 7–32.
Thiele, J. (1931). *Handbuch Syst. Weichtierkunde* **2,** 377–734.
Throckmorton, L. H. (1968). *Ann. Rev. Ent.* **13,** 99–114.
Tryon, G. W. (1884). "Structural and Systematic Conchology: An Introduction to the Study of Mollusca". Vol. 3. Tryon, Philadelphia.
Van Mol, J. J. (1967). *Mém. Acad. r. Belg. Cl. Sci.* 8° **37,** 1–168.
Van Mol, J. J. (1970). *Ann. Musee r. l'Afrique Cent. Tervuren.* (8°), **180,** 1–234.
Van Mol, J. J. (1972). *Annls Soc. r. zool. Belg.* **101,** 183–225.
Voris, H. E. (1971). *Syst. Zool.* **20,** 442–458.

Wächtler, W. (1934). *Zool. Anz.* **105,** 161–172.
Wächtler, W. (1935). *Jb. Akad. gemeinnütz. Wiss. Erfurt.* **52,** 107–135.
Walton, M. L. (1963). *Nautilus, Philad.*, **76,** 127–135.
Warburg, M. R. (1965). *Proc. malac. Soc. Lond.* **36,** 297–305.
Watson, H. (1915). *Ann. Natal Mus.* **3,** 107–267.
Watson, H. (1920). *Proc. malac. Soc. Lond.* **14,** 6–30.
Wurtz, C. B. (1955). *Proc. Acad. nat. Sci. Philad.* **107,** 99–143.
Yen, T. C. (1939). *Abh. der senckenb. naturforsch. Ges.* **444,** 1–234.
Zilch, A. (1959–1960). Gastropoda. Teil 8. Euthyneura. *Handbuch der Paläozoologië* **6** (2), 1–834.

Chapter 3

Experimental methods in molluscan systematics

G. M. DAVIS

Department of Malacology, Academy of Natural Sciences of Philadelphia, USA

I. Introduction

Simpson (1961) defined systematics as "the scientific study of the kinds and diversity of organisms and of any and all relationships among them" while taxonomy was "the theoretical study of classification, including its bases, principles, procedures and rules". The subject matter of taxonomy is classification, and, in turn, meaningful classification can only result from sound systematics.

A systematist is primarily involved in comparative studies, of any type, which yield data concerning the relationships of 2 or more organisms, no matter which aspect is being investigated. For example, is taxon *A* a new species in contrast with taxa *B* or *C*? Did taxon *A* evolve from ancestors common to *B* or *C*? Is a certain biochemical trait found in taxon *A* and not in taxa *B* or *C* due to several gene differences, a mutation, or part of a wider species polymorphism? How does one account for the fact that within the same species population *a* will transmit a human parasite, yet population *b* is refractive to transmission? What are the factors, ecological or genetic, which account for the fact that population *a* of a gastropod species living in a lake is large and ornately sculptured while river population *b* is small and smooth?

A single operation may be sufficient to answer a given question. For example, does taxon *A* belong to the genus *Helisoma* or to *Biomphalaria* of the pulmonate basommatophoran family Planorbidae? A single study involving the comparison of 2 or 3 anatomical features is sufficient to answer the question. In another example, does population *a* belong to the same species as population *b*? A detailed morphological study will not necessarily settle the question, as 2 populations may have identical or nearly identical morphologies, yet be different species. We are not simply involved in identification *per se*, but in assessing the degree of genetic affinity between taxa. In making such an assessment, we must be capable of selecting suitable methodologies or "tools", remembering that a stereotyped approach is to be avoided; no single method or technique will resolve every problem.

In a paper entitled "Experimental taxonomy: a review of some techniques and their applications", Wright (1966) discussed the development and importance of such techniques as chromatography, electrophoresis, and serology. Examples were drawn from biology as a whole and only a few papers dealing with Mollusca were mentioned. Nevertheless, the reader is encouraged to consult this work for earlier references to work outside the field of malacology.

In preparing this chapter it has been necessary to discuss molluscs other than pulmonates in order to make adequate assessments of certain techniques. Likewise, developmental trends within the plant and animal kingdoms are indicated where they provide insight into further uses of techniques, interpretation of results, or an understanding of the limitations of a particular approach.

It is anticipated that, at the time this paper is published, there will be several worthy additions to the literature. Nevertheless, although the analysis presented here was completed in 1973, it is expected that the general conclusions and scope of techniques presented will remain current for some years.

II. An integrated approach to molluscan systematics

It should be obvious that adequate definitions of taxa and assessments of their relationships depend on integrating all available information that pertains to them. At present the most valuable area of investigation for assessing relationships is detailed morphology. There are, however, those who suggest that morphological data are not satisfactory in describing molluscan taxa objectively, particularly at the species level. The most commonly stated reasons for this suggestion are variation in organ shape and dimensions can be very extensive, and there is frequently a lack of discrete characters that can be used for defining specific and infraspecific levels that correspond with observed physiological, ecological, or parasitological differences. As stated by Davis and Lindsay (1967), one finds adequate reasons for such opinions when one reviews papers of molluscan anatomy; for example: (i) "too much anatomy is presented as precise, although what is actually involved is organology indicative of family, subfamily, or generic levels when the specific level was intended"; (ii) much anatomical data are unusable because the figures are crudely drawn, and the lack of uniform orientation of organs in the drawings makes comparisons between different papers difficult or impossible; (iii) the body "architecture", that is the manner in which organs are arranged and interrelated, is inadequately portrayed for any interspecific comparisons to be made; stylized drawings cannot be used in place of precise drawings of organ shape. Even greater insight is derived when the anatomical relationships are studied in terms of organ function and integration of functions; only after this level of precision is reached do stylized or diagrammatic drawings have a value in comparing traits of interest; (iv) unfortunately, important taxonomic decisions are all too often based on very limited anatomical data. As the operculum, radula and penis are easy to study, these features are frequently given undue weight in assessing relationships and all too often no consideration is given to the whole reproductive system or other organ systems, which may contain greater information.

The need to use methods other than anatomical to assess relationships arises when: (i) few morphological characters are available to differentiate between taxa and it is impossible to cluster species on series of unique characters (Wilson, 1965); (ii) considering taxa below the level clearly defined by morphology, for example, where genera, but not species, are clearly defined by morphology; (iii) one or more morphological features are convergent; (iv) information on relative genetic relationships is desired in terms of percentage of shared alleles or genetic distances, data that cannot be determined from anatomy. Cytology, chromatography, electrophoresis and immunology are all techniques that could provide relevant information. Specialized techniques such as DNA hybridization and amino acid sequencing are not considered here because they have, so far, not been used in molluscan systematics and because the costs and facilities needed for them far exceed the capabilities of many institutions that can, however, readily support basic electrophoretic and immunological studies. Nor has a discussion been included of mathematical techniques associated with numerical taxonomy. Techniques such as multivariate analysis are used to assess the relationships between taxa or characters, but such computer mediated procedures must be regarded as tools to handle data. Of concern here are methods for creating data, that is, characters whether they be derived chromatographically or electrophoretically.

The use of cytological data strengthens taxonomic assessments made on the basis of conchology and functional morphology. Burch (1964), for example, demonstrated the occurrence of a polyploid series in the planorbid subfamily Bulininae. The polyploid series not only served to delineate species groups, but also it discriminated between schistosome and non-schistosome (*S. haematobium* (Bilharz)) susceptible snails. Davis (1969d) could discriminate between 2 sibling species of Japanese *Semisulcospira* in terms of shell traits (adult and embryo) only after the taxa had been separated (Burch, 1968a) on the basis of different chromosome numbers.

It is important to note that there are certain limitations to the use of cytological data, such as chromosome number, in characterizing taxa. Chromosome number may vary because of supernumerary elements, and taxa separated by small differences in chromosome number may be able to interbreed and are, therefore, the same species. Conversely, 2 or more distinct, reproductively isolated species may have the same chromosome number and seemingly similar karyotypes. In the Basommatophora, the haploid chromosome number of 18 is characteristic of over 60% of the taxa studied (Burch, 1965).

It is within a framework established by precise anatomical and cytological studies that biochemical data have their greatest value for assessing the relationship between taxa. It should be realized that, by themselves, electrophoretic profiles do not serve to define categories such as genera or species, nor does the number of taxon-specific antigen–antibody precipitin systems justify automatic placement of a taxon in a particular category. Electrophoretic data alone may be less objective than precise functional morphology when inter-taxa comparisons are made and phylogenetic relationships are considered, because of the problem of determining the homology of protein fractions from different taxa.

Four examples of integrated approaches to systematics are presented here, to illustrate the value of using several tools to unravel complex relationships. In the first, a series of studies are cited where relationships between species could not be adequately assessed by anatomical studies, but were clarified first by an analysis of chromosome number and then with greater precision by electrophoretic and immunological methods. In the second example, species could be defined on anatomical grounds, but infraspecific relationships in a polytypic species complex were delineated first through immunology and secondly by electrophoresis. The third example draws attention to a situation where convergent evolution of certain anatomical features produced taxonomic confusion; yet shell characters, chromosome number and immunological data clearly correlate and clarify relationships. The last indicates the type of misinterpretation of data that must be avoided when using experimental procedures, especially attempting to correlate category rank with degree of electrophoretic or immunological differences.

(i) The African basommatophoran genus *Bulinus* is of considerable medical importance because certain species transmit the human urinary schistosome *S. haematobium*. About 120 species of *Bulinus* have been described, although today this number has been reduced to some 30 species and several subspecies (Mandahl-Barth, 1957, 1958b; Wright, 1971). There are considerable problems in determining species and in assessing subgeneric relationships.

Mandahl-Barth (1957, 1958b) established 4 species groups based on shell sculpture and shape, radular teeth, male copulatory organ, and ridges on the mantle cavity roof. These are the *africanus*, *forskalii*, *tropicus* and *truncatus* groups. Species of the *africanus* and *truncatus* groups are involved in the transmission of human schistosomiasis; some *forskalii* group taxa are likewise involved. However, differences between groups are not always clearcut when morphological traits are studied.

Burch (1960, 1964) made a significant discovery when he found that a polyploid series existed in *Bulinus*, with the taxa of the non-schistosome-transmitting *tropicus* group diploid ($2n = 36$) and the schistosome-transmitting *truncatus* group tetraploid (2n = 72). Here was an objective tool for distinguishing between 2 groups that had been most difficult to separate.

Wright *et al.* (1967) studied *Bulinus nyassanus* (Smith) and *B. succinoides* (Smith) from Lake Malawi, Africa. The question arose as to the placement of these taxa in one or the other of the *tropicus* or *truncatus* species groups. The anatomy of *B. nyassanus* suggested relationships with the *tropicus* group. The haploid chromosome number was "probably 18" and thus the taxon was referred to the *tropicus* group. The electrophoretic pattern of egg proteins was unique and did not help in deciding group placement; immunoelectrophoretic results, however, clearly indicated placement in the *tropicus* group.

Shells of *B. succinoides* resembled some of the *truncatus* group. The presence of aphallic specimens and mantle pigment polymorphism also suggested affinities to the *truncatus* group, as did the general form of the lateral mesocones, which was definitely more like the arrowhead type generally attributed to *truncatus* group, rather than the simple triangle considered a trait characterizing the *tropicus* complex, yet the haploid chromosome number of *B. succinoides* was 18, a *tropicus* group trait. Electrophoretic analysis of the egg proteins revealed a *tropicus* group pattern and the immunoelectrophoretic results confirmed an affinity with the *tropicus* group.

Wright (1971) studied the morphology, chromatograms of body mucus, egg protein electrophoresis and comparative serology of Indian Ocean taxa of *Bulinus*. Serological data were of greater value in determining species-group affinities than morphology, chromatography, or electrophoresis. For example, *B. wrighti* Mandahl-Barth had mantle ridges of the *tropicus-truncatus* type and lacked the fluorescent material in the mucus characteristic of the *africanus* group; the serological data clearly showed affinity with the *tropicus* group. Wright considered both *B. wrighti* and *B. reticulatus* M.-B. to be unspecialized bulinids without close affinities to other taxa of *Bulinus* and thus considered them a separate complex, the *reticulatus* group.

Bulinus obtuspira (Smith) lacked distinctive morphological traits. Shell shape suggested relationship to the *tropicus* group, while microsculpture suggested the *africanus* group; mantle ridges suggested *tropicus-truncatus* affinities, but the sharply pointed posterior foot was similar to the *africanus* group taxa. The haploid chromosome number was 18 and thus there was no clear relationship with the *truncatus* group. Results from serological studies showed a closer affinity to the *africanus* group; however, it is clear from the data that if absorption studies were carried out there would probably be antigens unique to *B. obtuspira*, indicating a genetic divergence from the *africanus* group homologous reactions. This situation is worthy of further investigation.

Burch and Lindsay (1970, 1971) reviewed the taxonomic position of *B. natalensis* (Küster), which had a haploid chromosome number of 18, as associated with *tropicus* group. Mandahl-Barth (1957) at first placed this species in the synonymy of *B. tropicus tropicus* (Krauss), but later (1965) referred it to the *truncatus* group on the basis of the shape of the mesocone of the first lateral radular tooth. Immunological data confirmed the placement in the *tropicus* group. It is evident that the radula, shell and phallicaphallic traits used in the past to differentiate the *truncatus* and *tropicus* groups are of relatively little value; the shape of the mesocone particularly is not to be trusted.

Burch (1967) and Brown and Burch (1967) revealed further cytological complexity within *Bulinus* with the discovery of hexaploids and octoploids in snails from Ethiopia. Burch (1964) found octoploids in Aden. Brown and Wright (1972) initiated a multi-phased project to detect traits that might be used to discriminate between the taxa of the Ethiopian polyploid series, without making chromosome preparations. Assessments were made of shell, radula, copulatory organs, egg size, chromatography of surface mucus, electrophoresis of egg proteins and esterase isoenzymes.

There was continuous variation in shell sculpture and dimensions of the radular teeth. Octoploids and hexaploids achieved a larger maximum size in shell and egg than diploids and tetraploids. Overall, there were no morphological traits that served to characterize the diploid group, because of the great variation between populations. There was continuous variation in chromatograms and electrophoretic patterns of isoenzymes between the groups. Nevertheless, an electrophoretic analysis of proteins and chromosome number did correlate in part. The electrophoretic pattern and the chromosome number placed the tetraploids in the *truncatus* species group; hexaploids were similar in electrophoretic pattern to diploids, but octoploids had a unique electrophoretic pattern.

Immunological results showed that antigens of tetraploids were homologous with those of *B. truncatus* (Audovin). No immunological studies were conducted with hexaploids. When antigens of the octoploid were compared with an Ethiopian tetraploid in an Ouchterlony diffusion system, there was homology with the tetraploid and spur formation with the diploids. The diploid forms developed strong spurs with the tetraploids, but less strong with octoploids. Absorption of antitetraploid sera with diploid antigens cancelled all reactions with diploid antigens, but not with tetraploid and octoploid antigens. The data indicated a stronger genetic affinity between tetraploids and octoploids than between tetraploids and diploids. Nevertheless, it is evident that there were antigens in common between all 3 groups.

Wu (1972) used polyacrylamide electrophoresis to separate nonspecific esterases of parental and hybrid forms of the Ethiopian polyploid series. Diploid albino snails and polyploid pigmented snails were successfully crossed, although fertilization proceeded only from albino to pigmented snails; diploid–hexaploid crosses were more successful than diploid–octoploid crosses, but in all cases the F_1 were sterile. As the different parental stocks possessed taxon-specific esterase patterns and the hybrids showed a fusion of the 2 parental patterns, the electrophoretic analyses confirmed that the F_1 were indeed hybrids and not the product of self-fertilization by hermaphrodites. The results of his entire study indicated that although there were no significant anatomical differences between the taxa in the polyploid series, on the basis of crossing data, the hexaploid population belonged to the *tropicus* group, while the octoploids were allied to the *truncatus* group, the tetraploid group. These data, independently derived, agreed with the findings of Brown and Wright (1972).

Clearly, the above studies show that subgeneric delineation of genetic relatedness is best accomplished by use of cytology and immunology, while electrophoretic studies, particularly those analysing isoenzymes, are more useful in characterizing populations. Burch and Lindsay (1970, 1971) found by using the in-gel absorption test of Davis (1968a) that one could not differentiate between taxa of *Bulinus* with the same chromosome number. Brown and Wright (1972), however, gave results that were in contrast with those of Burch and Lindsay; they seemed to have found antigenic heterogeneity between some diploid populations. Burch and Lindsay used foot-muscle antigens while Brown and Wright used egg-protein antigens. There are undoubtedly differences between foot-muscle and egg-protein antigens and these differences should be assessed, nevertheless an examination of Brown and Wright's limited data suggests that heterogeneity between *B. natalensis* and other diploid populations is not yet proven. In this type of investigation, in the absence of a single purified antigen, absorption studies must be used to show relative degrees of homo- or heterogeneity. Results in double diffusion systems, without absorption and where spur formation occurs, are inadequate; when mixtures of antigens are used absorption must be undertaken. The protein concentration of antigens should be determined and the concentration of the heterologous antigen mixture should equal that of the homologous antigen mixture. Immunoelectrophoresis or a Y-shaped simple diffusion pattern should be used. In the absence of immunoelectrophoresis, the Ouchterlony diffusion system (Davis, 1968a) is preferred, as it is possible, in one gel, to observe the control reaction, check if absorption is complete, and determine the percentage of precipitin systems unique to the homologous species.

(ii) Wright (1973), in his review of the molluscan hosts of the Asian schistosome, *Schistosoma japonicum* (Katsurada), suggested that each geographic region of Asia had its own species of *Oncomelania*. For example, the intermediate host of *S. japonicum* on mainland China was *Oncomelania hupensis* Gredler, while that in Japan was *O. nosophora* Robson. He rejected the possibility that these were subspecies, by asserting that the subspecies concept was a myth, as discussed by Sokal (1964). Actually, it is evident that *O. hupensis* is a polytypic species, and it is reasonable to consider some segments of this species complex as subspecies.

Unlike the situation discussed for *Bulinus*, anatomical data have provided objective characters for separating species of hydrobiid genera such as *Oncomelania*. Two species of *Oncomelania* have been separated on the basis of morphology, these are *O. hupensis* and *O. minima* (Davis, 1967, 1968a, c, 1969c, 1971b). Wright (1973) in his review, omitted *O. minima*. The morphological ground-plan of taxa included in the *O. hupensis* species complex is the same. Accordingly, 6 taxa have been relegated to subspecies rank by Davis; these are *O. hupensis hupensis* from mainland China, *O. h. formosana* (Pilsbry and Hirase) and *O. h. chiui* from Taiwan, *O. h. quadrasi* (Möll.) from the Philippines, *O. h. nosophora* from Japan, and *O. h. lindoensis* (Davis and Carney) from Sulawesi.

O. h. chiui was originally described as *Tricula chiui*, but morphological studies showed the taxon was part of the *O. hupensis* complex (Davis, 1968a) and serological data clearly demonstrated that it was closely related to *O. h. formosana*. Electrophoretic data indicated that *O. chiui* evolved from stock that also gave rise to north-eastern populations of *O. formosana* (Davis, 1968a). *Tricula* and *Oncomelania* actually belong in separate subfamilies (Davis, 1968c).

All members of the *O. hupensis* polytypic species-complex have the same chromosome number, similar general ecology and reproductive strategies, and can readily interbreed with one another to produce F_1 and F_2 generations without loss of viability or fecundity (Davis, 1971b). Davis (1973) and Davis and Ruff (1973) have shown by breeding studies that size differences, pigmentation and sculptural traits were controlled by simple Mendelian genetic factors.

Taxa considered as subspecies of *O. hupensis* have a number of genetically controlled differences: (i) there are discrete differences in shell size and sculpture that serve to separate the taxa; however, there may be overlap in shell size, for example, 10 to 15% overlap between the smaller *O. h. quadrasi* of the Philippines and *O. h. formosana* of Taiwan; (ii) the size of the organs generally correlated with shell size; (iii) there are small but significant differences in reproductive potential as well as rates of growth in a controlled laboratory environment, (iv) the taxa differ in potential for being infected with the various strains of *S. japonicum*; (v) there are antigenic differences between the taxa.

O. h. chiui is not differentiated antigenically from *O. h. formosana*; therefore, if this alone was the criterion for ranking, then perhaps *O. h. chiui* could be relegated to the rank of race of *O. h. formosana*. However, *O. h. chiui* has a number of distinctive anatomical features that remain constant during breeding in the laboratory. It is significantly smaller than *O. h. formosana*, produces no varix or only a slight varix on the shell, has a significantly reduced number of gill filaments and a disproportionately enlarged section of the nervous system. In contrast, *O. h. quadrasi* on the basis of anatomy alone could easily be considered a race of *O. h. formosana*, yet it is readily differentiated immunologically.

While it is argued that each taxon in question could transmit a specific strain of *S. japonicum* and could thus be considered distinct, *O. h. chiui* in nature does not transmit any strain of *S. japonicum*, although experimentally the taxon can transmit all known strains. Genetic studies of Chi *et al.* (1971), Davis (1973) and Davis and Ruff (1973) have shown that infectivity appears to be controlled by genes operating at the infraspecific level.

Infectivity patterns, electrophoretic data, and small differences in shell size and colour serve to define races within each subspecies. There are races of *O. h. formosana*, morphologically indistinguishable from other races, which are resistant to infection with any strain of *S. japonicum*, and other races transmitting only the zoophilic Formosan strain. Electrophoretic data indicate considerable interpopulation differences.

(iii) The need for integration of approaches and avoidance of provincial thinking in terms of a "single method" approach to determining intertaxon relationships is made quite clear from studying lymnaeid systematics. Burch and Lindsay (1968) stated, concerning their immunological analysis of certain lymnaeid taxa: "We do not believe that a morphological study alone will give us . . . answers [concerning relationships] for the Lymnaeidae . . .". Walter (1968) added the following viewpoint: "I hope that everyone recognizes the possible potential value of immunological data where integrated with other information, especially morphological". Walter then pointed out, "It is neglect and misuse of this [morphological] approach which have instigated the confused, misdirected views that have prevailed until now . . .". He stated that ". . . unsatisfactory morphological work" has been done on lymnaeids and "surely, immunological and chromosomal data are relatively useless except in a concept of detailed, accurate morphological data." In rebuttal, Burch (1968b) pointed out:

> We do not envision a conflict with anatomical studies, nor have we implied this; in fact, our whole research effort is directed towards supplementing morphological data, especially in those groups where it is difficult to make systematic assessments because of the lack of obvious gross morphological characters, or because of the difficulty in determining the significance of observable morphological variation.

Walter is correct; more detailed and accurate morphological studies of lymnaeid taxa are badly needed. Nevertheless, certain lymnaeid shell types do not closely agree with anatomical groupings and the possibility of convergent or parallel evolution must be considered. In this situation, an understanding of the relationships between taxa cannot be assessed with certainty on the basis of morphological data alone. Inaba (1969) reviewed the cytological data published on the Lymnaeidae and it seems clear that lymnaeids may be grouped into units on the basis of their different chromosome numbers.

Hubendick (1951) showed that "*Lymnaea*" *luteola* Lam. of India had a multifolded prostate, as did many other lymnaeids including the holarctic *Lymnaea stagnalis* (L.) and "*L.*" *corvus* (Gmel.) from Poland. Walter (1968) held that anatomical reversions in lymnaeid evolution give the impression that the phylogenetic pattern is reticulate. He mentioned the similarity between *L. stagnalis* and *L. corvus* to make his point. He absolutely rejected all generic and subgeneric subdivisions in the family.

In order to elucidate relationships within the Lymnaeidae, Burch and Lindsay (1969) studied the electrophoretic patterns of foot-muscle proteins of *L. luteola*, *L. acuminata* (Lam.) from India, and *L. auricularia* (L.) from North America. These taxa have 17 pairs of chromosomes, indicative of the *Radix* species group. While *luteola* has a multifolded prostate, *acuminata* and *auricularia* do not. They compared the 3 taxa with *L. stagnalis jugularis* (Say) of North America that has 18 pairs of chromosomes (indicative of the genera *Lymnaea*, *Stagnicola* and *Fossaria*) and a multifolded prostate. The electrophoretic patterns of *luteola* and *auricularia* were so similar that the taxa could not be differentiated on this basis. "*L.*" *acuminata* had a 75% relationship to *luteola* and an 81% relationship to *auricularia*. The 3 species had only a 69% relationship to *stagnalis*. They concluded that the multifolded prostate was not a trait confirming a close relationship between *luteola* and *stagnalis*.

Further clarification of this problem resulted from a study by Burch *et al.* (1971) of the lymnaeid species *stagnalis* and *corvus* from Poland along with *palustris elodes* (Say) and *stagnalis jugularis* (Say) from North America. "*L.*" *palustris elodes* does not have a multifolded prostate. They used foot-muscle proteins in the micro-Ouchterlony immunological test (Davis, 1968a) and the polyacrylamide electrophoretic methods of Davis and Lindsay (1967) modified for general esterase analysis. Antigens of *corvus* and *palustris elodes* were homologous; Polish *stagnalis* antigens were homologous with United States *stagnalis jugularis*; *corvus* had antigens unrelated to *stagnalis jugularis*. Likewise, *stagnalis* had antigens unrelated to *palustris elodes*. There were strong differences between North American *palustris elodes* and *stagnalis jugularis*. Studies of esterases yielded data that correlated with the immunological data.

Studies of shell characters, chromosome number and immunological results have confirmed distinct species groupings in the Lymnaeidae. To include all lymnaeid species in 1 genus, *Lymnaea*, obscures relationships. It is evident that certain anatomical traits obscure relationships because of convergence or parallel development. It is therefore justifiable to place *corvus* and *palustris* (Müll.) in *Stagnicola*, and *auricularia* in the group *Radix*, and *stagnalis* in *Lymnaea*. The question is not one of arguing for recognition of genera or taxonomic structuring into a rigid genus–subgenus Linnaean system, but of elucidating the different species-groups in the Lymnaeidae.

A further problem in lymnaeid systematics solved by an integrated approach concerns the genera *Fossaria* and *Stagnicola*. The shells of *Fossaria* and *Stagnicola* can be distinguished. Nevertheless, the species of *bulimoides* (Lea) and *cubensis* (Pfr.) have *Fossaria*-type shells, but a radula bearing bicuspid lateral teeth, a trait Baker (1928) considered a prime character of the genus *Stagnicola*. The micro-Ouchterlony double-diffusion tests mentioned above showed antigenic homogeneity between *bulimoides* and *F. obrussa* Say (Burch and Lindsay, 1972) and between *cubensis* and *F. obrussa* (LoVerde and Burch, 1973). There was strong heterogeneity between *obrussa* and *cubensis* when tested against *Stagnicola palustris elodes* and *Stagnicola emarginata* Say. Once more the traits of shell and antigenic reactions correlate to define species groups, when anatomical data obscured relationships.

(iv) Seed (1971) pointed out the difficulties encountered in separating the marine bivalves *Mytilus edulis* (L) and *M. galloprovincialis* (Lam.) on the ground of shell morphology alone. Five areas of investigation were integrated in an attempt to elucidate reliable differentia for the taxa. These were reproductive cycles, laboratory growth rates, infection with the "pea crab" *Pinnotheres*, spectrophotometric and chromatographic analysis of mantle-edge carotenoids, and polyacrylamide electrophoresis of muscle myoglobin.

Contrary to Seed's conclusion that . . . "physiological and biochemical differences between the taxa strengthens the case for considering these mussels as distinct species of *Mytilus*," the data presented more readily infer intraspecific variability. It is incorrect to assert as Seed has done that . . . "Since protein patterns are generally species-specific, a single analysis of 1 animal should be sufficient for identification, whilst on morphological grounds this might only be possible after statistical analysis of a large sample has been made."

The taxa could not be distinguished on the basis of carotenoid differences and the electrophoretic difference was extremely minor. Such a small difference is more commonly seen between populations or races of a species when one surveys the findings of numerous papers dealing with this subject. Protein patterns are not necessarily species-specific; they may characterize individuals and be different between individuals, races, subspecies, or species. In some cases they differ only significantly between genera.

The differences in reproductive cycle, infection with the pea crab, and growth rates under identical conditions do indicate genetic divergence between the taxa. Conclusive data to show that this divergence is indicative of either race or species ranking have yet to be provided.

From these examples it should be evident that without a framework of precise anatomy and cytology one cannot proceed with meaningful biochemical investigations. However, as is evident from studies of the Lymnaeidae and the planorbid genus *Bulinus*, morphology alone may lead to confusion due to parallel or convergent evolution of certain morphological features. It must be stressed that reliance for systematic evaluation should not be placed on 1 or 2 traits; multiple traits, not correlated with each other, should be used.

Within a framework of precise anatomy and cytology, biochemical data have their useful place. Electrophoretic and serological procedures are extremely valuable tools for establishing relationships, the degrees of genetic homology, inferred or actual rates of evolution, polymorphism, and for selecting genetic markers. Immunological methods are more suitable for macromolecular comparisons where relationships to be assessed are between closely-related families, genera and species groups. Depending on the degree of macromolecular divergence, it may or may not be possible to distinguish between congeneric species. Electrophoretic procedures are suitable for analysing relationships between congeneric species and between infraspecific taxa. By these procedures one may select protein markers to be isolated for use as specific antigens or to be traced from population to population. Certain proteins may serve as species-specific markers while others may reflect polymorphisms. Increased polymorphism may be interpreted as reflecting a greater ability of a species to respond to selective pressures, to adapt, and to survive. By avoiding a stereotyped approach to systematic problems, and by reaching for the tool that will most efficiently provide insight into relationships, it is possible to achieve a much greater degree of objectivity in systematics, a sounder taxonomic structure, and consequently increased stability in classification.

III. Chromatography

A. Introduction

Chromatography is the separation of substances where success depends on the balance between stationary and mobile phases. The stationary phase is an adsorbent such as paper, silica gel or sephadex, while the mobile phase is the solvent. Each substance being tested, for example, an amino acid, peptide, carotenoid, vitamin, alkaloid, lipid, terpene or steroid has a characteristic affinity for the solvent and the adsorbent, thus producing a characteristic rate of movement. Movement results from capillary action, gravity, pressure and diffusion, or combinations of these. Detection of the separated fractions depends on the development of colour, for example, by staining with ninhydrin, or by recording substances spectrophotometrically and electronically.

Chromatographic methods have today become highly sophisticated. Paper partition chromatography flourished from the decade beginning in 1940, becoming a universal technique in the 1950s. Thin layer, gel, and gas chromatography methods emerged in the 1960s. Thin-layer methods are replacing paper partition procedures, because the uniform fine-grained nature of the former provides greater sensitivity and use less material. Using paper and thin-layer procedures, substances are detected simultaneously in 1 or 2 dimensions. In gel chromatography, however, substances are collected in different elution volumes as they pass out of a column. Gel chromatography has been called gel filtration, column chromatography, molecular sieve filtration and exclusion chromatography. Karmen (1963) discusses the biological implications of gas chromatography. Standard texts of value in these areas are those of Block *et al.* (1952), Allen (1961), Hais and Macek (1963), Giddings and Keller (1965), Stahl (1969), Work and Work (1969) and Zlatkis (1970). Florkin (1966), Florkin and Bricteux-Grégoire (1972), Campbell and Bishop (1970) and Schoffeniels and Gilles (1972) have reviewed amino acid work as it relates to nitrogen metabolism and osmoregulation.

The application of chromatographic tools for systematic work has been greatest in non-molluscan groups (see review of Wright, 1959, for early work). Techniques applied to molluscan studies have centred on paper partition methods for detecting amino acids, other ninhydrin-positive molecules, or fluorescent substances. Relatively few papers have been concerned with using chromatographic tools for solving problems in molluscan systematics.

B. Critical factors in comparing taxa

Amino acids frequently form the basis for comparing taxa, but it is necessary to distinguish between "bound" and "free" amino acids; the former are a structural part of peptides or proteins, while the latter are in "pools" in either tissues circulating haemolymph, excretory products or in the contents of the digestive tract. Amino acids may be "transient" for varying periods of time during a particular metabolic or autolytic process when they are temporarily separated from a molecule. Further, when comparing taxa care must be taken to distinguish between amino acids from the environment, and thus not involved in genetically directed processes of the organism, and those that form a part of the taxon-specific complement, whether as structural proteins or involved in a homeostatic mechanism, like osmoregulation.

Different experimental procedures may yield vastly different results, as alluded to above and suggested by scrutiny of Table I. Michejda (1958) simultaneously investigated fluorescent and ninhydrin techniques using paper as the adsorbent; he concluded that the best qualitative results were obtained using foot and possibly whole-organism extracts, and that ninhydrin techniques were better for quantitative analyses. Certainly more conservative results were obtained using ninhydrin. Michejda and Urbanski (1958), studying 9 species of Clausiliidae, found the fluorescent patterns were strikingly uniform when whole-organism extracts were used. The fluorescent patterns derived from foot-tissue extracts were of little value because differences between individuals were greater than differences between species. Moreover, results from fluorescent and ninhydrin methods often did not agree, for example, the fluorescent patterns differed for *Cepaea vindobonensis* (Fér.) from Poland and Bulgaria while the ninhydrin-developed patterns were identical. *Helicigona lapicida* (L.) had a different fluorescent pattern from *Arianta arbustorum* (L.), but the ninhydrin-developed patterns were identical.

Many workers have been concerned with detailed quantitative and qualitative analyses of amino acids during ontogeny, or associated with different organs, diets, environmental features and parasitism. The studies of Brahmachary and Bhattacharaya (1963b), Morrill (1963) and Friedl (1961) should be consulted. From these studies it is evident that changes in buffer and pH resulted in a different spectrum of amino acids resolved. Pratt and Auclair (1948) presented a list of the minimum quantities of 36 amino compounds necessary to give a visible colour using ninhydrin with transmitted light, and a 2-dimensional paper method. Glutamic acid and glycine could be detected at concentrations of only 0·1 μg, while 5 μg of phenylalanine and 25 μg of histidine monohydro-

Table I. Qualitative analyses of amino acid composition of the haemolymph of some pulmonate Basommatophora. All results based on specimens reared in laboratory

	Planorbarius corneus					*Bulinus nasutus*	*Bulinus tropicus*
	non-infected[4]	infected 6 days	hydrolysed	FAA, lab reared[2]	FAA, field collected	hydrolysed[2]	[5]
Alanine	+	+	+	+	+	+	+
Arganine	+	+	NT	—	—	NT	+
Aspargine	NT	NT	NT	NT	NT	NT	NT
Aspartic acid	+	+	+	+	+	+	+
Cystine	NT	NT	NT	+	+	NT	+
Glutamine	NT	NT	NT	NT	NT	NT	+
Glutamic acid	+	+	+	+	+	+	NT
Glycine	+	+	+	+	+	+	+
Histidine	+	+	NT	—	—	NT	+
Isoleucine	+	+	+	+[a]	+[a]	+	NT
Leucine	+	+	+	+[a]	+[a]	+	+
Lysine	+	+	+	+[a]	+[a]	+	+
Methionine	+[c]	—	NT	—	—	NT	NT
Ornithine	NT	NT	NT	NT	NT	NT	NT
Phenylalanine	+	+	+	+	—	+	+[ac]
Proline	+	NT	NT	—	—	NT	NT
Serine	+	+	NT	+[a]	+	NT	NT
Threonine	+	+	+	+	+	+	+
Tryptophan	NT	NT	NT	—	—	NT	NT
Tyrosine	+[c]	—	+	+	+	+	+[ac]
Valine	+	+	+	+	+	+	+

[a] Found in some samples.
[b] Found only once.
[c] Trace.

chloride were needed. Certainly problems related to concentration must be considered when making tissue squashes on filter paper or comparing marine and freshwater organisms. Duchateau *et al.* (1952) have shown that the amino acid concentration in marine molluscs is much greater than that in freshwater and land molluscs. Michejda (1958) found that tissue squashes on paper were useless in the case of freshwater snails as "it gives a dispersal pattern that is quite illegible". Starvation may also drastically reduce amino acid concentrations to a level where they are hard to detect. Amino acids can be obtained in a more concentrated form from homo-

except for *Bulinus tropicus*, where they were collected in the field, and *Planorbarius corneus*, where both types are present.

Lymnaea stagnalis				*Biomphalaria glabrata*			
non-hydrolysed[1]	hydrolysed	FAA, lab reared[2]	FAA, field collected	hydrolysed[2]	FAA	non-infected[3]	infected (0·75–14 h)
+	+	+	+	+	+	NT	NT
NT	NT	—	—	NT	—	NT	NT
+	+	NT	NT	NT	NT	NT	NT
—	+	+	+[a]	+	+	NT	NT
—	+[c]	+[a]	+[a]	NT	+	NT	NT
+	+	NT	NT	NT	NT	NT	NT
—	+	+	+	+	+	NT	NT
+	+	+	+	+	+	+	+
+	+	—	—	NT	—	NT	NT
+	+	+[a]	+[a]	+	+[a]	+	NT
+	+	+[a]	+[a]	+	+[a]	—	+
+	+	+	+[a]	+	+[a]	+	+
+	—	—	—	NT	—	NT	NT
+	+	NT	NT	NT	NT	NT	NT
+	+	+	—	+	+[b]	+	+
+[c]	+	—	—	NT	—	NT	NT
+	+	+	+	NT	+	NT	NT
+	+	+	+	+	+	—	+
+	+	—	—	NT	—	NT	NT
+	+	+	+[b]	+	+[b]	NT	NT
+	+	+	+	+	+	NT	NT

(¹) Friedl (1961), paper and column; (²) Targett (1962), resin column, paper; (³) Dusanic an Lewert (1963), paper; (⁴) Gilbertson *et al.* (1967), amino acid analyser; (⁵) Stiglig and Van Eeden (1970), paper; FAA = free amino acids; NT = not tested or discussed; + = present; — = absent.

genates of tissues by extraction and concentration in ethanol. However, the work of Kavanau (1958) with *Spisula soldissima* (Dillwyn) shows that great care must be taken in preparing such extracts, because autolytic degradation of proteins to free amino acids and peptides occurs rapidly and to such an extent that the basic free amino acid patterns can be drastically changed.

It is evident that during ontogeny, generally, there are initially qualitative changes in amino acids, while later differences are largely quantitative (Ramamoorthi, 1955, 1958; Brahmachary and Bhattacharya, 1963a;

Collyer, 1961). Unless an ontogenetic problem is being investigated, it is advisable to use mature organisms for inter-taxon comparisons.

Moreover, in comparing taxa in terms of amino acids the whole organism should not be used; rather, specific tissues should be selected taking care to exclude contamination from gut contents, tissues, or other organs which can greatly affect results (Michejda and Urbanski, 1958; Wright, 1959; Kiecol and Michejda, 1960). For example, Michejda and Urbanski (1958), in the most extensive chromatographic study of molluscs undertaken to date, found that the fluorescent patterns of whole-snail extract showed individual variability while those with foot extract were constant. The foot tissue must be cleared of other tissues as the latter yielded different chromatographic patterns. They reported that only the results with skin and foot extracts paralleled higher category arrangements established on anatomical and conchological data. Morrill (1963) presented data on amino acids from the head-foot, liver-ovotestis, albumen gland, and hermaphroditic duct of *Lymnaea palustris*. Of 20 amino acids, aspartic acid, hydroxyproline, lysine, proline, threonine, tryptophan, valine, aspargine, and ornithine were not ubiquitous. Tyrosine and glutamine were universal in only trace amounts. Tissues particularly suitable for use in taxonomic investigations are foot-columellar muscle, albumen gland, muciparous gland, pallial oviduct and prostate. The gonad is most difficult to dissect free from the surrounding digestive gland tissue and the latter should not be used, because of heterogeneity within the system and the high probability of contamination with environmental and food substances.

A number of studies have considered the qualitative and quantitative distribution of amino acids in the pulmonate haemolymph (Table I); for example, Friedl (1961), Targett (1962), Gilbertson *et al.* (1967), Dusanic and Lewert (1963), Cheng (1963), Stiglig and van Eeden (1970), Feng *et al* (1970) and Stadnichenko (1971). These workers have found that there is considerable variation in amino acid composition of haemolymph within and between populations both in field and laboratory-reared conditions. Parasitism has a profound effect by decreasing the quantities of circulating amino acids and ecological conditions could likewise affect amino acid composition.

The interpretation of the data presented in many studies is equivocal. For example, Kirk *et al.* (1954) reported that 3 different diets did not affect the paper chromatographic pattern of fluorescent substances from *Theba pisana* (Müll.), but they obtained only a single diffuse spot from a squashed-foot extract for this taxon. Collyer (1961) found that starvation caused gradual reduction and disappearance of certain ninhydrin-positive spots derived from foot-muscle squashes (on paper) of *Nassarius*, but had no effect on *Littorina littorea* (L.). Nevertheless, Emerson (1966) found that starving *Littorina planaxis* Phil. for 70 days reduced their free amino acid content from 6·3 to 2·9 μg mg of dry weight and that nearly all of the individual constituents were reduced by approximately the same factor. Taurine, however, was not reduced; taurine is found in high concentration in marine organisms and not generally detected in land and freshwater molluscs (Simpson *et al.*, 1959). Allen and Awapara (1960) showed that the brackish water bivalve *Rangia cuneata* (Sow.) did not lack taurine as previously thought, but lost the compound from its tissue very rapidly. In contrast, the marine bivalve *Mytilus edulus* (and other marine molluscs) retained taurine under osmotic stress; an involvement in osmoregulation has been postulated (Florkin, 1966).

Environmental changes in salinity, temperature, food and season have been correlated with quantitative changes in amino acids and other molecules (Michejda and Urbanski, 1958; Wright, 1959; Rao, 1963; Hillman, 1964; Uikar and Webb, 1970; Campbell, 1970).

A 2-dimensional paper chromatographic study of *Haliotis crocherodii* (Leach) and *Pachygrapsus crassipes* (marine crab), using ethanol extracts of muscle, showed that specimens from polluted water lacked the asparagine that was present in animals from clean water. Also, phenylalanine and 3 unidentified substances were present only in specimens of *H. crocherodii* from polluted water (Schafer, 1961). This could be correlated with observations that a significant source of amino acids in sea water is from photosynthesising plankton and organic decay, and molluscs can absorb these amino acids (Carefoot, 1967). Stephens and Schinske (1961) demonstrated that *L. littorea* can absorb 100% of the glycine in a solution (2mM concentration) in 24 h.

C. Distinguishing between infraspecific taxa

Results and opinions are divided regarding the use of chromatographic procedures for distinguishing between races or strains of a species. Kirk *et al.* (1954) could not distinguish between populations of *Theba pisana* in Australia utilizing fluorescent paper chromatograms of foot-tissue substance. However, as this species was introduced into Australia only about 80 years ago genetic divergence between populations in this short period of time might not be anticipated. As discussed earlier Michejda and Urbanski (1958) frequently found that the recognition of differences between populations depended on the technique used. Care must also be taken to test for variation between individuals of a population at 1 point in time and on a seasonal basis (Michejda and Urbanski, 1958; Wright, 1959).

Gilbertson *et al.* (1967) found that variation in the amino acids from the haemolymph of different samples from a single population of *Biomphalaria glabrata* (Say) was of the same order as variability between strains. Wright (1959), however, found that the magnitude of variation in fluorescent chromatograms of foot mucus between populations of *L. palustris* was equal to that recorded between species in the *peregra-auricularia-natalensis* group. In later studies, Wright (1964) analysed many populations of *Lymnaea peregra* (Müll.) and found that 2 of 4 fluorescent bands present on paper chromatograms were very variable even within the same individual at different times of the year. Accordingly, increased emphasis was placed on analysis of the other 2 bands, and he demonstrated that there were a number of chromatographically distinct populations of *L. peregra*. It appears that the greatest variability of both chromatographic patterns and morphological traits coincided in the same geographic area, north-west Europe. In contrast, variation between populations of *L. auricularia* occurred only in eastern Europe. Wright suggested that *L. peregra* and *L. auricularia* were derived respectively from western and eastern segments of a common ancestral stock and that the chromatographic results supported the hypothesis that genetic variation of these species is greatest in areas of their origin.

Hillman (1964) showed under controlled conditions a single reproducible difference between 2 populations (races) of the oyster *Crassostrea virginica* (Gmelin). He stated (1965) that all allopatric populations could be distinguished on the basis of qualitative analysis of ninhydrin-positive substances, but it must be remembered that genetic control of these substances has yet to be proved.

D. Distinguishing between species and higher taxa

Kirk *et al.* (1954) initiated the use of fluorescent paper chromatograms to distinguish between species of molluscs; each of 7 land-snail species examined had a distinct pattern. Unfortunately, the tests for possible effects of diet on the chromatograms were conducted only with *Theba pisana*, but even this work is inconclusive. Nevertheless, the complexity of the chromatographic patterns produced for each of the other taxa studied appeared to indicate real differences.

Collyer (1961) compared *Nassarius reticulatus* (L.) from Plymouth, England, with a population of clearly the same species from North Wales and a population from Essex, which was provisionally identified as *N. reticulatus*, but had some traits resembling *N. nitida* (Jeffreys). Chromatographic analyses of the 3 populations, where diet and other factors were controlled, revealed that there were no differences between the Plymouth and North Wales populations. There were, however, reproducible differences between these and the Essex population. The data support the contention that there were genetic differences between the populations. Collyer wisely did not speculate on the taxonomic ranking of the populations based on the chromatographic data available.

Michejda (1958) noted that fluorescent patterns of snail foot mucus were species-specific. Chromatograms of prosobranch species were "... more or less parallel to the [traditional] taxonomic relationship[s]. On the other hand, among Pulmonata, the dissimilarity between *Radix* and *Lymnaea* or *Galba* is striking, the more so that the patterns of *Galba* are quite similar to those of *Planorbarius*, which belongs to quite a different family."

Michejda and Urbanski (1958) studied 63 taxa, primarily land pulmonates, and made a number of pertinent observations. They suggested that a re-evaluation of the clausiliid subfamilies Clausilliinae and Cochlodininae might be necessary because of the striking uniformity in chromatograms derived from 9 species of 3 genera within these subfamilies. They found that the foot-extract patterns of species of the subgenera *Helicella* and *Xeropicta* did not differ greatly and that this might correlate with the fact that several subgenera of the genus *Helicella* are similar in shell shape and anatomy. In contrast, the chromatograms for *Monacha carthusiana* (Müll.) were very different and this substantiated the conclusions derived from conchological and morphological comparisons with taxa of *Helicella*. The morphological differences are so great that some taxonomists would exclude *Monacha* from the Helicellinae. Yet in other cases it is impossible to distinguish between species using foot extracts and fluorescent techniques; for example, between *Helicigona faustina* (Ross.), *H. cingulella* (Ziegl.), and *Trichia villosula* (Ziegl.) and *T. hispida* (L.) (Michejda and Urbanski (1958). Indeed, ninhydrin-developed patterns within the Helicidae were quite similar, even to the point of monotony, but it was concluded that this lack of variation might not be typical of all families.

Wright (1959) used fluorescent techniques with paper chromatography to investigate foot mucus and obtained the following results: (i) similarity in chromatograms for *L. stagnalis*, *L. truncatula* (Müll.), and *L. glabra* (Müll.) was considered to be misleading as these taxa were divergent morphologically; (ii) the relationship between the patterns derived from the *L. auricularia* group, *L. peregra* and *L. natalensis* Krauss, was supported by morphological data; (iii) the patterns derived from *L. palustris* were very distinct and reinforce the conclusion based on morphological data that there is great divergence between *L. palustris* and other lymnaeid taxa studied.

Collyer (1961) found that paper chromatograms of ninhydrin-positive molecules of 2 geographically separated populations of *Nassarius reticulatus* were identical but that their patterns differed from that of *N. nitida*.

Okumura *et al.* (1966) made a quantitative analysis of 20 amino acids in 3 marine prosobranchs, *Neptunea poycostata*, *Buccinum verkruzeni* (Dantz. and Fischer), *Fusitriton oregonensis* (Redfield). Taurine represented about 50% of the total for each species. Although they stated that no marked differences were found in the amino acid composition of the proteins in the 3 species, that is in hydrolysates, their data showed methionine present in *Neptunea* and not in the others. In a tissue extract, *Buccinum* lacked cystine, which was present in the other taxa.

Wada (1966) studied the amino acids from hydrolysates of the conchiolin matrix obtained from decalcified shells of cephalopods, gastropods and pelecypods using an amino acid analyser. He stated that the conchiolin of any 1 species appeared to have a specific amino acid pattern and that the 3 classes could be distinguished on the basis of the amino acid composition. Moreover, a quantitative investigation of amino acids in hydrolysates of shell organic proteins, using ion exchange chromatography, demonstrated that the co-variant groups of amino acids were correlated with phylogenetic arrangement of the Mollusca (Degens *et al.*, 1967; Degens, 1967). There was also a significant decrease in amino sugars in the sequence: Amphineura > Cephalopoda > Gastropoda > Pelecypoda. The influence of temperature and salinity fluctuations on the secretion of the organic matrix of the shell could also be considered, indeed these factors may have had a pronounced influence on the evolution of invertebrates. Also see Hunt (1971).

Campbell and Speeg (1968) determined the whole body amino acid concentrations (free and bound) for *Otala lactea* (Müll.) and *Helix aspersa* Müll., together with the free amino acid content of the blood, digestive gland, kidney, lung, foot, reproductive and albumen glands of *O. lactea*. These determinations, part of a study of arginine biosynthesis and metabolism in terrestrial snails, indicated no significant differences between species in terms of whole-body amino acids. Also see Pujol *et al.* (1970).

These results indicate that there is a risk in assessing relationships solely on chromatographic data, for one must be prepared to find similarity in chromatograms derived from divergent taxa. Data from chromatography are, however, of value as supporting evidence to morphological studies.

E. Evaluation of chromatographic tools for systematic studies

It is accepted that the primary structure of proteins is genetically controlled, for a mutation can lead to amino acid loss or substitution. Evolution of species appears to be accompanied by a divergence in molecular structure and the hypothesis discussed by Zuckerkandl (1965) that the greater the morphological difference between taxa, the greater the molecular difference, seems to be supported. This divergence is inter- and intramolecular and may be expressed in different metabolic pathways. Careful analysis of amino acids in selected tissues or proteins has, therefore, enormous potential for solving phylogenetic problems, as exemplified in the review by Degens (1967) and studies of the evolution of haemoglobin (Zuckerkandl, 1965).

The techniques of thin-layer gel and gas chromatography are sophisticated enough to allow precise qualitative and quantitative determinations of amino acids and ninhydrin-positive substances, as well as numerous other classes of compounds, even when they are present in very low concentrations. The question then is no longer whether or not chromatography tools are potentially valuable tools in systematic studies, they are. The problem is how to use these techniques and with what purposes in mind. It is the question whether a single stereotyped method or a combination of methods will provide the best assessment of the relationships between taxa. Already experience has shown that the latter approach, that is a combination, is preferable.

There are over 300 naturally occurring amino acids (Bell, 1971) and of these, 20 are common to all living organisms. Of the remainder, the vast majority are from plants. At least 13 usual amino acids, as well as additional ones in trace amounts, are readily demonstrated free in the haemolymph of any gastropod. Accordingly, it is inadvisable to perform qualitative analyses of common free amino acids for taxonomic purposes in systematic studies of animals. Quantitative studies of free amino acids are likewise not advised because of the problem of proving genetic control of quantity; also the influence of diet, parasitism, ontogeny and other factors on quantity may be considerable.

In special cases it is possible to hydrolyse structural proteins or other molecules and analyse for amino acid content. This technique is particularly important in assessing genetically-controlled change in individual molecules of interest. In most cases, however, studies of variation in proteins are best investigated using electrophoretic and immunological procedures.

The most promising use of chromatography in systematic studies is in the investigation of the presence of unusual amino acids or related amines that are taxon-specific and indicative of distinct metabolic pathways. Success with this approach is particularly evident in plants where unusual amino acids abound. The distribution of such amines as tyramine, octopamine and tetramine in the Mollusca are cases in point (Florkin, 1966). For example, tyramine is found in quantity in *Octopus macropus* Risso, but is lacking in *Sepia officinalis* L. Studies using fluorescent compounds have yielded relatively stable results. However, as shown by Michejda and Urbanski (1958), who compared fluorescent and ninhydrin-paper procedures, at times ninhydrin-developed chromatograms were more conservative at the species or population levels.

It is essential to note that the best results obtained to date for differentiating taxa are those where chromatographic data were correlated with morphological data. Examples of a few problems where chromatographic tools are of potential value are:

(i) Morphological heterogeneity in a population may indicate possible hybridization. Chromatographic analyses may reveal compounds that are genetically controlled, as to type and quantity, and these may differentiate the parental types giving rise to the hybrids. It might then be possible to determine the extent of hybridization and follow the patterns of gene flow;

(ii) In a study where the lack of definitive morphological data prevents a conclusive assessment of the relationships of several taxa, chromatographic analyses coupled with other lines of investigation may serve to prove or disprove genetic homogeneity;

(iii) Genetic heterogeneity between populations of a distinct species as revealed by chromatographic tools may give clues to the centre of evolution of the species and the rate of genetic change. It is essential that numerous populations be studied throughout the species range in order to fully establish the distribution of variants;

(iv) Species-specific molecules may be of aid in discriminating between taxa where the morphology is too variable to permit certain identification without cytogenetic or laboratory-breeding experiments, for example, in the recognition of sibling species;

(v) Chromatography can be used to identify different biochemical pathways that may reflect differences that are phylogenetically important;

(vi) One may be able to study the rate of molecular and gene-group evolution, as has been possible with haemoglobin.

IV. Electrophoresis

A. Introduction

Electrophoresis is the separation of molecules by use of a constant direct electric current; in this context the molecules are usually proteins. The molecules are either positively, negatively, or neutrally charged, depending on their structure, but the net charge may be changed by the use of buffers. This charge affects the rate and, at times, the direction of movement of molecules in an electric field. Various inert materials can be used for supporting the molecules while the current is applied. Paper is rarely considered now, but starch and agar-gels, cellulose acetate strips, and polyacrylamide gels in columns or slabs are commonly used. Polyacrylamide methods, as developed by Ornstein (1962, 1964) and Davis (1964), have emerged in the past 10 years as the most sensitive means of separating proteins (Nerenberg, 1966; Chrambach and Rodbard, 1971). Polyacrylamide gels are unique in that the pore size can be adjusted to impose frictional resistance on molecules of a selected size; this increases the separatory capabilities of the gel. Only 200 to 300 μg protein are needed for each experiment and the actual run time, current on to current off, can be as little as 40 min; proteins are readily fixed and stained, and gel columns are easily stored for long periods of time.

Enzymes and other proteins are structured according to a genetically controlled program. Therefore, many of the problems in dealing with amino acids, which may be controlled genetically to varying degrees, are eliminated. One should note, however, that amino acids may be separated electrophoretically and specifically stained for. While one expects to find at least 17 to 20 amino acids in tissues of any animal, the combinations of these allow for the formation of innumerable proteins. Accordingly, the probability of finding taxon-specific biochemical markers among proteins is greatly enhanced compared to qualitative analyses of free amino acids.

The amino acid sequence of many proteins is frequently different; such variation is genetically controlled and a single mutation resulting in an amino acid substitution may result in changing the net charge of the molecule, thus rendering the mutant molecule detectable by electrophoresis. Studies with abnormal haemoglobins in human populations have clearly shown the effects of mutations causing amino acid substitutions, deletions and additions (Livingstone, 1967).

Systematists can use electrophoresis to determine the degree of genetic variability within populations and to estimate the extent of genetic divergence between populations, thereby permitting an assessment of relationships between sibling or closely-related species. Electrophoretic tools make it possible to determine correlations between such factors as shell and genetic variability. Practical uses are innumerable.

B. The development phase

Early studies aimed at exploring the possibility of differentiating taxa were those of Dubuisson and Roubert (1947) using muscle extracts from snail, frog and rabbit; Deutsch and McShan (1949) using blood from amphibians, reptiles, fish and snail, and Woods *et al.* (1958) using blood from Crustacea and 2 molluscs. The results suggested "that starch-gel electrophoresis of serum proteins may be useful in certain racial studies, taxonomic problems and considerations of biochemical individuality". Many of the earlier investigations were concerned, however, with the detailed examination of a particular metabolic pathway or substances in a single species; for example, Pellegrini and Stagni (1957), glycogen in *Helix pomatia* L.; Yagi *et al.* (1957), haemoglobin of *Anadara inflata* Reeve; Wright and Ross (1959 and 1963), studying blood proteins of *Biomphalaria glabrata* and Cheng and Sanders (1962), blood of *Viviparus malleatus*. Other relevant papers are those of Masseyeff *et al.* (1963), Morrill *et al.* (1964), Norris and Morrill (1964), and Waehneldt *et al.* (1973).

Ontogenic changes were reported; Wright and Ross (1963) found a decrease in blood-protein density with increasing age in *Biomphalaria glabrata* and suggested that results with planorbid eggs were most satisfactory because qualitative changes did not occur. Davis and Lindsay (1964) found that blood proteins of *Helix pomatia* decreased quantitatively, but not qualitatively, during ontogeny. A study of the haemolymph proteins in the slug *Agriolimax reticulatus* (Müll.) was undertaken by Bailey (1971) for the purpose of finding fractions that might correlate with reproductive development. Of the 4 techniques used, the best results were obtained with polyacrylamide electrophoresis techniques of Davis and Lindsay (1967), with poor results from agar-gel, starch-gel or isoelectric focusing on 5% polyacrylamide. The problems with the latter techniques were attributable to extremely low concentrations of protein available for any experiment.

Several early electrophoretic studies of molluscs were undertaken on a systematic basis. Targett (1963) did not obtain satisfactory results in comparing species; Manwell (1963) could discriminate between the arcid clams *Noetia ponderosa* (Say) and *Anadara transversa* Say on the basis of electrophoretic data, and Cheng (1964) concluded from his study of several species of snails and clams that "species" yielded specific electrophoretic patterns and that undoubtedly much more extensive surveys of the serum proteins of molluscs must be made before useful taxonomic information will be obtained. Gysels (1966) could distinguish only 1 fast-migrating band in the haemolymph of prosobranch species that was of any importance in separating taxa, although in pulmonates there were several components which he stated were highly species-specific. He concluded that electrophoretic analysis of molluscan serum proteins ". . . may furnish valuable information on a taxonomic lower level . . ."

C. Selecting methods and presenting results

Electrophoresis has been used to (i) isolate 1 or more proteins for biochemical studies; (ii) establish a taxon-specific "fingerprint", that is, a linear series of protein bands characterizing either an individual or population on the basis of unique proteins or varying combinations of occurrence and quantity of proteins, and (iii) determine the genotype of an individual, where the electrophoretic result is the phenotype (e.g. one or more enzyme bands may be the result of a single allele). Systematists tend to concentrate on the results from studies of types (ii) and (iii); in developing these approaches they may pool tissue extracts from several individuals and determine the "fingerprint" or zymogram for a population, or determine the protein patterns for individuals using either non-specific proteins or enzymes. In the former the systematist is looking for proteins that are taxon-specific and that serve to identify a population, species, or genus. The contention is that those taxa with greatest overall similarity in electrophoretic patterns are more closely related than taxa with molecular divergence. In the latter, the emphasis is on the individual and the distribution of alleles or polymorphisms throughout a population. It should be obvious that studies of populations based on individuals have greater potential for systematics than studies based on populations where extracts of individuals are pooled. Nevertheless, both approaches depend for success on factors that include the correct selection of experimental procedures, careful choice of tissues or fluids that will be the source of proteins and knowledge of biological factors that can affect the results. For example, in considering the protein separating medium, it is evident that polyacrylamide gel is superlative for detailed investigations of non-specific proteins and starch-gel for population studies of variation in enzymes. Starch-gels are readily cut into a number of slices that can be analysed for different enzyme systems. Paper electrophoresis is not recommended; it is particularly insensitive as shown by a comparison of the work of Sibley and Johnsgard (1959), using paper, and Lush (1961) using starch-gel. The former found no significant differences between 23 breeds of domestic fowl (*Gallus gallus*) based on egg-white protein, and concluded that egg-white proteins were conservative, but the latter found differences between individual hens of 16 small closed populations; the data suggested that variation was due to 2 alleles segregating at each locus for each of 3 variant classes. Kobayashi (1964) could find no significant difference between taxa as widely divergent as the clam *Corbicula leana* and the snails *Viviparus intertextus* Say and *Cipangopaludina malleata* (Reeve) using paper electrophoresis of extrapallial fluid and blood of 14 species of molluscs.

Polyacrylamide gel provides better results than cellulose acetate for separating proteins, as shown by a comparison of the works of Targett (1963) and Wright and Ross (1963) with that of Michelson (1966b), where all used snail haemolymph. The former demonstrated 1 to 8 components, while the latter demonstrated 9 to 12. Cheng (1969) also using polyacrylamide, found 15 fractions in the blood of *Helisoma duryi normale* Pilsbry. Because of the great sensitivity of polyacrylamide gel electrophoresis, one must be aware of factors that can drastically alter results. Experimental error can be induced by both non-biological and biological factors.

Non-biological factors affecting reproducibility of results frequently involve technical aspects, such as cleanliness of glassware, chemical purity, and factors such as differential heating in a gel, improper polymerization due to poor technique, and current fluctuations. Many chemicals of supposedly best grade are, in fact, dirty; passing solutions through a bacteriological filter does help to remove contaminants. For these reasons a standard control should be used with each experiment. A pooled serum can serve as a control, where deviation from the standard pattern indicates experimental error. From 8–50%, average of 30%, of experiments thus controlled cannot be used because of procedural error (Davis and Takada, 1969; Kobayashi, 1964; Davis and Lindsay, 1967).

Biological factors involve varying protein concentration in tissues or blood (Cheng, 1963; Wright and Ross, 1963; Gilbertson *et al.*, 1967; Cheng, 1969), differences due to ontogenic processes, protein differences in different tissues, physiological effects and parasitism. Protein concentration may vary $\pm$ 28%, even using the same standardized techniques. It may be necessary, therefore, to establish the protein concentration of haemolymph and tissue extracts before undertaking electrophoresis in order to dilute or concentrate the solution to a level that will yield uniform results. It is a simple matter to use the Folin-phenol reagent test (Daughaday *et al.*, 1952), which can be adapted to ultra-micro procedures; a standard of crystalline bovine albumen is suggested. Dilute solutions may be concentrated by forced evaporation or vacuum dialysis (Clausen, 1969).

There are ontogenic changes, both quantitative and qualitative, in proteins (Norris and Morrill, 1964; Decleir and Richard, 1970); Wright and Ross (1963, 1965, 1966) used planorbid eggs of a fairly uniform age and thus they should have obtained comparable results. Instead, there was extreme infraspecific variation in the "fingerprints" obtained. There are additional problems with using eggs when one considers that many species of molluscs do not readily produce eggs in the laboratory and that eggs of numerous species are very minute and are laid only a few at a time. Accordingly, organs from mature individuals should be used whenever possible.

The choice of tissues from which to extract proteins is most important. The whole organism should not be used for several reasons: (i) the digestive system contains bacteria and food contaminants that may alter the taxon-specific protein patterns; (ii) the mixture of proteins is too complex to obtain meaningful results, and (iii) it is much more possible to isolate a "key" protein and assess its function when one knows the organ from which it was derived.

Different tissues may yield different results. Schwabl and Murray (1970) found in polyacrylamide gel studies of *Partula* that tissues other than the muciparous gland, which is part of the female reproductive tract, yielded patterns either too variable from individual to individual or too constant throughout a species group; unfortunately these findings were not documented with data. Davis and Lindsay (1964, 1967) advocated the use of foot-muscle extracts because: (i) the foot is a characteristic organ of the phylum Mollusca; (ii) more components can be separated from foot-tissue extracts, than from haemolymph alone; (iii) foot tissue is dense and more proteins can be extracted per unit weight, compared with other organs of the body, and (iv) foot tissue is readily obtained even from small molluscs. Tissues should be carefully checked for parasites, for example the digestive gland often harbours sporocysts and cercariae; tissue of the foot is easily checked. Other tissues of potential value for comparing taxa are mantle, albumen gland, pallial oviduct and prostate gland.

The physiological state of the animal is also important. Saleuddin *et al.* (1970) used cellulose acetate and polyacrylamide electrophoresis to demonstrate that during regeneration of damaged shell 2 additional protein bands were isolated from the mantle of *Ampullarius glaucus* L. and that these bands corresponded to 2 fractions found in extracts of the normal digestive gland. Likewise, Stadnichenko (1970) found that female *Viviparus contectus* (Millet) brooding embryos had a heightened synthesis of protein in the digestive gland, which was reflected by an increase of blood-soluble proteins. Differences between males and females were established for 3 fractions; altogether there were 6 fractions in males and 7 in females.

Parasitism can have a great effect on electrophoretic results. For example, it can cause a depletion of blood proteins as found by Cheng (1969), Pascoe (1970) and Lee and Cheng (1972). Coles (1969a) studied 7 enzyme systems from the digestive glands of *Lymnaea*, *Bulinus* and *Pila* and subsequently found that infection of *Biomphalaria sudanica* with plagiorchoid daughter sporocysts correlated with a reduction in alkaline phosphatase, a result contrary to those of Cheng (1964) studying *Helisoma trivolvis*, infected with *Echinoparyphium*, and James and Bowers (1967) studying *Cardium edule* (L.), infected with *Cercariae bucephalopsis haimaena*. Conversely, Pascoe (1970) found that parasitism was associated with the appearance of a fraction not found in uninfected snails. Also see Pino *et al.* (1962) and Coles (1969b). The same phenomenon was found by Lee and Cheng (1972) and Michelson and Dubois (1973) studying the effect of *Schistosoma mansoni* Sambon on the haemolymph proteins of *Biomphalaria glabrata*.

The uninitiated should be aware that there can be considerable variability between experiments. Even with pooled extracts from animals of the same population, not all fractions will necessarily be found in each experimental separation of proteins. A combination of genetic factors and experimental error is involved. Morrill *et al.* (1964) noted that the band patterns of albumen from capsules of different egg masses of *L. palustris* frequently differed qualitatively; of 11 fractions separated in starch-gel, only 6 developed consistently. Davis and Lindsay (1967) studied proteins from foot-muscle extracts of 4 individuals of *Helix pomatia*; of 20 fractions, 4 were not found consistently. R_F values for proteins separated from foot-muscle extracts of 2 populations of *Semisulcospira libertina* (Gould) are given in Fig. 1 and Table II (Davis, 1969a). Note that the percentage of "runs" where a band was present was recorded; very rarely encountered fractions can be documented aside from the principal analysis (Table III).

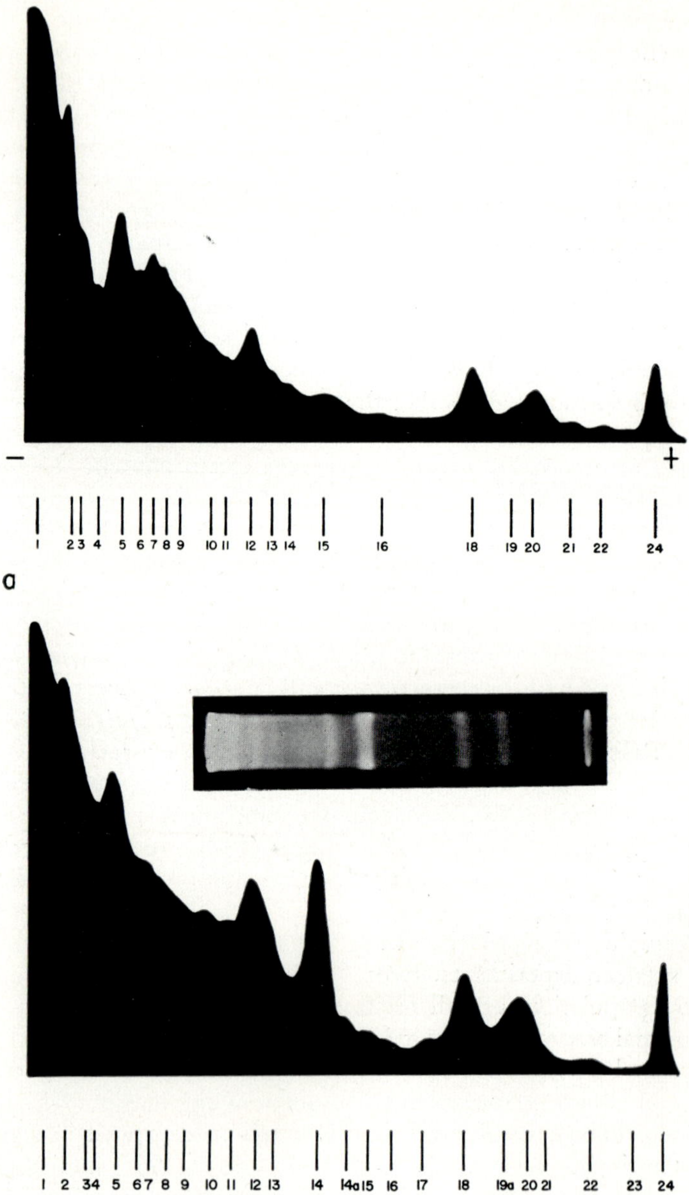

Fig. 1. Densitometric tracings of results from electrophoresis, of proteins from foot-muscle extracts of Japanese *Semisulcospira libertina* using polyacrylamide techniques. Samples from Nahari, Shikoku (a and b) and Shimoda, Honshu (c and d). Prints of gels are shown and the protein fractions are numbered below each tracing (from Davis, 1969a).

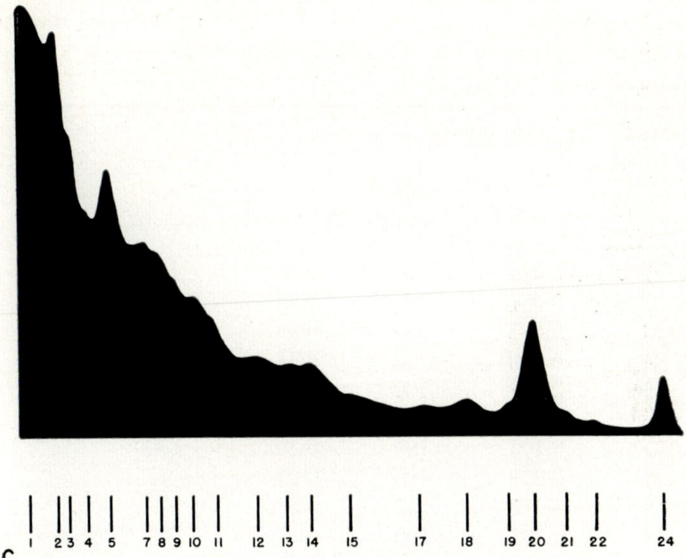
1 2 3 4 5 7 8 9 10 11 12 13 14 15 17 18 19 20 21 22 24

c

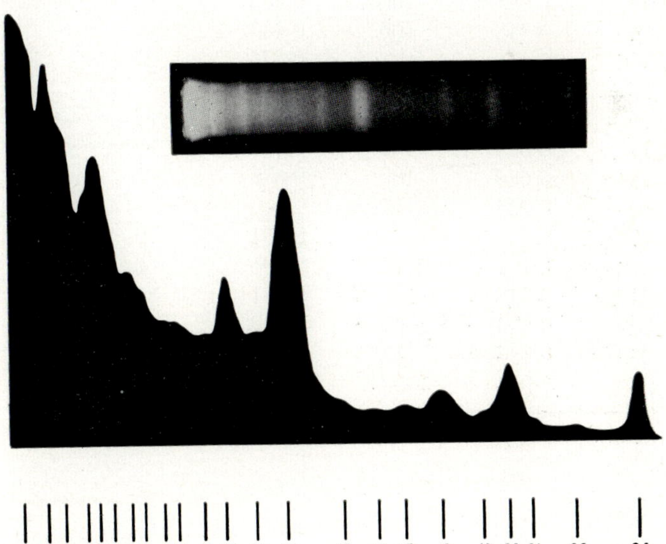
1 2 3 4 5 6 7 8 9 10 11 12 13 14 15 16 17 18 19 20 21 22 24

d

Table II. Congruence in the occurrence of protein fractions separated electrophoretically from fresh foot muscle extracts of *Semisulcospira libertina* from Shimoda and Nahari (from Davis 1969a).

Band number	Shimoda		Nahari		Significant difference (SD)
	R_F value	% runs present	R_F value	% runs present	
1	0·015[a]	100	0·019[a]	100	
2	0·067	100	0·059	100	
3	0·092	83	0·088	100	
4	0·110	92	0·111	53	
5[a]	0·144	100	0·138	100	
6	0·175	33	0·170	63	
7[a]	0·198	100	0·192	100	
8[a]	0·215	100	0·213	87	
9	0·250	42	0·240	73	
10	0·285	100	0·276	63	
11	0·312	33	0·305	67	
12[a]	0·345	100	0·345	100	
13	0·391	100	0·388	93	
14[a]	0·448	100	0·433	100	
14[a]	—	—	0·476	27	SD
15					
16	0·578	92	0·556	87	SD
17	0·629	100	0·622	87	
18[a]	0·680	100	0·687	100	
19	0·725	67	—	—	SD
19[a]	—	—	0·751	73	SD
20[a]	0·794	100	0·786	100	
21	0·835	100	0·831	80	
22	0·890	92	0·882	87	
23	0·936	17	0·937	47	
24	1·000	1000	1·000	100	

[a] Dense, distinct bands as shown in Fig. 1.

Table III. Congruence in the occurrence and location of 9 rare fractions of proteins extracted from foot muscle of laboratory reared and field collected *Oncomelania hupensis nosophora* (from Davis and Takada 1969).

Band	Field collected			Laboratory reared	
number	Mixed	Females	Males	Females	Males
1a[a]	0·029 (20)[b]	—	0·024 (27)	0·032 (63)	0·019 (66)
1b	—	—	0·052	—	0·053
4a	—	—	—	0·155	—
9a	—	—	—	0·312	—
10a	0·391	0·383	—	—	0·387
17a	—	—	—	—	0·586
18a	0·650	—	—	—	—
20a	—	0·815	—	—	—
23a	0·967	—	—	0·950	—

[a] Only bands 1a and 1b were encountered in 2 or more experiments involving each of the 5 conditions studied.

[b] The percentage of experiments in which the fraction was resolved is enclosed within parentheses.

In comparing taxa, emphasis is placed first on the prominent stable fractions found 90–100% of the time, for such fractions are the basis of the taxon-specific densitometer profile. Results should not be presented simply as photographs of gels, schematic drawings of gels, or densitometer profiles where fractions are not properly identified by number and R_F value. The R_F value is defined as the ratio of the distance from the origin to the band compared to the distance from origin to the front band. Reproducibility and comparisons with data of other workers depend on knowing the numerical or coded designation of the bands, or the R_F identification of the fractions. In isoenzyme work a band consistently found in the reference population is frequently used to indicate the relative position of other bands in both the reference and other populations.

When using polyacrylamide techniques and when the distance from origin to front is no more than $\pm$ 1 mm from a set distance, for example, 35 mm, the R_F values can be so accurately determined that when a fraction from taxon *a* differs from that of taxon *b* by a R_F of 0·018, the difference can be recognized. This can be tested by mixing extracts from the 2 taxa to see if the 2 fractions can be distinguished in the same gel; this principle is illustrated in Fig. 2.

The densitometer profile is a machine translation of the differentially-spaced bands of varying density (see Fig. 1). Slight variations in density may have a pronounced effect on the densitometric profile (Davis and

Lindsay, 1967). A study of the 17 densitometer profiles presented by Cheng (1969) of proteins separated from the blood of *Helisoma duryi normale* Pilsbry will illustrate the effect of changes in protein density on the densitometric profile; such variation could be attributed to experimental technique where the protein concentrations, naturally quite variable, were not adjusted.

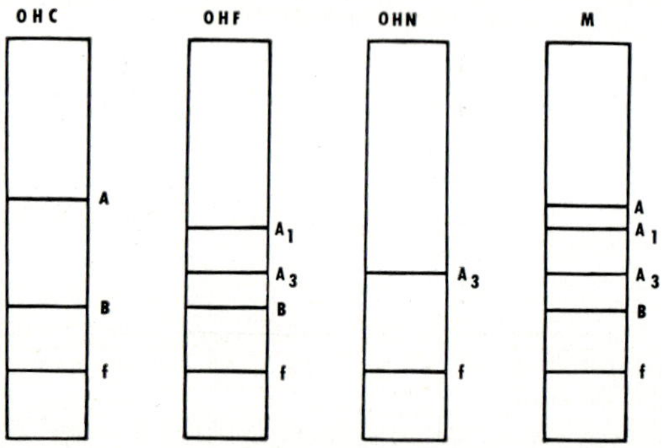

Fig. 2. Schematic presentation of selected protein bands from mid-gel (polyacrylamide electrophoresis) to the front (f) to demonstrate the identity of fractions A and A_1. The individual patterns for *Oncomelania hupensis chiui* (OHC) from Alilao, Taiwan, *O. h. formosana* (OHF) from Pu Yen, Taiwan and *O. h. nosophora* (OHN) from Yamanashi, Japan are shown. Foot muscle extracts of the 3 taxa are mixed and separated (M), and the result is a composite of the patterns of the 3 taxa. Fractions A and $A_1$are distinct (derived from Davis, 1968a, and Davis, unpublished).

In polyacrylamide electrophoresis (using 7·5% polyacrylamide gel) it is apparent that many of the characteristic differences between taxa are usually found from mid-gel to the front. The region near the origin is usually packed with a series of large molecules that only migrate slowly (Davis and Lindsay, 1967). Sodeman and Meuwissen (1966) reported that the area near the origin was "technically difficult to interpret". Schwabl and Murray (1970), studying *Partula*, showed great variability amongst fractions near the origin and taxon-specific bands from mid-gel to the front.

D. Non-specific proteins

The series of investigations summarized below illustrate the use of electrophoretic data to supplement morphological data in the identification of species, species groups and even genera. The researchers hoped to find protein markers or series of markers in extracts where individuals or tissues from individuals were pooled. In some cases populations of the same species were studied to assess if genetic differences between the populations correlate with environmental factors or with different capacities to transmit a parasite.

During 1960 to 1970 the phase of general electrophoretic surveys of diverse animal assemblages passed and it was realized that electrophoretic methods could yield valuable data regarding the relationships of closely-related taxa. Yet one still found in some papers, e.g. Chaetum and Darlington (1971) examining *Triodopsis*, such general statements as: electrophoretic examination of "plasma proteins" yielded similarities and differences between taxa and thus could be of use in systematic studies, but no supporting data were given. Moreover, the impact of the *Drosophila* work of Hubby and Lewontin (1966) and Lewontin and Hubby (1966) had not reached the malacological world (see pp. 139–142) and thus there was no realization that alleles at a single locus could cause considerable variation in protein banding when extracts were derived from single individuals.

1. *Distinguishing between infraspecific taxa*

Wright and Ross (1965, 1966) studied a series of populations and species of African *Bulinus* and *Biomphalaria* using egg proteins and cellulose acetate as the supporting medium. They discovered a great amount of interpopulation variability, but the variation between populations of *B. globosus* (Morelet) was so great it encompassed variation seen between different species of the same genus. No distinct species-specific densitometer profiles were observed, which is disappointing, as species concepts in these genera are diffuse and great uncertainty prevails concerning identifications at the species level. Although Wright and Ross (1965) presented data for populations of *B. africanus* (Krauss) from Nelspruit and Richmond (S. Africa), they suspected that the former should be classified as *B. globosus*. Further, there was closer resemblance between profiles of *B. forskalii* (Ehrenb.) from Badja Kunda and *B. cernicus* (Morelet) from Cresonville than between 2 populations of *B. cernicus* from Mauritius (1966). Unfortunately, their work is weakened because (i) no band identifications are given, (ii) variability between experiments is documented in only 1 example and (iii) curved, distorted bands in at least 6 gels indicate poor experimental conditions and these distortions are reflected in the densitometer tracings. Accordingly, one should compare the densitometer patterns with caution.

Davis and Lindsay (1967), using foot tissue and polyacrylamide gels, demonstrated distinct differences between 4 populations of *Pomatiopsis lapidaria* (Say). These differences did not interrupt the species-specific densitometer profile seen for each population. Variability in profiles due to experimental or genetic conditions was documented. Davis (1968a) demonstrated differences between populations of *Oncomelania hupensis formosana* in Taiwan; snails from central Taiwan at Pu Yen were distinct from populations from the north-east at I-lan County. This supported earlier studies that showed that each of the subspecies of *O. hupensis* had a distinct fingerprint (Davis and Lindsay, 1964; Davis, 1967; Davis and Takada, 1969).

In a detailed investigation of the genus *Semisulcospira*, it was possible to link electrophoretic data with morphological data to produce a taxonomic revision of the group (Davis, 1968b, 1969a, 1972a). He demonstrated that: (i) *S. trachea* (West.) was a synonym of *S. libertina*; (ii) a population of *S. libertina* transmitting the human lung fluke *Paragonimus westermani*, was closer genetically to the topotype population several hundred miles away than was *S. trachea* from less than 100 miles away; (iii) there were only slight genetic differences between populations of *S. libertina* from Okinawa and Houshu, Japan.

Davis (1971a) could find no significant differences between electrophoretic profiles of 2 populations of Malayan *Brotia costula* from vastly different environments.

2. Comparing species and higher categories

Morrill *et al.* (1964) found that albumen and polysaccharides of *Lymnaea palustris*, *L. stagnalis* and *L. columella* were species-specific.

In a series of studies on *Bulinus* and *Biomphalaria* Wright and Ross (1965, 1966) found that it was possible to distinguish the *B. truncatus* species-group from 3 other species-groups on the basis of a pronounced peak in the densitometer pattern caused by 4 or more closely-associated bands. Initially it was thought that the species of *Biomphalaria* had a number of cathodal-migrating fractions not present in *Bulinus*, but later species of *Bulinus* with such fractions were found. In a study of 48 populations of *Bulinus* of the *africanus* species-group, only those identified as *B. nasutus* (Mart.) had a species-specific egg-protein pattern; but the "fingerprints" of *B. nasutus* and *B. abyssinicus* (Mart.) were similar enough to suspect a close relationship between these species. The patterns for *B. forskalii*, *B. beccarii* (Paladilhe), and *B. camerunensis* (M.-B.) were very similar, while "fingerprints" of *B. globosus* populations from Kenya belonging to a different species-group were quite distinct.

In contrast to the work of Wright and Ross, Rosé *et al.* (1966), using polyacrylamide gel, found species-specific proteins for species of *Bulinus* and *Biomphalaria*. They found greater similarity between *Bulinus truncatus* and *B. rolfsii* compared with *B. tropicus* and *B. forskalii*. *B. forskalii* had a unique pattern.

As part of a detailed study on the population genetics and systematics of *Partula* Schwabl and Murray (1970) analysed proteins from the muciparous gland using polyacrylamide techniques. Fractions of similar mobility were found for all species of *Partula* studied (= group-specific proteins). Animals of the same species, collected from different populations, usually yielded the same "fingerprint" and this was consistent with laboratory reared individuals. *P. suturalis* Pfeiffer and *P. dendroica* Crampton, although geographically isolated, showed no divergence in pattern and this suggested that they may be conspecific; while *P. tohiveana* Crampton and *P. suturalis* that were sympatric, had distinct "fingerprints". Animals with morphological features that suggested they were hybrids had "hybrid fingerprints" and by mixing tissues of the 2 species (as in Fig. 2) this hypothesis was confirmed. *P. suturalis* and *P. aurantia* Crampton likewise hybridized and did so to such an extent that "ultimate merging of the two entities seems likely.".

Wium-Anderson (1970) could differentiate *Littorina striata* King and Brod. from *L. littorea* and *L. rudis* (Maton) on the basis of migrational differences of haemoglobin, but regarding protein fractions, she stated: "the protein staining of *L. rudis* showed great variation in the protein pattern from one locality to another . . . contrary to the pattern of *L. littorea*. The combination of bands in *L. rudis* appears entirely random, and there is a great variation from one individual to the other in each locality, with the exception of individuals from Heligoland and from Greenland (*L. rudis groenlandica* Menke)." Unfortunately, Wium-Anderson presented insufficient information for taxon-specific protein profiles to be determined and it is impossible to substantiate the statement that *L. rudis groenlandica* should be considered a distinct species. Apparently there is a great amount of genetic variability in *L. rudis*, but from the available data the magnitude of this cannot be assessed. It is evident that there is much greater variability in so-called "rare bands". It has been demonstrated several times (Davis, 1968b, 1969a; Davis and Takada, 1969) that even when tissues of several individuals of a population are mixed, over a number of experiments the band frequency is variable for some components and particularly in "rare bands". This variability may be attributed to genetic differences between populations and to experimental error. Comparatively greater variation in shell morphology and protein patterns of *L. rudis*, in contrast to *L. littorea* and *L. striata*, can be correlated with a brackish water environment and ovoviviparity, both of which are barriers to dispersal. See also More *et al.* (1971a, b).

E. Enzyme analyses and molecular genetics

Historically, Markert and Møller (1959) demonstrated that many enzymes exist in multimolecular forms called isozymes. Isozymes (isoenzymes) are currently defined as proteins having the same catalytic function, but with different structures formed by various polypeptides controlled by different genetic loci. They are detected by electrophoretic procedures. After 1959 researchers explored the potential of isoenzyme analyses for gaining insight into biological phenomena. Early emphasis was placed on changes associated with ontogeny.

1. Molluscan development

A number of studies have considered protein differentiation and development during ontogeny of *Ilyanassa obsoleta*. Morrill (1961) studied various enzymes using starch-gel electrophoresis and found that enzyme complexity increased during development. In a more detailed investigation Morrill and Norris (1965) analysed eggs, embryos, and veligers for the effect of developmental sequence on the occurrence of isoenzymes of nonspecific esterases, phosphatases (alkaline and acid), alanyl napthylamidases, leucyl napthylamidases, alpha glucosidase, beta glucuronidase, and sulphatase. Certain fractions were present from days 0 to 10; others disappeared after day 4; yet others only appeared after day 3. The changes between days 3 and 5 coincided with differentiation of organ primordia. Presence of unique preorganogenesis bands contrasted with findings for *Lymnaea* (Norris and Morrill, 1964).

Meizel and Markert (1967) studied the malate dehydrogenase (MDH) isoenzymes from *Ilyanassa obsoleta* (Say) for use as markers for other experiments concerning the development of embryos. The organs of the adult all had the same MDH isoenzymes and these could be classified into 2 categories, mitochondrial and supernatant. The extensive experimental methodology is rigorously explained. Freeman (1971; polyacrylamide gel) studied the appearance of alkaline phosphatase and non-specific esterases during embryogenesis of the same species. The effect of removal of the polar lobe on enzyme synthesis was assessed; certain enzyme fractions were reduced or failed to develop.

Morrill and Dow (1962) and Norris and Morrill (1964) have studied 9 hydrolitic enzymes from the adult organs and developing embryos of *L. palustris*. Forty-two enzyme bands were detected in extracts from 16 adult organs. The number of bands varied from 32 in the digestive gland to 10 in the muciparous gland. Five "adult" isoenzymes were found in the 0- to 2-day embryo, but from the beginning of organogenesis in the third day, additional bands appeared. Enzymes present in the early stages were thought to be associated with yolk and albumen metabolism. Goldberg and Cather (1963) used polyacrylamide gels to study lactate dehydrogenase (LDH) isoenzymes during the development of *Argobuccinum oregonense*.

2. Esterases

These enzymes, which hydrolyse carboxylic esters are extensively used in systematic studies. Indeed, non-specific esterases have been used more than any other class of enzymes for "fingerprinting" taxa and in genetic studies. The advantage of using esterases is that sharp resolution is obtained with starch-gel or polyacrylamide electrophoresis, they are easy to detect and are readily separated from extracts of almost all organs. Esterases also appear to be more polymorphic than any other enzyme systems; for example, the *esterase*-5 locus in *Drosophila* has 12 alleles

(Prakash *et al.*, 1969), while 1 esterase in the butterfly *Colias eurytheme* has 13 alleles (Burns and Johnson, 1967) compared to the 2 to 4 alleles that are typical of other systems. The problem of analysing non-specific esterases is that different groups of esterases can be involved; there are 4 main categories: arylesterase, acetylesterase, cholinesterase and carboxylesterase. Therefore when comparing different taxa it is impossible to be sure that the same groups of esterases are being considered in each taxon. Nevertheless, the advantages outweigh the disadvantages.

Haites *et al.* (1972) found that, with few exceptions, the various tissues of *H. aspersa* were alike in isoenzyme content and that there was less tissue differentiation of isoenzymes than in mammals. These data were obtained during an analysis of carboxylesterases, cholinesterases and arylesterases from 8 different groups of animals. *H. aspersa* had 18 isoenzymes, surpassed only by the king prawn, an arthropod, which had 29 isoenzymes. The foot muscle of *H. aspersa* had 8 carboxylesterases and 3 cholinesterases, while arylesterases were dominant in the digestive gland. There were similar isoenzymes in the digestive gland, reproductive system and intestine. Nine of the carboxylesterases were similar to subgroup I carboxylesterases found in sheep, horse, ox and possum (Holmes *et al.*, 1968).

3. Systematic studies

Many of the earlier studies concentrated on establishing isozyme "fingerprints" for taxa, in order to facilitate identification or assess relationships. In many instances tissues of individuals were pooled and comparisons were undertaken without realizing that 1 or several bands could result from a single locus. This approach is of little value compared to the study of individuals and the assessment of genetic relationships using sets of loci and alleles. That such studies were continued into the 1970s is surprising since the approach of molecular genetics was clearly established by 1970.

The results from the early studies were rather equivocal; for example Reid (1968) concluded from an analysis of 12 species of marine bivalves that the value of isozyme patterns for delineating pathways of speciation was doubtful, but they could be of value for differentiating below the species level. Similarly, Reid and Dunnill (1969) in an examination of 8 species of the marine bivalve *Macoma* found that patterns of variation were species-specific, but variation between individuals was sufficient to confuse clear-cut delineation of species. Their data indicated that a number of polymorphisms were present, for example, in *M. incongrua* 20% of the individuals were polymorphic at 1 of 3 loci governing digestive gland esterases; they did not analyse these polymorphisms.

Other workers have drawn conclusions which were certainly not justified by the results presented from studies of isoenzymes, for example Jones (1971) using malate dehydrogenase isoenzyme patterns to assess affinities between species of *Littorina*, and Sick (1972) studying populations of *Urosalpinx cinerea* (Say) presented the data inadequately for any clear deductions to be reached. Jelnes *et al.* (1971) used esterase isoenzymes from the digestive gland of 4 species of *Cardium*. The results did not justify the conclusion that the 4 species studied were valid species, and that no subspeciation in 2 of the taxa was evident. What could be deduced was that the 4 taxa had different isoenzyme patterns; *C. hauniense* exhibited a considerable degree of polymorphism, and that *C. edule* exhibited a very low degree of polymorphism between individuals and between the 2 populations where isoenzyme patterns were shown. These examples illustrate the tendency of many workers to over extrapolate from the basic data.

Grossu and various co-workers have used electrophoretic techniques to assist in elucidating a number of difficult taxonomic problems. In attempting to separate 2 species of *Limax*, Grossu and Tesio (1971) used general esterases and proteins from foot muscle and intestine, together with proteins from the haemolymph. They reported no intrapopulation differences and readily distinguished each species on the basis of specific patterns and this was considered adequate evidence that they were considering 2 distinct species. In a later study, Grossu and Tesio (1972) examined species of the clausiliid genus *Alopia* from the Carpathian Mountains in Rumania in terms of anatomy and esterase isoenzymes. Initially the specific status of the taxa being studied was confused, because of the occurrence of both dextral and sinistral shells. *Alopia lactea* Bielz. was originally considered amphidromic, that is, sinistral and dextral, but later the 2 morphs were separated as different species. Two morphologically similar populations corresponding to *A. livida* Mke. had the same isoenzyme pattern when extracts of the whole-snail were compared. Differences in both anatomy and electrophoretic pattern indicated that a third population belonged to a different species from *A. livida*, and this was considered to be *A. lactea*. These taxa could not be distinguished on the basis of shell. It was concluded that (i) shell sizes are unreliable characters for the separation of subspecies, as populations of *A. livida* with different shell sizes have similar anatomy and isoenzyme patterns; (ii) dextral snails referred to *A. lactea*, once considered amphidromic, had a distinctive internal anatomy and isoenzyme pattern, although shell character traits converge with some alpine forms of *A. livida*; (iii) *A. lactea* differed anatomically and electrophoretically from sinistral *A. nixa* K. and *A. canesceus* Parr.

In these papers Grossu and his co-workers state that the isozyme patterns are species-specific. This is a surprising result as esterase analyses of individuals of a population are generally undertaken to determine the degree of variation within a population; consequently, variation both within and between populations would be anticipated. This work should be tested to determine if indeed populations of *Limax* or clausiliid genera are so conservative that there is no interpopulation variation.

Medical malacologists have commenced to assess the value of isoenzyme patterns in distinguishing populations of snails with the potential to transmit the human blood fluke *Schistosoma*. There are also hopes of finding diagnostic markers that would be useful in identifying taxa in such genera as *Bulinus* and *Biomphalaria* where species concepts are badly tangled (see pp. 103–106, 133–134).

Wright *et al.* (1966) examined various enzyme systems from the maximum of 10 different organs or tissues in *B. truncatus* and *Biomphalaria sudanica* (Martens). The best results, in terms of number of closely-developed clear bands, were obtained with non-specific esterases from the digestive gland and kidney. In studying digestive gland esterases, the clearest results were obtained from starved and fully mature specimens, but even so, considerable variation occurred in *B. sudanica* when specimens of similar age and degree of starvation were compared; such variation was not found in *B. truncatus*. Wright and File (1968) used esterases from the digestive gland in a study of species representing the 4 species-groups of the genus *Bulinus*. No group-specific patterns were found and interpopulation differences were greater than those between species of the same group. In a similar study Malek and File (1971) examined individuals of 5 species of *Biomphalaria*, 2 species of *Fossaria*, *Pseudosuccinea columella* (Say) and *Stagnicola palustris*, and again there was pronounced variation between individuals and populations. The intra- and interpopulation variation for *Biomphalaria glabrata* was greater than for any of the other species. Certain patterns even considered the result of genetic homozygosity, and one pattern was considered to reflect the heterozygous state; one appeared to be correlated with poor susceptibility to *Schistosoma mansoni*. Such studies contrast markedly with that of Ukoli (1972) who found population-specific patterns for 3 species of *Biomphalaria* with only minor intrapopulation or interspecific differences.

Coles (1969a, b, 1970) studied a number of different enzyme systems from *Bulinus* and *Biomphalaria* using pooled extracts of the digestive glands to reduce variation. Populations and some species could be identified by specific patterns depending on the enzyme system used, for example, dehydrogenases indicated differences between species but not populations. It was also found that morphologically distinct species, such as *Biomphalaria sudanica* and *B. choanomphala* (Mart.), could have similar electrophoretic patterns. Burch and Lindsay (1967) found that the polyploid species of *Bulinus*, e.g. *B. truncatus*, were characterized by 1 particularly dense, fast-moving esterase band, while diploid species, such as *B. tropicus*, had 2 dense bands. See also Michelson and Dubois (1971).

4. *Molecular genetics*

Since the pioneering work of Hubby and Lewontin (1966) and Lewontin and Hubby (1966) the field of molecular genetics has developed considerably (Gottlieb, 1971). The discussion in this section is restricted primarily to enzymes, although any molecule could be used that can be shown to be coded for by 1 or more genes. Enzymes that vary because of the effect of different alleles at the same locus are called allozymes (Prakash *et al.*, 1969). Allozyme patterns produced by electrophoresis may be considered a phenotypic display, where variation in banding can be directly equated to variation in genetic coding for a variant protein; thus it is possible to recognize heterozygotes, as bands coded by each allele will be present (Fig. 3). This contrasts markedly with the situation in classical genetics, where the genetics of a phenotype can frequently only be recognized by breeding. Nevertheless, the importance of breeding data in the interpretation of electrophoretic data cannot be over-emphasized; the problems are clearly illustrated by the studies of Manwell and Baker (1968, 1970) on 2 *Cepaea* species. Although they discuss hybrid isozymes,

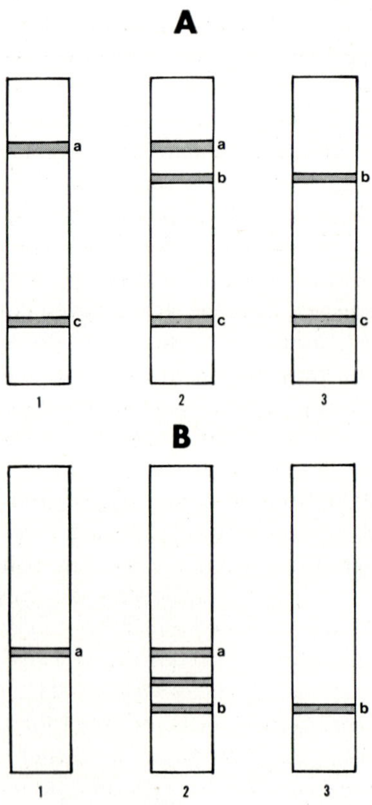

Fig. 3. Schematic representation of 6 gels using electrophoretic techniques; the bands represent the results of an analysis of esterases. Each allele controls a single enzyme of characteristic mobility. In example A one sees a simple 2 allele situation, where band c is monomorphic. Individuals 1 and 3 are homozygous for bands a and b respectively. Individual 2 is heterozygous, i.e. with both bands a and b. In B, again a 2 allele situation is shown, but the heterozygous individual, 2, has a hybrid band. The hybrid band is due to a multimeric enzyme where the intermediate band is caused by a polypeptide subunit.

gene frequencies and polymorphisms, on the basis of studies on isozymes of esterases, acid phosphatases and dehydrogenases, the work should be viewed with caution for 3 reasons Whole-snail extracts were used, including all gut contents; conclusions regarding genetics and gene frequencies were not verified by breeding experiments and there was a failure to clearly recognize a mixture of species of *C. nemoralis* (L.) and *C. hortensis* (Müll.), especially in the juvenile stage (Cain, personal communication). While the inspection of the electrophoretic patterns can give a great deal of information on the genetics of the bands, with some systems such as esterases, the heterozygotes often do not have easily recognized patterns of zones and thus grossly misleading conclusions can be made by visual inspection alone (Oxford, 1973). Manwell and Baker (1968) interpret 1 of the *Cepaea* esterase systems as controlled by "at least 7 alleles". In 1970 they describe this multiallelic set of esterases as being a problem. In fact, these esterase systems appear, in the light of Oxford's work, based on breeding experiments, to be coded by 4 linked loci; a fifth zone seems to be coded for by a locus closely linked to the others.

Haemoglobin from the haemolymph of *Anadara trapezia* was studied (Nicol and O'Gower, 1967; O'Gower and Nicol, 1968). It was shown that 2 fractions were apparently controlled by one pair of alleles; frequencies approximated the Hardy-Weinberg distribution. There was a distinct latitudinal cline in haemoglobin polymorphism correlated with the East Australian current, which governs a water temperature gradient.

Wium-Anderson (1970) demonstrated variations in haemoglobin and proteins in 3 species of *Littorina*. There was a cline in the gene frequency governing the haemoglobin expression in *L. rudis* along the west coast of Jutland, observed by applying the Hardy-Weinberg law to the latter. Also see Milkman and Beaty (1970).

Gooch *et al.* (1972) studied the isoenzymes of malic dehydrogenase, lactate dehydrogenase, tetrazolium oxidase and leucine aminopeptidase from *Nassarius* [= *Ilyanassa*] *obsoletus*. The gene frequencies governing specific protein patterns were determined for populations ranging along the eastern coast of the United States from North Carolina to Cape Cod, Massachusetts; 6 gene loci were characterized, 4 were monomorphic and 2 polymorphic. The data indicated that natural selection did not have a differential action on age classes in the populations; genotype evaluations agreed with the expected Hardy-Weinberg distribution and frequencies of alleles were extremely uniform along the coast. Thus the evidence suggested that there was very little genetic differentiation of populations along the coastline examined and this could be attributed possibly to extensive gene flow.

Gooch and Schopf (1972) tested the hypothesis that the deep sea represents the most stable and homogeneous habitat of such magnitude; on the basis of this premise it was postulated that organisms living there should be genetically less heterogeneous than organisms living in a more unstable environment. Eight species of marine invertebrates, including the marine bivalve *Nuculana pontonia* (Dall), were examined, using several enzyme systems. They found from 30–47% of the 74 loci observed were polymorphic. These data are of the same order of magnitude found for organisms in heterogeneous environments. Therefore, the hypothesis that the deep-sea fauna was genetically impoverished was not substantiated and one suspects that the deep sea is not as stable and homogeneous as previously considered. It is interesting to compare this study with that of Levinton (1973) on genetic variation in 6 marine bivalves living intertidally at various depths within the sediment and sublittoral in different depths of water. Phosphohexose isomerase and leucine aminopeptidase systems were examined. There was a decrease in the number of absolute and effective alleles, that is, a decrease in heterozygosity, with increasing depth, suggesting that environmental variability could be correlated with genetic heterogeneity.

Koehn and Mitton (1972) studied leucine aminopeptidase (LAP) and malate dehydrogenase (MDH) isoenzymes from *Mytilus edulus* and *Modiolus demissus*. LAP isoenzyme determinations were likewise made for *Mercenaria mercenaria*, *Mya arenaria* and *Pecten irradians*. Both taxa (*Mytilus* and *Modiolus*) were polymorphic as regards LAP and both had 3 common and 2 rare alleles. The alleles of the 2 species were not electrophoretically alike. Maintenance of these molecular changes due to selective pressure is suggested by the fact that the ranking order of relative abundance of each allele (same electrophoretic mobility) is almost the same for both species. With change in micro-environment the species responded with like genetic change, suggesting that natural selection is effecting a similar evolutionary result relative to the LAP locus.

Mitton and Koehn (1973) analysed the isoenzymes of aminopeptidase and LAP of *Mytilus edulis* to assess potential inter-locus interactions that might show possible inter-locus dependency between genotypes. They demonstrated such dependency, and because dependency varied with the position of the organisms in the intertidal zone, they concluded that epistasis was a function of both non-homologous alleles and environment factors.

Koehn *et al.* (1973) studied the genetic differences of isoenzyme loci of *Modiolus demissus* from different microhabitats, the high- and low-intertidal zones. The smallest animals had a heterozygote deficiency; heterozygosity increased with age. These genetic differences most likely resulted from the differential survival of heterozygotes.

F. Concluding evaluation

Electrophoresis can be used both to locate genetic markers and "fingerprint" a population or species; consequently the technique is widely used in the study of infraspecific variability, population genetics and systematics. Many natural populations exhibit considerable variation in electrophoretic profiles, indeed there can be greater variation between populations of a single species than between species. Nevertheless, success in the use of electrophoresis depends on a careful evaluation of experimental procedures, methods of sampling populations, plus a judicious choice of tissues used. Comparisons should be made between animals of comparable age and physiological condition, and when laboratory populations are used the influence of inbreeding on the taxon-specific "fingerprint" must be determined by comparisons with field-collected material. A control should always be used during electrophoresis; blood serum provides an excellent substance for this purpose and if the pattern obtained varies, experimental error should be suspected.

The difficulties in assessing the relationships between taxa increase as the genetic distance between the taxa increases. The problem is that proteins with the same R_F value derived from 2 different taxa are not necessarily homologous; indeed, the greater phylogenetic distance between taxa, the lower probability that proteins with similar R_F are homologous. A study by Jones (1972) on variation in myoglobin, lactate- and malate dehydrogenases of 12 species of *Littorina* illustrates some of the problems that can be encountered in this type of study. He attempted to define geminate (= cognate or analogous) species using polyacrylamide techniques and unpooled extracts of the snail heads. Protein concentrations were not determined and no indication was given of the enzyme patterns and the number of isoenzymes isolated. Few specimens were studied (5–9) per species. Band-to-band comparisons were made; the degree of relatedness was calculated by dividing the number of bands in common between 2 species by the sum of the total number of bands in each, less the number in common. Subsequently, cross-comparative matrices were made using the percentages of similarity; these matrices were then used to construct dendrograms from which relationships between the taxa were derived. There was little agreement in the results obtained from the 3 different classes of molecules and none agreed with the taxonomic arrangement of Rosewater (1970) at either the subgeneric or species levels. There were no consistent similarities between species pairs, where one species came from the Pacific Coast and the other from the Atlantic, indeed, the closest association appeared to exist between taxa from the same coast. Available data show that polymorphisms of allozymes involve from 15–40% of the loci examined. There may be greater variation between populations of one species than between species. It is also obvious that the number of alleles observed at a locus can be dependent on sample size; certainly larger samples than were used in Jones' study would have to be tested to have an adequate grasp of infraspecific polymorphism. The work of Koehn and Mitton (1972), Levinton (1973) and others have demonstrated significant correlations between environmental factors and geographic allozyme heterogeneity or variability; it is evident in some cases that increased polymorphic variation is a response to more variable environmental conditions. Moreover, 2 species, under the same shift in environmental stresses, may demonstrate a similar shift in alleles (Koehn and Mitton, 1972). From the above, it should be evident that just comparing bands from few individuals of species currently relegated to several subgenera will not clarify genetic relationships. It is quite possible for 2 bands of the same R_F value, but from different species, to be produced by different gene loci. Polymorphism within 1 locus can result in the appearance of a band within the pattern of a different locus of another species. Jones' work would have greater validity if numerous individuals of many populations of each species were studied and if loci were compared, not single bands. However, as shown by Oxford (1973), in the absence of breeding studies one cannot be sure that certain bands are controlled by a locus designated by inspection.

It has been shown that "fingerprints" based on isoenzymes or non-specific proteins can be used to characterize races, species and genera. By pooling extracts of tissues of several individuals of a population it is possible to decrease the effects of individual genotypes and thus reduce the importance of intrapopulation variation. This method is useful for recognizing a taxon-specific protein or establishing a specific densitometer profile for use in identifying taxa. A species-specific "fingerprint" must be tested by analysing numerous individuals from different populations, but it is always possible that 2 or more species may have the same "fingerprint", especially when utilizing non-specific proteins. In other words, if 2 populations have similar densitometer profiles one may not assume that the populations belong to the same species. Two sympatric species may be reproductively isolated for several reasons, yet have similar proteins in the extracts of the tissues chosen for study.

A number of researchers have avoided use of non-specific proteins, because there are 18–25 bands that can be resolved from the extract of almost any tissue. Nevertheless, if these substances are considered, the question arises as to how to deal with so many bands. It is possible to select only prominent fractions that are resolved in 90–100% of the tests conducted on a population (Davis, 1971, 1972); these are the more stable non-variable proteins. Or, one may record only those fractions occurring in 60% or more of the analyses; genetic variability increases considerably below this level, and it is advisable to use studies of allozymes to analyse the amount of genetic variability implied by these low-frequency protein bands.

There are acknowledged limitations to the use of electrophoretic techniques. Considerable attention has been focused on the debate whether allozyme polymorphisms are selectively neutral or not representative of traits that are sensitive to natural selection. Gottlieb (1971) has reviewed the arguments and suggests that the evidence strongly supported the case that allozyme polymorphisms are not selectively neutral.

There is a tendency to underestimate or produce a bias in the estimate of genetic variability when using electrophoretic techniques (Gottlieb, 1971), e.g. (i) amino acid changes in a protein may go undetected, because only about 30% of nucleotide substitutions dictate amino acid substitutions that effect a net change in the electrical charge; (ii) it is possible for 2 proteins to move to the same position in the gel although they have different amino acid sequences; (iii) a small proportion of the genome is analysed, as genes regulating water-insoluble proteins or membrane-bound proteins will not be detected; (iv) enzymes involved in gluclose metabolism are significantly less variable than those acting on substrates such as esters (Gillespie and Kojima, 1968; Kojima *et al.*, 1970). In spite of these limitations, electrophoresis has the potential for being a powerful tool when applied to systematics, but it is important that the deficiencies or difficulties associated with the technique are clearly understood.

V. Systematic serology

A. Introduction

Systematic serology is that discipline of biology that makes use of immunological methods to assess the degree of correspondence between antigens derived from different organisms, as an aid in determining the genetic relationships between taxa. Immunological methods rely on the nature and types of interactions of antigens and antibodies. The general subject is discussed by Boyden (1964) in his "Perspectives in Systematic Serology". The reader, in order to gain perspective, is referred to a number of papers dealing with a taxonomic serological approach with various groups of animals in "Taxonomic Biochemistry and Serology" (Leone, 1964). Other references of value for understanding principles and techniques in immunology are those of Campbell *et al.* (1963), Ackroyd (1964), Boyden (1966), Cawley (1969), and Taliaferro and Humphrey, (1961).

While immunology had its origin in the late nineteenth century, systematic serology has had a phenomenal growth since 1950. In the 1950s, Ouchterlony double-diffusion methods and immunoelectrophoretic procedures became well established. An introduction to these and other pertinent techniques can be obtained by studying the standard texts of Crowle (1961), Kabat and Meyer (1964) and Clausen (1969). The most important technique relies on the precipitin reaction in either a tube or a gel. The agglutination reaction has been widely used in the field of microbiology, but not in malacology. Agglutination will be considered only with regard to the topic of molluscan humoral antibodies.

B. Tube precipitin studies

Both Bordett (1899) and Uhlenhuth (1901) found that the precipitin reaction was not absolutely specific because an antiserum against one protein might react with another closely-related protein; it is this phenomenon that makes it possible to assess degrees of relatedness among organisms. Uhlenhuth (1901) stated that the precipitin reactions were quantitatively specific, that is, an antiserum will react more strongly with the homologous antigen than with the heterologous antigen, thus allowing for a quantitative analysis of relationship.

Erhardt (1931) presented a very comprehensive paper describing the use of the precipitin reaction to determine relationships among taxa of many invertebrate and vertebrate groups; the bibliography to this work is extensive and covers papers that date back to 1903. Cross-reactions among gastropods were described, but antigastropod serum did not react with chitons, although that for bivalves reacted with gastropods. The titre for the homologous reaction involving *Arion empiricorum Fer.* (= *A. ater* L.) was 1:3200, and 1:100 for both *Limax maximus* (L.) and *Lymnaea stagnalis*. Makino (1934) used precipitin (floculation) methods as well as complement fixation and anaphylactic reactions to differentiate genera of Mollusca and to group them.

Boyd (1937) could not demonstrate inter-phylar precipitin reactions in comparing Mollusca and Arthropoda using purified haemocyanin as an antigen, but did demonstrate intra-ordinal relationships between *Viviparus* and *Busycon*. Leone (1954) likewise could not demonstrate any relationship between Annelids, Arthropoda or Mollusca, but demonstrated inter-class precipitin reactions between Gastropoda and Cephalopoda and Gastropoda and Bivalvia with an index of 1–3% relationship (10–12% using turbidity measurements).

Wilhelmi (1944) was also interested in the phylogenetic derivation of the Mollusca and approached the subject serologically. He was correct, for his time, in stating that of several types of serological tests available, the precipitin test was the most suitable for determining the interrelationships between organisms. He found, as did Wolfe before him (1935), that to increase titres with heterologous antigens it was necessary to extend the injection series to produce antisera of less specificity. He concluded that the Mollusca were serologically more closely related to the Annelida than to any other invertebrate phylum tested. He found that there was over 27% relationship between 2 species, *Busycon carica* (Gmel.) and *B. canaliculatum* (L.), about 2% relationship between the classes Gastropoda and Bivalvia, and interphyla relationships were less than 1%.

Hiscock (1949) felt that the use of a specific tissue or secretion as a source of antigen would be preferable to mixtures of antigens extracted from whole animals, as used by Wilhelmi (1944) and Makino (1934). Hiscock used foot mucus from 5 marine prosobranchs and the marine basommatophoran *Siphonaria*. The interfacial ring test titre with the homologous reaction using *Austrocochlea torri* (Gray) was 1/1024 and with heterologous antigens, 1/8 or 1/2 indicating great specificity, at least in discerning families.

Tomita and Koizumi (1951) used the interfacial ring test of Ascoli (1902) and Uhlenhuth (1903) to study the relationship between 5 species of bivalves. They showed that the greatest affinity was between 2 species of the same genus, and absorption studies revealed taxon-specific antigens; the relationships between *Ostrea* and *Anadara* were closer than between *Pecten* and either *Anadara* or *Ostrea*. *Anodonta* showed no affinities with the other species, all belonging to a different order. Numachi (1959, 1962) used the techniques of the above-mentioned authors in attempts to discriminate between closely-related species and races of oysters. He found race-specific and species-specific antigens by absorption studies and could readily distinguish between genera as classically defined. Unfortunately, the concentration of antigens in these studies was not given.

Michelson (1966a) could not separate species of a genus of freshwater planorbid pulmonates using the interfacial ring test. Results were not changed by using a serial dilution of the antigen. The concentration of antigen was not determined.

Boyden (1943, 1964) discussed the problems of documenting relationships between taxa using only a single arbitrary dilution of antigen to react with a single volume of antiserum, as done by Erhardt and Wilhelmi. He demonstrated the value of testing the whole range of antigen–antibody reactions from antigen excess to antibody excess. With excess of antigen there was a higher estimate of similarity among related antigens. Certainly this procedure should be followed when quantifying the amounts of precipitate found during an analysis of the relative degrees of relatedness between taxa.

C. General considerations for in-gel procedures

While tube-precipitin procedures have proved quite valuable in systematics, in-gel procedures have a much wider application. Ouchterlony (1958) reviewed the development, theory and use of in-gel methods for immunological analysis. These techniques overcame the difficulties of dealing with complex mixtures of antigens, by separating the different reactions using either simple diffusion or electrophoresis in a gel. Immunoelectrophoresis (Grabar and Williams, 1955) is particularly sensitive, with frequently more than double the number of antigens detected from the same antigen mixture than are found using Ouchterlony double diffusion methods. In-gel methods, therefore, enable one to detect single precipitin systems of interest, isolate them, and later identify purified antigens. Antigens that are very nearly alike may be recognized by reactions of partial identity, that is, spur formation. Table IV lists (in

Table IV. The number of in-gel precipitin arcs obtained by various workers using different antigens from gastropods (G), pelecypods (P) and squid (S).

Family	Author	Antigen source	Number of precipitin systems: Ouchterlony	Immunoelectrophoresis
Lymnaeidae, Planorbidae (G)	Tran Van Ky *et al.* (1962)	lymphilized whole body	—	15–16
Ostreidae (P)	Numachi (1962)	extract of D-shaped larvae	3–6	—
Haliotidae (G)	Denuce and Cushing (1963)	haemolymph	—	2–3 a
Sepiidae, Ommatostrephidae (S)	Huneeus-Cox (1964)	axoplasm, haemolymph	6, 3	7 a + d
Lymnaeidae (G)	Morrill *et al.* (1964)	egg albumen	14	19a
Lymnaeidae, Planorbidae (G)	Rose *et al.* (1966)	whole body	—	18–20 ar
Planorbidae (G)	Michelson (1966a)	haemolymph	—	5 a + d
Planorbidae (G)	Michelson (1966b)	haemolymph	1–5	—
Ostreidae (P)	Li *et al.* (1967)	haemolymph	6	—
Planorbidae (G)	Wright and Klein (1967)	egg extract	c. 4–5	—
		whole snail	c. 4–6	—
		albumen gland	3	—
		foot muscle	2	—
Planorbidae (G)	Wright *et al.* (1967)	egg	—	a 7 a + c
Unionidae (P)	Gibb *et al.* (1968)	intestine	5–7	4
		foot	4–6	10
		gonad	4–7	6
Hydrobiidae (G)	Davis (1968a)	foot muscle	4–9	3–4 a + d
Pleuroceridae (G)	Davis (1968b)	foot muscle	4–5	7–8 a
Pleuroceridae (G)	Davis (1969a)	lymphilized foot muscle	5	5–10 a
Planorbidae (G)	Burch and Lindsay (1970)	foot muscle	5–6	—
Pleuroceridae (G)	David and Suzuki (1971)	lymphilized foot muscle	5–6 R	4–12 R, a
			3–5 A	4–11 A, a
Pleuroceridae (G)	Davis (1971a)	lymphilized foot muscle	4–7	8–11 R; 11–6 A, a
Pleuroceridae (G)	Davis (1972a, b)	lymphilized foot muscle	—	6– 9 A, a

R = rabbit sera; A = mouse ascites fluid; a = agar; a + c = agar and cellulose acetate; a + d = agar and disc-gel; ar = agarose.

chronological order) those workers who have used in-gel procedures and their results in terms of numbers of precipitin arcs obtained.

In assessing the results of serology for systematic purposes several important factors must be considered, which pertain to the reproducibility of the work and to the level at which conclusions may be drawn; Boyden (1964) and Crowle (1961) discuss some of the factors.

1. *Antigen concentration*

When comparing taxa, antigens should be extracted from the same tissues or fluids and whole-body extracts should not be used. Likewise, animals should be of the same age and physiological conditions. Whatever the source of the antigens, they should be characterized in terms of weight or amount per volume; this can be done as mg protein per ml relative to crystallized bovine albumen, or more precisely (and much less rapidly) in terms of mg N per ml as determined by Kjeldahl procedures. Standardization in these terms does not necessarily mean that the actual weights of antigens in the extracts from different taxa will be precisely the same. Some antigens are carbohydrates or mucoproteins and some proteins will not be effective antigens. A more precise method that has been used is to lyophilize the extract, remove the lipids, and use the residue in terms of dry weight as antigen. One problem with this method is the potential denaturization of molecules that may be good antigens in freshly-prepared extract.

Lack of standardization may lead to considerable error. Consider the possible error when one extract contains 2·0 mg protein per ml and another has 10·0, and diffusion experiments are conducted without considering such differences. One result could be the formation of concentration artifact where a precipitin pattern of non-identity occurs, even though the antigens are actually homologous.

Wright and Klein (1967) stated that the antigen concentration in their diffusion tests affected the nature and number of precipitin bands, and that optimal concentrations of test antigen resulted from using 20 mg wet weight tissue per ml of extracting saline. Unfortunately, concentrations of antigen were not given. Li *et al.* (1967) found, using haemolymph antigen concentrations of 1·35 mg ml^{-1} that concentration or dilution of the test antigen did not improve results, and Davis (1968b, 1969a) found that dilution of test antigen below 5·0–6·0 mg ml^{-1} did not improve results. The optimal concentration of antigen to be used in agar-diffusion experiments appears to range from about 5·0–10·0 mg protein ml^{-1}.

2. *Injection series*

In reporting an experiment the types of animals used, site of injection, and amounts of antigen injected must be carefully considered and factually described. The number of animals used must be stated. Control serum must be obtained, tested against antigen and extracting solutions, and the results recorded.

It is well known that the more prolonged the injection series, the greater the extent of cross-reactions with taxa further removed phylogenetically. The most specific antibody would be that obtained after 1 injection with a minimal dose of antigen; as one injects more antigen, antibodies are formed against more and different antigens. Recently Davis and Suzuki (1971) obtained excellent high titre anti-snail mouse ascitic fluid by injecting a total of 2·0 mg protein (1·0 mg in 1 injection, intraperitoneally), which was more specific than hyperimmune rabbit sera, where rabbits were injected intravenously 14 times with 2·0 mg per injection. In these experiments, it was obvious that individual mice and rabbits may behave quite differently in terms of producing usable antiserum. In a few individual cases the titer was too low and precipitin systems could not be demonstrated. In some cases there was a sufficient titre, but only one precipitin system resulted. Davis (1968a) used 5 rabbits to obtain 1 rabbit where the antiserum yielded highly-specific antibodies. Accordingly, it is necessary to use sufficient animals to be sure that the best possible antiserum is obtained for the extract used and injection methods employed.

Greater specificity theoretically results from using the least amount of antigen and fewest injections that can result in high-titer antisera. Davis (1971b) found that the optimal dosage of extract per injection to stimulate the production of hyperimmune mouse ascitic fluid was 0·50–0·25 mg protein. Two doses of antigen were given; the doses were administered 4 weeks apart. From 6–9 precipitin systems were demonstrated using immunoelectrophoretic procedures. Two strong precipitin systems resulted from using ascites stimulated by the lowest dosages (0·008 mg protein); however, these 2 systems were the least taxon-specific when comparing distantly-related genera.

3. *Titre*

It is necessary to assess the quality of an antiserum to avoid wasting valuable antigen as well as time and effort with poor antiserum. The interfacial ring test is one way to assess in a serial dilution of homologous antigen, the end point where a precipitin band is no longer observable in the glass capillary tube. Michelson (1966a) stated that a titre of 8×10^{-3} or greater was adequate for his studies. Davis (1968b) found that titers of 1:128 ($7{\cdot}8 \times 10^{-3}$) or 1:256 ($3{\cdot}9 \times 10^{-3}$) at 30 min (23 ± 2 °C) usually yielded strong precipitin systems while lesser titres indicated sera unusable for micro-Ouchterlony or immunoelectrophoretic tests. Nevertheless, Davis and Suzuki (1971) found the interfacial ring test to be unreliable for titre determination of mouse ascitic fluid. They utilized a 5-well micro-Ouchterlony diffusion test where the homologous antigen was adjusted to a concentration of 6·0 mg ml^{-1} protein. The ascitic fluid was saved for experimentation if 2 or more distinct precipitin systems developed using the fluid.

To determine the optimal concentrations for in-gel studies first the extract, then the antisera, are diluted for use in micro-Ouchterlony diffusion tests. As stated above, optimal results have been obtained using extracts with a concentration of 5·0 to 6·0 mg ml^{-1}, and hyperimmune sera with a titre of 1:256 or 1:128 (Davis, 1968b, 1969a; Davis and Suzuki, 1971). As a routine procedure, homologous and heterologous antigens are diluted in 2-fold series and tested simultaneously in a 7-hole micro-Ouchterlony system to determine if dilution causes a change in

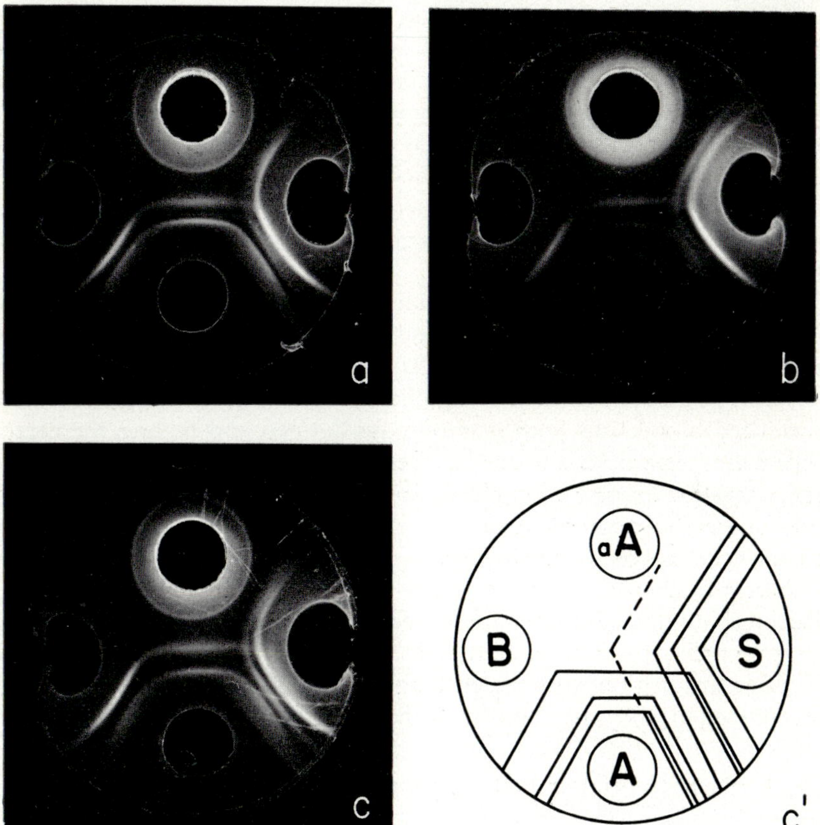

Fig. 4. Results of in-gel absorption using a micro-Ouchterlony double diffusion system. Three different hyperimmune sera were used (a–c); clarification of c is shown in c′. In these gels, the heterologous antigens (*Brotia costula episcopalis* from Malaya) were placed in a well aA and allowed to diffuse into the gel for 16 hours after which any excess was withdrawn. Subsequently heterologous antigens were placed in well B, and homologous antigens (*Semisulcospira libertina*, Japan) in well S. Unabsorbed serum was then placed in well A and partially absorbed serum into well aA. As the partially absorbed antiserum diffused from well aA towards B, A and S, it reached previously diffused heterologous antigens thus completing the absorption and leaving a ring of precipitate around well aA. Proof of complete absorption is the absence of bands between wells aA and B. Bands between wells aA and S indicate antigen–antibody systems "unique" to the homologous reaction (from Davis, 1971a).

results compared with the standard procedure (Davis, 1969a, Plate 5, Fig. 9c).

4. Absorption

One of the most important manoeuvres in systematic serology is to add heterologous antigen to an antiserum to precipitate all antibodies that can react with the heterologous antigens. Remaining antibodies are thus unique to the homologous system and the absorbed serum can be used to discriminate between taxa. Michelson (1966a) accomplished absorption by mixing predetermined volumes of antiserum and heterologous antigen (haemolymph), maintaining the mixture at 5 °C for 18–24 hours, and subsequently removing the precipitate by centrifugation. Mixing in volumes of heterologous antigen was continued until no further precipitation was detected. The problem with this technique is that one may excessively dilute the antisera when absorbing with an heterologous antigen mixture derived from a species closely allied to the homologous species.

Alternative methods are available. Davis (1968a, b) used an in-gel method shown in Fig. 4. A method yielding a relatively undiluted absorbed antiserum was used by Davis (1969a) that involved partially absorbing the serum with lyophilized heterologous antigen, coating sheep red blood cells with heterologous antigen and mixing the coated cells with the partially absorbed antiserum, and then removing the cells by centrifugation leaving a completely absorbed antiserum. Absorptions have been successfully carried out using lyophilized antigen alone or with only coated cells (Davis, unpublished).

Absorption procedures are called for when one compares 2 taxa by Ouchterlony diffusion or immunoelectrophoretic procedures and finds 1 or more precipitin systems seemingly unique to the homologous reaction. If, after absorption with heterologous antigens, these systems can be demonstrated, they may be considered taxon-specific.

D. Differentiating between infraspecific taxa

It is often impossible to distinguish between infraspecific taxa of a species using qualitative analysis. For example, Davis (1969a) could not discriminate between 2 populations of *Semisulcospira*, only 1 of which transmits the human lung fluke *Paragonimus westermani*. Burch and Lindsay (1970) could not distinguish between *B. tropicus tropicus* and *B. tropicus alluaudi* using anti-*B. tropicus tropicus* sera. Using anti-*B. tropicus* they likewise could not recognize *B. natalensis*. However, a few studies have successfully distinguished subspecies or populations.

Davis (1968a) used in-gel absorption procedures in micro-Ouchterlony diffusion systems to demonstrate that Taiwanese *Oncomelania hupensis chiui* was more closely related to *O. h. formosana* than to other subspecies of *O. hupensis*. Other workers have used these methods to distinguish between disease-resistant and susceptible races of the oyster *Crassostrea virginica* (Gmel.) from the Atlantic Coast of Canada (Li *et al.* 1967).

Usually it will not be possible to discriminate between infraspecific taxa using crude antigen mixtures and immunodiffusion methods. Nevertheless, it should be feasible to discern antigenic differences, if they occur below the species level, using purified antigens and a sensitive technique such as micro-complement fixation, where it is possible to estimate the amino acid difference between related antigenic proteins (Prager and Wilson, 1971a, b). This quantitative approach with purified antigens has never been applied to molluscs.

E. Differentiating between species and higher categories

A number of workers have investigated the relationships and systematics of the Basommatophora, particularly the families Planorbidae and Lymnaeidae, using serological techniques. An examination of these studies provides an insight into the development of this approach to more general problems of molluscan systematics.

Tran Van Ky *et al.* (1962) prepared antisera against lyophilized whole-body extracts of 4 freshwater pulmonate snails: *Biomphalaria* (= *Australorbis*) *glabrata*, *Planorbis corneus* (L.), *Lymnaea palustris* and *L. stagnalis*. Immunoelectrophoretic analysis of antigens from 5 species, the 4 above plus *Lymnaea limosa*, revealed that the 2 planorbid species were more closely related to each other than to the lymnaeid species, with the converse also true. There was, in the absence of absorption studies, 20–33% correspondence between species in the same family.

Morrill *et al.* (1964), using anti-*L. palustris* egg albumen antigens of *L. palustris*, *L. stagnalis* and *L. columella* showed that there was only a 33% correspondence of *L. palustris* with *L. stagnalis* and a 47% correspondence with *L. columella*; absorption studies were not undertaken. Burch and Lindsay (1968) presented a preliminary report stating that there were distinct immunological differences between the genera *Lymnaea*, *Bulimnea*, *Fossaria*, *Pseudosuccinea*, *Radix* and *Stagnicola*.

Rosé *et al.* (1966) could not distinguish between species and subspecies of the *Bulinus tropicus* and *B. truncatus* groups with immunoelectrophoretic procedures, using both absorbed and unabsorbed antisera. Nor could *Lymnaea stagnalis* be differentiated from *L. palustris*. However, *B. forskalii* could be distinguished from taxa of the *B. truncatus* and *B. tropicus* groups. There was a 25–28% correspondence between the families Planorbidae and Lymnaeidae, which confirmed the findings of Tran Van Ky *et al.* (1962). Burch and Lindsay (1970), however, in their immunological study of *Bulinus* s.s. could distinguish between the *tropicus* and *truncatus* groups using in-gel absorption with micro-Ouchterlony double-diffusion techniques. Antisera were produced against foot muscle antigens of laboratory-reared cultures of *B. tropicus tropicus* with a chromosome number of n = 18, *B. coulboisi* (Bgt) n = 36, and *B.* sp. n = 72. There was a distinct correlation between the immunological results and chromosome number. No species with a haploid chromosome number of 18, that is members of the *tropicus* species-group species, was antigenically related to species with a chromosome number of n = 72. Strong precipitin arcs unique to the homologous reaction were found each time a species of 1 group was tested against a species of another chromosome number group. These results have an important medical application as taxa of the *tropicus* group are not susceptible to infection with the human urinary schistosome, while those of the *truncatus* group are susceptible.

Michelson (1966a) used haemolymph antigens from 9 genera, 17 species, and 26 strains of the families Planorbidae and Lymnaeidae. Ouchterlony diffusion procedures were successful in differentiating genera. Some antisera were specific enough to enable differentiation between some congeneric species, but in no case could populations of a species be differentiated. Michelson (1966b) coupled polyacrylamide electrophoresis with agar-gel diffusion (diffusion of antigens separated from polyacrylamide into agar), where the homologous system involved the haemolymph of the planorbid *Biomphalaria glabrata* and the heterologous antigens were from *Helisoma anceps* (Mke) and *H. caribaeum* (d'Orb.). In absorption studies it was shown that there was a 1-system correspondence between heterologous and homologous systems showing a 20% relationship.

Wright and Klein (1967), in the study of planorbids, found that anti-egg sera were more specific than antisera against prostate, uterine gland, foot or blood, while results with whole-snail were too unreliable. *Bulinus globosus* and *B. tropicus* could not be distinguished by muscle antigens, but could be by use of albumen gland antigens. Unfortunately, the reactions with both sets of antigens were very poor with only 2 precipitin systems resulting. Contrast this with the results of Burch and Lindsay (1970), who demonstrated 5 to 6 precipitin systems in the homologous reaction with foot-muscle from taxa of *Bulinus*. Wright and Klein's results showed a close antigenic relationship between *B. glabrata* and *B. sudanica* (100%) and *B. pfeifferi* Krauss (75%); there was a 50% correspondence with *B. centrimetralis* Lutz and 25% with *B. sericinus*. Their results show great antigenic similarity between *B. tropicus* and *B. truncatus*, while there were pronounced differences between antigens of these 2 species-groups and those of the *B. africanus* and *B. forskalii* groups. These results confirmed the findings of Rosé *et al.* (1966), but the results of Burch and Lindsay (1970) using a more sophisticated method enabled the *tropicus* group to be differentiated from the *truncatus* group using foot-muscle antigens.

An analysis of these data of all authors studying taxa of the family Planorbidae reveals that serological procedures are readily used to separate genera. Refined methodology and several different antisera are necessary to differentiate congeneric species. Indeed, it may be difficult to differentiate between some species. Serological procedures seem insufficient for discrimination at the infraspecific level in the Planorbidae.

Davis and Suzuki (1971) compared taxa of prosobranch gastropods representing 4 superfamilies and 7 families; the homologous reaction was employed with foot-muscle antigens of *Semisulcospira libertina* belonging to the family Pleuroceridae using 8 different pools of hyperimmune mouse ascites fluid and 5 different rabbit antisera. It was found that: (i) the closest relationships were between species of the same genus and it was difficult to distinguish congeneric species belonging to the Pleuroceridae; (ii) the results substantiate the standard classification in terms of the relationships between orders and superfamilies; (iii) although Morrison (1964) considered the cerithiacean families Pleuroceridae and Thiaridae to have a polyphyletic origin from the Cerithiidae and Planaxidae respectively, serological evidence indicated that the Pleuroceridae were more closely related to the marine Potamididae, than to the Cerithiidae, and that representatives of the Thiaridae and Potamididae were more closely allied to the Pleuroceridae than were representatives of the Cerithiidae. There was no evidence to substantiate a polyphyletic origin of the Pleuroceridae and Thiaridae.

Davis and Suzuki (1971) using unabsorbed sera (immunoelectrophoresis) and Davis (1971a) using in-gel absorption of sera, showed approximately a 50% relationship between the thiarid taxon *Brotia costula episcopalis* and the pleurocerid species *Semisulcospira libertina*, and a 41–53% correspondence between 4 thiarid taxa and *S. libertina*. This level of correspondence is indicative of only subfamily or generic differences, when compared with the findings of Tran Van Ky *et al.* (1962) and Rosé (1966) who studied taxa of the pulmonate families Lymnaeidae and Planorbidae. It is possible that the distinctions between families Thiaridae and Pleuroceridae are indeed superficial and the taxa more closely allied than placement in 2 separate families would suggest. The rates of evolution and divergence of antigenic macromolecules appears to have been very conservative over millions of years (Davis and Suzuki, 1971; Davis, 1971a) in contrast with other genetically-controlled systems that are used to support separation of these "melanid" taxa into different families.

Davis (1972b) attempted to distinguish between the 2 species *Semisulcospira libertina* and *S. niponica* using immunoelectrophoresis and an anti-foot-muscle serum produced using low concentrations of antigen to increase the chances of a highly-specific reaction. There was no immunoelectrophoretic distinction between the taxa with the methods used, although on the basis of evidence from cytology (Burch, 1968a), embryo developmental patterns and adult shell (Davis, 1969d) the 2 taxa cannot be confused; they are distinct species. Rutkowski and Davis (1970) had already pointed out that the taxa could not be separated on a macromolecular basis using electrophoresis and immunology.

Also see Owen *et al.* (1971) and Pietrazak *et al.* (1973).

F. Molluscan humoral antibodies

The literature dealing with transplantation immunity and histo-incompatibility phenomena has been reviewed by Cheng and Galloway (1970) and Cheng (1970). Cheng and Sanders (1962) have given an extensive review of the internal defence mechanisms that were known amongst molluscs; these included phagocytosis, encapsulation, leucocytosis and humoral immunity. The last category is of particular interest as it evokes the concept of antibody and with it the question of humoral immune mechanisms. To date, no molecules exhibiting vertebrate immunoglobin properties have been found in the Mollusca, however, Cantacuzène (1915) reported that acquired agglutinins and precipitins were demonstrable in the pul-

Table V. Haemagglutination properties of some gastropod body substances.

	Ag non specific	Ag human group A	Ag human group B	Other
Subclass Prosobranchia				
Order Archaeogastropoda				
Family Trochidae: *Tegula lividomacula*-	—	—	—	NR
Livona pica-	—	—	—	NR
Family Neritidae: *Nerita tessalata*-	—	—	X	—
Nerita peloronata-	X	—	—	—
Order Mesogastropoda				
Family Viviparidae: *Viviparus japonica*-	—	—	—	NR
Viviparus malleatus:	—	—	—	AgRaBC
Family Ampullariidae: *Marisa cornuarietis*-	—	—	—	NR
Pomacea bridgesii diffusa-	—	X	X	—
Family Littorinidae: *Littorina littorea*-	—	—	—	NR
Littorina angulifera-	—	—	—	NR
Littorina nebulosa-	—	—	—	NR
Otder Neogastropoda				
Family Muricidae: *Thais patula*-	—	—	—	NR
Family Fasciolariidae: *Fasciolaria tulipa*-	—	—	—	NR
Subclass Pulmonata				
Order Basommatophora				
Family Planorbidae: *Planorbarius corneus*-	—	—	—	NR
Helisoma anceps-	—	—	—	haemolysis
Family Physidae: *Physa heterostropha*-	X±	—	—	—
Order Stylommatophora				
Family Pleurodontidae: *Carscolus marginella*-	—	—	—	NR
Family Helicidae: *Otala lactea*-	—	X(s)	—	—
Helix pomatia;	—	X(w)	—	—
Family Bulimulidae: *Bulimulus quadalupensis*-	X	—	—	—

- Boyd *et al.* (1966).
: Cheng and Sanders (1962)
; Prokop *et al.* (1965).

BC = blood cells; NR = non-reactive; Ag = agglutinates; Ra = rabbit; (s) = strong; (w) = weak; ± = very weak reaction; X = reaction.

monate *Helix pomatia*. Tyler (1946) produced a report of haemagglutinins in the body fluids of 6 species of non-pulmonate molluscs which agglutinated the blood cells of marine invertebrates.

There is an extensive series of studies on the presence of human and other vertebrate haemagglutinins in molluscan tissues; the results of some of these investigations using vertebrate erythrocytes are given in Table V. This includes the results of the early studies where a high-titre anti-A component was reported from the snail *Otala lactea* (Body and Brown, 1965), with a weaker anti-A component from *Cepaea hortensis* (Prokop *et al.*, 1965). Boyd *et al.* (1966) studied agglutination of human O, A and B blood cells using body components of 17 gastropods of a wide taxonomic range. Gilbertson and Etges (1967) demonstrated haemagglutinins against human A, B and O groups in the haemolymph of the planorbids *Biomphalaria glabrata* and *B. sudanica*, but not in *B. pfeifferi* or *B. stramineus*. Only *B. sudanica* had a pronounced titre (1:8). The agglutination factor was not present in all populations of *B. glabrata*, but 2 with it present were from the same general locality collected at different times and reared in different laboratories, attesting to the genetic control of this factor. Other pertinent papers are those of Prokop *et al.* (1967), Dietz (1966), Kim *et al.* (1966), Prokop and Rackwitz (1965), Uhlenbruck and Prokop (1966), and Uhlenbruck *et al.* (1966). Kothbauer (1970) investigated an agglutinin anti-AHP from the albumen gland of *Helix pomatia* which appeared to serve as protection for living eggs and a preservative for dead eggs.

Rudolph (1972) assessed the presence or absence of haemagglutinins in the haemolymph, saline extracts of whole body, and albumen gland of various species of Basommatophora and Stylommatophora, including 10 belonging to the genus *Bulinus*. Red blood cells were tested from man (type A, B, O), *Rana pipiens*, rat, gerbil, rabbit, dog, cat and Rhesus monkey. The preliminary results showed that the presence or absence of haemolymph haemagglutinins in the various species of *Bulinus* varied between populations and between individuals from the same population.

Various questions are raised regarding the presence of these agglutinins: (i) are these components that appear to be genetically controlled used for defence purposes?; (ii) are the molecules causing agglutination primarily programmed for functions not related to defence mechanisms, and thus are their haemagglutination or haemolysin properties fortuitous?; (iii) can studies of haemagglutinins be used in systematic analyses? These questions can only be answered by large-scale comparative studies, by determining the chemistry and functional significance of receptor molecules and sites, and by determining if receptor molecules correlate with basic types of defence strategies.

Cheng and Sanders (1962) showed that the agglutinating property of the serum from *Viviparus malleatus* was associated with proteins found in each of the 5 fractions they separated electrophoretically. Bhatia *et al.* (1967) purified a potent anti-A agglutinin for pig from *Otala lactea* and later (Bhatia *et al.*, 1968) employed column chromatography to establish a sediment constant of 4·2 for this component. They determined its molecular weight as 42,000.

A_1 human blood group specific agglutinins have been reported in marine and freshwater bivalves (Johnson, 1964; Tripp, 1966; McDade and Tripp, 1966; Li and Fleming, 1967; Gibb *et al.*, 1967, 1968). Acton and Evans (1968, 1969) studied the potential immune response of *Crassostrea virginica*. The oyster cleared secondary injections of T_2 bacteriophage more rapidly than primary injections. The response was determined to be cellularly mediated because long-term exposure to T_2 did not elicit a detectable antibody. Oysters injected with red blood cells likewise did not produce an immune response. An oyster haemagglutinin was isolated by density-gradient ultracentrifugation. Electrophoresis of oyster haemolymph using cellulose actate revealed only 1 band, but immunoelectrophoresis gave 3 bands. The major subunit after total reduction and alkylation had no physical-chemical relationship to any known vertebrate immunoglobin.

Khalap *et al.* (1970, 1971) investigated the haemaglutinins from sponges and various pulmonate snails; their interest was the receptor sites on red blood cells, which they analysed by haemagglutination inhibition procedures using a range of substances and in-gel precipitation and precipitation-inhibition procedures with snail and sponge extracts where the agglutinators were absorbed. The snails used were *Hedleyella falconeri* (Reeve), *Achatina granulata* (Krauss), *Achatina fulica hamillei* Petit, *Theba pisana*, *Helicella virgata* (da Costa) (banded and unbanded), *Cepaea hortensis*, *Hygromia striolata* (C. Pfr), *Helix aspersa* and *Otala lactea*. The results did not appear to reflect any phylogenetically derived pattern; different haemagglutination systems from the same biological source may yield different results with red cells of different species.

Insufficient data have been gathered to determine the role of agglutinins within the structure, physiology, and biochemical pathways of the Mollusca. The fact that erythrocyte agglutinins and lysins are rather uncommon and have no clear taxonomic distribution raises questions as to the origin of these molecules and their possible homologies. This area of study is of potential value to systematics, as the presence or absence of these molecules coupled with their selectivity and potency in agglutinating specific erythrocyte types may give insight into the relationships of various taxa. Khothbayer and Schnitzler (1971) reported on the usefulness of haemagglutinators in *Cepaea* for systematic studies. Extracts of albumen gland were used in immunoelectrophoretic analyses with anti-*H. pomatia* and anti-*C. nemoralis* sera produced in rabbits. Data from previous studies on agglutinators in *C. hortensis*, *C. nemoralis* and *C. vindobonensis* (Fér.) were also reviewed. Receptors reported were *N*-acetyl-neuramininic acid (NAMA), *N*-acetyl-D-galactosamine, and *N*-acetyl-D-glucosamine. *C. vindobonensis* had only 1 agglutinator and this perhaps correlates with lack of a NAMA. The other 2 taxa had 2 agglutinators and the NAMA receptor. The results suggested that *C. vindobonensis* had a closer relationship to *H. pomatia* than to the other species of *Cepaea*. The haemagglutination data clearly indicate that *C. vindobonensis* differs considerably from the other species of *Cepaea*. However, a quote from Tyler (1946) still pertains: "There is no reason, as yet, for considering this trait (haemagglutination reaction) to be of more general significance than any other in any application that might be made to phylogenetic problems."

G. Concluding evaluation

Serological methods are more powerful tools than those of electrophoresis for assessing relationships above the species level, as it is possible to test homologies between proteins. While the key fractions seen in electrophoretic profiles of 2 taxa are assumed to be the same when they have identical R_F values, their identity can be tested serologically. Immunodiffusion and immunoelectrophoresis employing mixtures of antigens extracted from tissues yield excellent results above the species level, but there are more sensitive quantitative methods using purified antigens that can be used to assess relationships at all levels. Sensitive techniques, such as micro-complement fixation (Wasserman and Levine, 1961), permit the calculation of immunological distances that can be correlated with the number of amino acid substitutions in the antigen (Arnheim *et al.*, 1969). However, such techniques have not yet been used in molluscan systematics and the following discussion pertains to studies using combinations of antigenic and immunodiffusion procedures.

It is impossible using immunodiffusion methods to demonstrate genetic variability at the population or interpopulations levels similar to that which is feasible with electrophoresis; for example, the amount of genetic differentiation that separates congeneric species, as seen from allozyme analyses using starch-gel electrophoresis, may not be indicated by immunodiffusion methods, as the antigens of the 2 species appear homologous. Accordingly, an intertaxon difference of even 1 recognizable antigen indicates a major difference, that is, a considerably greater degree of genetic divergence than that seen in the change in electrophoretic mobility of a protein caused by a single amino acid substitution. Therefore, as genetic differences between populations of a single species are usually not detected by immunodiffusion, any differences which are found frequently indicate that one is considering, at least, separate species.

Results from qualitative in-gel studies suggest that it is impossible to resolve any antigenic correspondence between phyla, less than 1% between classes, and less than 4% between orders. There may be from 5–20% similarity between superfamilies in the same order, 20–50% between families, 40–70% between genera, and 60–100% between species. These assumptions are based on the demonstration of 8 or more precipitin systems using specific antisera against tissue extracts or whole haemolymph.

To date, few papers have been published that explore the possibilities of serology as applied to molluscan systematics. Future refinements hold great promise for assessing rates of evolution, testing systematic relationships suggested by evidence from conchology and morphology and even identifying species. Studies of the evolution of primates are producing such results, for example Moore and Goodman (1968) and Goodman *et al.* (1969) have coupled immunodiffusion techniques with a computer programme to solve a systematic problem; using modified Ouchterlony plates, spur formation phenomena, purified antigen and set theory, they have been able to construct phylogenetic trees and suggest a timetable for the divergence of primate groups. Other investigations have produced similar results in assessing rates of evolution in the Amphibia (Salthe and Kaplan, 1966); evolution of primate albumens (Sarich and Wilson, 1966, 1967) and molecular evolution in reptiles (Gorman *et al.*, 1971). The tools and procedures are available to malacologists to assess relationships between contemporaneous taxa if they wish to use them. The technique of lyophilization, which has recently become commonplace and routine, enables easy and convenient preservation of tissue extracts, which may later be used for immunological studies. Extracts lyophilized in 1 country are easily shipped elsewhere, avoiding problems of quarantine or death during transit. Consequently, lyophilized material may very well become incorporated into museum collections, together with shells and alcohol-preserved bodies.

VI. Conclusions

It is evident that molecules that serve to characterize populations, species, or higher taxa can be separated from the tissues or fluids of individuals and demonstrated by 3 basic types of procedure: chromatography, electrophoresis, immunology. Proteins are particularly suitable for examination, as they are coded for by specific genes and the effects of mutations in these genes can frequently be detected by electrophoresis and serology. The genotype of individual animals can be deduced by analysing the phenotype; and the homology of proteins between populations can be tested by immunological methods. It is possible, therefore, to assess the genetic relationships between individuals, populations and species in a precise manner.

Such assessments are done with contemporaneous species, excepting amino acid analysis of shell proteins from fossil shells, and yet a network analysis of the genetic distance between species does not provide a definitive answer to the question—how did they evolve? There may be no simple way of stating the limits of species or genera by examining electrophoretic or immunological data. Similarly taxonomic relationships and a stable classification cannot be based only on genetic relationships derived from a study of 1 or even a series of proteins. Certainly phylogenetic pathways cannot be discussed solely on the basis of genetic relationships of contemporaneous species.

The limits of species and genera, the establishment of a classification, and the awesome task of reconstructing phylogenies depend on utilizing all available data including information on morphology, ecological requirements and biogeography. Phylogenetic speculations must be based on a good fossil record that can provide an indication of the direction of evolution and of faunal changes correlated with distinct geological events. Thus the reader is reminded of the first section of this paper where the necessity of an integrated approach to systematic analyses is pointed out.

Needless to say, pulmonate systematics and molluscan systematics as a whole could profit from a greater use of the experimental methods discussed here, especially those involving electrophoresis and immunology. The value of these experimental methods has been pointed out, yet their enormous potential to assist in unravelling problems in pulmonate systematics has yet to be exploited.

VII. References

Ackroyd, J. F. (Ed.) (1964). "Immunological Methods". F. A. Davis, Philadelphia.

Acton, R. T. and Evans, E. E. (1968). *In vitro* 3, Dept. Microbiol., Univ. Alabama Med. School, *Lerner Marine Lab. Coll. Reprints* **9,** No. 286.

Acton, R. T., Evans, E. E. and Bennett, J. C. (1969). *Comp. Biochem. Physiol.* **29,** 149–160.

Allen, K. (1961). *Am. Zool.* **1,** 253–261.

Allen, K. and Awapara, J. (1960). *Biol. Bull. mar. biol. Lab. Woods Hole* **118,** 173–182.

Arnheim, N., Prager, E. and Wilson, A. (1969). *J. Biol. Chem.* **244,** 2085–2094.

Ascoli, M. (1902). *Münch. Med. Wochsehr.* **49,** 398–401.

Bailey, T. G. (1971). *Comp. Biochem. Physiol.* **40C,** 699–707.

Baker, F. C. (1928). "The Fresh Water Mollusca of Wisconsin" Pt. I. Gastropoda. Wisconsin Academy of Science, Arts and Letters.

Bedford, J. and Reid, M. S. (1969). *Comp. Biochem. Physiol.* **29,** 659–664.

Bell, E. A. (1971). *In* "Chemotaxonomy of the Leguminosae" (J. Harbourne,

D. Boulter and B. Turner, Eds) pp. 179–206. Academic Press, London and New York.

Bhatia, H. M., Boyd, W. C. and Brown, R. (1967). *Transfusion, Philad.* **7,** 53–59.

Bhatia, H. M., Kim, Y. C. and Boyd, W. C. (1968). *Vox Sang.* **14,** 170–178.

Block, R. J., Le Strange, R. and Zweig, G. (1952). "Paper Chromatography, a Laboratory Manual". Academic Press, New York and London.

Bordett, J. (1899). *Annls Inst. Pasteur, Paris* **13,** 225–250.

Boyd, W. C. (1937). *Biol. Bull. mar. biol. Lab. Woods Hole* **73,** 181–183.

Boyd, W. C. and Brown, R. (1965). *Nature, Lond.* **208,** 593–594.

Boyd, W. C., Brown, R. and Boyd, L. G. (1966). *J. Immunol.* **96,** 301–303.

Boyden, A. (1943). *Am. Nat.* **77,** 234–255.

Boyden, A. A. (1964). *In* "Taxonomic Biochemistry and Serology" (C. A. Leone, Ed.) Vol. 10, pp. 75–99. Ronald Press, New York.

Boyden, S. U. (1966). *Adv. Immunol.* **5,** 1–28.

Brahmachary, R. and Bhattacharya, A. (1963a). *Experientia* **19,** 143–144.

Brahmachary, R. and Bhattacharya, A. (1963b). *Experientia* **19,** 225–226.

Brown, D. S. and Burch, J. B. (1967). *Malacologia* **6,** 189–198.

Brown, D. S. and Wright, C. A. (1972). *J. Zool. Lond.* **167,** 97–132.

Burch, J. B. (1960). *Z. Tropenmed. Parasit.* **11,** 449–452.

Burch, J. B. (1964). *Malacologia* **1,** 387–400.

Burch, J. B. (1965). *Proc. 1st Europ. malac. Congr.* 1962, 215–241.

Burch, J. B. (1967). *Ethiopian Med. J.* **5,** 245–257.

Burch, J. B. (1968a). *J. Conch. Lond.* **107,** 3–51.

Burch, J. B. (1968b). *Bull. Am. malac. Un.* **35,** 25–28.

Burch, J. B. and Lindsay, G. K. (1967). *Bull. Am. malac. Un.* **34,** 39–40.

Burch, J. B. and Lindsay, G. K. (1968). *Bull. Am. malac. Un.* **35,** 87–33.

Burch, J. B. and Lindsay, G. K. (1969). *Malac. Rev.* **2,** 135.

Burch, J. B. and Lindsay, G. K. (1970). *Malac. Rev.* **3,** 1–18.

Burch, J. B. and Lindsay, G. K. (1971). *Zoologica Afr.* **6,** 39–44.

Burch, J. B. and Lindsay, G. K. (1972). *Malac. Rev.* **5,** 16–17.

Burch, J. B., Lindsay, G. K. and LoVerde, P. T. (1971). *Zool. Zh.* **50,** 1158–1168.

Burns, J. M. and Johnson, F. M. (1967). *Science, N.Y.* **156,** 93.

Campbell, D. H., Garney, J. S., Cremer, N. E. and Sussdorf, D. H. (1963). "Methods in Immunology". W. A. Benjamin, New York.

Campbell, J. W. and Bishop, S. H. (1970). *In* "Comparative Biochemistry of Nitrogen Metabolism. I. The Invertebrates" (J. W. Campbell, Ed.) pp. 103–206. Academic Press, New York and London.

Campbell, J. W. and Speeg, K. V. (1968). *Comp. Biochem. Physiol.* **25,** 3–32.

Campbell, S. A. (1970). *Comp. Biochem. Physiol.* **32,** 97–115.

Cantacuzène, J. (1915). *C.r. Séanc. Soc. Biol.* **79,** 528–530.

Carefoot, T. H. (1967). *Comp. Biochem. Physiol.* **21,** 627.

Cawley, L. P. (1969). "Electrophoresis and Immunoelectrophoresis", Little, Brown and Co., Boston.

Chaetum, E. P. and Fullington, R. W. (1971). *Bull. Dallas Mus. nat. Hist.* **1,** 74.

Cheng, T. C. (1963). *Ann. N.Y. Acad. Sci.* **113,** 289–321.

Cheng, T. C. (1964a). *In* "Taxonomic Biochemistry and Serology" (C. A. Leone, Ed.) Vol. 10, pp. 659–666. Ronald Press, New York.

Cheng, T. C. (1964b). *Parasitology* **54,** 73–79.
Cheng, T. C. (1969). *J. invert. Pathol.* **14,** 60–81.
Cheng, T. C. (1970). *Transplant. Proc.* **2,** 226–230.
Cheng, T. C. and Galloway, P. C. (1970). *J. invert. Pathol.* **15,** 117–192.
Cheng, T. C. and Sanders, B. G. (1962). *Proc. Pa. Acad. Sci.* **36,** 72–83.
Chi, L. W., Wagner, E. D. and Wold, N. (1971). *Am. J. trop. Med. Hyg.* **20,** 89–94.
Chrambach, A. and Rodbard, D. (1971). *Science N.Y.* **172,** 440–451.
Clausen, J. (1969). *In* "Laboratory Techniques in Biochemistry and Molecular Biology" (T. S. Work and E. Work, Eds) Vol. 1, pp. 397–572. American Elsevier, New York.
Coles, G. C. (1969a). *Comp. Biochem. Physiol.* **29,** 403–411.
Coles, G. C. (1969b). *Comp. Biochem. Physiol.* **31,** 1–14.
Coles, G. C. (1970). *Parasitology* **61,** 19–25.
Collyer, D. M. (1961). *J. mar. biol. Ass. U.K.* **41,** 683–693.
Crowle, A. J. (1961). "Immunodiffusion". Academic Press, New York and London.
Daughaday, W. H., Lowry, O. H., Rosebrough, N. J. and Fields, W. S. (1952). *J. Lab. clin. Med.* **39,** 663–665.
Davis, B. J. (1964). *Ann. N.Y. Acad. Sci.* **121,** 404–427.
Davis, G. M. (1967). *Malacologia* **6,** 1–143.
Davis, G. M. (1968a). *Malacologia* **7,** 17–70.
Davis, G. M. (1968b). *Proc. Symp. Mollusca, Mar. biol. Assoc. India* **1,** 16–35.
Davis, G. M. (1968c). *Arch. Molluskenk.* **98,** 291–317.
Davis, G. M. (1969a). *Jap. J. Parasit.* **18,** 93–119.
Davis, G. M. (1969b). *In* "Proc. Fourth S.E. Asian Seminar on Parasitology" (C. Harinasuta, Ed.) 93–103, Bangkok.
Davis, G. M. (1969c). *Venus, Kyoto* **28,** 1–36.
Davis, G. M. (1969d). *Malacologia* **7,** 211–294.
Davis, G. M. (1971a). *Proc. Acad. nat. Sci. Philad.* **123,** 53–86.
Davis, G. M. (1971b). *In* "Culturing *Biomphalaria* and *Oncomelania* (Gastropoda) for Large-scale Studies of Schistosomiasis" (Alexander Hubert, Ed.). *Bio-med. Rep.* 406, *Med. Lab. No.* **19,** 85–161.
Davis, G. M. (1972a). *Proc. malac. Soc. Lond.* **40,** 5–32.
Davis, G. M. (1972b). *Comp. Biochem. Physiol.* **41A,** 339–347.
Davis, G. M. (1973). *Bull. Am. malac. Un.* **39,** 37.
Davis, G. M. and Carney, W. P. (1973). *Proc. Acad. nat. Sci. Philad.* **125,** 1–34.
Davis, G. M. and Lindsay, G. K. (1964). *Bull. Am. malac. Un.* **31,** 20–21.
Davis, G. M. and Lindsay, G. K. (1967). *Malacologia* **5,** 311–334.
Davis, G. M. and Ruff, M. (1973). *Malac. Rev.* **6,** 181–197.
Davis, G. M. and Suzuki, S. (1971). *Veliger* **13,** 207–225.
Davis, G. M. and Takada, T. (1969). *Exp. Parasit.* **31,** 193–201.
Decleir, W. and Richard, A. (1970). *Comp. Biochem. Physiol.* **34,** 203–211.
Degens, E. T. (1967). *Medd. dansk geol. Foren.* **17,** 112–124.
Degens, E. T., Spencer, D. W. and Parker, R. H. (1967). *Comp. Biochem. Physiol.* **20,** 111–138.
Deutsch, H. F. and McShan, W. H. (1949). *J. biol. Chem.* **180,** 219–234.

Dietz, G. (1966). *Z. ges. Hyg.* **12,** 70.
Dubuisson, M. and Roubert, L. (1947). *C.r. Séanc. Soc. Biol.* **141,** 802–805.
Duchâteau, G., Sartlet, M., Camien, M. and Florkin, M. (1952). *Arch. int. Physiol. Biochem.* **50,** 124–125.
Dupaul, W. D. and Webb, K. L. (1970). *Comp. Biochem. Physiol.* **32,** 785–801.
Dusanic, D. G. and Lewert, R. M. (1963). *J. infec. Dis.* **112,** 243–246.
Emerson, D. (1966). *Proc. S.Dak. Acad. Sci.* **45,** 196–201.
Erhardt, A. (1931). *Ergebn. Fortschr. Zool.* **7,** 279–377.
Feng, S. Y., Khairallah, E. A. and Canzonier, W. J. (1970). *Comp. Biochem. Physiol.* **34,** 547–556.
Florkin, M. (1966). *In* "Physiology of Mollusca" (K. M. Wilbur and C. M. Yonge, Eds) Vol. 2, pp. 309–351. Academic Press, New York and London.
Florkin, M. and Bricteux-Grégoire, S. (1972). *In* "Chemical Zoology", (M. Florkin and B. Scheer, Eds) Vol. 8, pp. 301–348. Academic Press, New York and London.
Freeman, S. B. (1971). *J. Embryol. exp. Morphol.* **25,** 339–349.
Friedl, F. E. (1961). *J. Parasit.* **47,** 773–776.
Gibb, G., Zahn, I. and Scheibe, E. (1967). *Z. Immun., Allerg. Klin. Immunol.* **133,** 385–393.
Gibb, B., Becker, I., Zahn, I. and Scheibe, E. (1968). *Z. Immun., Allerg. Klin. Immunol.* **134,** 417–423.
Giddings, J. C. and Keller, R. A. (Eds) (1965). "Advances in Chromatography" Vols 1–7. Marcel Dekker, New York.
Gilbertson, D. and Etges, F. J. (1967). *Ann. trop. Med. Parasit.* **61,** 144–147.
Gilbertson, D., Etges, F. J. and Ogle, J. (1967). *J. Parasit.* **53,** 565–568.
Gillespie, J. A. and Kojima, K. (1962). *Proc. natn. Acad. Sci. U.S.A.* **61,** 582–585.
Goldberg, E. and Cather, J. (1963). *J. cell. comp. Physiol.* **5,** 31–38.
Gooch, J. L. and Schopf, T. J. M. (1972). *Evolution, Lancaster, Pa.* **26,** 545–552.
Gooch, J. L., Smith, B. S. and Knupp, D. (1972). *Biol. Bull. mar. biol. Lab. Woods Hole* **142,** 36–48.
Goodman, M. (1964). *In* "Taxonomic Biochemistry and Serology" (C. A. Leone, Ed.) pp. 467–485. Ronald Press, New York.
Goodman, M. (1969). U.S. Government Clearing House Document ARL-TR-69-10; 6571st Aeromedical Research Laboratory, Holloman Air Force Base, N.M., iii, 49 p.
Gorman, G. C., Wilson, A. C. and Nakanishi, M. (1971). *Syst. Zool.* **20,** 167–185.
Gottlieb, L. D. (1971). *Bioscience* **21,** 939–944.
Grabar, P. and Williams, C. A. (1955). *Biochim. biophys. Acta* **17,** 67–74.
Grossu, A. V. and Tesio, C. (1971). *Atti Soc. ital. Sci. nat.* **112/3,** 289–300.
Grossu, A. V. and Tesio, C. (1972). *Ann. Univ. Bucuresti* (*Biol. Anim.*) **21,** 43–48.
Gysels, H. (1966). *Biol. Jahrb. Dodon.* **34,** 23–27.
Hais, I. M. and Macek, K. (Eds) (1963). "Paper Chromatography". Academic Press, New York and London.
Haites, N., Don, M. and Masters, C. J. (1972). *Comp. Biochem. Physiol.* **42C,** 303–332.

Hillman, R. E. (1964). *Syst. Zool.* **13,** 12–18.
Hillman, R. E. (1965). *Chesapeake Sci.* **6,** 115–116.
Hiscock I. D. (1949). *Aust. J. Sci.* **11,** 209.
Holmes, R. S., Masters, C. J. and Webb, E. C. (1968). *Comp. Biochem. Physiol.* **26,** 837–852.
Hubby, J. L. and Lewontin, R. C. (1966). *Genetics, Princeton* **54,** 577–594.
Hubendick, B. (1951). *K. svenska Vetensk Akad. Handl.* **3,** 1–223.
Hunt, S. (1971). *Comp. Biochem. Physiol.* **40,** 37–46.
Inaba, A. (1969). *Malacologia* **7,** 143–168.
James, B. L. and Bowers, E. A. (1967). *Parasitology* **57,** 67–77.
Jelnes, J. E., Petersen, G. H. and Russell, P. J. C. (1971). *Ophelia* **9,** 15–20.
Johnson, H. M. (1964). *Science, N.Y.* **146,** 542–549.
Jones, M. L. (1971). *Assoc. Inst. Mar. Lab. Carib.* **9,** 26.
Jones, M. L. (1972). *Int. Congr. Zool.* **17,** 1–10.
Kabat, E. A. and Mayer, M. M. (1964). "Experimental Immunochemistry", 2nd edition. C. C. Thomas, New York.
Karmen, A. (1963). *Science, N.Y.* **142,** 163–172.
Kavanau, J. L. (1958). *In* "Chemical Basis for Development" (W. McElroy and B. Alass, Eds) pp. 433–447. Johns Hopkins Press, Baltimore.
Khalap, S., Thompson, T. E. and Gold, E. R. (1970). *Vox Sang.* **18,** 501–526.
Khalap, S., Thompson, T. E. and Gold, E. R. (1971). *Vox Sang.* **20,** 150–173.
Kiecol, T. and Michejda, J. (1960). *Bull. Soc. Amis Sci. Lett. Poznan* (D) **1,** 47–50.
Kim, Z., Uhlenbruck, G., Prokop, O. and Schlesinger, D. (1966). *Z. Immun. Forsch.* **130,** 290.
Kirk, R. L., Main, A. R. and Beyer, F. G. (1954). *Biochem. J.* **57,** 440–442.
Kobayashi, S. (1964). *Biol. Bull. mar. biol. Lab. Woods Hole* **126,** 414–422.
Koehn, R. K. and Mitton, J. B. (1972). *Am. Na.* **106,** 47–56.
Koehn, R. K., Turano, F. J. and Mitton, J. B. (1973). *Evolution, Lancaster, Pa.* **27,** 100–105.
Kojima, J., Gillespie, J. and Tobari, Y. N. (1970). *Biochem. Gen.* **4,** 627–637.
Kothbauer, H. (1970). *Oecologia* **6,** 48–57.
Kothbauer, H. and Schnitzler, St. (1971). *Zeitsch. Zool. Syst. Evol. Forsch.* **10,** 133–136.
Lee, O. F. and Cheng, T. C. (1972). *Exp. Parasit.* **31,** 203–216.
Leone, C. A. (1954). *Physiol. Zoöl.* **27,** 317–325.
Leone, C. A. (Ed.) (1964). "Taxonomic Biochemistry and Serology". Ronald Press, New York.
Levinton, J. (1973). *Science, N.Y.* **180,** 75–76.
Lewontin, R. C. and Hubby, J. L. (1966). *Genetics, Princeton,* **54,** 595–609.
Li, M. F. and Flemming, C. W. (1967). *Can. J. Zool.* **45,** 1225–1234.
Li, M. F., Flemming, C. W. and Stewart, J. E. (1967). *J. Fish. Res. Bd Can.* **24,** 443–446.
Livingstone, F. B. (1967). "Abnormal Hemoglobins in Human Populations". Aldine, Chicago.
Lo Verde, P. and Burch, J. B. (1973). *Malac. Rev.* **6,** 44–45.
Lush, I. E. (1961). *Nature, Lond.* **981,** 981–984.
McDade, J. and Tripp, M. R. (1966). *J. invert. Pathol.* **9,** 523–530.

Makino, K. (1934). *Zeitschr. Immun.* **81,** 316–335.
Malek, E. A. and File, S. K. (1971). *Bull. Wld Hlth Org.* **45,** 819–825.
Mandahl-Barth, G. (1957). *Bull. Wld Hlth Org.* **17,** 1–65.
Mandahl-Barth, G. (1958a). *Bull. Wld Hlth Org.* **22,** 565–573.
Mandahl-Barth, G. (1958b). *Wld Hlth Org. Monog.* Ser. 37, 1–89.
Mandahl-Barth, G. (1965). *Bull. Wld Hlth Org.* **33,** 33–34.
Manwell, C. (1963). *Comp. Biochem. Physiol.* **8,** 209.
Manwell, C. and Baker, C. M. A. (1968). *Comp. Biochem. Physiol.* **26,** 195–209.
Manwell, C. and Baker, C. M. A. (1970). "Molecular Biology and the Origin of Species". University of Washington Press, Seattle.
Markert, C. L. and Møller, F. (1959). *Proc. natn. Acad. Sci. U.S.A.* **45,** 753–763.
Masseyeff, R., Gombert, J., Tanguy, U. and Neuzil, E. (1963). *Bull. Soc. Chim. biol.* **45,** 1133–1143.
Mattinson, G. (1965). *Can. J. Zool.* **43,** 1–12.
Meizel, S. and Markert, C. (1967). *Archs Biochem. Biophys.* **22,** 753–765.
Michejda, J. (1958). *Bull. Soc. Amis Sci. Lett. Poznan* (B) **14,** 341–344.
Michejda, J. and Turbánska, E. (1958). *Bull. Soc. Amis Sci. Lett. Poznan* (B) **14,** 359–365.
Michejda, J. and Urbanski, J. (1958). *Bull. Soc. Amis Sci. Lett. Poznan* (B) **14,** 345–358.
Michejda, J. W. and Walski, M. (1966). *Bull. Soc. Amis Sci. Lett. Poznan* (D) **7D,** 115–126.
Michelson, E. H. (1966a). *J. Parasit.* **52,** 466–472.
Michelson, E. H. (1966b). *Ann. trop. Med. Parasit.* **60,** 280–287.
Michelson, E. H. and Dubois, L. (1971). *Comp. Biochem. Physiol.* **38C,** 263–268.
Michelson, E. H. and Dubois, L. (1973). *Comp. Biochem. Physiol.* **44C,** 763–767.
Milkman, R. and Beaty, L. (1970). *Biol. Bull. mar. biol. Lab. Woods Hole* **139,** 430.
Mitton, J. B. and Koehn, R. K. (1973). *Genetics, Princeton* **73,** 427–496.
Moore, G. W. and Goodman, M. (1968). *Bull. math. Biophys.* **30,** 279–289.
More, P., More, M. T., Monnet, R. and Poisbeau, J. (1971a). *C.r. hebd. Séanc. Acad. Sci., Paris* (D) **273,** 222–225.
More, P., More, M. T., Monnet, R. and Poisbeau, J. (1971b). *C.r. hebd. Séanc. Acad. Sci., Paris* (D) **273,** 904–906.
Morrill, J. B. (1961). *Am. Zool.* **1,** 464.
Morrill, J. B. (1963). *Acta Embryol. Morphol. exp.* **6,** 339–343.
Morrill, J. B. and Dow, E. N. (1962). *Biol. Bull.* **123,** 463–464.
Morrill, J. B. and Norris, E. (1965). *Acta Embryol. Morphol. exp.* **8,** 232–238.
Morrill, J. B., Norris, E. and Smith, S. D. (1964). *Acta Embryol. Morphol. exp.* **7,** 155–166.
Morrison, J. P. E. (1954). *Proc. U.S. natn. Mus.* **103,** 357–394.
Nerenberg, S. T. (1966). "Electrophoresis—a Practical Laboratory Manual" Vol. 8, 272 p. F. A. Davis, Philadelphia.
Norris, E. and Morrill, J. B. (1964). *Acta Embryol. Morphol. exp.* **7,** 29–41.
Nicol, P. and O'Gower, A. K. (1967). *Nature, Lond.* **216,** 684.
Numachi, K. (1959). *Tohoku J. agric. Res.* **10,** 313–319.
Numachi, K. (1962). *Am. Nat.* **97,** 211–217.
O'Gower, A. K. and Nicol, P. I. (1968). *Heredity, Lond.* **23,** 485–492.

Okumura, A., Murata, K., Takagi, M. and Oishi, K. (1966). *Bull. Fac. Fish Hokkaido Univ.* **17,** 147–151.
Ornstein, L. (1962). "Theory of Disc Electrophoresis", Preprint Canalco Corp., Bethesda.
Ornstein, L. (1964). *Ann. N.Y. Acad. Sci.* **121,** 321–349.
Ouchterlony, O. (1958). *Prog. Allergy* **5,** 1–78.
Owen, B., McLean, J. H. and Meyer, R. J. (1971). *Bull. Los Ang. Cty Mus. nat. Hist.* **9,** 1–37.
Oxford, G. S. (1973). *Heredity, Lond.* **30,** 127–139.
Pascoe, D. (1970). *Acta parasit. pol.* **18,** 359–368.
Pellegrini, R. and Stagni, A. (1957). *Boll. Soc. ital. Biol. sper.* **33,** 1150–1152.
Pietrzak, J. E., Bates, J. M. and Scott, R. M. (1973). *Biol. Bull. mar. biol. Lab. Woods Hole* **144,** 391–399.
Pino, E. C., Annexy-Martínez, A. M., Pérez-Vilar, M. and Clintrón-Rivera, A. A. (1962). *Exp. Parasit.* **22,** 288–294.
Prager, E. M. and Wilson, A. C. (1971a). *J. Biol. Chem.* **246,** 5978–5989.
Prager, E. M. and Wilson, A. C. (1976). *J. Biol. Chem.* **246,** 7010–7017.
Prakash, S., Lewontin, R. C. and Hubby, J. L. (1969). *Genetics, Princeton* **61,** 241-258.
Pratt, J. J. and Auclair, J. L. (1948). *Science, N.Y.* **108,** 213–214.
Prokop, O. and Rackwitz, A. (1965). *Acta biol. med. germ.* **15,** 191.
Prokop, O., Rackwitz, A. and Schlesinger, D. (1965). *J. forensic Med.* **12,** 108.
Prokop, O., Uhlenbruck, G. and Schnitzler St. (1967). *Bibl. Haematol. Humboldt Univ., Berlin* **27,** 103–114.
Pujol, J. P., Rolland, M., Lasry, S. and Vinet, S. (1970). *Comp. Biochem. Physiol.* **34,** 193–201.
Ramamoorthi, K. (1955). *Curr. Sci.* **24,** 339–340.
Ramamoorthi, K. (1952). *J. anim. Morphol. Physiol.* **5,** 34–42.
Rao, K. P. (1963). *Proc. Ind. Acad. Sci.* (B) **57,** 290–296.
Reid, G. B. (1968). *Comp. Biochem. Physiol.* **24,** 727–744.
Reid, G. B. and Dunnill, R. M. (1969). *Comp. Biochem. Physiol.* **29,** 601–610.
Rosé, F., Osteux, F., Rosé, G. and Tran Van Ky, P. (1966). *Bull. Soc. zool. Fr.* **91,** 345–352.
Rosewater, J. (1970). *Indo-Pac. Moll.* **2,** 417–506.
Rudolph, Paul H. (1972). *Malac. Rev.* **6,** 48–49, 1 tbl.
Rutkowski, R. and Davis, G. M. (1970). *Jap. J. Parasit.* **19,** 171–181.
Saleuddin, A. S. M., Miranda, E., Losada, F. and Wilbur, K. (1970). *Can. J. Zool.* **48,** 495–499.
Salthe, S. N. and Kaplan, N. O. (1966). *Evolution, Lancaster, Pa.* **30,** 603–616.
Sarich, V. M. and Wilson, A. C. (1966). *Science N.Y.* **150,** 1563–1566.
Sarich, V. M. and Wilson, A. C. (1967). *Proc. natn. Acad. Sci. U.S.A.* **58,** 142–148.
Schafer, R. D. (1961). *Pacif. Sci.* **15,** 49–55.
Schoffeniels, E. and Gilles, R. (1972). *In* "Chemical Zoology" (M. Florkin and B. T. Scheer, Eds) Vol. 7, pp. 393–420. Academic Press, New York and London.
Schwabl, G. and Murray, J. (1970). *Evolution, Lancaster, Pa.* **24,** 424–430.

Seed, R. (1971). *Cah. Biol. mar.* **12,** 291–322.
Shick, J. M. (1972). *Mar. Biol.* **13,** 276–283.
Sibley, C. G. and Johnsgard, P. (1959). *Am. Nat.* **93,** 107–115.
Simpson, G. G. (1961). "Principles of Animal Taxonomy". Columbia University Press, New York and London.
Sodeman, W. A. and Meuwissen, H. E. T. (1966). *J. Parasit.* **52,** 23–25.
Sokol, R. R. (1964). *In* "Taxonomic Biochemistry and Serology" (C. A. Leone, Ed.) pp. 33–48. Ronald Press, New York.
Stadnichenko, A. P. (1970). *Zool. Zh.* **49,** 680–684.
Stadnichenko, A. P. (1971). *Vest. Zool.* **2,** 25–30.
Stahl, E. (Ed.), (1969). "Thin-Layer Chromatography" Vol. 24. Springer-Verlag Inc., New York.
Stephens, G. C. and Schinske, R. A. (1961). *Limnol. Oceanogr.* **6,** 175.
Stiglig, I. S. and van Eeden, J. A. (1970). *Wetensk. Bydraes, Reeks B: Natuurweten, South Africa* **13,** 1–4.
Taliaferro, W. H. and Humphrey, J. A. (Eds), (1961). "Advances in Immunology" Vols 1–3. Academic Press, New York and London.
Targett, G. A. T. (1962). *Ann. trop. Med. Parasit.* **56,** 61–66.
Targett, G. A. T. (1963). *Expl. Parasit.* **14,** 143–151.
Tomita, G. and Koizumi, S. (1951). *Sci. Rep., Tohoku Univ.* **19,** 113–117.
Tran Van Ky, P., Rosé, F. and Laude, F. (1968). *C.r. hebd. Séanc. Acad. Sci., Paris* (D) **225,** 366–370.
Tripp, M. R. (1966). *J. invert. Pathol.* **8,** 478–484.
Tyler, A. (1946). *Biol. Bull. mar. biol. Lab. Woods Hole* **90,** 213–219.
Uhlenbruck, G. and Prokop, O. (1966). *Vox Sang.* **11,** 519.
Uhlenbruck, G., Prokop, O. and Haferland, W. (1966). *Zentbl. Bakt. Parasitkde* **199,** 271.
Uhlenhuth, P. (1901). *Dt. med. Wschr.* **27,** 260–261.
Uhlenhuth, T. (1903). *Dt. med. Wschr.* **29,** 309.
Ukoli, F. M. A. (1972). *Malac. Rev.* **6,** 47–42.
Virkar, R. A. and Webb, K. L. (1970). *Comp. Biochem. Physiol.* **32,** 775–723.
Wada, K. (1966). *Bull. Jap. Soc. sci. Fish.* **32,** 253–259.
Waehneldt, T. V., Osborne, N. N. and Nuehoff, V. (1973). *Comp. Biochem. Physiol.* **44,** 1043–1050.
Walter, H. J. (1968). *Bull. Am. malac. Un.* **35,** 23–25.
Wasserman, E. and Levine, L. (1961). *J. Immunol.* **87,** 290–295.
Wilhelmi, R. W. (1944), *Biol. Bull.* **87,** 96–105.
Wilson, E. O. (1965). *Syst. Zool.* **14,** 214–220.
Wium-Andersen, G. (1970). *Ophelia* **8,** 267–273.
Wolfe, A. R. (1935). *J. Immunol.* **29,** 1–11.
Woods, K. R., Paulsen, E. C., Engle, R. L. and Pert, J. H. (1958). *Science, N.Y.* **127,** 519–520.
Work, T. S. and Work, E. (Eds) (1969). "Laboratory Techniques in Biochemistry and Molecular Biology", Vol. 1, pp. 1–972. North Holland Publishing Co., Amsterdam.
Wright, C. A. (1959). *J. Linn. Soc. (Zool.)* **44,** 222–237.
Wright, C. A. (1964). *Proc. zool. Soc. Lond.* **142,** 371–378.

Wright, C. A. (1966). *In* "General and Experimental Zoology" (W. J. L. Felts and R. J. Harrison, Eds) Vol. 2, pp. 1–42. Academic Press, New York and London.

Wright, C. A. (1971). *Phil. Trans. R. Soc.* **260B,** 299–313.

Wright, C. A. and File, S. K. (1968). *Comp. Biochem. Physiol.* **27,** 871–874.

Wright, C. A. and Klein, J. (1967). *J. Zool., Lond.* **151,** 429–495.

Wright, C. A. and Ross, G. C. (1959). *Trans. R. Soc. trop. Med. Hyg.* **53,** 308.

Wright, C. A. and Ross, G. C. (1963). *Ann. trop. Med. Parasit.* **57,** 47–51.

Wright, C. A. and Ross, G. C. (1965). *Bull. Wld Hlth Org.* **32,** 709–712.

Wright, C. A. and Ross, G. C. (1966). *Bull. Wld Hlth Org.* **35,** 727–731.

Wright, C. A., File, S. K. and Ross, G. C. (1966). *Ann. trop. Med. Parasit.* **60,** 522–525.

Wright, C. A., Klein, J. and Eccles, D. H. (1967). *J. Zool., Lond.* **151,** 199–209.

Wright, W. H. (1973). *In* "Epidemiology and Control of Schistosomiasis (Bilharziasis)" (N. Ansari, Ed.), pp. 32–249. University Park Press, Baltimore.

Wu, S. K. (1978). *Malac. Rev.* **5,** 95–164.

Yagi, Y., Mishima, T., Tsujimara, T., Sato, K. and Egami, F. (1957). *J. Biochem., Tokyo,* **44,** 1.

Zlatkis, A. (Ed.) (1970). "Advances in Chromatography", Vol. 7, pp. 1–434. Ulrichs Int. Dir. Per. List.

Zuckerkandl, E. (1965). *Sci. Am.* **212,** 110–118.

Chapter 4

Chromosomes of pulmonate molluscs

C. M. PATTERSON and J. B. BURCH

Museum of Zoology, University of Michigan, Ann Arbor, USA

I. Introduction

The first studies of chromosomes of pulmonate gastropod molluscs date from the late nineteenth and early twentieth centuries. These initial reports mostly only recorded chromosome numbers, and recent studies have shown most of them to be incorrect. Errors in the early literature were undoubtedly due to inferior optical equipment and the use of the paraffin section technique, which shrinks the cells and chromosomes substantially (Burch, 1960a). Unreliable information from the old reports is not included in the sections below. Most researchers now have adequate optical equipment and employ various modern techniques of chromosome preparation, which ensures accuracy. Accordingly, in recent years a considerable amount of information has accumulated on the chromosomes of pulmonate snails. These recent studies have been concerned mainly with cytotaxonomy, but several papers have included aspects of cytogenetics. In this chapter on chromosomes of pulmonate snails, we shall (i) describe the chromosome cycle, (ii) summarize current information on pulmonate chromosomes, (iii) discuss this information in light of present schemes of systematics and evolution in the various taxa, (iv) include a complete bibliography on chromosomes of pulmonates, and (v), as a supplement, present a short section on techniques.

II. The chromosome cycle

The chromosome cycle is basically the same in all pulmonate snails that we have examined. A detailed description of the chromosomes in the spermatogenesis of *Catinelle vermeta* (Say), a snail with few chromosomes, is described below, and several of the stages are illustrated in Figs 1 and 2. For additional details see Patterson and Burch (1966).

Spermatogonial mitosis

Spermatogonial cells undergo a number of mitotic divisions prior to the meiotic divisions. (Burch, 1960a, observed the spermatogonial divisions in North American *Stagnicola* always to be 6. Whether or not the number of divisions in other pulmonate snails is as rigidly controlled, and if so whether numbers other than 6 prevail in certain species or groups is not known.) In adequate preparations of spermatogonial divisions, the diploid number of chromosomes can be seen in late prophase, metaphase or anaphase cells and the exact number can be counted.

The chromosomes of early mitotic prophase appear as rather fuzzy, diffuse strands with extremely irregular margins. The strands are long with slightly darker staining sections along their length. The 12 mitotic chromosomes of *C. vermeta* can be seen as distinct entities and actually can be counted at this early stage in favourable preparations. The mid-prophase chromosomes are somewhat shorter, more deeply stained and have smoother margins than those of early prophase as a result of increased chromosome contraction. The coiled nature of the strands is particularly evident at this mid-prophase stage. The area of the centromere appears as a lightly stained or non-stained portion of each strand and

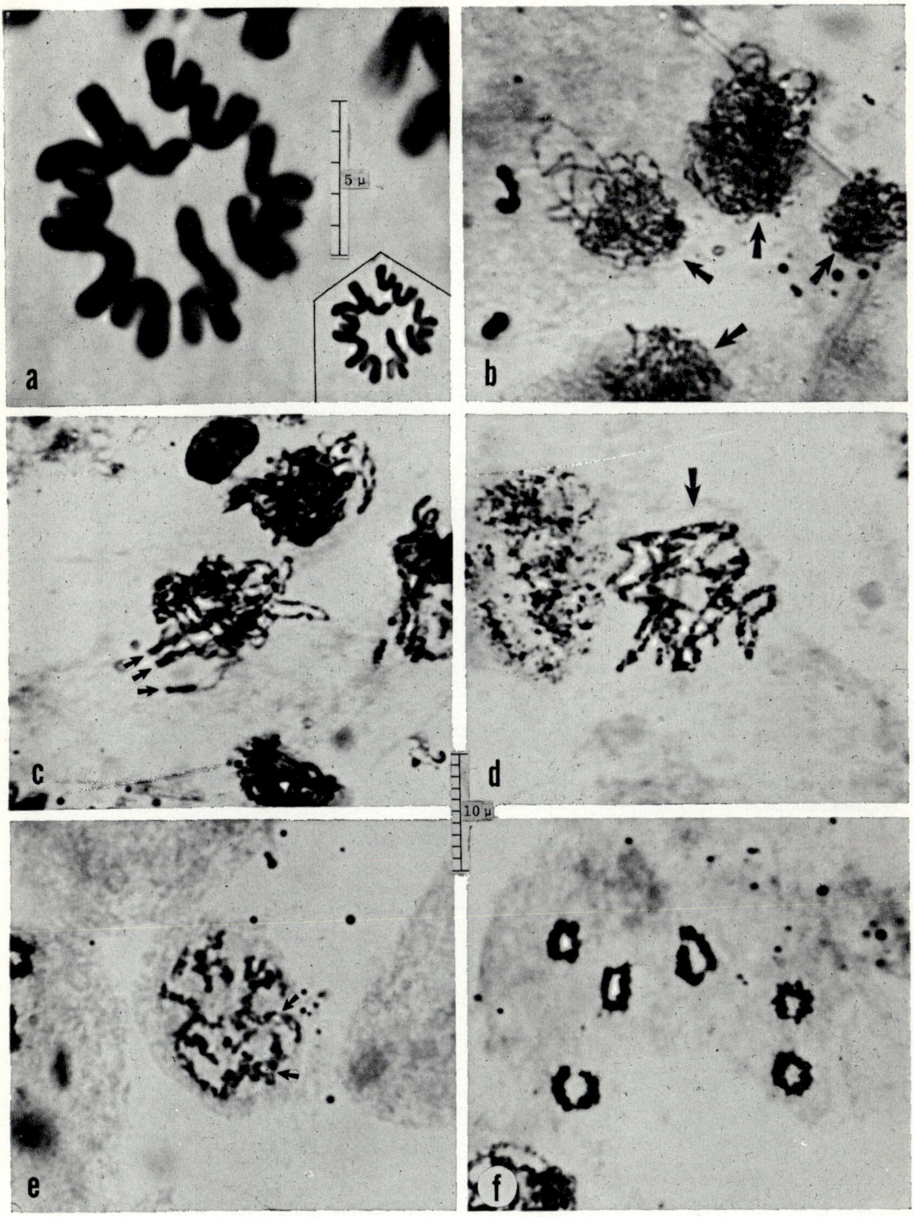

Fig. 1. (a) Spermatogonial metaphase; the inset shows the chromosomes at the same magnification as (b)–(f) and Fig. 2; (b) four leptotene nuclei; (c) zygonema: the arrows indicate synapsis at the polarized ends, while most of the rest of the strands are not yet paired; (d) pachytene bivalents; (e) early diplonema; arrows indicate places where the chromosomes are beginning to "open out"; (f) late diakinesis showing 6 ring-shaped bivalents. The measurement line in the centre is for all pictures except the enlarged chromosomes in Fig. 1(a).

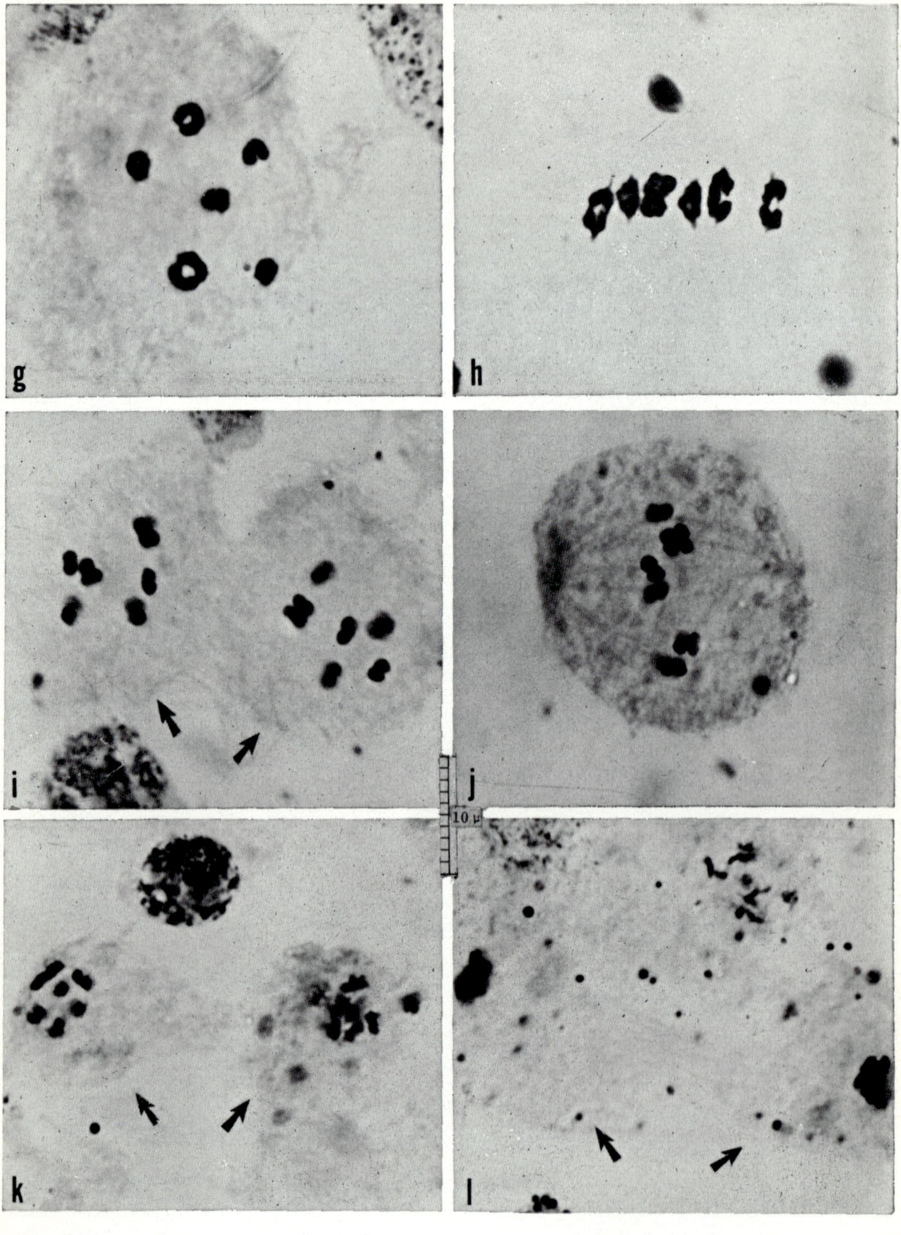

Fig. 2. (g) Metaphase I bivalents; (h) metaphase–anaphase I bivalents in side view; (i) prometaphase II dyads 2 sister cells; (j) metaphase II dyads (note spindle fibres); (k) telophase II monads in 2 sister cells; (l) late telophase II. The measurement line in the centre is for all pictures in Fig. 2.

this area produces the primary constriction of the chromosome. All 12 chromosomes can easily be counted. In late prophase, the chromosomes are even more condensed, deeply stained, shorter and have smoother margins than those of mid-prophase. The coiled nature of the strands and the differential staining of the centromeric regions are not as apparent because of the increased contraction. However, constrictions can be seen to indicate the centric area in some chromosomes. The 12 chromosomes can be distinguished with particular ease at the late prophase stage.

Metaphase chromosomes are greatly contracted, deeply stained, have smooth margins and show no evidence of coiled structure as they do in prophase. The centromeric regions are indicated only by constrictions which are located at the "bend" of the chromosomes, where the centromere is attached to the bipolar spindle apparatus. The karyotype of *C. vermeta* includes 1 pair of conspicuously large chromosomes, 4 pairs which are intermediate in size and 1 smaller pair. All of the chromosomes appear to be almost medianly constricted (metacentric) but several may be submedianly constricted to a slight degree. A secondary constriction is clearly evident near the end of an arm of one of the large chromosomes; it is probably present in its homologue, but is not always as clear. Spermatogonial metaphase chromosomes range in length from 5·6 μ for the largest pair to 3 μ for the smallest pair.

Anaphase chromosomes have much the same shape and density of stain as those of the preceding metaphase, but are usually somewhat smaller. The chromosomes of the 2 anaphasic groups are pulled from their centromeres towards opposite poles and sometimes sister chromosome arms may encounter difficulty in separating and lag behind the other chromosomes in the complement. The secondary constriction on one of the largest chromosomes is usually again evident in sister chromosome arms that lag in anaphase.

During telophase, the chromosomes become more tightly grouped within the nucleus and resemble a rather solid ball of chromatin. Cytokinesis follows and results in the formation of 2 equivalent daughter nuclei each with the diploid complement of chromosomes.

Leptonema

After the last pre-meiotic division, the nucleus enlarges to form the first meiotic prophase nucleus. The pre-leptotene nucleus contains localized accumulations of chromatin that are rather "granular" in appearance and only partially formed into chromatin strands. In leptonema the chromosomes appear as maximally extended single strands which are lightly stained, with more darkly stained bead-like chromomeres along their length. In late leptonema, the free ends of the chromosomes are polarized, i.e. associated at one side of the nucleus, forming the "bouquet stage".

Zygonema

The pairing (synapsis) of homologous chromosomes begins in zygonema. The chromosomes begin to synapse at the polarized ends, while the remainder of the strands are yet unpaired in the "bouquet" formation. The zygotene chromosomes appear to be somewhat more contracted and darkly stained than leptotene chromosomes and the chromomeres are still clearly evident.

Pachynema

Pairing of homologous strands is completed in zygonema forming pachytene chromosomes of double thickness and haploid in number. The "bouquet" persists into early pachynema and the separate homologues can still be discerned. Pairing seems to be chromomere-by-chromomere along the length of the homologous strands. Late pachytene chromosomes (bivalents) are much shorter than leptotene or zygotene chromosomes, due to the increased contraction of the strands, which begins in leptonema and continues throughout prophase I. In favourable preparations, the 6 bivalents can be counted in late pachynema.

Diplonema

In early diplonema, the homologues begin to "repel" one another, causing the chromosomes to separate slightly; but since they remain connected at the points of crossing-over (chiasmata), open areas or loops are formed along their length. Early diplotene chromosomes still stain rather deeply; however, the chromosomes of mid to late diplonema appear more diffuse, poorly stained and with irregularly characterized margins.

Diakinesis

As terminalization of chiasmata begins and contraction continues, the chromosomes soon form the ring, rod, cross or multiple loop-shaped figures characteristic of diakinesis. In early diakinesis, the chromosomes are still quite diffuse and have irregular margins. Due to the increased contraction, mid to late diakinetic chromosomes are smaller, more deeply stained and have somewhat smoother margins. Cells in which the 6 bivalents form 6 ring-shaped figures or 5 ring- and 1 rod-shaped figure occur in about equal frequency and are most common. Less frequently, 4 ring- and 2 rod-shaped figures are observed and occasionally cross and multiple loop-shaped bivalents are present. Ring-shaped bivalents have 2 remaining chiasmata and rod-shaped figures presumably have 1 terminal chiasma remaining. Of the other figures observed, the cross has 1 non-terminal chiasma and the multiple loop has 3 remaining chiasmata, producing 2 loops.

Metaphase I

The continued contraction throughout prophase I forms metaphase I bivalents that are very condensed with deep stain and smooth marginal outlines. The bivalents form either ring, half-ring or rod-shaped figures in polar view, depending upon the number and location of remaining chiasmata. Cells with 6 rings or those with 5 rings and 1 half-ring are most frequently observed.

Anaphase–telophase I

Homologous centromeres, with their chromatids of varying genic constitution, separate at anaphase I and the dyads move towards opposite poles. The chromosomes in each anaphasic group are about half the size of metaphase I bivalents, but density of stain and smoothness in outline are much the same. The centromeric regions of the dyads can be seen as constrictions in some chromosomes and may appear more lightly stained. Two dyad members of a bivalent are sometimes late in separating and they therefore lag in the centre of the cell after the other dyads have passed to their respective poles. Cytokinesis follows and spindle formation takes place as the chromosomes seem to immediately enter the second meiotic division without undergoing an observable period of interkinesis.

Metaphase II

During pro-metaphase II spindle formation is completed and the dyads begin to move towards the equator of the cell. The dyads appear to be just as highly contracted, or even more so, as in the previous anaphase. Centromeric regions are indicated only by constrictions, with no evidence of differential staining. In metaphase II, dyads appear as single or double "dumb-bell"-shaped figures with the centromeric constrictions clearly evident. The chromosomes of metaphase II are highly contracted, deeply stained and have smooth marginal outlines. Six dyads are easily counted, with the largest and smallest of the complement measuring $2{\cdot}6\,\mu$ and $2{\cdot}1\,\mu$ respectively in length, thus being smaller than the metaphase I bivalents.

Anaphase II

In anaphase II, each dyad separates to form 2 monads or daughter chromosomes, which pass to opposite poles. The anaphase II chromosomes often appear to have slightly more irregular margins than metaphase II dyads.

Telophase II

When the chromosomes reach their respective poles, telophase II is initiated and cytokinesis begins to take place. The haploid complement now consists of 1 large chromosome and 5 smaller chromosomes randomly oriented within the nucleus. In late telophase II, the chromosomes are often grouped together near the pole, appearing as a solid mass of chromatin.

Spermiogenesis

Upon completion of telophase II and cytokinesis, the chromosomes pass into an interphasic-like state, the chromatin characteristically becoming somewhat diffuse and irregular in shape. Thus begins the formation of young spermatids, each with its conspicuous nuclear membrane. All 6 irregularly shaped chromatin masses can still be clearly discerned in some of the young spermatids. During spermiogenesis, the chromatin becomes progressively more disorganized and nuclear condensation occurs, until very small, deeply stained, nearly mature sperm cells are formed. The mature sperm head measures 2·5 μ in diameter and is about one-third the size of the young spermatid.

III. Chromosome numbers

A. Review of chromosome numbers

Most reports on chromosomes of molluscs have dealt chiefly or only with chromosome numbers, but in spite of the increasing number of studies on molluscan chromosomes in recent years, chromosome numbers are known for less than 0·5% of the species of Recent molluscs. The class Gastropoda has received the most attention, although the great majority of species have still not yet been studied. Nevertheless, enough representative species have now been investigated to present some broad evolutionary concepts concerning their chromosomes. Among the gastropods, chromosome numbers have been determined for species of 23 streptoneuran families (c. 19%) and 73 euthyneuran (Opisthobranchia and Pulmonata) families (c. 42%).* In the euthyneuran Pulmonata (Basommatophora + Stylommatophora), there are reliable reports for at least 1 species in 46 (c. 61%) of the families.

In non-pulmonate streptoneuran gastropods, chromosome numbers range from $n = 7$ to $n = 36$. In the Archaeogastropoda, the sole chromosome number of the Patellacea is $n = 9$, occurring in 9 species belonging to

* The higher classification used throughout this article follows Zilch (1959–1960) as modified by Taylor and Sohl (1962), except for the inclusion of the names Archaeopulmonata and Branchiopulmonata for the lower Basommatophora and the higher limnic Basommatophora respectively, and including the Lancidae within the Lymnaeidae.

the families Acmaeidae and Patellidae. In the other superfamilies (Pleurotomariacea, Fissurellacea and Neritacea), the numbers range from n = 11–18, except for the Trochacea, which has n = 18 in all species but *Stomatella lyrata*, which has n = 21. In the Mesogastropoda, the chromosome numbers range from n = 7 in *Viviparus contectus* Millett to n = 18 in one member of the Littorinidae and to n = 20 in several of the Pleuroceridae. Chromosome numbers are higher in some of the Thiaridae (n = 45–60), but these numbers are considered to be polyploid. In the more advanced Neogastropoda, the numbers range from n = 28 to n = 36, except for some populations of *Nucella lapillus* (L.), which have been reported to have 13 pairs of chromosomes.

In the euthyneuran opisthobranchs, the numbers range from n = 12 in the Pleurobranchidae and n = 13 in all of the nudibranchs (44 species examined) to n = 17–18 in the rest, except for the aberrant *Bosellia* with n = 7.

The range of chromosome numbers in the pulmonates, i.e. in the 2 major groups of basommatophorans and in the stylommatophoran superfamilies, are shown in Fig. 3, and are discussed in the following in greater detail.

The order Basommatophora comprises 12 families arranged into 2 groups, the primitive "lower" largely marine Archaeopulmonata and the so-called "higher limnic Basommatophora", sometimes placed in the ill-named suborder Branchiopulmonata. Chromosome numbers are known for 125 species in 10 of the families. In the Archaeopulmonata the chromosome numbers of 21 species from 6 families have been determined. In the Siphonariidae all 7 species studied have 16 pairs of chromosomes. Haploid chromosome numbers range from n = 17 to n = 19 in the 9 species of the Ellobiidae studied. Different chromosome numbers may occur in related species of the same genus, as here exemplified by the genera *Pythia* (n = 17, 18) and *Melampus* (n = 18, 19). Interestingly, one individual of *Melampus coffeus* (L.) in an otherwise diploid population was found by Natarajan and Burch (1966) to be tetraploid (n = 38). The single representative from each of the remaining 4 archaeopulmonate families studied (Amphibolidae, Chilinidae, Latiidae and Acroloxidae) has the chromosome number n = 18, the number most frequently found in the "higher limnic Basommatophora".

In the higher Basommatophora, several species of each of the 4 families (Lymnaeidae, Physidae, Planorbidae and Ancylidae) have been investigated. All 7 species of the family Physidae studied have 18 pairs of chromosomes. Chromosome numbers in the world-wide family Lymnaeidae range from 16–19 haploid among the 43 species studied. The 3 species of *Austropeplea* have the lowest chromosome number, n = 16. *Radix* is characterized by 17 pairs of chromosomes and all of the species

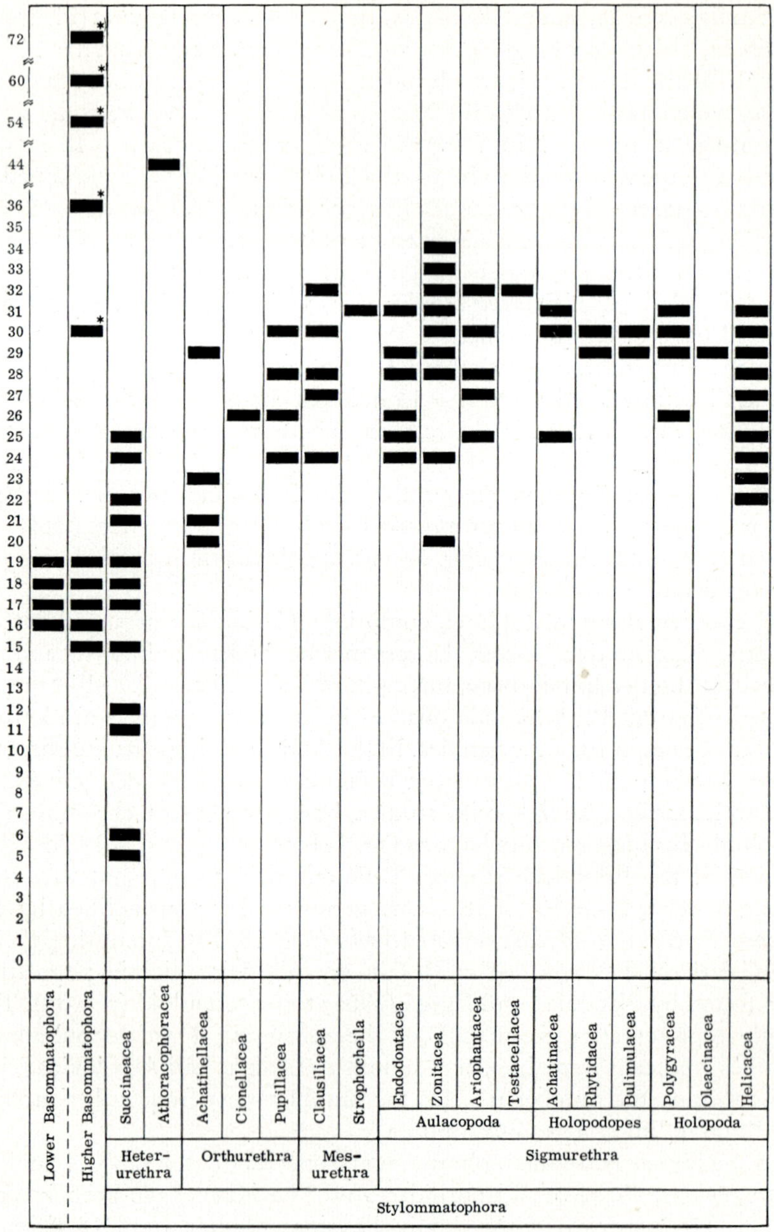

Fig. 3. Haploid chromosome numbers in the Pulmonata. The bars with an asterisk represent polyploid numbers.

in the genera *Lymnaea*, *Stagnicola*, *Acella*, *Pseudosuccinea*, *Bulimnea*, *Erinna*, *Pseudisidora* and *Lanx* have 18 pairs; *Fossaria modicella* (Say), *F. parva* (Lea) and *F. truncatula* (Müll.) also have 18 pairs of chromosomes, but *F. rustica* (Lea) has 19 pairs.

Haploid chromosome numbers range from n = 18 to n = 72 in the 46 species studied in the family Planorbidae. Their base chromosome number is x = 18. In the subfamily Bulininae (comprising the genera *Bulinus* and *Indoplanorbis*), chromosome numbers range from n = 18 to n = 72. Polyploid species have n = 36, n = 54, or n = 72. Supernumerary chromosomes (see below under Cytogenetics) have been reported in 2 species, *Bulinus forskalii* (Ehrenb.) and *B. natalensis* (Küst.). In the subfamily Planorbinae, polyploidy has also been found. *Gyraulus parvus* (Say) has the haploid chromosome number n = 36 while the other 6 *Gyraulus* species studied and *Anisus crassilabrum* (Morel.), *A. vortex* (L.) and *Armiger crista* (L.) have the chromosome number n = 18. Additional chromosomes have also been reported for individuals of a population of *Gyraulus deflectus* (Say). The haploid chromosome number 18 occurs in 5 species (*Helisoma anceps* (Menke), *H. subcrenatum* (Carp.), *H. trivolvis* (Say), *Planorbarius corneus* (L.) and *Promenetus exacuous* (Say)) of the subfamily Helisomatinae while 2 species of *Planorbula* have 19 pairs of chromosomes. All 5 species of *Biomphalaria* (subfamily Biomphalariinae) investigated have the number n = 18. Three members of the subfamily Segmentininae, species of the genera *Segmentina*, *Helicorbis* and *Hippeutis*, and one species of the Camptocerinae (*Camptoceras hirasei* Walker) also have the chromosome number n = 18.

Ten species of the Ancylidae have chromosome numbers ranging from n = 15 in the North American *Rhodacmea* to n = 60 in the related Eurasian genus *Ancylus*, with polyploidy undoubtedly involved. Numbers between these 2 extremes are n = 17, 18 and 30. The n = 30 for the Ethiopian *Ancylus* species is obviously related to the higher polyploid number n = 60, but the relationship of the two n = 30 *Ferrissia* is not clear.

Most land snails belong to the large order Stylommatophora, which contains 59 recent families. Chromosome numbers have been reported for species in 35 of these families. The greatest range in haploid chromosome numbers (n = 5–44) of any stylommatophoran group occurs among members of the Heterurethra, an aberrant suborder containing 3 families, the Succineidae, Aillyidae and Athoracophoridae. In 48 species of the family Succineidae, the haploid chromosome numbers range from n = 5 to n = 25. Species of a primitive succineid genus, *Catinella*, have the lowest number of chromosomes yet found in the Mollusca, n = 5 and n = 6 in the 9 species studied. *Succinella oblonga* (Drap.) has been reported by Butot and Kiauta (1964) to have 12 pairs of chromosomes, but the 2 specimens examined in another population by Rainer (1967) had 10, 11 and 12 bivalents (?) in meiosis. Haploid numbers range from n = 15 to n = 19 among the 10 *Oxyloma* species studied, and in *Succinea* the range is n = 17–25, except for 1 Tanzanian species which has only 11 pairs of chromosomes. The aberrant slug-like *Omalonyx felina* Guppy has 21 pairs of chromosomes. *Aneitea* and *Triboniophorus* of the Anthoracophoridae have the surprisingly high chromosome number of n = 44, the highest yet found in the Stylommatophora.

Haploid chromosome numbers range from n = 20 to n = 30 among the 24 species studied in 7 families of the suborder Orthurethra. The lowest numbers (n = 20–23) occur in 4 species of the Achatinellidae, while the highest haploid chromosome numbers (n = 30) are found among members of the family Chondrinidae. The intermediate numbers are n = 24 in the Enidae, n = 26 in the Cionellidae and Pyramidulidae, n = 28 in the Valloniidae and n = 29 in the Partulidae.

There is cytological information for 3 of the 6 families in the suborder Mesurethra, with chromosome numbers ranging from 24 to 32 haploid. *Cerion incanum* (Binney) of the Ceriidae has 27 pairs of chromosomes. *Strophocheilus oblongus* (Müll.) of the Strophocheilidae has 31 pairs, and the haploid numbers 24, 28, 30 and 32 occur in the Clausillidae. In the latter family, 12 species have the chromosome number of n = 24, 3 species have n = 28, 3 species have a haploid number of 30 or "about" 30, and 1 species has 32.

The largest suborder of the Stylommatophora, the Sigmurethra, contains 3 infraorders (Holopodopes, Aulacopoda and Holopoda), with cytological information available for 28 families. The chromosome numbers of 15 species from 6 of the 14 Recent families of the Holopodopes range from n = 25 to n = 32. All species (except *Lamellaxis mauritianus* (Pfeif.) with n = 25) fall into the chromosome number range n = 29–32.

The infraorder Aulacopoda contains 11 families with chromosome numbers of n = 20 to n = 34 reported for 38 species in 11 families. In the superfamily Endodontacea, data are available for 3 families, the Philomycidae, Endodontidae and Arionidae. The haploid number is 24 for 2 Japanese philomycids, 29 for the European *Discus rotundatus* (Müll.), and 31 for both *D. patulus* (Deshayes) and *Anguispira alternata* (Say) of North America. The haploid numbers of the Arionidae range from 25–29. Slugs of the aulacopod superfamily Zonitacea have haploid chromosome numbers which range from 24–31 in the Limacidae and 33–34 in the Milacidae. In the limacid slugs each genus has a different but constant chromosome number (*Lehmannia*, n = 24; *Deroceras*, n = 30; *Limax*, n = 31). In the milacid slugs the haploid number of *Milax gracilis* (Leydig) is 33 and that of *M. sowerbyi* (Fér.) is 34. *Milax gagates* (Drap.), on the other hand, is reported to have both 33 and 34 pairs of chromosomes. Shelled zonitaceids of the family Zonitidae have haploid numbers which range from

28–30, except for the European *Vitrea diaphana* (Stud.), which has only 20 pairs of chromosomes. The 1 member of the family Vitrinidae studied, *Cryptaegis pilsbryi* Clapp (called "*Vitrinopsis* sp" in Burch, 1967a and Patterson, 1969), has 28 pairs of chromosomes. In the superfamily Ariophantacea, the greatest number of species with known chromosome numbers are in the Ariophantidae, where the numbers range from 25 pairs in *Mariaella dussumieri* Gray to 32 in *Cryptozona bistrialis* Pfeif. The 2 trochomorphids studied have 28 and 30 pairs of chromosomes respectively, and the single helicarionid, *Helixarion solomonensis* (called "*Helicarion* sp." in Burch, 1967a and Patterson, 1969), has 28 pairs. A chromosome number is available for only 1 species of the Testacellacea, n = 32 for *Testacella haliotidea* Drap.

Information on chromosome numbers is available for 7 of the 10 families in the Holopoda. Haploid chromosome numbers range from n = 26 to n = 31 in 20 species of the family Polygyridae; 16 of them have the chromosome number n = 29. The family Bradybaenidae is also conservative in regard to chromosome numbers; all 18 species studied have either 28 or 29 pairs of chromosomes. The 30 species studied in the family Camaenidae have haploid numbers ranging from n = 27 to n = 29. All but 2 species have the latter number. The 1 species of the Helminthoglyptidae studied (*Monadenia fidelis* (Gray)) has the haploid number n = 29, the same number found in the single oleacinid studied, *Euglandina rosea* (Férussac). Five of 6 species of the Oreohelicidae investigated have 32 pairs of chromosomes. The sixth species (*Radiocentrum clappi* (Ferriss)) has 31 pairs.

The greatest number (64) of stylommatophoran species and subspecies studied cytologically belong to the holopod family Helicidae. Haploid chromosome numbers range from a low of 21 to a high of 31. Any one particular haploid chromosome number does not seem to consistently characterize any of the subfamilies. For example, the haploid numbers 27 and 30 occur in 3 different subfamilies and n = 26 and n = 23 each occur in 2 different subfamilies.

Whether or not the Soleolifera should be included with the Pulmonata or left with the Opisthobranchia is not yet settled, but in the eventuality that they should remain with the pulmonates, we are including what little is known about their chromosome numbers. Five species from 2 families have been studied. In the Veronicellidae, *Veronicella floridana* (Leidy) has 16 pairs of chromosomes and *Laevicaulis alte* (Fér.) has 17 pairs. In the Onchidiidae, *Onchidella kurodai* has 17 pairs and *O. evelinae* (Marcus and Burch) and *Onchidium verraculatum* have 18 pairs of chromosomes.

B. Chromosome numbers and systematics

The chromosome numbers of animal species are, in general, uniform and constant, each species having a characteristic chromosome number. This number for any species may be the same as that found in related species, or it may differ. In pulmonate molluscs, the chromosome numbers vary from the 5 pairs found in the land snail *Catinella rotundata* (Gould) of Hawaii to the 60 pairs found in the European *Ancylus fluviatilis* Müll. and the 72 pairs in the Ethiopian freshwater planorbid *Bulinus octoploidus* Burch. This is a very wide range of chromosome numbers, but the span is reduced (n = 5–44) when the polyploid numbers found in *Ancylus* and *Bulinus* are not considered. *Ancylus* is a polyploid (octoploid), its high chromosome number being derived from the base number x = 15. *B. octoploidus* is also an octoploid, its chromosome number being derived from the base number x = 18. Its sibling species, *B. hexaploidus* Burch, with 54 pairs of chromosomes and *B. truncatus* (Audouin) with 36 pairs, are also polyploids, i.e. hexaploids and tetraploids respectively. Discounting these cases of polyploidy in *Ancylus* and *Bulinus*, the range of chromosome numbers in pulmonates is still rather great, from n = 5 to n = 44 pairs in the land slug family Athoracophoridae. But, in spite of the wide divergence in chromosome numbers in pulmonate snails as a whole, their chromosome numbers in general are quite conservative within the various pulmonate taxa, seldom varying more than by 1 or 2 bivalents.

Changes in chromosome numbers in animals may occur by (i) chromosomal fusion; (ii) fragmentation; (iii) mitotic or meiotic nondisjunction; (iv) translocations involving a supernumerary fragment serving as a donor of a centromere and 2 telomeres and (v) duplication of entire chromosomal complements through polyploidy. Chromosomal fusion produces a decrease in chromosome number while the other conditions result in an increase in chromosome number. Changes in the chromosomal complement that involve the loss or gain of incomplete chromosomal sets occur by aneuploidy in contrast to polyploidy. Because most variation in molluscan chromosome numbers within a taxon is seldom greater than ± 1 or 2 bivalents, mechanisms for addition or deletion of chromosomes must operate at a low or inefficient level. Nevertheless, nearly all changes in chromosome numbers that did occur and are now evident undoubtedly represent aneuploid conditions which became stabilized. Furthermore, basic diploid species can tolerate duplications or additions of chromosomes more easily than loss of chromosomes (see Husted and Burch, 1946); therefore, variation in chromosome number within a taxon probably more usually reflects addition rather than loss of chromosomes. When the variation in chromosome number between lower taxonomic categories exceeds 1 or 2 bivalents, i.e. 3 or 4 or more (but less than an additional full haploid complement of chromosomes), this probably represents successive aneuploid conditions, each change followed by a period of stability in the new chromosome number. Accordingly, we may expect to find more primitive molluscs to have lower chromosome numbers and species derived from them to have a greater number of chromosomes. When an unexpectedly low chromosome number is found in an advanced group, this number either represents an unchanged ancestral condition or it has resulted from successive reductions in the chromosome number. Subsequent chromosome number increases may be expected in either case, which would be primary increases in the former case and secondary increases in the latter. However, in referring to chromosome number change as an increase or decrease, care must be taken in assessing which groups are morphologically primitive or advanced. It often appears that the determination of whether or not a taxon is "primitive" or "specialized" ("advanced") is quite subjective and has not always been determined by careful and detailed analysis of many characters. The lack of sufficiently critical morphological studies and careful analyses relating to phylogeny pose serious problems for the molluscan cytotaxonomist. Until the chromosome numbers of more species are determined and related to reliable information on molluscan comparative morphology, systematics and phylogeny, we can only present suggestions concerning the cytotaxonomic relationships in the Mollusca.

A general conservativeness with regard to chromosomal change is evident in most mollusc groups. Many higher taxa have characteristic chromosome numbers, with only a few species having chromosome numbers that deviate widely from the "basic" haploid number for their particular groups. In the prosobranch Archaeogastropoda, for example, all members of the Patellacea for which there is information have the haploid chromosome number n = 9; in the Mesogastropoda all members of the rissoacean families Hydrobiidae and Bithyniidae studied have 16 to 18 pairs of chromosomes; the opisthobranch order Nudibranchia is characterized by the single haploid number n = 13 (44 species), and the other opisthobranch orders are also conservative, with all species studied (except 1) having n = 16–18; the basommatophoran Lymnaeidae all have n = 16–19 (n = 18 is the most prevalent, with n = 16 for *Austropeplea* and n = 17 for *Radix* being characteristic of and restricted to their respective genera), all Physidae have n = 18, nearly all planorbids have n = 18 or a polyploid multiple of n = 18; of the Stylommatophora studied, all Partulidae have n = 29, Chondrinidae n = 30, Enidae n = 24, Bradybaenidae either n = 28 or n = 29, all but 2 of the 30 species of Camaenidae studied have n = 29 and in the Polygyridae 17 (out of 21) polygyrid species have n = 29. There are other examples (e.g. see Patterson, 1969), but these will suffice.

It has not been possible in many cases to correlate low chromosome numbers with species groups considered to be "primitive" by taxonomists and morphologists, but in other cases such a correlation can be made and deserves mention. In the cytologically most conservative groups, chromosome numbers *per se* may have no bearing on phylogenetic schemes because chromosomal changes may not have occurred.

The range in haploid chromosome numbers among non-polyploid members of the order Basommatophora is n = 15–19; polyploid species extend this range to n = 72. Classification of the "lower" Basommatophora or Archaeopulmonata is, at this time, still uncertain. The Archaeopulmonata contain 8 families of rather diverse gastropods grouped together mainly because they have characteristics which include them in the pulmonates, yet they do not appear to be very closely related to the 4 families of "higher limnic Basommatophora", the Lymnaeidae, Physidae, Planorbidae and Ancylidae (= the Branchiopulmonata). The diversity of the archaeopulmonates can be exemplified by comparing the limpet-like Siphonariidae, the operculate Amphibolidae, the peculiar strand-zone Ellobiidae, and the continental ellobiid Carychiinae. We can only speculate about their relationships to each other and to other pulmonate snails, although it seems clear that the Chilinidae and Latiidae are closely related. It does seem justified to follow the suggestion of Morton (1955) in separating the order Basommatophora into the more primitive ("lower") basommatophoran species and the more specialized ("higher") limnic Basommatophora (Morton's suborders Archaeopulmonata and Branchiopulmonata). The question arises, however, as to which groups belong to these respective suborders. Originally, Morton (1955) included the following families in the Archaeopulmonata: Ellobiidae, Otinidae, Chilinidae, Latiidae, Amphibolidae, Gadiniidae and Siphonariidae. The Lymnaeidae, Physidae, Planorbidae and Ancylidae belonged to the Branchiopulmonata (Morton considered *Acroloxus* to be a member of the Ancylidae). Bondeson (1950), Burch (1961a, 1962) and Hubendick (1962) have convincingly argued for familial status of *Acroloxus*, and Hubendick (1962) placed the Acroloxidae with the lower Basommatophora. However, Van Mol (1967) included only the Ellobiidae and Otinidae in the Archaeopulmonata, which he raised to an order. Similarly, disagreements are prevalent as to which pulmonate familial group is the most primitive, the arguments mainly centring about the families Siphonariidae and Ellobiidae. In Hyman's (1967) view *Siphonaria* is probably the most primitive pulmonate. The lower chromosome numbers of the Siphonariidae may well indicate their primitive position in the Archaeopulmonata, as changes in chromosome numbers in the Euthyneura usually involve an increase rather than a decrease in numbers and as higher numbers are nearly always found in the phylogenetically more advanced taxa. The Siphonariidae have the lowest chromosome number characterizing any single basommatophoran family group. The 7 widely distributed species of *Siphonaria* so far studied have only 16 pairs of chromosomes.

In the Ellobiidae, the genera *Pythia*, *Ovatella* and *Cassidula*, together with *Ophicardelus*, are placed in the primitive subfamily Pythiinae by Morton (1955), *Pythia* being considered the most primitive genus. *Melampus*, and presumably *Detracia*, are considered by Morton to belong to one of the most advanced groups of the family. Haploid chromosome numbers in the Ellobiidae range from n = 17–19. Three species of the Pythiinae have the lowest chromosome numbers in the Ellobiidae, n = 17, and the highest number, n = 19, occurs in *Melampus* which belongs to one of the most advanced ellobiid groups. However, 2 other species of *Melampus* have n = 18, the chromosome number commonly observed in the other Basommatophora. Morton (1955) considers *Pythia* to be structurally the most primitive ellobiid even though some of its species may occupy terrestrial habitats characteristic of more advanced or specialized pulmonates. It is of interest to note that *Pythia plicata* (Fér.) collected from an estuarine environment has 17 pairs of chromosomes (Natarajan, 1958a), whereas *Pythia scarabaeus* (L.) which lives in terrestrial habitats often far removed from the seashore has 18 pairs of chromosomes (Burch, 1967a).

The haploid chromosome number n = 18 is reported for each of the species studied in the archaeopulmonate families Amphibolidae, Chilinidae, Latiidae and Acroloxidae. Chromosome numbers, *per se*, are of little help in establishing the phylogenetic positions of the above families within the Archaeopulmonata since all have 18 pairs of chromosomes, the basic haploid chromosome number characterizing the Branchiopulmonata. Characteristics of egg morphology and chromosome cytology convinced Bondesen (1950) and Burch (1962) that *Acroloxus lacustris* (L.) should be removed from the Ancyiidae and placed in its own family, the Acroloxidae. Natarajan and Burch (1966) included the Acroloxidae in the Archaeopulmonata. Burch and Patterson (1963) pointed out that *Acroloxus* is quite distinct from *Latia* in contrast to the view of Hubendick (1962).

Most of the 43 species of the Lymnaeidae studied have 18 pairs of chromosomes, but those of the widely distributed genus *Radix* have 17 pairs, and the Australasian *Austropeplea* has only 16. Chromosome numbers and other considerations have led to the opinion (Patterson, 1969; Burch, 1965b, 1967a; Inaba 1969) that *Austropeplea* is the most primitive living lymnaeid group, representing a stock from which *Radix* evolved, which in turn gave rise to the other lymnaeids. Inaba's (1969) interesting hypothesis concerning the geological and evolutionary history of early lymnaeids deserves repeating:

The hypothetical pro-lymnaeid ancestor probably appeared in the Paleozoic era. The 16 (n) group seems to be an archaic form, perhaps appearing in the late Paleozoic or early Mesozoic eras. The first to diverge from this stock was *Radix* (n = 17). In the Jurassic era *Fossaria* (n = 18) appeared. *Stagnicola* (n = 18) probably originated from *Radix* stock in the Cretaceous era. Soon after this divergence, *Acella* (n = 18) branched from the *Stagnicola* stem. In the Paleocene of the Cenozoic era, *Lymnaea* (n = 18) branched from the *Radix* stem. *Pleurolimaea* (fossil) branched in this time from the *Acella* stem, and many fossil groups of *Radix*, and *Polyrhytis* from *Stagnicola*, appeared in the Pliocene. From the middle of the late Pliocene, *Hinkleyia* and *Bakerilymnaea* branched from the *Stagnicola* stem. At almost the same time, *Bulimnea* and *Pseudosuccinea* appeared, probably from the *Radix* stem. Recently a 19 (n) species originated within the *Fossaria* group. The origin of *Pectinidens* and its relationships to *Fossaria* is obscure.

The conservativeness in chromosome numbers in the Lymnaeidae and the distinctiveness of the genus *Radix* in having 17 pairs of chromosomes can be used to show that the similar neritiform shell shapes peculiar to the lymnaeids of Lake Biwa, Japan and Kauai Island, Hawaii, are related to species of their own areas rather than having a common ancestry. In both places there is a graded series of species whose shells range from spired to neritiform—in Japan: *Radix japonica* (Jay) (spired), R. (*Biwakoia*) *hamadai* (Habe) (intermediate) and R. (*B.*) *onychia* (Westerlund) (neritiform); in Hawaii: "*Pseudisidora*" *rubella* (Lea) (spired), *Erinna aulacospira* Ancey (intermediate), and *E. newcombi* (H. and A. Adams) (neritiform). *Radix* alone occurs in Japan (other than recently introduced species), and each of the species, including the neritiform one, has the characteristic *Radix* chromosome number, n = 17 (Burch *et al.*, 1964). The Hawaiian species have the chromosome number n = 18 (Burch and Patterson, 1971b), the number characterizing all the North American genera. Therefore, the Japanese species would clearly seem to be related to the other Eurasian species of *Radix*, and the Hawaiian species to lymnaeids of the eastern rim of the Pacific (the Western Hemisphere) (Fig. 4). The neritiform shell then does not show genetic relation, but in both cases adaptation to turbulent water habitats, and is an example of parallel evolution (Burch, 1968b; Burch and Habe, 1969).

Fossaria rustica with 19 pairs of chromosomes is not easy to distinguish morphologically from other species of *Fossaria* with 18 pairs of chromosomes (e.g. *F. obrussa*), but since it has been found in Michigan, Ohio, and Idaho it is not just an isolated, cytologically aberrant population. Its cytological distinction from the common *F. obrussa* would argue against opinions that there is only 1 fossarid-type species (*F. humilis* (Say)) in most of North America.

Little can be said about cytological relationships in the Physidae except

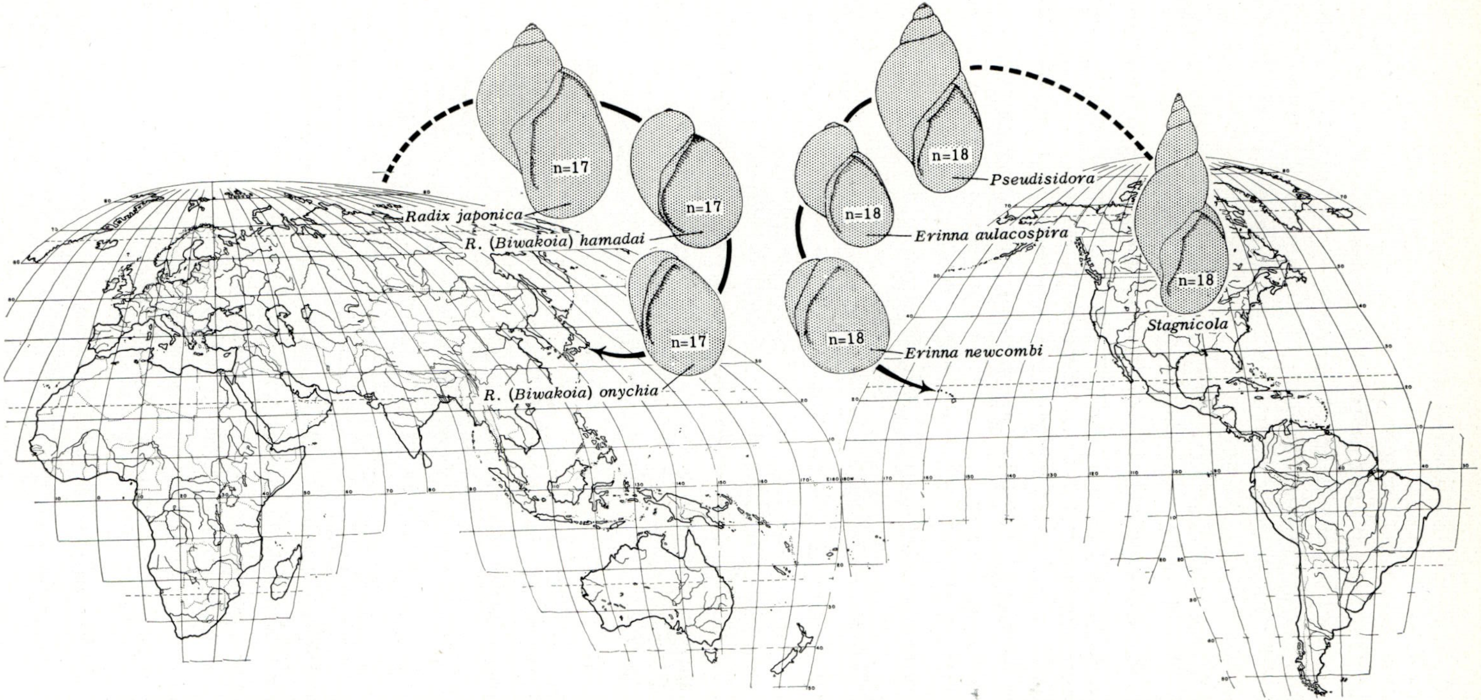

Fig. 4. Convergence of shell characters leading to the neritiform lymnaeids peculiar to Lake Biwa, Japan (*Radix* (*Biwakoia*) *onychia*) and Kauai Island, Hawaii (*Erinna newcombi*). Each of the geographic areas has a related species illustrating an intermediate stage in spire reduction (Japan: *R.* (*B.*) *hamadai*; Hawaii: *E. aulacospira*). Each of the neritiform species has evolved from higher spired forms in their region.

that all 7 members of this conservative group have 18 pairs of chromosomes, and in this way do not differ from species of most other higher limnic basommatophoran families. If all other species of Physidae have this same number of chromosomes, then this family will continue to be the cytologically most conservative among the higher limnic Basommatophora.

The Planorbidae are characterized by the basic haploid chromosome number x = 18; all of the genera have 18 pairs of chromosomes (or a multiple of 18), except the 2 species of *Planorbula* which have 19 pairs, and 1 specimen of a population of *B. natalensis* and a population of *Gyraulus tokyoensis* Mori which had 1–3 extra bivalents.

Members of several gastropod genera have been the subject of cytological studies in recent years due to their importance as intermediate hosts of human parasitic diseases, and snails of the planorbid genus *Bulinus*, intermediate hosts of *Schistosoma haematobium*, have received considerable attention (see Brown and Burch, 1967; Brown *et al.*, 1967; Brown *et al.*, 1971; Burch, 1960c, 1963, 1964a, 1966, 1967b, c, d, 1972, 1973; Burch and Lindsay, 1966, 1969, 1970, 1971; Claugher, 1971; Natarajan, Burch and Gismann, 1965; Wu, 1972; Wu and Burch, 1973). Of particular interest is the subgenus *Isidora* (referred to as *Bulinus* s.s. in some of our previous publications) because the cytological condition of polyploidy occurring among its species is associated with the susceptibility of the snails to schistosome infection.

Species of the subgenus *Isidora* occur in most of the African continent and comprise 2 species groups, the "*tropicus*" and "*truncatus*" groups. Members of the "*truncatus*" group occur in the northern portion of the continent and are tetraploid (n = 36). They are the actual or potential intermediate hosts of *Schistosoma haematobium* in Africa north of the Sahara, in some areas of West Africa and in adjacent regions of Asia Minor. Members of the more southerly "*tropicus*" (n = 18) group are not implicated in the transmission of schistosomiasis haematobia. Thus, there is a correlation between polyploidy and the ability of tetraploid species of the "*truncatus*" group to transmit schistosomiasis haematobia (Burch, 1964a, 1966, 1967c). Since there is great difficulty in distinguishing the species morphologically, knowledge of chromosome numbers is of value in discerning which field populations are capable of transmitting schistosomiasis. Higher polyploids have been found in Ethiopia: hexaploids with 108 chromosomes and octoploids with 144 chromosomes (Fig. 6). Neither of these have been implicated in the transmission of schistosomiasis, although the octoploids were experimentally infectable in the laboratory (Lo, 1972).

Species of the genus *Biomphalaria* serve as intermediate hosts of schistosomiasis caused by *Schistosoma mansoni* in Africa and South America.

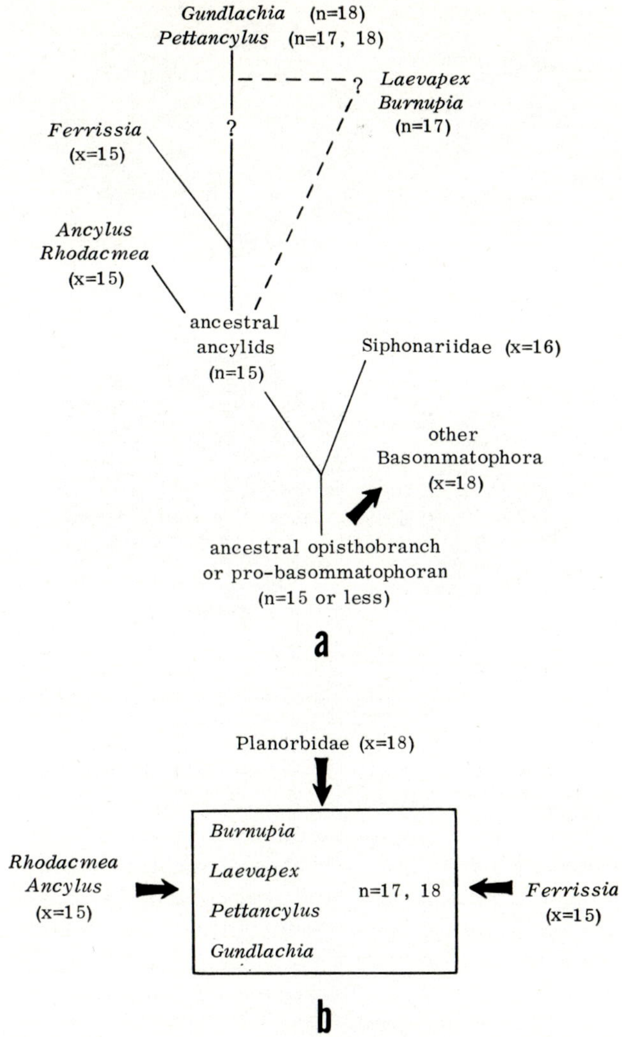

Fig. 5. Two possible evolutionary trends in the derivation of the Ancylidae. An increase in basic chromosome number through aneuploidy is shown in (a), and a decrease in (b). (From Patterson, 1969.)

Chromosome numbers have been determined for 7 species of *Biomphalaria*, but since all species have the haploid number n = 18, chromosome numbers have not been helpful in determining systematic relationships in this genus.

The freshwater limpet family Ancylidae has the chromosome numbers n = 15, 30 and 60, which seem to form a polyploid series, and n = 17 or

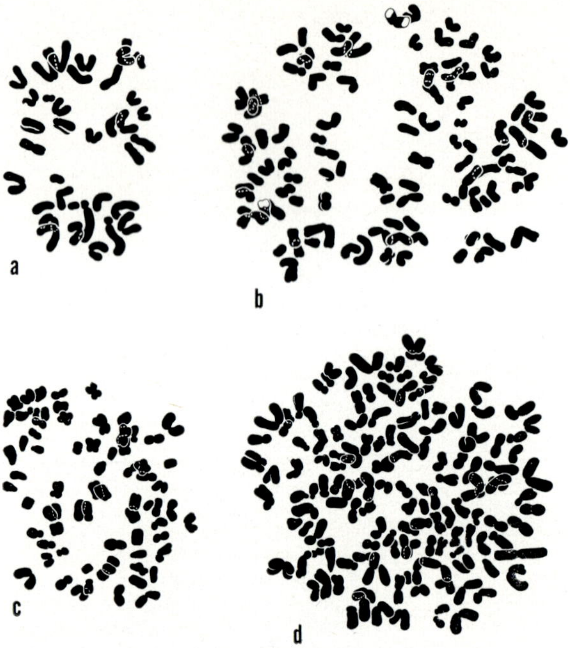

Fig. 6. Mitotic metaphase chromosomes of (a) *Bulinus tropicus* (2n = 36); (b) *B. coulboisi* (2n = 72); (c) *B. hexaploidus* (2n = 108); and (d) *B. octoploidus* (2n = 144).

18. *Rhodacmea cahawbensis* (n = 15) is related to the genus *Ancylus* (n = 30, n = 60) (Burch, 1962, 1974) and so x = 15 would seem to be the base number for the higher chromosome numbers of *Ancylus*.

Two species of *Ferrissia* also have 30 pairs of chromosomes and may be tetraploids derived from an ancestral species with 15 pairs of chromosomes. *Laevapex fuscus* (Adams) from North America and the widely separated *Burnupia* sp from Africa both have 17 pairs of chromosomes, and Burch (1962) believes them to be related. *Pettancylus nipponica* (Kuroda) also has 17 pairs of chromosomes, but does not seem to be closely related on morphological grounds to either *Laevapex* or *Burnupia*. Only 2 of the 10 species of Ancylidae studied (*Pettancylus japonica* (Burch) of Japan and *P.* sp of Tanzania) have the haploid number n = 18 which is characteristic of most other Branchiopulmonata.

Pelseneer (1901) contended that the Ancylidae were derived from the Planorbidae, but Burch (1962) followed by Hubendick (1964) thought that the ancestral ancylid was perhaps more primitive than any recent or higher limnic basommatophoran. Two possible evolutionary trends in the derivation of the Ancylidae are shown in Fig. 5. If the primitive ancylids were derived from a probasommatophoran or an opisthobranch ancestor with less than 15 pairs of chromosomes, there would have been an increase in haploid chromosome numbers by aneuploidy and polyploidy. However, if a planorbid with 18 pairs of chromosomes served as the ancestor, there would have been successive loss of bivalents (as many as 3) resulting in the n = 17 and n = 15 representatives and a subsequent increase through polyploidy to derive the n = 30 and n = 60 individuals from the n = 15 stock. The former is perhaps the "easiest" derivation since gain in chromosome numbers are easier to account for than reductions of chromosome numbers.

Nearly all pulmonate land snails are members of the order Stylommatophora. A commonly held view is that the ancestors of the stylommatophoran land snails were the basommatophoran Ellobiidae. The Stylommatophora have been divided into 4 suborders according to the structural arrangement of the kidney and ureter. The Orthurethra are considered by Pilsbry (1900a) to be the most primitive suborder because the kidney and ureter resemble those of the Basommatophora. He apparently considered the Heterurethra to be somewhat more advanced than the Orthurethra, and he considered the Sigmurethra to be the most specialized suborder. Pilsbry (1900b) implies that both the Heterurethra and Sigmurethra evolved separately from the Orthurethra. Boettger (1955) follows this same scheme. However, Baker (1955) thought that the ancestral group of the Sigmurethra had heterurethrous pallial organs and an aulacopod foot. Baker (1955, 1961) proposed the establishment of the suborder Mesurethra to distinguish the cerionoids and clausiliids from other stylommatophoran snails. The Mesurethra are believed to be phylogenetically close to the Orthurethra (Baker, 1961; Pilsbry, 1946).

Mörch (1865) united the trachaeopulmonate slugs *Janella* and *Aneitea* with the succineids *Omalonyx* and *Succinea* in the group "Elasmognatha" on the basis of jaw structure. However, Pilsbry (1948) believed that the tracheopulmonates (Athoracophoridae) were so different from the Succineidae that they should be retained in the separate suborder, Trachaeopulmonata, leaving the Succineidae alone in the Elasmognatha (= Heterurethra). Baker (1955) placed the trachaeopulmonate slugs in the same group with succineids, the Heterurethra. Likewise, Van Mol (1967) unites these 2 groups, retaining Mörch's name Elasmognatha. However, Van Mol's classification differs importantly from that of other authors in that he considers the Elasmognatha (= Heterurethra) to be the most primitive stylommatophoran suborder, the Orthurethra to be derived from an heterurethran ancestor and the Sigmurethra from an orthurethran predecessor. If evolutionary change in the Euthyneura has usually been accompanied by an increase in chromosome number as suggested by Burch (1965b), then the contention of Van Mol that a heterurethran snail is ancestral to the other stylommatophoran groups is strengthened by the low chromosome numbers of some succineids. Burch and Patterson (1969), from morphological considerations (structure and method of inversion of the tentacles, body surface pattern, pedal grooves and male genitalia) other than those used by Van Mol, suggested that the Succineidae would seem to be an ideal ancestral type to the Athoracophoridae on one hand and to the remaining Stylommatophora on the other.

The greatest range of haploid chromosome numbers in the stylommatophoran land snail suborders (Fig. 3) occurs in the aberrant suborder Heterurethra (n = 5–44). Its most prominent family, the Succineidae, is of particular interest cytologically because the lowest chromosome numbers found in molluscs (n = 5 and 6) occur in several of its species. In other species of the family haploid chromosome numbers range from n = 11–25. The Succineidae contain 2 subfamilies, the Catinellinae which lack a penial sheath and the Succineinae which have a penial sheath, and their chromosome numbers fall into 2 groups corresponding to these morphological divisions. Members of the Catinellinae have the chromosome number n = 5 or n = 6 in all 5 species studied while members of the Succineinae have haploid chromosome numbers ranging from n = 11 to n = 25. There appears to be a consistent correlation between low haploid chromosome numbers in the most primitive group, Catinellinae, and higher haploid numbers in the more advanced group, Succineinae. It will be interesting to observe whether further cytological data bear out this relationship.

Butot and Kiauta (1964) reported 8 specimens from populations of *Succinella oblonga* (Drap.) from North Holland and Belgium to have 12 pairs of chromosomes. Rainer (1967) examined 2 specimens of a population from Baden (Switzerland) and found the haploid chromosome number to be predominantly n = 11 in one specimen, and a variation in the other of n = 10–12, with n = 12 predominant. He concluded that n = 11 is probably the normal haploid chromosome number of *S. oblonga* and that a connection to *Succinea putris* (L.) (n = 22) through polyploidy was conceivable. However, it seems probable that the population Rainer sampled was aberrant and that the normal haploid chromosome number of *S. oblonga* is n = 12 as Butot and Kiauta (1964) reported. Furthermore, a connection to *S. putris* through polyploidy seems unlikely since there are species with intermediate numbers of n = 11, 17, 18, 19 and 21 in the genus *Succinea*, and an explanation of the increase in chromosome number through aneuploidy is more likely.

The 2 species of athoracophorid slugs with 44 pairs of chromosomes have by far the highest chromosome number known for any stylommatophoran land snail (Burch and Patterson, 1971a), the next highest number reported being n = 34 for 2 milacid slugs of the genus *Milax* (Beeson, 1960). The fact that *Aneitea* sp of the New Hebrides and *Triboniophorus graeffei* (Humbert) of Australia have the same chromosome number although they are separated by 1000 miles of ocean indicates that the Athoracophoridae exhibit the same cytological conservativeness characteristic of most other snail groups. Some of the anatomical peculiarities of the Athoracophoridae have recently been discussed by Burch (1968c) and in regard to the related Succineidae by Burch and Patterson (1969). Although these 2 heterurethran snail families obviously seem to be related, the wide gap in their chromosome numbers, if they are the result of aneuploidy, might well indicate that the 2 families have diverged considerably, and over a long period of time. On the other hand, even though polyploidy has not been demonstrated in the Stylommatophora, the high chromosome number of the athoracophorid slugs could well be the result of polyploidy. If such were the case, then the aberrant chromosome numbers found in the Athoracophoridae would have little bearing on their possible phylogenetic remoteness. It is evident that cytological studies should be extended to include more species of the Athoracophoridae, especially the species of New Caledonia and New Zealand.

Haploid chromosome numbers range from n = 20 to n = 30 in the suborder Orthurethra. Pilsbry (1900a) considered the Achatinellidae to be one of the most primitive orthurethran groups, but Solem (1959) placed this family in a more advanced phylogenetic position. It may be of significance that members of this family have the lowest haploid chromosome numbers (n = 20–23) in the Orthurethra, which may strengthen Pilsbry's contention.

In the suborder Mesurethra nearly all of the species studied cytologically belong to the Clausiliidae. Zilch (1959–1960) does not recognize the Mesurethra, but includes the members of this group in various superfamilies in the suborder Sigmurethra. The current classification of the Clausiliidae is very complicated; there are 10 subfamilies with various tribes. Species for which we have cytological data have been placed in 5 different subfamilies, but their haploid chromosome numbers do not correspond to their present classification into these groups. A critical morphological–cytological study of the Clausiliidae would probably lead to revisions in the current classification.

The largest stylommatophoran suborder, the Sigmurethra, is divided into 3 infraorders, the Holopodopes, Aulacopoda and Holopoda. Baker (1955, 1962) proposed the infraorder Holopodopes and included in it several primitive "Achatinidae like" genera and families that had a less distinct aulacopod foot and had similar kidneys. The chromosome numbers of members of the Holopodopes so far investigated are n = 25, 29, 30, 31 and 32. However, cytological information for this group is currently too fragmentary to draw any conclusions about cytotaxonomic relationships.

The lowest (n = 20) and the highest (n = 34) chromosome numbers in the Sigmurethra occur in members of the infraorder Aulacopoda, a morphologically diverse group considered to be somewhat more advanced than the Holopodopes. The family Endodontidae seems to be the most primitive aulacopod group (Solem, 1959). Chromosome numbers in this family (n = 29–31) are higher than those reported for members of other aulacopod families. The lowest chromosome number in the Aulacopoda occurs in *Vitrea diaphana* (Zonitidae), but the other 4 species studied in this family have 28–31 pairs of chromosomes. There is also a considerable amount of variation in chromosome numbers in the families Arionidae, Limacidae and Ariophantidae. In the slug family Limacidae each genus is characterized by a different chromosome number, but in the Arionidae, the chromosome number varies in the genus *Arion* from n = 25–29. Nevertheless, the different chromosome numbers in these slugs correspond to subgeneric groupings based on gross morphology (Burch, 1965b).

The Philomycidae exhibit several morphological characters that differentiate them from other slugs. Pilsbry (1948) thought that this family was perhaps an early branch of the endodontid stock from which the Arionidae also arose. It is of interest that both philomycid species examined have 24 pairs of chromosomes, the lowest haploid chromosome number reported for the 2 endodontacean slug families (Philomycidae and Arionidae). Pilsbry considers the limacid slugs (including Limacidae and Milacidae) to have arisen independently from different ancestors than those of the endodontacean slugs. The chromosome number in the Limacidae and Milacidae are higher than those of the endodontacean slugs (except for *Lehmannia*). *Lehmannia* is a form intermediate to *Limax* and *Deroceras*, which is given a subgeneric ranking in *Limax* by Pilsbry (1940). However, its specialized penial structure and chromosome number suggest that perhaps *Lehmannia* deserves full generic ranking, and some recent authors have treated it in this way.

The infraorder Holopoda includes some morphologically rather specialized groups of land snails. Their chromosome numbers range from n = 21–31 in the Helicidae, to n = 31 in a mosaic polygyrid specimen, i.e. one with extra chromosomes, and to n = 31 and n = 32 in the Oreohelicidae. Most members of the Polygyridae studied have 29 pairs of chromosomes, but both species of *Allogona* have the haploid number n = 26. Pilsbry (1940) held the view that polygyrid genera which retain an epiphallus and flagellum are primitive and that these structures are lost in the more specialized forms. *Allogona* has retained both an epiphallus and flagellum and is therefore considered more primitive; it also has a lower chromosome number than the other genera (Husted and Burch, 1953; Burch, 1965b).

The higher chromosome numbers in the Oreohelicidae (n = 31 and n = 32) strengthens the taxonomic separation, based on morphological criteria, of this group from the Camaenidae (n = 27–29), where they were once placed. The higher chromosome numbers of the oreohelicids may indicate that they are more advanced and specialized than the camaenids (Babrakzai, *et al.*, 1974).

The variation in chromosome numbers from n = 21–31 in the Helicidae is rather wide compared to that found in most other stylommatophoran families. Several of the chromosome numbers of this apparently specialized land snail family are relatively low (n = 21, 22, 23). Recently, Rainer (1967) determined the chromosome numbers of 19 species and subspecies of helicid snails and discussed some of their cytotaxonomic relationships. He studied representatives of 5 helicid subfamilies and 2 tribes of the subfamily Helicellinae. The tribe Helicelleae includes the genera *Candidula* (n = 27), *Helicella* (n = 26), *Trochoidea* (n = 26) and *Cochlicella* (n = 23). In his opinion *Cochlicella* should be assigned to the tribe Monacheae in which 2 species of *Monacha* have also been found to have 23 pairs of chromosomes. Should morphological similarities between the members of *Cochlicella* and *Monacha* warrant it, Rainer's contention may be justified. Rainer also has suggested a division of the genus *Cepaea* into 2 subgenera, based on cytology. One subgenus would contain the species with n = 22 and the other subgenus the species with n = 25. Such a suggestion, however, does not seem justified unless there exist correlative morphological characters (Patterson, 1969). It is interesting to note that Rainer found the European *Euparypha* (=*Theba*) *pisana* (Müller) to have 28 pairs of chromosomes while Laws (1965) reported n = 30 for species introduced into Australia.

Rainer (1967) remarks that in general the cytological findings correlate well with the modern classification of the Stylommatophora. However, there are too many exceptions yet to such a correlation with current classification of the Stylommatophora to justify so broad a statement. Also, within and among a great many taxa, chromosome numbers, *per se*, neither support nor detract from present systematic schemes (Patterson, 1969).

The systematic position of the Soleolifera or Systellommatophora is still being debated. Plate (1893), Fretter (1943), Boettger (1955) and Hyman (1967) contend that they are more reasonably placed with or near the opisthobranch orders, and such an arrangement is followed by Zilch (1959–60) and Taylor and Sohl (1962). Others, such as Hoffmann (1925, 1929), Pilsbry (1948) and Van Mol (1967), place them with the Pulmonata. From the meagre amount of cytological information available on these molluscs, chromosome numbers *per se*, which are n = 16, 17 and 18, neither support nor weaken either position. Their chromosome numbers overlap with those of the opisthobranch orders Cephalaspidea (n = 17, 18), Sacoglossa (n = 7, 17), Entomotaeniata (n = 17) and Anaspidea (n = 17), as well as with those of the Basommatophora (n = 15, 16, 17, 18, 19, excluding polyploid numbers) and the stylommatophoran Succineidae (n = 5, 6, 11, 12, 15, 17, 18, 19, 21, 22, 24, 25) (see Patterson, 1969). All other Stylommatophora have higher chromosome numbers than the Soleolifera.

The utility of cytological information in taxonomic discrimination at the generic and specific levels has been demonstrated in several cases. For example, the lymnaeid species "*Limnaea*" *ollula* Gould of the Orient was considered to be a member of the genus *Fossaria*. However, Burch *et al.* (1964) found this species to have the chromosome number n = 16, a number quite distinct from other fossariids (n = 18, 19). It was provisionally assigned to the radicine group, and later H. J. Walter (personal communication) found this species to have a radicine-like anatomy. Hubendick (1951) considered all North American fossarids, excluding those of the southernmost region of the Gulf coast states, to be only 1 species, "*Lymnaea*" (=*Fossaria*) *humilis*. The chromosome number of *F. rustica* (n = 19), a species with a wide geographic range, demonstrates its distinctiveness from other fossariids (n = 18) and supports those who contend that there are several species of the genus in North America. In addition to these lymnaeids, cytological information has been helpful in discerning various planorbid species. Burch (1965b) reported *Gyraulus circumstriatus* (Tryon) to have 18 pairs of chromosomes, and the closely related and often indistinguishable form, *G. parvus* (Say), to have 36 pairs. These 2 species have sometimes been considered synonymous by malacologists, but their difference in chromosome number demonstrates their distinctness (i.e. one is a normal diploid, the other a tetraploid). Chromosome numbers have been especially helpful in unravelling the taxonomic problems in the subgenus *Isidora* of *Bulinus*. Burch (1964a, 1967b, c, 1969b, 1972), Burch and Huber (1966), Brown and Burch (1967) and Brown and Wright (1972) found Ethiopian bulinine snails to have chromosome numbers of n = 18, 36, 54 and 72. The diploid and tetraploid populations occur at lower altitudes and the hexaploids and octoploids occur at higher altitudes. Mandahl-Barth (1965) earlier had considered only 1 species of *Isidora* to occur in Ethiopia, but it would seem that each of these different polyploid populations represent distinct biological species. There are then at least 4 species of the subgenus *Isidora* in Ethiopia.

IV. Polyploidy

Polyploidy is the exact multiplication of the entire chromosomal complement of an organism. The degree of polyploidy is indicated by the number of multiples of the basic haploid number. For example, if the haploid number of the snail is 18, a closely related species with a diploid number of $2n = 144$ is an octoploid (8×18). There are 2 types of polyploidy, autopolyploidy and allopolyploidy. An autopolyploid condition arises when more than 2 haploid chromosomes sets of a single species participate in zygote formation. Thus, autopolyploidy is the multiplication of 1 basic genome. For example, an autotriploid can be produced by the union of an unreduced (2n) gamete with a normal (n) gamete, or a tetraploid can result from an aberrant mitotic division where the chromosomes duplicate and divide but the nucleus fails to divide, thereby producing a 4n condition. Polyploids produced artificially by such drugs as colchicine are of this type. Allopolyploids, on the other hand, have a chromosomal complement composed of 2 or more genomes from different species or populations. The fact that the 2 species can cross implies at least partial homology between the chromosomes of the parents. The pattern of association of chromosomes should differ with the species involved in the crosses. Polyploids formed from 2 species having no cytologically detectable evidence of homology between genomes would show only bivalent formation while autoploids with 4 completely homologous genomes would show multivalents or bivalents that are formed at random from the members of the group of 4 homologues. Between these 2 extremes are cases of constituent genomes with intermediate degrees of homology where the homologous segments are probably interspersed with others along the length of the chromosomes. All molluscan polyploids appear to be allopolyploids.

Because of the inability of a newly derived polyploid to breed with its sibling diploids, it is reproductively isolated from them. Accordingly, polyploidy is a method, and the only clearly established one, of instantaneous speciation (Mayr, 1963). Its importance and predominance in producing new species of plants is well known. Cytologists generally agree that its significance in animal speciation is slight (Muller, 1925; White, 1954, 1973). Nearly all proven cases occur in permanently hermaphroditic species that are capable of self-fertilization or of sexual reproduction, or in those species that have abandoned sexual reproduction in favour of parthenogenesis. Reasons given for the rarity of polyploidy among animals are the disturbance of the sex chromosome–autosome balance, the improbability of a newly arisen polyploid individual finding a polyploid mate, or developmental difficulties in cellular differentiation.

In hermaphroditic animals, such as pulmonate snails, there are no sex chromosomes (White, 1954; Burch, 1960a). A newly arisen polyploid need not find a mate, but through self-fertilization it can produce its own strain, if there is cytological compatibility. It would seem then, that the only barriers to polyploidy in pulmonates would be the infrequency of accidents of chromosomal doubling, the infrequency of self-fertilization in many groups, the infrequency of the production of viable interspecies hybrids, and possible developmental difficulties. That polyploidy has been found in less than 3% of the euthyneuran species examined (Burch, 1965b) indicates that 1, some, or all of these factors are important in reducing its occurrence (Burch and Huber, 1966). One likely factor is lack of self-fertilization, an ability which is probably limited to a few groups only of euthyneuran snails. In the Planorbidae, where self-fertilization is known to occur and in which many specimens of some populations are aphallic, about 11% of species examined to date are polyploid. On the other hand, in the Lymnaeidae, where self-fertilization is also known to occur, none of the 43 species studied are polyploids. Therefore one or more of the other factors limiting polyploidy must be operating in lymnaeid snails.

Most changes in chromosome numbers in pulmonate molluscs have obviously been brought about through aneuploidy. When polyploidy occurs, it operates at or below the species level and probably has been of little significance in the derivation of higher taxa.

In the Planorbidae, polyploidy has been found in 2 different subfamilies. In the subfamily Planorbinae, *Gyraulus parvus* (n = 36) has twice the number of chromosomes characteristic of its family (n = 18) and found in its nearly indistinguishable sibling species, *G. circumstriatus* (Burch, 1960b, 1965b). All other species of the Planorbinae have the haploid number n = 18. A study of *G. parvus* and *G. circumstriatus* should prove interesting. What are their geographic ranges? Do they overlap in distribution? Are there any gross morphological distinctions between the 2? Are they physiologically different? Do they occupy different ecological niches? How did *G. parvus* arise—by autopolyploidy, allopolyploidy, or by yet another method?

A bit more is known about the polyploids in the subfamily Bulininae. The genus *Bulinus* contains 3 groups which are in general easily separated on shell characters: the *Physopsis* group or subgenus, the *Bulinus* s. s. group (formerly often called *Pyrgophysa*) and the *Isidora* group (formerly known as *Bulinus* s.s.). The polyploids, as well as some diploids, belong to the latter group. Although the taxonomy of the genus is still quite unsatisfactory, and many nominal species must still be dealt with, broad outlines of geographical distribution of the various polyploids and their diploid siblings are reasonably well known. The diploids range from South Africa northward to the highlands of Ethiopia and from the coastal areas of the Indian Ocean (and Madagascar) westward perhaps to West Africa. The tetraploids occur on several islands of the Mediterranean, have been reported from the Mediterranean area of south-western Europe, are found in North Africa, from Morocco to Tunisia, in Egypt and the Middle East until Persia, and southward through the Sudan, Ethiopia and to the

Great Lakes region. In West Africa they are found south of the Sahara, e.g. in Ghana. The hexaploids and octoploids are found in the Ethiopian highlands (the octoploid perhaps occurs also across the Red Sea in Aden).

Cytologically, the various polyploids behave as diploids, i.e. all chromosomes of the first meiotic division pair as bivalents. Therefore Burch and Huber (1966) concluded that most likely these bulinine polyploids are allopolyploids, i.e. are the result of hybridization followed by chromosome doubling.

Wu (1972) and Brown and Wright (1972) have studied various aspects of these polyploid populations in considerable detail, but because of the different techniques and populations used, it is somewhat difficult at this stage to directly compare their results. Both studies showed that it is difficult to separate the ploidy groups morphologically, because clear-cut diagnostic characters either do not exist, or population variability extends beyond ploidy boundaries. Wu (1972) concluded that the hexaploids had more characters in common with the diploids he examined than did either of the other polyploid forms, and that the tetraploids and octoploids were more similar to each other than to the hexaploids or diploids. This is what one might expect if the octoploids were derived from tetraploids and the hexaploids from hybridization of tetraploids with diploids. Brown and Wright (1972) found a similarity in egg proteins between the hexaploids and diploids, but reported digestive gland esterases of octoploids to be more similar to those of diploids, and octoploids to have more egg-protein antigens in common with diploids than with tetraploids.

Polyploidy is also apparent in some of the Ancylidae, but here the evidence is still rather sketchy. The high chromosome number ($n = 60$) of the Eurasian–North African *Ancylus fluviatilis* would obviously seem to be derived by polyploidy, and finding a species in Ethiopia with half the number found in *A. fluviatilis* confirms this view. However, the origin of the number $n = 30$ is not yet clear; the rather close relationship of *Ancylus* to the American *Rhodacmea* with 15 pairs of chromosomes would indicate that $n = 15$ is the basic and ancestral number. Perhaps the 30 pairs of chromosomes found in *Ferrissia* were derived from the same basic number. The derivation of a base number $x = 15$ in the Ancylidae is difficult to explain other than by a reduction in number from the $x = 18$ in the Planorbidae, from which the ancylids were supposedly evolved. However, the separation of the ancylids and planorbids probably occurred very long ago, and from ancestors morphologically and cytologically quite unlike present day representatives of their families. A study of the ancylids in ancient lakes might be revealing, and especially pertinent might be the findings from a study of the aberrant *Ancylastrum* of Tasmania, a freshwater limpet with apparently strong affinities to the Planorbidae.

Natarajan and Burch (1966) found the ellobiid *Melampus coffeus* to have the haploid number n = 19, but in a nearby population with the same chromosome number they found a tetraploid individual with 38 pairs of chromosomes. This population was somewhat deviating morphologically; it may have consisted of hybrids from which the tetraploid individual arose by allopolyploidy.

There has been some discussion concerning the possibility of polyploidy occurring within the land snail family Succineidae (Rainer, 1967). However, Patterson (1971b) has pointed out that there is no correlation between chromosome numbers and morphological similarity of the species having multiples of existing haploid numbers. In fact, they are so different morphologically that a derivation through ploidy is not indicated for any succineid taxa.

The sperm nurse cells, or Sertoli cells, of pulmonate snails have greatly enlarged nuclei with large amounts of chromatin. The increased chromatin material of these cells as compared to that found in ordinary cells has not been carefully studied, but it may have arisen by a form of endomitotic polyploidy, i.e. by numerous divisions of the chromosomes without cytokinesis, which produces a highly polyploid nucleus. We have seen some evidence of progressive multiplication of chromatin masses in *Catinella vermeta* by successive factors of 2, as would be expected if the chromatin duplication was by endomitosis resulting in endopolyploidy. The role of the increased amount of chromatin material in the functions of the nurse cells would be an interesting subject to investigate.

V. Chromosome morphology and karyotype analyses

The morphology of mitotic metaphase chromosomes has long been recognized as a character useful in species recognition, and in some groups of animals where techniques have been developed for readily producing dividing cells and easily handling their chromosomes, cytotaxonomy has become a very useful adjunct to more classical systematics. However, mollusc cells as yet are not so easily manipulated, and workers are few; therefore the use of chromosome morphology for studying relationships and evolution of molluscs is at a very primitive stage. What little informa-

---

Fig. 7. Mitotic colchicine-mitosis chromosomes of (a) *Helix pomatia* and (b) *Catinella vermeta*, aligned according to position of the primary constrictions and in order of decreasing lengths. Chromosomes in lines (1) and (2) are medianly constricted or nearly so; chromosomes in line (3) are submedianly or subterminally constricted.

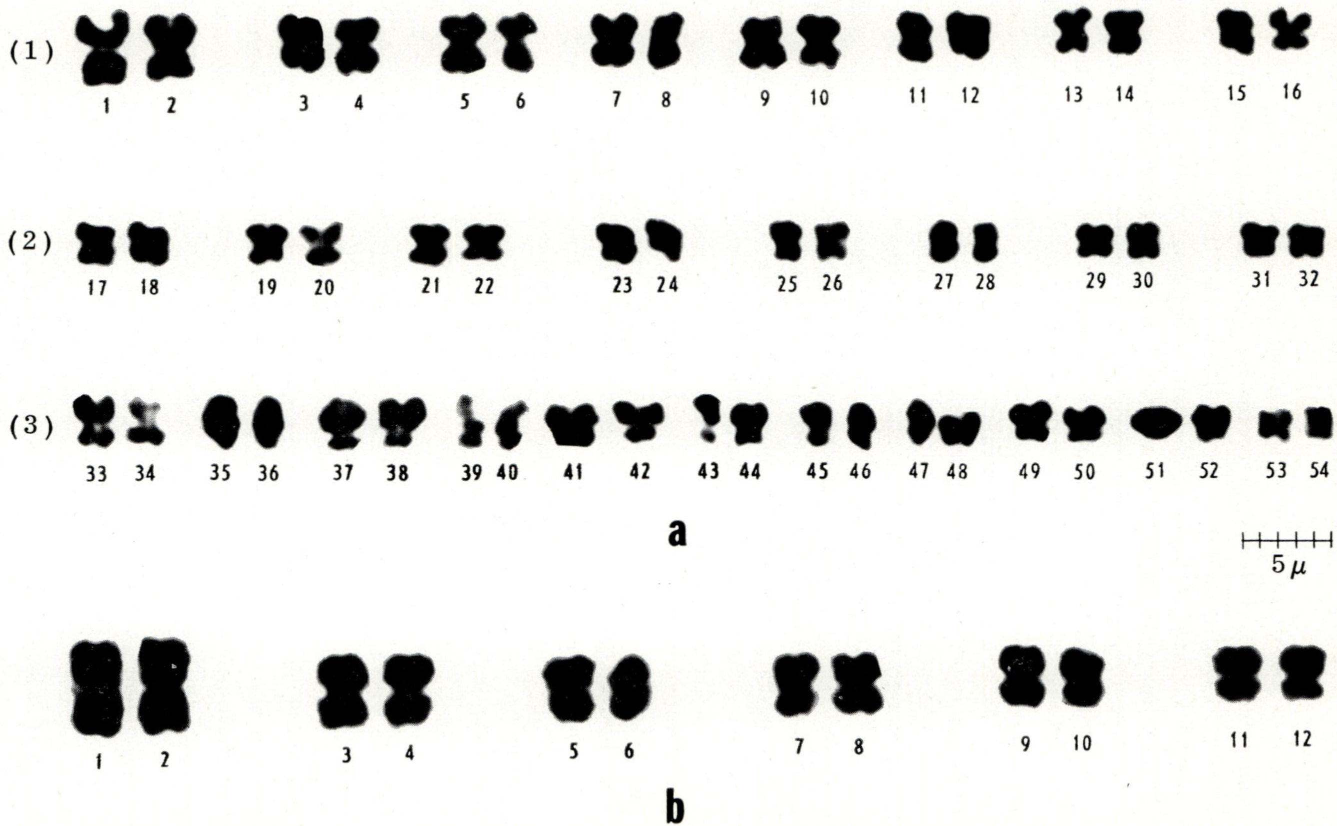

(1)
1 2 3 4 5 6 7 8 9 10 11 12 13 14 15 16
(2)
17 18 19 20 21 22 23 24 25 26 27 28 29 30 31 32
(3)
33 34 35 36 37 38 39 40 41 42 43 44 45 46 47 48 49 50 51 52 53 54
a
5 μ
1 2 3 4 5 6 7 8 9 10 11 12
b

tion is available is mostly at random on the karyotypes of a few isolated species.

Chromosome morphology of pulmonate snails has been discussed by Burch (1960a, d). The shapes of mitotic metaphase chromosomes in aquatic and land pulmonates are similar in that they are elongate and monocentric, with median, submedian or terminal centromeres which form a "primary constriction" in the highly contracted chromosomes. On the spindle in polar view the medianly and submedianly constricted chromosomes, especially the larger ones, appear as V's and J's respectively, because, due to the crowding on the spindle, their arms extend peripherally away from their centromeres, which are attached to the spindle. The orientation and distribution of chromosomes on the spindle and as seen in the mitotic metaphase plate is probably random. Sometimes homologues may be observed to be close to each other, but just as often they appear to be some distance apart or on opposite sides of the cell. Jacob (1959) designated homologues by their close proximity on the assumption of a regular or determined arrangement of chromosomes within the nucleus. We have not observed such an arrangement.

At mitotic anaphase the chromosomes have the same V- and J-shaped appearance (as seen in mitotic metaphase chromosomes) while they are being pulled through the cytoplasm *en route* to their respective poles. Chromosomes with terminal or nearly terminal centromeres appear as straight rods.

In addition to the primary constriction in the chromosome which marks the location of the centromere, there may also occur secondary constrictions marking the locations of the nucleoli.

Details on the morphological and behavioural aspects of chromosomes during the mitotic and meiotic cycles have been given for the land pulmonate *Catinella vermeta* by Patterson and Burch (1966).

The karyotype of a species is usually represented by a drawing or photograph in which the homologous pairs of mitotic chromosomes, or 1 representative of each pair, are arranged according to their lengths and the positions of their centromeres (Figs 7–9). Karyotypes have been presented so far for only a few pulmonate species: *Laevapex fusca* and *Acroloxus lacustris* (Burch, 1962), *Catinella vermeta* (Patterson and Burch, 1966; Burch, 1968a), *C. arenaria* (Butot and Kiauta, 1967), *C. texana* Hubricht, *Succinea greeri* Tryon, *S. urbana* Hubricht and *S. grosvenori* Lea (Natarajan, Hubricht and Burch, 1966), *Melampus bidentatus lineatus* Say (Natarajan and Burch, 1966), *Helix aspersa* Müller (Rainer, 1967), *H. pomatia* L. (Burch, 1968a) (Fig. 7), *Stagnicola elodes wyomingensis* Baker, *S. exilis* (Lea) and *S. catascopium* (Say) (Inaba, 1969) (Fig. 8), *Bulinus tropicus* (Krauss) (Patterson, 1971a) (Fig. 9), *B. forskalii*, *B. globosus* (Morelet), *B. liratus* (Tristram), *B. obtusispira* (Smith) and *B.*

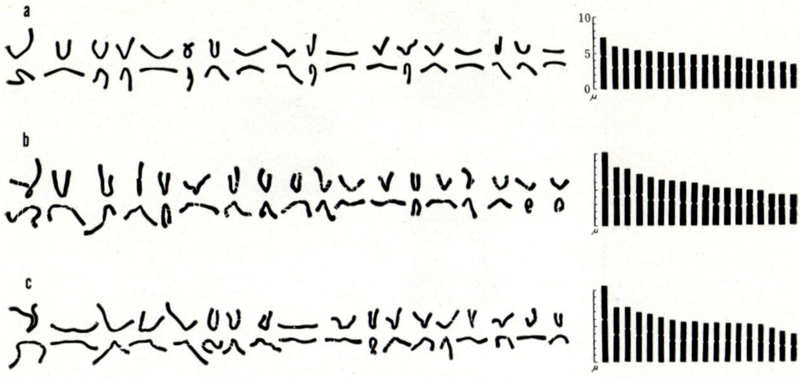

Fig. 8. Serial arrangement and schematic representation of the karyotypes of 3 *Stagnicola* species. (a) *S. elodes wyomingensis*; (b) *S. exilis*; (c) *S. catascopium*. (From Inaba, 1969.)

truncatus (Claugher, 1971), and *Radiocentrum clappi*, *R. chiricahuana* (Pilsbry), *Oreohelix anchana* Gregg, *O. concentrata* Dall, *O. subrudis* Reeve and *O. houghi* Marshall (Babrakzai, Miller and Ward, 1974).

As techniques are designed and perfected for production and easier handling of mollusc chromosomes (see Burch and Cuadros, 1965; Burch, 1968a; Patterson, 1971a), the karyotype will undoubtedly gain in importance as a tool for biosystematic analyses of pulmonate snails.

VI. Cytogenetics

A. Chiasma frequency

During the meiotic cycle, homologous chromosomes become associated in pairs during zygonema, after which genetic interchange takes place through the process of crossing-over. The points of such cross-overs can be identified and are called chiasmata. The homologous chromosomes are thus attached to each other, but during diplomena the free portions are repelled, making cross-like figures if there is a single chiasma which is not near the ends of the homologues, a rod-shaped figure if the single chiasma is near the ends, a ring if there are 2 chiasmata near opposite ends, and multiple loop-shaped figures if there are more than 2 chiasmata. Thus, the shapes of such bivalents depend on the number and position of the chiasmata. As meiosis proceeds, the chiasmata progressively slip towards the terminal ends of the chromosomes, a process called "terminalization". As terminalization of chiasmata begins and chromosomal contraction continues through diplonema into diakinesis, the cross, rod, ring or

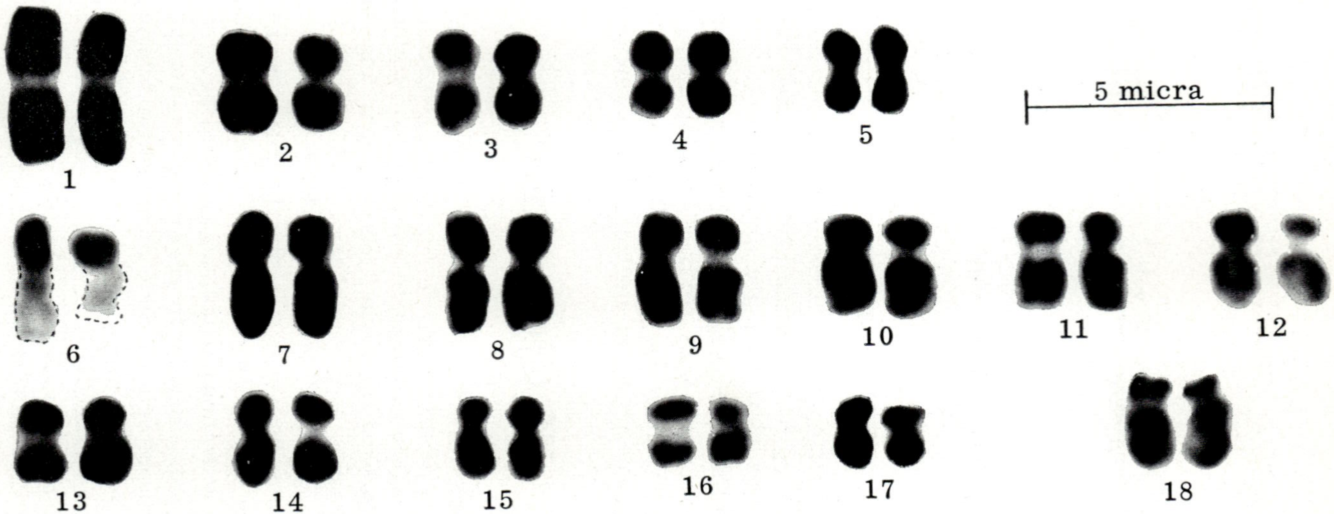

Fig. 9. Karyotype of *Bulinus tropicus* composed of 5 metacentric pairs (1–5, top row), 12 submetacentric pairs (6–17, bottom 2 rows), and 1 subacrocentric pair (18, lower right). Pair 6 is especially distinct because 1 arm of each chromosome is very lightly stained (heterochromatic). (From Patterson, 1971a.)

multiple loop-shaped figures are more easily defined. The average frequency of chiasmata per cell has been used for some time as a species specific character in various animal and plant groups, but has been applied very little to molluscs. However, we have noted differences in chiasma frequencies between several species of *Catinella* whose chromosomes otherwise were very similar (Natarajan *et al.*, 1966).

B. Paracentric inversions

Changes in gene sequence in a chromosome occur when a segment of the chromosome becomes inverted. Changes in position of the centromere on a chromosome can be brought about by an inversion of a segment of the chromosome containing the centromere. Due to the small size of molluscan chromosomes, inversions are difficult to observe, although their detection may become more common as chromosome banding techniques are perfected for molluscs. However, inversions may be detected by other means, one of which is the observation of chromosomal bridges and acentric fragments during meiotic anaphase in organisms heterozygous for the inversion. A cross-over within a paracentric inversion will result in a bridge at anaphase accompanied by the acentric fragment. Bridges and fragments have been reported in the stylommatophoran *Allogona profunda* (Say) (Husted and Burch, 1953) and in the basommatophoran *Gyralus deflectus* (Burch, 1959). Undoubtedly they will be reported in other pulmonates as cytogenetic studies become more common in these molluscs.

Sometimes mitotic chromosomes have difficulty in separating at anaphase and tend to "stick" together after the chromosomes have nearly reached their respective poles. Such "sticky chromosomes" (see Patterson and Burch, 1966) form a bridge in mitotic cells that resembles the bridge during anaphase following a cross-over within a paracentric inversion, except for the lack of the acentric fragment. Also, sometimes 1 or more bivalents have difficulty terminalizing in a dividing meiotic cell and the dyads may lag behind the others in anaphase I movement (see Patterson and Burch, 1966). Care must be taken to not confuse these with the acentric fragment characteristic of an inversion heterozygote.

C. Extra or supernumerary chromosomes

Chromosomes in excess of the normal complement have occasionally been reported in molluscs. When more than 2 homologous chromosomes are present in the complement, multivalent associations may occur. Trivalents and quadrivalents have been most recently reported by Burch (1967e) for several species of the prosobranch *Semisulcospira*. Nonhomologous chromosomes do not have pairing partners and hence are usually observed as univalents in meiotic metaphase I cells (Burch, 1960a, 1967e). Univalents can be derived by non-disjunction, failure of meiotic pairing, or as a consequence of aberrant meiotic pairing as reported in 1 cell of *Catinella vermeta* which had a quadrivalent, 3 bivalents and 2 univalents (Patterson and Burch, 1966). The occurrence of heteromorphic bivalents has been described in *Triodopsis* (Husted and Burch, 1946).

Supernumerary chromosomes have usually been defined as those that are additional to the karyotype and that are non-homologous to members of the normal complement (White, 1954). These additional chromosomes, when present, appear in variable frequency among individuals of a wild population and may not necessarily characterize all populations of a species. In fact, supernumerary chromosomes have not been observed commonly in those molluscs that have been studied cytologically. This may be due to the fact that supernumerary chromosomes often have deleterious effects on fertility and viability. Supernumerary chromosomes have been reported to occur in 4 pulmonate species, although chromosome number variation has been reported in several other species (Ford, 1962; Rainer, 1967). In the latter cases, extra or supernumerary chromosomes may be involved, although the possibility exists that some of these observations were made on faulty cell preparations. Since a number of cells in many squash preparations are not perfect for study in that all the chromosomes are not clearly separated from each other, it is sometimes difficult to distinguish some of the chromosomes as precise entities. Also, some bivalents are broken by the pressure of the squash. Thus, cells with seemingly aberrant chromosome numbers are not necessarily anomalous.

Individuals of 1 population of *Gyraulus deflectus* (normally $n = 18$) were found by Burch (1960a) to have 3 extra elements which were associated variously as 3 univalents, a univalent and a bivalent, or 1 trivalent at diakinesis. These supernumeraries stained more deeply than the regular bivalent chromosomes of the complement.

Supernumerary chromosomes were found in 2 populations of *Bulinus natalensis* (Burch, 1964a). The number of additional chromosomes varied from 1–3 among the individuals of 1 population and 1 extra chromosome was present in some cells of individuals of the other population. These varying chromosome numbers may have been the result of hybrid origins of the 2 populations.

Husted and Burch (1946) found individuals of *Triodopsis fraudulenta* (normally $2n = 58$) to have chromosome numbers of $2n = 58$, 59, 60, 61(?) or 62 and the bivalents to number 29, or 30, or $n = 29$ with an additional univalent. The authors suggested that the extra chromosomes were perhaps duplications of other chromosomes of the complement. If this were the case, the chromosomes would be at least partially homologous, thereby not qualifying as supernumeraries in the strict sense. Nearly

40% of the specimens of *T. fraudulenta* examined were found to have extra chromosomes. One individual was also found to be mosaic, i.e. to have cells with 62, 64 and 66 chromosomes (2n). Husted and Burch (1946) noted that some individuals of *Mesomphix oxycoccus* had 29 pairs of chromosomes, while others had 30 pairs.

The most thorough study of supernumerary chromosomes of a mollusc is that of Evans (1960), in which he investigated their occurrence in wild populations of *Helix pomatia* L. From 1–6 extra chromosomes were found in cells containing the basic diploid complement. All of the extra chromosomes were the same in the 2 populations studied in that they were submedianly constricted and smaller than the chromosomes of the basic complement. When only 1 extra chromosome was present it usually did not pair or come to lie on the metaphase plate of the first meiotic division, but it divided mitotically at the second meiotic division and was always included in daughter nuclei. If 2 additional chromosomes were present, they were usually associated as bivalents in meiosis I, and more than 2 extra chromosomes appeared as univalents or bivalents in the first meiotic division. Multivalent associations (consisting of more than 2 homologues) such as have been reported in other gastropods were not observed. Supernumerary bivalents seemed to behave normally during meiosis. In the view of Evans (1960) the supernumerary chromosomes might have arisen from chromosome change occurring after a period of genotype instability due to inbreeding of these normally cross-fertilizing hermaphrodites.

D. Hybridization studies

Hybridization studies combined with a cytological investigation are almost non-existent in pulmonate molluscs. Wu (1972) hybridized various populations and species of the planorbid genus *Bulinus* and some limited observations were made on the cytology of the hybrids. In his experiments double recessive albinism in a diploid population was used as a genetic marker. Thus, he could determine from the presence or absence of eye pigmentation in the first generation offspring of the acting female albino whether cross-fertilization by the mating partner had occurred or whether reproduction was restricted to self-fertilization. Since albinism is inherited in a simple Mendelian fashion, cross-fertilization of the pigmented parent acting as a female could be determined from the ratios of pigmented to albino offspring in the second filial generation. In hybridization between ploidy groups where a second filial generation was not produced, hybrids were identified by a comparison of their foot muscle esterase patterns with those of their parents.

Wu (1972) found that crosses were most successful between breeding individuals from different diploid populations that were geographically closest, with genetic exchange decreasing as geographical separation increased. Diploids mated to various polyploids were generally not successful in producing offspring, but in a few instances F_1 progeny were produced. For example, in a cross between an albino diploid *Bulinus tropicus* mated with a pigmented tetraploid *B. truncatus sericinus*, the tetraploid produced 19 F_1 progeny which developed and lived for 5 months. During that time they mated with their siblings but were sterile. No eggs were produced. In a similar case, in 4 experiments where the albino diploid was mated with the pigmented hexaploid *Bulinus hexaploidus*, the latter produced many vigorous F_1 progeny, which were sterile upon sibling mating. Histological examination of the ovotestis in these hybrids revealed that spermatogenesis was not completed—no sperm were produced, only spermatocytes. In a cross between the albino diploid and the octoploid *B. octoploidus*, the octoploid produced eggs for 7 weeks. The resulting hybrid progeny were sterile, although some of them did lay eggs which, however, proved to be unembryonated. Histological examination of the ovotesticular tubules of the F_1 snails revealed that no sperm were produced. A cytological examination of the hybrids revealed that they had the expected diploid chromosome number, $2n = 90$ ($n = 18 + n = 72$) (Burch, unpublished). In male gametogenesis, meiosis was started but not completed, always being arrested at anaphase I because of the inability of the chromosomes to separate.

VII. Methods for molluscan chromosome preparations

Most cytotaxonomic studies of molluscs have been based on chromosome squash preparations of ovotestis tissue or embryos. The basic chromosome squash procedure is described in LaCour (1941). However, over the years many people have expressed difficulty in successfully making molluscan chromosome preparations. Therefore, we would like to describe the squash technique employed in our studies in some detail with the hope that this will aid those researchers not yet experienced in handling molluscan chromosomes.

Chromosome squash technique using ovotestis tissue

There are 2 important initial considerations for achieving success in preparing chromosomes for cytological study: there must be active cell division occurring in the reproductive tissue and the tissue must be suitably fixed for cytological study.

Fixation. The fixative which we now routinely use is a solution of 3 parts by volume 95% ethanol (or methanol) and 1 part glacial acetic acid, both reagent grade. It is preferable to remove the ovotestis prior to fixation. However, whole animals may be placed directly into the fixative. If the latter method is chosen, then it is best to remove the shell or at least crack or pierce it in the apical region prior to fixation in order to assure rapid penetration of the fixative into the gonadal tissue. At least 10:1 parts by volume of fixative to tissue should be used. If the tissue is to be stored for future use, then the fixative should be changed after 24 hours. Tissue fixed in this way may keep for as long as a year under refrigeration, or for a shorter time if not refrigerated.

Stain. Several suitable chromatin stains are available. Acetic-orcein has been most commonly used, and is simple to prepare. The stain is made by placing 2·2 g of synthetic orcein in 100 ml glacial acetic acid and boiling gently for 10 min, stirring frequently. The mixture is then cooled, and filtered to remove undissolved stain. The undiluted stain seems to work best. However, the stain will work if diluted to 45%, but it may take longer to fully stain the chromatin. This stain will give excellent results for several years and, in fact, we prefer to use aged stain. Subsequent filtering may be necessary. It is most convenient to keep some stain in a dropper bottle.

Supplies. The following supplies are suggested: 2 fine dissecting forceps (Watchmaker's forceps), spear-head dissecting needles, clean slides and No. 1 or 1·5 coverslips, a coverslip sealant for temporary mounts (i.e. 1 part gum mastic:1 part paraffin, nail polish, dental wax, petroleum jelly, etc.), absorbent paper and 45% acetic acid in a dropper bottle.

Squash procedure. A small piece of ovotestis tissue (c. 1 mm^3) is placed in a tiny vessel or on a depression slide, to which is added 1 or 2 drops of stain. The tissue is allowed to stain for at least 15 min. After staining, the tissue is moved to a clean slide and divided into about 6 pieces of equal size (using spear-head needles). The tissue should never be allowed to dry, and should be handled rapidly so as to minimize drying of the stain on the slide. One to two drops of 45% acetic acid is then added to rinse off excess stain. This procedure is repeated as necessary. After rinsing, the pieces of stained tissue are moved equidistantly apart, a drop of 45% acetic acid added, and then covered with a coverslip, taking care to avoid air bubbles around the tissue. The slide is placed between absorbent paper (a bibulous paper packet is most convenient) and squashed with thumb pressure, being careful not to let the coverslip slip on the slide. Finally, a temporary sealant is applied to the edges of the coverslip, or the coverslip may be attached with a permanent mounting medium (see below).

Alternatively, the tissue may be placed directly on the microscope slide, cut into small pieces, stained for about 5 min, rinsed with 45% acetic acid and squashed as above. In this case, as the stain dries on the slide it must be continuously wiped off with a piece of absorbent paper.

Any foreign particulate matter on the slide or coverslip may prevent a monolayer squash or may entrap air. This is especially true of precipitated or dried stain and fibrous material. Pressure must be exerted to force the chromosomes of a cell to spread apart adequately for counting and to produce a monolayer of cells. Smaller pieces of tissue squash most easily.

Each investigator should be intimately familiar with the morphological appearance of chromosomes in the various stages of both mitotic and meiotic cell division (see Patterson and Burch, 1965). Good quality optical equipment must be used, a green filter is often helpful in viewing the chromosomes which are stained a reddish colour. Chromosome counts should be made from as many cells as possible, in order to be certain of the exact chromosome number.

Some investigators have injected animals with a chemical spindle inhibitor prior to fixation. Chromosomes are associated with microtubules (spindle fibres) during cell division. Because of this association, it is often difficult to achieve squash preparations that do not have a high degree of chromosome overlap, especially in species characterized by high chromosome numbers. The 3 most commonly used inhibitors are colchicine, colcemide and vinblastine phosphate (Velban). Stern (1974) injected living specimens of the land snail *Euglandina rosea* near the apical whorl with 0·15 ml of Velban having a concentration of 1 mg ml^{-1}. The ovotestis was removed 16 hours after inhibitor treatment and chromosome preparations were made using the acetic-orcein squash technique. Babrakzai and Miller (1974) injected 0·1 ml per gram body weight of 10^{-3} M colchicine into living *Ashmunella*. After 6–8 hours, the ovotestes were removed to a hypotonic 0·65% sodium citrate solution containing 10^{-4} M colchicine for 12–15 min before fixation. Chromosome squashes were prepared using the squash technique and several different stains.

Chromosomes for karyotypes. Karyotypes can be determined from studying mitotic metaphase chromosomes. In most reproductively active snails gonial cells are in mitosis for relatively short time periods. Thus, very rarely are a large number of mitotic cells found. In an effort to obtain a good sampling of cells for karyotype analysis, Burch (1965) developed a tissue culture technique for land snails. The method involves placing ovotestes in a culture medium suitable for cell growth and adding a spindle inhibitor such as colchicine to arrest cell division at metaphase. Acetic-orcein chromosome squash preparations are made after 42–48 hours of culture. If the snails are laying eggs, embryos are also an excellent source of a large number of mitotic cells. Techniques for obtaining mitotic metaphase chromosomes from embryos which are suitable for karyotype analyses have been published independently by Claugher (1971) and Patterson (1971), both using embryos of the freshwater pulmonate, *Bulinus*. Both methods utilize a spindle inhibitor, a hypotonic treatment and the acetic-orcein squash technique.

Making slides permanent. If it is desirable to make temporary chromosome preparations permanent, this may be accomplished by first removing the temporary sealant with a razor blade and then removing the coverslip by soaking the slide in a 1:1 mixture of 45% acetic acid and 95% ethanol. The coverslip can also be removed, usually more easily than by soaking, by freezing the slide on a block of dry ice and raising the coverslip by gently prying under one end with a single-edged razor blade. After being separated, the slides and coverslips are dehydrated in 95% and 100% ethanol and the same coverslip reapplied with a suitable mounting medium (i.e. Permount or Euparal). If balsam is used as the mounting medium, the slides must be passed through xylene before mounting the coverslip.

Coverslips may be wiped with a silicon paper prior to making the squash preparation—this will help keep tissue from adhering to the coverslip when it is removed for making permanent mounts.

VIII. Acknowledgements

We wish to thank Anne Gismann for critically reading the manuscript, and to acknowledge the assistance of the Smithsonian Institution project (FR6-50008) at Ain Shams University, Cairo, for help in preparing the manuscript.

IX. References

Babrakzai, N. and Miller, W. B. (1974). *Malac. Rev.* **7,** 37–38.

Babrakzai, N., Miller, W. B. and Ward, O. G. (1974). *Bull. Am. Malac. Union* **1974,** 4–11.

*Baker, H. B. (1955). *Nautilus, Philad.* **68,** 109–112.

*Baker, H. B. (1961). *Nautilus, Philad.* **75,** 33–36.

* References not dealing with chromosomes.

*Baker, H. B. (1962). *Nautilus, Philad.* **75,** 116–121.
Baltzer, F. (1913). *Arch. Zellforsch.* **11,** 151–169.
Beeson, G. E. (1960). *Nature, Lond.* **108,** 1–56.
*Boettger, C. R. (1955). *Verh. d. zool. Ges.* **1954,** 253–280.
*Bondeson, P. (1950). *Nature Jutl.* **3,** 1–208.
Bonham, K. (1955). *Growth* **19,** 9–18.
Brown, D. S. (1976). *J. nat. Hist.* **10,** 257–267.
Brown, D. S. and Burch, J. B. (1967). *Malacologia* **6,** 189–198.
Brown, D. S. and Wright, C. A. (1972). *J. Zool., Lond.* **167,** 97–132.
Brown, D. S., Schutte, C. H. J., Burch, J. B. and Natarajan, R. (1967). *Malacologia* **6,** 175–188.
Brown, D. S., Oberholzer, G. and van Eeden, J. A. (1971). *Malacologia* **11,** 141–170.
Burch, J. B. (1959). *Dissert. Abst.* **20,** 1487–1488.
Burch, J. B. (1960a). *Nucleus, Calcutta* **3,** 177–208.
Burch, J. B. (1960b). *Nature, Lond.* **186,** 497–498.
Burch, J. B. (1960c). *Z. Tropenmed. Parasitol.* **11,** 449–452.
Burch, J. B. (1960d). *Trans. Am. microsc. Soc.* **79,** 451–461.
Burch, J. B. (1961a). *Amer. malac. Union ann. Reps.* **28,** 15–16.
Burch, J. B. (1961b). *Basteria* **25,** 45–52.
Burch, J. B. (1962). *Malacologia* **1,** 55–72.
Burch, J. B. (1963). *Am. malac. Union. ann. Reps* **30,** 15–16.
Burch, J. B. (1964a). *Malacologia* **1,** 387–400.
Burch, J. B. (1964b). *Occ. Paps. Mus. Zool. Univ. Michigan* **638,** 1–8.
Burch, J. B. (1964c). *Acta biol. Acad. Sci. Hung.* **15,** 87–94.
Burch, J. B. (1965a). *Malacologia* **2,** 253–258.
Burch, J. B. (1965b). *Proc. first Europ. malac. Congr.* **1962,** 215–241.
Burch, J. B. (1966). *Malacologia* **5,** 25–27.
Burch, J. B. (1967a). *Venus* **25,** 118–135.
Burch, J. B. (1967b). *Malacologia* **5,** 127–135.
Burch, J. B. (1967c). *Ethiopian Med. J.* **5,** 245–257.
Burch, J. B. (1967d). *Ann. Rep. Proc. Papua N. Guinea Sci. Soc.* **1967,** 26–28.
Burch, J. B. (1967e). *J. Conchyliol.* **107,** 3–51.
Burch, J. B. (1968a). *Venus, Kyoto* **27,** 20–27.
*Burch, J. B. (1968b). *Malac. Rev.* **1,** 15–30.
*Burch, J. B. (1968c). *J. Malac. Soc. Austr.* **11,** 62–67.
Burch, J. B. (1969a). *Proc. Symp. Moll. Mar. biol. Ass. India Pt.* **2,** 616–625.
Burch, J. B. (1969b). *Malac. Rev.* **2,** 113–114.
Burch, J. B. (1972). *Malac. Rev.* **5,** 7–8.
Burch, J. B. (1973). *Malac. Rev.* **6,** 43.
*Burch, J. B. (1974). *Malac. Rev.* **7,** 127–132.
Burch, J. B., Basch, P. R. and Bush, L. L. (1960). *Rev. portug. Zool. Biol. Ger.* **2,** 199–204.
Burch, J. B. and Bush, L. L. (1960). *J. Conchyliol.* **100,** 49–54.
*Burch, J. B. and Cuadros, C. (1965). *Nature, Lond.* **206,** 637–638.
Burch, J. B. and Habe, T. (1969). *Bull. Nat. Sci. Mus., Tokyo* **12,** 583–591.
Burch, J. B. and Heard, W. H. (1962). *Acta. biol. Acad. Sci. Hung.* **12,** 305–312.

Burch, J. B. and Huber, J. M. (1966). *Malacologia* **5,** 41–43.
Burch, J. B. and Kim, D. C. (1962). *Bull. natn. Inst. Hlth, Korea* **5,** 181–186.
Burch, J. B. and Lindsay, G. K. (1966). *Am. malac. Union ann. Reps* **33,** 37–38.
Burch, J. B. and Lindsay, G. K. (1969). *Malacologia* **9,** 37–38.
Burch, J. B. and Lindsay, G. K. (1971). *Zool. Afr.* **6,** 39–44.
Burch, J. B. and Natarajan, R. (1965). *Bull. Inst. Zool. Acad. Sinica* **4,** 11–17.
Burch, J. B. and Natarajan, R. (1966). *Am. malac. Union. ann. Reps* **33,** 39–40.
Burch, J. B. and Natarajan, R. (1967). *Pacific Sci.* **21,** 252–259.
Burch, J. B. and Patterson, C. M. (1963). *Malacologia* **1,** 313–319.
Burch, J. B. and Patterson, C. M. (1965). *Amer. biol. Teacher* **27,** 203–207.
*Burch, J. B. and Patterson, C. M. (1969). *Malacologia* **9,** 259–260.
Burch, J. B. and Patterson, C. M. (1971a). *Malac. Rev.* **4,** 25–27.
Burch, J. B. and Patterson, C. M. (1971b). *Malac. Rev.* **4,** 209–210.
Burch, J. B., Williams, J. E., Hishinuma, Y. and Natarajan, R. (1964). *Malacologia* **1,** 403–415.
Burch, J. B., Patterson, C. M. and Natarajan, R. (1966). *Venus* **24,** 342–353.
Butot, L. J. M. and Kiauta, B. (1964). *Corresp. Ned. malac. Ver.* **111,** 1163–1164.
Butot, L. J. M. and Kiauta, B. (1967). *Beaufortia, Amsterdam* **14,** 157–164.
Butot, L. J. M. and Kiauta, B. (1970). *Chromosome Inf. Serv.* **11,** 9–11.
Claugher, D. (1971). *Bull. Wld Hlth Org.* **45,** 855–858.
Evans, H. J. (1960). *Heredity* **15,** 129–138.
Ford, J. M. (1962). "The Chromosomes of Northwest Pulmonate Snails". Ph.D. thesis, Oregon State University.
Franzen, D. S. (1966). *Nautilus, Philad.* **80,** 59–69.
Franzen, D. S. (1971). *Nautilus, Philad.* **84,** 131–142.
Franzen, D. S. (1973). *Nautilus, Philad.* **87,** 66–87.
*Fretter, V. (1943). *J. mar. biol. Assoc. U.K.* **25,** 685–720.
*Hoffmann, H. (1925). *Z. Naturwiss, Jena* **61,** 1–374.
*Hoffmann, H. (1929). *Zool. Jahrb.* (*Syst.*) **57,** 253–302.
*Hubendick, B. (1951). *Kungl. svenska Vet.-akad. Handl.* (4) **3,** 1–223.
*Hubendick, B. (1962). *Göteborgs Kungl. Vetensk. Samh. Vit. Handl.* (B) **9,** 1–68.
*Hubendick, B. (1964). *Göteborgs Kungl. Vetensk. Samh. Vit. Handl.* (B)**9,** 1–72.
Husted, L. and Burch, P. R. (1946). *Amer. Nat.* **80,** 410–429.
Husted, L. and Burch, P. R. (1953). *Va. J. Sci.* **4,** 62–64.
*Hyman, L. (1967). "The Invertebrates". McGraw-Hill, New York and London.
Inaba, A. (1945). *Venus, Kyoto* **14,** 97–102.
Inaba, A. (1953). *J. Sci. Hiroshima Univ.* **14,** 221–228.
Inaba, A. (1959). *J. Sci. Hiroshima Univ.* **18,** 71–93.
Inaba, A. (1961a). *Dobutsugaki Zasshi* **70,** 24.
Inaba, A. (1961b). *Venus, Kyoto* **21,** 402–413.
Inaba, A. (1962). *Dobutsugaki Zasshi* **71,** 397.
Inaba, A. (1965). *Venus, Kyoto* **23,** 223–228.
Inaba, A. (1969). *Malacologia* **7,** 143–168.
Jacob, J. (1959). *Cytologia* **24,** 427–497.
Kawabe, M. (1947). *Kromosomo* **3–4,** 133–134.
Koyama, T. (1955). *Venus, Kyoto* **18,** 259–265.

La Cour, L. (1941). *Stain Techn.* **16,** 169–174.
Laws, H. M. (1965). *Rec. S. Aust. Mus.* **15,** 79–87.
Laws, H. M. (1966). *Rec. S. Aust. Mus.* **15,** 257–260.
Laws, H. M. (1967). *Rec. S. Aust. Mus.* **15,** 495–496.
Laws, H. M. (1971). *Malacologia* **11,** 217–224.
Laws, H. M. (1973a). *Cytologia* **38,** 229–235.
Laws, H. M. (1973b). *J. malac. Soc. Aust.* **2,** 389–392.
Le Calvez, J. and Certain, P. (1950). *C. R. Acad. Sci. Paris* **231,** 794–795.
*Lo, C. T. (1972). *Malacologia* **11,** 225–220.
Lo, C. T. and Chang, K. M. (1975). *Bull. Chin. Malac. Soc.* **2,** 95–101.
*Mandahl-Barth, G. (1965). *Bull. Wld Hlth Org.* **33,** 33–44.
*Mayr, E. (1963). "Animal Species and Evolution". Belknap Press, Harvard University Press.
Meyer, K. O. (1955). *Arch Molluskenk.* **84,** 1–43.
*Mörch, O. A. L. (1865). *J. Conch., Paris* **13,** 262–283.
*Morton, J. E. (1955). *Proc. zool. Soc. Lond.* **125,** 127–162.
Muller, H. J. (1925). *Am. Nat.* **59,** 346–353.
Natarajan, R. (1958a). *J. zool. Soc. India* **10,** 103–107.
Natarajan, R. (1958b). *Curr. Sci.* **27,** 311–312.
Natarajan, R. (1959). *J. zool. Soc. India* **11,** 30–33.
Natarajan, R. (1960). *J. zool. Soc. India* **12,** 69–79.
Natarajan, R. (1965). *Am. malac. Union ann. Reps.* **32,** 15–16.
Natarajan, R. and Burch, J. B. (1966). *Cytologia* **31,** 109–116.
Natarajan, R., Burch, J. B. and Gismann, A. (1965). *Malacologia* **2,** 239–251.
Natarajan, R., Hubricht, L. and Burch, J. B. (1966). *Acta biol. hung.* **17,** 105–120.
Nishikawa, S. (1967). *J. Shimonoseki Univ. Fish.* **16,** 17–21.
Olazarri, J. (1973). *Neotropica* **19,** 50–52.
Patterson, C. M. (1964). *Am. malac. Union ann. Reps.* **31,** 21–23.
Patterson, C. M. (1967). *Am. malac. Union ann. Reps.* **34,** 31–32.
Patterson, C. M. (1968a). *Proc. Symp. Moll. Mar. biol. Ass. India* **1,** 46–50.
Patterson, C. M. (1968b). *Malac. Rev.* **1,** 1–13.
Patterson, C. M. (1969). *Proc. Symp. Moll., Mar. biol. Ass. India* **2,** 635–686.
Patterson, C. M. (1970). *Malac. Rev.* **3,** 25–36.
Patterson, C. M. (1971a). *Malac. Rev.* **4,** 27.
Patterson, C. M. (1971b). *Malac. Rev.* **4,** 131–202.
Patterson, C. M. (1973). *Malac. Rev.* **6,** 141–150.
Patterson, C. M. and Burch, J. B. (1966). *Malacologia* **3,** 309–325.
*Pelseneer, P. (1901). *Mém. Acad. R. Belg.* **54,** 1–76.
*Pilsbry, H. A. (1900a). *Proc. Acad. nat. Sci. Philad.* **1900,** 561–567.
*Pilsbry, H. A. (1900b). *Proc. Acad. nat. Sci. Philad.* **1900,** 568–581.
*Pilsbry, H. A. (1940). *Acad. nat. Sci. Philad., Monogr.* **1,** 575–994.
*Pilsbry, H. A. (1946). *Acad. nat. Sci. Philad., Monogr.* **2,** 1–520.
*Pilsbry, H. A. (1948). *Acad. nat. Sci. Philad., Monogr.* **2,** 521–1113.
*Plate, L. H. (1893). *Zool. Jb. Anat.* **7,** 93–234.
Perrot, J. L. (1930). *Rev. suisse Zool.* **37,** 397–434.
Perrot, J. L. (1934). *Rev. suisse Zool.* **41,** 693–697.
Perrot, J. L. and Perrot, M. (1938). *C.R. Séanc. Soc. Phys. Hist. nat. Genéve* **53,** 98–93.

Perrot, M. (1938a). *Rev. suisse Zool.* **45,** 487–566.

Perrot, M. (1938b). *C.R. Séanc. Soc. Phys. Hist. nat. Genève* **45,** 90–92.

Rainer, M. (1967). *Malacologia* **5,** 341–373.

Scvortzoff, E. (1966). "Chromosome numbers of the land snails of the genus *Partula* that inhabit the island of Moorea". Masters Thesis, University of Virginia.

*Solem, A. (1959). *Fieldiana Zool.* **43,** 1–359.

Stern, E. M. (1974). *Nautilus, Philad.* **88,** 29–30.

Taylor, D. W. and Sohl, N. F. (1962). *Malacologia* **1,** 7–32.

Taylor, D. W., Walter, H. J. and Burch, J. B. (1963). *Malacologia* **1,** 7–32.

Thaler, E. (1963). *Ost. Akad. Wissensch.* **24,** 1–19.

*Van Mol, J.-J. (1967). *Mém. Acad. r. Belg. Cl. Sci.* **8°** **37,** 1–162.

*White, M. J. D. (1954). "Animal Cytology and Evolution" 2nd edition. Cambridge University Press.

White, M. J. D. (1973). "Animal Cytology and Evolution" 3rd edition. Cambridge University Press.

Wu, S. K. (1972). *Malac. Rev.* **5,** 95–164.

Wu, S. K. (1973) *Malac. Rev.* **6,** 203.

Wu, S. K. and Burch, J. B. (1973). *Malac. Rev.* **6,** 62–63.

*Zilch, A. (1959–1960). "Gastropoda, Vol. 2, Euthyneura". *In* Schindewolf, *Handl. Paläozool.* **6,** 1–834. Bornträger, Berlin.

Chapter 5

Genetic variation and natural selection in pulmonate molluscs

B. CLARKE, W. ARTHUR, D. T. HORSLEY and D. T. PARKIN

Department of Genetics, University of Nottingham, UK

I. Introduction

The pulmonate molluscs have several advantages for studies of microevolutionary genetics. They are often diverse within and between populations; they are slow-moving and easy to catch; they are generally tolerant of laboratory conditions, and they are easily marked. These advantages have led to a widespread interest in the pulmonates among practitioners of ecological and population genetics, and have generated a correspondingly extensive literature. Because the literature is scattered, it seems worthwhile to attempt a review.

In this chapter we propose successively to describe examples of the genetic variation found within populations of pulmonates, to examine the evidence that this variation is subject to natural selection, and to enquire about the factors that may act to maintain the variation in a state of dynamic equilibrium. We will then review the variation between populations, and assess the contribution of studies on pulmonates to our knowledge of the process of speciation. We will not consider variation above the level of the species.

The message of this chapter is a cautionary one. The great advantages of pulmonates for genetic experiment and observation have carried with them some corresponding hazards and pitfalls. The literature is encumbered with many potentially important works that have foundered because of too little regard for these dangers. As we shall see, the greatest problem has been the difficulty of adequate sampling from natural populations.

II. Variation within populations

The existence of variation within a population implies that the individuals differ in some measurable way. The differences may be continuous, in the sense that individuals can manifest any value within a range of measurement (as, for example, variations in shell-length or body-weight). Alternatively, the differences may be discontinuous, in the sense that each individual can only manifest one of a finite set of discrete values (as, for example, some variations in shell pattern or in the electrophoretic mobility of proteins). Statistically, the composition of a population with respect to continuous variables can usually be summarized by the mean and variance of the measurement. With respect to discontinuous variables, on the other hand, it must be described by the frequencies of the various types. Although the 2 categories of variation are not strictly exclusive, it is useful to treat them as if they were, because different methods of analysis are appropriate to each.

A. Continuous variation

All organisms show continuous variation within populations, and there are thousands of documented cases among the pulmonates. The early literature has been admirably summarized by Diver (1939). Later studies, among many others, are those of Mayr and Rosen (1956) on *Cerion*, Sacchi (1957) on *Helicella*, and Cook and O'Donald (1971) on *Cepaea*, which deal with metric variations in shell form. Continuous variation in other characters has been observed by Aubertin (1927, genital variation in *Cepaea*), Oldham (1934, shell-weight in *Arianta*), Diver and Anderson-Kottö (1938, mixed broodedness in *Lymnaea*), Wolda (1965, 1967, egg-production in *Cepaea*) and Richards (1973, tendency to aestivation, susceptibility to *Schistosoma*, and pigmentation in *Biomphalaria*).

With a few honourable exceptions, most of the studies on continuous variation have involved no attempt to determine the inheritance (or otherwise) of the characters under study. Consequently for the student of evolution they have little value.

Work on the inheritance of continuously-variable characters often lacks the precision of that dealing with discrete characters because of the relatively larger number of genes involved, and because there are usually direct effects of the environment. There are particular problems in experiments with pulmonates, because they have inconveniently long life cycles. Investigations have therefore usually taken the form of estimating the degree of genetic control over the variation, rather than attempting to locate the genes concerned.

A second level of imprecision has been introduced because virtually all studies have failed to distinguish the inheritance of differences within populations from the inheritance of differences between them. The theory of quantitative genetics gives us no encouragement that the genetics of the two classes of variation will necessarily be the same. Nevertheless, there is, as far as we know, no single study that has attempted to compare the results of intra- and inter-population crosses. Lacking such a comparison we must proceed with the uneasy assumption that the results from studies of variation between populations of pulmonates can be applied to variation within them.

Two types of approach have been made to the problem of detecting genetic elements in continuous variation. Some workers have transferred differing individuals or populations to the laboratory, and have observed whether or not their progeny converge to a common form. If they do not, the observation is presumed to indicate a genetic component in the variation. This presumption may be incorrect if cytoplasmically transmitted modifications (such as the passage of infectious microorganisms) are involved. The second approach has been to set up laboratory crosses, and to estimate heritability, usually from a regression of the mean value of the offspring on the mean value of the 2 parents. The advantage of this method is that it gives a numerical estimate of the (additive) genetic component of the variation, and allows us to state precisely the errors involved. Furthermore, other things being equal, the heritability is a good predictor of short-term responses to selection, and is therefore potentially useful to the student of microevolution. The major disadvantage of the method is that it tends to overestimate the natural heritability because laboratory environments are almost certainly more homogeneous than natural ones, and will tend therefore to minimize the environmental component of the variation. On the other hand, the heritability itself underestimates the total extent of genetic determination because it does not measure the components of variation due to dominance or to interaction between genes.

"Laboratory transplant" studies have been undertaken with *Lymnaea peregra* (Boycott, 1938), *Lymnaea stagnalis* (Roszkowski, 1914; Piaget, 1928; Arthur, 1973), *Lymnaea palustris* (Forbes and Crampton, 1942) and *Limicolaria martensiana* (Owen, 1965). They have produced diverse results.

Of the cases in which the variation appeared to be entirely environmental, the work by Owen (1965) on shell-size differences between populations of *L. martensiana* is worthy of mention. A survey of wild populations showed that those living on calcareous volcanic ash had larger and heavier shells than those living on ferralitic soils. Owen took a sample of newly-hatched snails from a population on volcanic ash and grew half the sample in a container with ferralitic soil, and the other half in a container with volcanic ash. After 6 months the latter group had very much larger and heavier shells than the former, indicating a strong environmental component in the variation. Unfortunately, the control experiment, using a sample from ferralitic soils, was not carried out. Owen's results resemble those of Oldham (1934) who found that a high level of calcium in the diet slightly increased the shell size of *Arianta arbustorum* and greatly increased the shell-weight (4- or 5-fold, on the average).

Another case in which the environment seems to play a primary role is that of the variation between some northern Irish populations of *L. stagnalis* studied by Arthur (1973). Young snails (about 5 mm in length) were grown in the laboratory in water taken from the environment of a second population. Within 2 months they became significantly different in shape (shell-length/aperture-length) from their parents, the mean value shifting so as to coincide almost exactly with the mean of the population whose environment had been simulated in the laboratory.

Roszkowski's (1914) work, also on the shell-shape of *L. stagnalis*, is important because snails transferred to the laboratory as eggs were modified in a manner similar to those in Arthur's experiments. However, those taken after hatching remained immutable throughout their life-span. This emphasizes the care that is necessary in concluding that variation is genetic in origin; a short contact with a particular environment may have a lasting effect.

A number of studies have found characters that remain constant through several generations of laboratory culture. Cook and Peake (1960) quote Stelfox's observation that the large-shelled western Irish race of *Cepaea nemoralis* breeds true in the laboratory. Forbes and Crampton (1942) report that differences between populations of *L. palustris* in longevity, and in various parameters of reproduction, persisted after several generations in the laboratory. The persistence of differences in shell-size was less clear.

One of the more thorough studies of this type is reported by Boycott (1938), who worked with the varieties of *L. peregra* known as var. *involuta*, *burnetti*, *praetenuis* and *lacustris*, all of which have shell shapes very different from the normal type. In the laboratory the offspring of var. *praetenuis* and *lacustris* reverted to normality in a single generation. On the other hand the F_5 generation of *burnetti* was still identical to the parent populations. In the case of var. *involuta*, the offspring of populations from some Irish lakes were still distinct after 5 generations, but the offspring of populations from other lakes reverted to normality in the first generation. There appears to be a genetic fixation in some populations of characters that in others are direct responses to the environment. However, the role of cytoplasmically transmitted factors has not been excluded. A further note of caution is introduced by the work of Wolda (1972). He established a series of experimental colonies of *C. nemoralis* in fenced outdoor enclosures, and found changes of shell-size in successive years that were probably due to natural selection acting upon genetic variation in juvenile growth-rates. The role of selection has not been formally excluded in the majority of transplant studies, and differences in the responses of populations to laboratory conditions might be the result of differences in the amount of genetic variance available for selection. Although laboratory transplant experiments may be misleading, they do illustrate the dangers of extrapolating the results obtained from one population to the variation observed in another.

There are few cases in which specific crosses have been set up to test the inheritance of continuously-varying characters. Richards has used a combination of crossing and selection on *Biomphalaria glabrata* to demonstrate genetic elements in the control of apertural lamellae associated with a tendency to aestivation (Richards, 1968), tentacle and eye malformations (Richards, 1969a), mantle pigmentation (Richards, 1969b), pulmonary occlusion, the presence of head bulb, shell scalariformity, polyembryony and haemolymph pigment (Richards, 1970, 1971) and juvenile susceptibility to *Schistosoma mansoni* (Richards, 1970; Richards and Merritt, 1972). Wolda (1969) has demonstrated that the fusion of adjacent bands in *C. nemoralis* is subject to multifactorial control, as are the finer shades of body-colour. He also found evidence for a genetic component in the variation of growth rate in the same species (Wolda, 1970).

The only studies in which heritabilities have been estimated are those of Cook (1965, 1967) on shell-size in *Arianta* and *Cepaea* and those of Murray and Clarke (1968a) on shell-size in *Partula*.

In Cook's (1965) work on *A. arbustorum* 3 (non-independent) parameters—shell-breadth, shell-height and width of aperture—were measured in 119 families. In 43 families the number of whorls was also measured. The regressions of offspring on midparent values were: breadth = 0·70; height = 0·54; aperture width = 0·63; and number of whorls = 0·53.

The regression coefficients are direct estimates of heritability (Falconer, 1960), so that approximately 60% of the variation in these characters can be attributed to additive genetic effects. In a smaller-scale experiment on shell-size in *C. nemoralis*, Cook (1967) also found a heritability of about 60%.

Murray and Clarke (1968a) have carried out a similar study on *Partula suturalis* and *P. taeniata* and have found the following values: *P. suturalis*—length = 0·81 ± 0·42, width = (0·53 ± 0·13); *P. taeniata*—length = 0·36 ± 0·17, width = 0·40 ± 0·14.

In one case (shell-width in *P. suturalis*) the values for the F_1 and F_2 generations were significantly different, so that combining the two for an average, bracketed above, is not strictly valid.

Although he did not estimate heritability, the data of Wolda (1969) on the inheritance of variations in the shade of the body among dark-bodied *C. nemoralis* suggest a value of about 40%.

Summarizing the work on continuous variation within populations of pulmonates we are obliged to say that, while its existence cannot be disputed, and while both genetic and environmental factors contribute to it, the relative importance of these factors in particular cases has rarely been investigated. The studies of heritability mentioned above have used methods that tell us more about the inheritance of variation between populations than about variation within them.

B. Discontinuous variation

When 2 or more discrete varieties coexist within a single population, the population is said to exhibit polymorphism. For varieties that are genetically determined, Ford (1945) has restricted the term to cases in which the rarer variety has a frequency too great to be the result of repeated mutation balanced by a selective disadvantage of the mutant form. Since, in practice, it is difficult or impossible to determine what that frequency may be, the term is now conventionally restricted to the cases in which the commonest variety has a frequency less than 99%.

The most striking polymorphisms among the pulmonates are those involving the colour and banding patterns of the shell and body. However, polymorphisms in shell-coiling, in the electrophoretic mobility of proteins, and in the numbers of chromosomes have also been observed.

1. *Polymorphisms in external colouration*

Some groups of pulmonates have large numbers of species that are polymorphic for colour and pattern, others have few. For example, this kind of polymorphism appears to be rare among the Basommatophora but common among the Stylommatophora. Within the latter order it seems to be particularly widespread among the Achatinellidae, Partulidae, Achatinidae and Helicidae, and relatively rare among the Vallonidae, Pupillidae, Succineidae, Zonitidae and Polygyridae. There also appear to be regional differences in the incidence of polymorphism. About 30% of the British land pulmonates are polymorphic for colour and pattern. Among North American species the proportion is much lower (less than 10%). Some island faunas, such as those of Polynesia and the Caribbean, seem to be particularly rich in polymorphic species; others, such as that of the Galapagos, seem to be relatively poor.

It is sad to relate that we cannot yet provide any clear explanation for these apparent regularities, indeed we cannot argue with more than moderate conviction that the regularities exist at all. The uncertainty is largely due to the habitat-descriptions and nomenclature of the specimens enshrined (or entombed) in our museums. Until recently, too many museum taxonomists have been describing and cataloguing haphazardly-collected single specimens, and calling them "types". The value of random sampling, the concept of polymorphism, even the nature of biological species, have been too little appreciated. Gould (1969), in an attempt to raise the taxonomy of the zonitid *Poecilozonites* from a morass of complex verbiage, points out that the major sin has been the naming of mere shapes and colours without regard to the question whether they co-exist in a single biological population. It is often found, for example, that a "subspecies" *alba* refers to pale-coloured shells otherwise resembling the "type", suggesting that these forms are merely morphs from a single population. The overuse of the trinomial (as, for example, by Welch, 1954) without information about the genetics or small-scale distribution of the variants concerned, makes our task as reviewers more difficult. Only a small part of the literature deals with adequately-documented examples of genetic polymorphism (*sensu* Ford, 1945). It is surprising that such an important component of natural variation should have received so little attention from museum taxonomists, and it is clear that there is a great potential for studies in "comparative polymorphology", and for attempts to define the rules that determine whether a particular species or group should have a high or low incidence of polymorphism.

In the absence of any such systematic and objective studies, it may be worthwhile to offer a few unsystematic and subjective propositions that might serve as starting points for future research. These propositions, which certainly cannot be dignified by the title of rules, are based upon our common experience of pulmonate variation. They are listed below as a series of bold statements, but must be regarded as very tentative and uncertain:

(i) aquatic species are less likely to be polymorphic for colour and pattern than terrestrial ones;

(ii) terrestrial species living in colder and drier environments are less likely to be polymorphic for colour and pattern than those living in warmer and wetter environments;

(iii) species living always on the ground are less likely to be polymorphic for colour and pattern than those that periodically, or permanently, climb above it;

(iv) species inhabiting dense woodland are less likely to be polymorphic for colour and pattern than those inhabiting more open places;

(v) species living on continental land-masses are less likely to be polymorphic for colour and pattern than those living in archipelagos;

(vi) species whose major predators hunt by smell are less likely to be polymorphic for colour and pattern than those whose major predators hunt by sight;

(vii) very small and very large species are less likely to be polymorphic for colour and pattern than those of intermediate size;

(viii) species whose average density is relatively low are less likely to be polymorphic for colour and pattern than those whose average density is relatively high;

(ix) inbreeding species are less likely to be polymorphic than outbreeding ones.

Several of the variables mentioned in these propositions are likely to be correlated with each other, so that groups of propositions may stand or fall together.

Colour and pattern polymorphisms among the pulmonates seem to be of 3 general types: (i) polymorphism of the mantle or body, (ii) polymorphism of shell colour, either a series of different hues (as in *Cepaea*) or the simple presence and absence of a particular pigment (as in *Arianta*), and (iii) polymorphism of banding or striation on the shell.

The colours of land pulmonates tend to be in the range of white, yellow, brown, red and orange (all of which, excepting white, have the property of absorbing ultraviolet light). A very few arboreal species (in the genera *Achatinella*, *Cochlostyla* and *Papuina*) have green shells. Those of aquatic species tend to be greenish-brown. What little is known about the nature and composition of pulmonate pigments has been reviewed by Comfort (1951a, b).

The commonest pattern of shell-banding among the pulmonates seems to be the single-banded condition, in which a single dark brown or black band runs along the centre of the whorl. Within the Helicidae the banding patterns appear to be variations on the pentataeniate theme (5 bands running parallel along the whorls). These bands are conventionally numbered 1 to 5 from the top of the whorl downwards, and the banding pattern is represented by the numbers of the bands that are present, with zeros representing the bands that are absent. Thus a 5-banded shell is designated 12345, and one with only the central band is designated 00300.

Table I briefly describes some of the better-known colour and pattern polymorphisms among pulmonate families other than the Helicidae, which are treated in Table II. Data on the genetics, where it has been

Table I. Colour polymorphism in the non-helicid pulmonates.

Family	Description	Source
Achatinellidae	*Achatinella mustelina* Mighels, *A. apexfulva* (Dixon) and *A. bulimoides* Swainson, very variable shell colour and banding. Genetics not known	Welch (1938, 1954, 1958)
Partulidae	*Partula taeniata* Mörch, an allelic shell-colour series grading from purplish brown to white, darker colours usually dominant to lighter. Three banding patterns segregate with, and are dominant to, unbanded. The colour and banding loci are linked	Murray and Clarke (1966, 1976a)
	Partula suturalis Pfeiffer, several shell patterns controlled by a series of at least 6 functional "alleles". Many other *Partula* species are polymorphic for shell-colour and banding	Murray and Clarke (1966, 1976b) Crampton (1916, 1925, 1932)
Clausiliidae	*Marpessa laminata* (Montagu), some populations polymorphic for shell-colour (dark brown v. pale). Genetics not known	Clarke (unpublished)
Cerionidae	*Cerion* spp, extensive variation in shell-colour, poorly understood	Mayr and Rosen (1956)

Table I—continued

Family	Description	Source
Achatinidae	*Limicolaria flammulata* (Pfeiffer), brown shell dominant to grey, streaked shell dominant to unstreaked. The loci are linked	Barker (1968, 1969)
	Limicolaria aurora Jay, pink shell dominant to grey, streaked shell dominant to unstreaked. The loci are linked	
	Limicolaria martensiana (Smith), polymorphic for shell-colour and streaking. Genetics not known	Owen (1963)
Endodontidae	*Discus rotundatus* (Müll.), polymorphic for "spotting" on the shell	Williamson (1957)
Bulimulidae	*Liguus fasciatus* (Müll.), highly polymorphic for shell-colour and banding. Pink spire segregates with white, and unbanded segregates with banded. Other species of *Liguus* are also highly polymorphic for shell-colour and banding. Genetics not known	Pilsbry (1912, 1945) Young (1960)
	Bulimulus guadalupensis Brug., polymorphic for shell-colour and banding. Genetics not known	Heatwole and Clarke (unpublished)
Arionidae	*Arion ater* (L.), black body-colour dominant to brown, banded body dominant to unbanded, fully pigmented body dominant to streaked, which is dominant to albino. Last 2 loci closely linked	Williamson (1959)
	Arion hortensis (Fér.), grey-blue body-colour dominant to pink, which is dominant to white	Abeloos (1945)
	Arion subfuscus Drap., dark mottled body-colour dominant to creamy buff	Getz (1962)
Zonitidae	*Poecilozonites bermudensis* (Pfeiffer) and *P. cupula* Gulick, polymorphic for presence and absence of shell-banding, and for flammulation. Genetics not known	Gould (1969)
Limacidae	*Lehmannia marginata* (Müll.), polymorphic for banding of body v. spotting, and for body colour. Genetics not known	Greenwood (1966)
Bradybaenidae	*Bradybaena similaris* (Fér.), brown shell dominant to yellow, single banded dominant to unbanded. The loci are linked	Komai and Emura (1955)

Table II. Colour polymorphism in the Helicidae.

Species	Description	Source
Helix aspersa Müll.	Red ground-colour of juvenile shell dominant to yellow; banded 12345 dominant to 10005; full pigment dominant to reduced pigment (*exalbida*). The third locus is epistatic to the first two	Cook (1969) and Cain (1971)
Polymita picta (Born)	Dramatically polymorphic for colour, banding and striation. Shell, bands and striations can be brown, red, white or black. Other species of *Polymita* also highly polymorphic, genetics not known	Torre y Huerta (1950)
Arianta arbustorum (L.)	Brown shell dominant to yellow, presence of single central band dominant to absence; mottled shell dominant to unmottled; pigmented band dominant to unpigmented; white opaque stripe dominant to absence of stripe. First two loci linked	Cook and King (1966)
Cepaea nemoralis (L.)	Highly polymorphic. An allelic shell-colour series from brown to pale yellow, darker colours dominant to paler; unbanded dominant to banded; punctate bands dominant to unmodified bands; "spread" bands dominant to unmodified; normal bands and lip dominant to hyalozonate bands and white lip. All these loci linked. Six other loci (2 controlling suppression of particular bands, and others controlling orange bands, darkening bands, colour of body, and quantity of dermal pigment) are known	Many authors, reviewed by Murray (1975)
Cepaea hortensis (Müll.)	Polymorphism very similar to that of *C. nemoralis*. Four linked loci known (controlling shell-colour, presence or absence of bands, pigmentation of bands and punctate bands)	Many authors reviewed by Murray (1975)
Cepaea vindobonensis (Fér.)	Less variable than *C. nemoralis* or *C. hortensis*, but still polymorphic for presence or absence of particular bands; intensity of band pigmentation, and band-fusion. Genetics not known	Jones (1971, 1973)
Cepaea sylvatica (Drap.)	Apparently very similar to *C. vindobonensis*. Genetics not known	Schilder and Schilder (1957)

Table II—continued

Species	Description	Source
Pseudotachea splendida (Drap.)	Polymorphic for shell-colour (pink v. yellow) and for presence and absence of particular bands. Genetics not known. *P. coquandi* has similar polymorphism	Sacchi (1956)
Monacha cantiana (Montagu)	Dark mantle-colour probably dominant to light	Cain (1971)
Hygromia striolata (C. Pfeiff.)	Dark mantle-colour (heavily spotted) segregates with light. Also polymorphic for shell-colour *H. hispida* has a similar polymorphism	Cain (1959)
Theba pisana (Müll.)	Complicated polymorphism for shell-banding, perhaps variation on a pentataeniate pattern. Genetics not known	Sacchi (1955a)
Helicella itala (L.)	Polymorphic for pigmentation of bands; dark, light or absent (hyalozonate), also for blotching of bands. Genetics not known	Lanyon (1966)
Cernuella virgata (Da Costa) [*Helicella virgata* (Da Costa)]	Polymorphic for pigmentation of bands; dark, light or absent (hyalozonate); also an unbanded (brown speckled) phenotype. Genetics not known	Clarke (unpublished)
Candidula intersecta (Poiret)	[*H. caperata* (Montagu)] and *Candidula gigaxii* (Pfeiff.) [*H. heripensis* (Pfeiff.)] have similar polymorphisms, but less clear-cut	
Cochlicella acuta (Müll.)	Amber shell-colour segregates with colourless; five-banded pattern (12345) dominant to single banded (00040), which is dominant to unbanded. The loci are linked	Lewis (1968; 1975)

studied, are also included in the tables. For a detailed discussion of molluscan genetics, we refer to the admirable review by Murray (1975).

Some generalities emerge from the tables. First, it seems that the alleles producing darker shells and bodies are dominant to those producing lighter ones (with the single exception of the *apex* allele in *P. suturalis*; see Murray and Clarke, 1976b). Second, the alleles controlling the presence of bands are usually dominant to those controlling the absence of bands. Exceptions are the alleles for the unbanded condition, and for the lack of individual bands, in *C. nemoralis* and *C. hortensis*. Even in these species, however, a lack of pigment in the bands (the hyalozonate condition) is recessive to its presence (see Murray, 1975). Third, the loci for shell-colour and shell-striation or banding are almost invariably linked. This is true for genera as diverse as *Arianta*, *Bradybaena*, *Cepaea*, *Cochlicella*, *Limicolaria* and *Partula*. The association, which has been recorded for every species in which genetic tests for linkage have been carried out, is too frequent to be attributable to chance. Possible explanations are discussed below.

2. *Polymorphisms for shell-coiling*

Most pulmonates are dextrally coiled. When the shell is held with the apex upward and the aperture towards the observer, the aperture is on the right-hand side. Some species (such as *Vertigo pusilla*) and a few genera (such as *Physa*) are uniformly sinistral. Rarely a species may be truly polymorphic for coil, with both dextral and sinistral individuals coexisting in the same population.

One such species in *L. peregra*, which is normally dextral but is polymorphic in a few scattered populations. Diver and his colleagues investigated the genetics of the polymorphism, and found that there is a major locus determining the direction of coiling, with dextrality dominant to sinistrality. However, the inheritance is peculiar because the locus shows a delayed expression. The phenotype of an individual is determined by the genotype of its mother, so that broods are normally uniform in coil (Boycott and Diver, 1923; Sturtevant, 1923; Diver *et al.*, 1925; Boycott *et al.*, 1930). A number of other genes are capable of modifying the homozygous sinistral genotype so that mixed broods of dextral and sinistral young can be produced (Diver and Andersson-Kottö, 1938).

In *P. suturalis*, which also has populations polymorphic for coil, the mode of inheritance is similar except that sinistrality is dominant to dextrality (Murray and Clarke, 1966, 1976b).

In each of the 2 species studied the commonest form (dextral in *L. stagnalis*, sinistral in *P. suturalis*) is dominant to the rarer one. This leads us to wonder if the dominance may have evolved. It would be interesting to examine the genetics of coiling in a range of other species, some that are predominantly dextral and some that are predominantly sinistral. Several species of *Partula* as well as *P. suturalis* may sometimes be polymorphic for coil (Crampton, 1916) as may several species of *Achatinella* (Welch, 1938, 1954, 1958). This kind of polymorphism has also been reported in *Fruticicola lantzi* (Gause and Smaragdova, 1940).

3. *Protein polymorphism*

In recent years large numbers of genetic polymorphisms in enzymes and other proteins have been detected by the use of gel electrophoresis. These polymorphisms are present in a wide range of species from *E. coli* to man, but there have been relatively few studies on the pulmonates.

The species most thoroughly investigated has been *C. nemoralis*, in which polymorphisms have been reported for 6-phosphogluconate dehydrogenase, malate dehydrogenase, isocitrate dehydrogenase, acid phosphatase and haemocyanin (Manwell and Baker, 1968), various esterases (Oxford, 1973a, b, c) leucine aminopeptidase and phosphoglucomutase (Brussard and McCracken, 1974) and phosphoglucose isomerase and tetrazolium oxidase (Johnson, 1976). Manwell and Baker (1968) also found polymorphisms for 6 phosphogluconate dehydrogenase, isocitrate dehydrogenase, malate dehydrogenase, acid phosphatase and haemocyanin in *C. hortensis*.

Schwabl and Murray (1970) describe polymorphisms for the proteins of the muciparous glands in *Partula suturalis* and *P. aurantia*. Burnet (1972) reports polymorphisms for esterase, leucine aminopeptidase, malate dehydrogenase and tetrazolium oxidase in *Arion ater rufus*. Levan and Fredga (1972) report an esterase polymorphism in *A. arbustorum*. Selander and Kaufman (1975) found polymorphisms for malate dehydrogenase, leucine aminopeptidase, glutamicoxaloacetic transaminase and phosphoglucomutase in *H. aspersa*, and Woodruff (1975a) found polymorphisms for malate dehydrogenase, leucine aminopeptidase, glutamicoxaloacetic transaminase, phosphoglucose isomerase and a non-specific esterase in *Cerion bendalli*.

Although it is reasonable to suppose that these polymorphisms are genetically determined, because similar ones are known to be inherited in other species, and the proportions of the putative genotypes usually fit the Hardy-Weinberg expectations, there are only 2 cases in which the genetics has been formally investigated. Oxford (1973a, b, c), studying the esterases of *C. nemoralis*, found an extremely complex pattern of inheritance which has not yet been clearly resolved. Brussard and McCracken (1974), using the same species, found that the loci controlling phosphoglucomutase (*Pgm*-2) and leucine aminopeptidase (*Lap*-2) display simple Mendelian inheritance, with 2 and 3 alleles respectively.

It appears that some of the pulmonates are highly polymorphic at the enzyme level. Manwell and Baker (1968) suggest that approximately 60% of enzyme loci in *C. nemoralis* are polymorphic. This estimate was based on a small sample of enzymes and an inadequate knowledge of the genetics, but subsequent studies have strengthened the view that the proportion of polymorphic loci in *Cepaea* is well above the average value for invertebrates. That this high level of heterozygosity is not common to all pulmonates has been clearly shown by the work of Selander and Kaufman (1973) on *Rumina decollata*. This species, which is predominantly self-fertilizing, is native to the Mediterranean region of Europe. It was introduced into North America during the early part of the nineteenth century, and has now spread throughout most of the southern United States, from the Carolinas to California. Selander and Kaufman found that, with respect to 26 enzyme "loci", individuals from 33 widely separated North American localities were all homozygous and identical to each other. There were no polymorphic enzymes. In Europe the species consists of several genetically different homogeneous strains that occasionally hybridize. One strain differed from another in approximately 50% of the enzymes studied (Selander and Kaufman, 1975). Thus, in a sense, some European populations are polymorphic, although the frequency of heterozygotes is usually very low.

4. Chromosomal polymorphism

We know only one possible case of chromosomal polymorphism among the pulmonates. It appears that *H. pomatia* is polymorphic for the presence or absence of accessory (B) chromosomes (Evans, 1960). They are smaller than the normal chromosomes but behave regularly during meiosis. Data on their distribution within and between populations are not available.

Since pulmonate chromosomes are usually small, numerous and difficult to see, it is quite possible that inversion polymorphisms could exist undetected. The presence of "supergenes" for the colour and pattern polymorphisms (see below) leads us to wonder if they may be associated with chromosomal inversions.

III. The evidence for natural selection

In this section we will discuss the evidence that variation in the pulmonates is subject to natural selection. The evidence for selective interaction between loci, and for selective mechanisms capable of maintaining genetic polymorphisms, will be discussed in subsequent sections.

A. Selection acting on continuous variation

Natural selection is the differential survival or reproduction of phenotypes. It is only effective in bringing about evolutionary change if the component of the phenotypic difference that is subject to selection is correlated with an underlying genetic difference. With respect to continuous variation, we have already noted that the evidence for correlation between phenotype and genotype among the pulmonates is sparse but genuine. We shall see that the evidence for natural selection on this type of variation is equally sparse, but less certainly genuine. Even in those studies where the case for selection (as defined above) is strongest, there is no evidence that it has any significance to the evolution of the organism concerned. Because heritabilities are never 100% (i.e. because all continuous phenotypic variation has environmental as well as genetic components) it is possible that the differential survival or reproduction is related only to the environmentally-induced component of the differences between phenotypes. Its effect on the genetic component can only be demonstrated by examining subsequent generations. To our knowledge, this has never been done.

There are 3 kinds of selection on continuously varying quantitative characters:

(i) stabilizing selection, as a result of which individuals with values near the mean survive or reproduce better than those that are more extreme;

(ii) directional selection, as a result of which individuals with values on one side of the mean survive or reproduce better than those with values on the other;

(iii) disruptive selection, as a result of which individuals with values near the mean survive or reproduce less well than those that are more extreme.

Over several generations stabilizing selection, if it acts on the genetic component of the variation, will maintain the mean value but reduce the variance. Directional selection will change the mean, with or without a change of variance. Disruptive selection will increase the variance, with or without a change of mean.

1. *Stabilizing selection*

Two studies are often quoted as classic examples of stabilizing selection. These are by Weldon (1902) and di Cesnola (1907) on *Clausilia laminata* and *Arianta arbustorum* respectively. In both cases certain measures on the shell spire appeared to have greater variability in young snails than in the upper (young) whorls of the adults. The means did not significantly differ. The results were interpreted as evidence for the selective elimination of extreme forms among the young.

It is here that we encounter, for the first time, the problems of sampling mentioned in the introduction. Di Cesnola collected his sample of *Arianta* over an area roughly 50 times greater than the panmictic unit of the related and similar snail *C. nemoralis* (as determined by Lamotte, 1951). Even if the panmictic unit of *Arianta* is larger than that of *Cepaea*, several different Mendelian populations were probably included in the sample. In particular, if juveniles were more difficult to find (as is often the case), they may have been taken from a larger area than the adults. Since clinal variations in the shell-measurements of *Cepaea* have been observed over distances approximately equal to the diameter of the panmictic unit (Wolda, 1969), this could well account for the larger variability of the young snails. Of Weldon's sampling we only know that he collected "in the great beech-wood known as Der Holm", so a similar argument may apply. Stabilizing selection may have occurred, but the data of Weldon and di Cesnola do not unequivocally show it.

2. Directional selection

At first sight, the best evidence for directional selection comes from the work of Cook and O'Donald (1971). Five samples of *C. nemoralis* were kept in cardboard boxes through the winter in an unheated room. In 4 of them the survivors were larger (in shell-size) than the dead snails. The results of the 5 experiments, when combined by the method of summing normal deviates, were highly significant ($p < 0{\cdot}001$). However, the sizes of the collecting areas were not recorded. If these samples came from areas within which there were local differences in shell-size and if in 4 of the cases the larger snails happened to come from subdivisions of the sampling areas where the condition of the animals was more favourable to survival, the observed association between shell-size and mortality could be produced. The probability of such an association being due to chance would then be approximately 0·2 (1-tailed). The discrepancy between this value and the highly significant result of pooling normal deviates indicates the potential dangers of sampling from phenotypically heterogeneous populations. The most rigorous way to conduct a selection experiment of this sort would be to take a large series of samples from different populations, each from a very small area, and to make sure that paired groups of snails (large v. small) from each sample were included in the experiment. The sampling area should indeed be small, no larger than a few square metres. It has not generally been appreciated that gene flow and the movement of individuals can produce significant differences of gene frequency even within a panmictic unit, as can local differences in the direction or intensity of natural selection. Clarke (1962b) has found significant changes of phenotype frequency over distances of 5 m in colonies of *C. nemoralis*.

Wolda (1963, 1967, 1973) has found in 4 different populations of *C. nemoralis* a positive relation between shell-size, oviposition frequency and clutch size. It seems very likely that the latter variables are related to fitness. Wolda's finding is supported by some unpublished studies of Clarke and Murray on *P. taeniata* and *P. suturalis*. In both these species, which are ovoviviparous, there is a positive association between shell-size and the number of embryonic young *in utero*. The trend occurs in more than 40 separate populations from each species, so it is likely to be a real phenomenon.

Bantock and Bayley (1973) have claimed to find visual selection for shell-size in *C. nemoralis* and *C. hortensis*. They put out marked snails of different sizes and observed predation by wild thrushes. They concluded that smaller individuals of *C. nemoralis*, and larger individuals of *C. hortensis*, were eaten by the birds. However, *C. nemoralis* came from several localities. Snails of different origin may have differed both in size and behaviour (and hence in the likelihood of their being predated). *C. hortensis* are described as coming from one "locality", but several colonies. Because of the hazards already mentioned, and because their results are not statistically significant, Bantock and Bayley's case is weak.

There are a few studies in which the values of quantitative characters among natural populations have been found to be correlated with environmental variables. Welch (1938, 1958) found a negative relation between size and altitude in *Achatinella mustelina* and *Achatinella bulimoides*. Cain and Sheppard (1952) report an association between body-colour in *C. nemoralis* and the background of the habitat in which the snails live. Cain and Sheppard explain this association in terms of visual selection by predators. In several species dark colours of shell and body are apparently associated with environments that are relatively cool and damp (see Sacchi, 1955b). In none of these cases have experiments been carried out to test possible selective mechanisms, nor is it certain that the associations are not due to direct effects of the environment on the characters involved.

We know of no studies on disruptive selection in the pulmonates.

B. Selection acting on discontinuous variation

From the viewpoint of the evolutionary geneticist, discrete variants have 2 great advantages; they are usually inherited in a simple Mendelian manner, and they are usually stable in development. Their phenotypes are relatively immune to direct modification by the environment. Consequently if a genuine association is found between the frequency of a particular variety and some environmental factor, we can be reasonably sure that natural selection is acting. Furthermore, if selection does act, it will almost certainly have an effect on subsequent generations.

1. *Selection on polymorphism for shell-colour and pattern*

Attempts to detect natural selection acting on shell-polymorphisms have taken 3 forms. First, there have been experiments and observations on putative agents of selection. Second, there have been studies relating the spatial distribution of phenotype frequencies to particular environmental factors. Third, there have been observations on temporal changes in these frequencies.

(a) *Selective agents.* The most obvious candidate as an agent of natural selection on colour and pattern polymorphisms is predation. Many animals eat snails, and a large proportion of them hunt by sight (Kleiner,

1936; Wild and Lawson, 1937; Morris, 1954; Lim Boo Liat, 1966). In Britain the song-thrush (*Turdus ericetorum*) is a major predator of pulmonates, and has the convenient habit of breaking the shells on stones (anvils), so that it is possible to compare the broken shells with the population from which they were taken. Such comparisons have been made by several workers (Lamotte, 1950; Sheppard, 1951; Cain and Sheppard, 1954; Sacchi, 1956; Murray, 1962; Wolda, 1963; Arnold, 1966; Carter, 1968a; Cain and Currey, 1968; Parkin, 1971) not all of whom have appreciated the pitfalls of sampling. The difficulty lies in establishing the proportions of morphs in the population(s) from which the thrushes are taking snails. With an organism in which frequencies can change in less than 10 m, a small discrepancy in the area sampled between the collector and the thrush can give spurious evidence of visual selection. Similarly, if the thrushes periodically change the areas where they hunt, this can give the impression that there are temporal changes in the direction or intensity of selection. For example, the increased proportions of pink *C. hortensis* found on thrush stones in the summer of 1956 by Goodhart (1958a), and attributed by him to a seasonal change in the direction of selective predation, might equally well be due to the thrushes having changed their hunting grounds. The same criticism applies to the work of Cameron (1969c) on the same species. The patterns of changing selection found by the 2 workers are consistent with each other, but inconsistent with the results of Wolda (1963) and Carter (1968a) on *C. nemoralis* (see below).

Some studies of visual selection (Sheppard, 1951; Murray, 1962; Wolda, 1963; Arnold, 1966; Carter, 1968a; Cain and Currey, 1968) are not subject to the criticism outlined above because they dealt with the predation of snails that had previously been marked. All of them examined the polymorphism in *C. nemoralis*, and all agreed in finding differential selection on the shell-colour morphs (yellow, pink and brown shells). Two studies (Sheppard, 1951; Wolda, 1963) followed the pattern of predation over extended periods, and found that the thrushes changed from preferring yellow during the early spring to preferring pinks during the late spring. Sheppard (1951) interpreted this change as a consequence of the changing background. Pinks, he argued, were more cryptic against a brown background of dead leaves and dry stems, but lost this advantage when new green growth predominated. Wolda (1963), however, continued his observations into the summer, and found that again the yellows were at a disadvantage. Since he could detect no further change in the background he questioned Sheppard's interpretation, and suggested that either the thrushes were changing their colour preferences as a result of experiences with alternative prey, or the selective predation was the result of behavioural differences between the shell-colour morphs, differences that changed with the season. Carter (1968a) who repeated Sheppard's experiments, also found yellows to be preferentially eaten during the summer, but argued that this could still be a consequence of changing background. By the early summer, bracken and rosebay willowherb had grown up to form a dense cover, under which the background at ground level was once more brown. Carter noted that the description of Wolda's study-area suggests that it was densely covered by mixed herbage, so that the same argument might apply.

Although the relation of visual selection to the background is not conclusive, the direct evidence for its action on the shell-colour polymorphism of *C. nemoralis* is very strong. The direct evidence for selection on the banding polymorphism, however, is less so. Several studies (Sheppard, 1951; Murray, 1962; Carter, 1968a; Cain and Currey, 1968) did not find it. Two of those that have claimed to do so (Lamotte, 1950; Cain and Sheppard, 1954) did not use marked snails, and are open to the objection that the live populations sampled may not have been exactly the same as those in which the thrushes were hunting.

Wolda (1963), using marked shells, found significant seasonal changes in the proportions of banded individuals taken by the thrushes. The most striking results were for the 00300 phenotype. However, his experimental colony was started with samples from 4 different populations, and only one of them (Haren) had appreciable frequencies of 00300. Differences in the behaviour of snails from different localities could explain Wolda's results. A similar criticism can be applied to the work of Bantock (1971) on *C. hortensis*, in which he claims to find evidence of visual selection on marked snails.

Parkin (1971) studied the predation of *A. arbustorum* by thrushes and rodents. He found, in 6 separate regions, that predated material from woods contained an excess of yellow snails compared with live material from the same places. He also found, in 4 out of 5 regions, that predated material from non-woodland sites contained fewer yellows than the live populations. These results are formally significant, even counting regions rather than snails. However, the possibility remains that Parkin's results were due to thrushes or rodents carrying snails from non-woodland sites into the woods, and vice-versa.

The only other predator to have been studied experimentally is the glow-worm. O'Donald (1968) offered to individual glow-worms a choice of 2 morphs of *C. nemoralis*. In one series of experiments the insects attacked significantly more unbanded brown shells than banded yellows. In another series they attacked more unbanded than banded yellows. O'Donald suggests that this type of selection might explain a sharp cline in the frequency of browns that occurred on the dunes where he had observed predation by glow-worms. However, his data do not exclude the possibility that the apparent selection on the basis of phenotype is a result rather than a cause of the clines. He does not give information about the exact localities from which the snails were collected. If the snails of different colours and patterns came from different localities, they would probably also differ in their behaviour, and consequently in their susceptibility to attack. Once more, there is a problem of sampling.

The second major candidate as a selective agent is heat. The pulmonates are poikilothermic and consequently are much affected by the ambient temperature, and by the intensity and duration of solar radiation. Schropp (1931) nicely demonstrated the importance of colour to the thermal balance of an irradiated object. He showed that a piece of black card attained an equilibrium temperature 40 °C higher than an adjacent white one, when they were both placed in bright sunlight. Watt (1968), using butterflies of the genus *Colias*, found that dark morphs attained a higher equilibrium temperature than pale ones, and reached it more quickly. He also found that dark butterflies flew earlier in the day, but in the hottest weather they sought shelter from the sun. They appeared to gain an advantage in cooler conditions, but to suffer a disadvantage at high temperatures. This is reflected in their distribution. Dark forms are commoner in the north, and at higher altitudes; pale forms are commoner in the south, and at lower altitudes.

Experimental studies on the thermal properties of colour and pattern morphs among the pulmonates are few. Richardson (1972), using empty shells of *C. nemoralis*, showed that pink banded shells reached a temperature 1 to 2 °C higher than that attained by yellow unbanded shells under the same conditions of radiation. The difference was most obvious when the banded shells had bands that were fused together to produce an effectively black shell. Heath (1975), in a more extensive study that used both living snails and empty shells, found qualitatively similar results. Darker shells were almost always hotter than lighter ones when they were exposed to sunlight. Jones (1973) showed that individuals of *C. vindobonensis* with fully-pigmented bands reached a higher temperature when exposed to sunlight than individuals with faintly-pigmented bands. Emberton and Bradbury (1960) found that the banded parts of the shell of *C. nemoralis* transmitted more visible light than the unbanded parts, and that darker shells (brown and pink) transmitted less light than paler (yellow) shells. They did not investigate the relative amount of absorption and reflection.

As well as the studies on the thermal properties of shells, there have been several investigations in which the relative viabilities of different morphs were examined after the snails had been subjected to various regimes of temperature and humidity. Due to sampling problems and the likelihood of linkage disequilibrium (see below), it is by no means certain that the differences in survival observed in these experiments are causally related to the variations in colour and pattern. They may be due to other genes accidentally (or otherwise) associated with the alleles at the colour and pattern loci. This is particularly likely when the results of different experiments conflict with each other.

Boettger (1954) performed a few selection experiments with *C. hortensis* and *C. nemoralis* by subjecting the snails to the heat from a strong electric bulb. He observed, particularly with *C. hortensis*, that yellow unbanded individuals seemed to survive better than yellow 5-banded individuals. However, he does not record precise details of the populations from which the snails were taken, and once more we are uncertain that the phenomenon is relevant to selection on the banding locus itself. The same criticism applies to the work of Lamotte (1959) who studied the mortality of various morphs of *C. nemoralis* after they were subjected to 40 h heating at 40 °C from a light source. The interpretation of Lamotte's results is made the more difficult because the numbers of snails used in his experiments are not fully recorded. In the single case where numbers are given, his finding that there was a greater mortality of pinks compared to yellows, and of 5-banded compared to unbanded, is statistically significant, and due almost entirely to an exceptionally low mortality of yellow unbanded snails. A second experiment, for which numbers are not given, also showed a greater mortality of pinks (34% of the pinks died, 26% of the yellows) and of 5-bandeds (29% of the 5-banded died, 26% of the unbanded) but did not show the exceptionally low mortality of yellow unbanded snails. Lamotte also claims that yellow and unbanded individuals are more resistant to cold. In an experiment in which the snails were exposed to –8 °C for 3 h, 46% of the yellows died and 50% of the pinks; 56% of the 5-bandeds died and 46% of the unbandeds. Without knowing the numbers involved it is difficult to assess whether these differences have any meaning.

Because the work of Boettger and Lamotte has often been quoted in support of particular theories of climatic selection (see below), it is important to stress that their conclusions are infirm. The uncertainty is increased because Sedlmair (1956) has obtained results apparently at variance with them. She placed snails in cloth bags out of doors in summer for a month, and found that pink 00345 *C. nemoralis* survived significantly better than yellow 00345. With *C. hortensis*, under the same conditions, she found that yellow unbanded survived significantly better than yellow 12345. Because the snails were not in direct sunlight it is unclear to what extent the thermal properties of the shells were involved in these differences. Once more the experiments are subject to the criticism that the snails used might not have been homogeneous in their origins. The *C. nemoralis* came from a sampling area of 4 × 400 m, and the *C. hortensis* from an area of 40 × 400 m. Since the diameter of the panmictic unit in *C.*

nemoralis is approximately 50–60 m (Lamotte, 1951), each sample could have included as many as 8 different populations. Sedlmair also reported differences in the behaviour of the morphs, finding that yellow unbanded *C. hortensis* and yellow *C. nemoralis* preferred warmer and drier resting positions in temperature and humidity gradients, but these differences might again have been artifacts of sampling.

It is worth pointing out that the most frequent results of such "survival" experiments (that paler shells do better in extreme conditions) are exactly what would be expected if they were indeed artifacts. We know that *Cepaea* populations often have a higher proportion of dark shells in habitats shaded by tall vegetation and a higher proportion of pale shells in more open places (see below). The microclimate of places shaded by trees, bushes or high herbage is likely to be more uniform than the microclimate of places not so shaded. If the populations occupying each type of habitat are genetically or physiologically adapted to their environments (regardless of colour or banding), we would expect to find that in a sample including 2 or more different habitats the paler shells would be better able to survive in extreme environments. Although there is a correspondence between what is found by experiment and what is observed in the field, we do not yet know which is cause and which is effect.

Further evidence for "climatic" selection in *Cepaea* has recently been reported by Richardson (1974). He studied a sand-dune colony of *C. nemoralis* occupying an area of 40 × 40 m and found a number of snails dead, apparently from overheating. In 2 successive years the dead snails showed an excess of pinks and 5-bandeds compared to those surviving. At another site on the same dune, with similar dimensions, a similar result was obtained. The results overall were significant, and this work provides the strongest evidence so far obtained for differential selection unrelated to predation. A very small element of doubt remains whether the live and the dead snails came from exactly the same areas, and a larger uncertainty whether the deaths were indeed due directly to overheating.

Bantock (1974) set up cages (made of fine mesh netting) into which he introduced *C. nemoralis*. After 14 weeks, from the end of May to the beginning of September, he scored the morphs that had died and those that had survived. There were highly significant differences between them. Brown unbanded survived best, pink 5-banded and yellow unbanded fared worst. Single banded (00300) survived better than 5-banded. Bantock interpreted these findings on the result of differential resistance to cold, because the cages were established in an area known to experience temperature inversions. We note that, if his interpretation is correct, Bantock's data are inconsistent with the claims of Lamotte (1959) mentioned above, and that the snails were collected from an area 270 m in length, which could include as many as 5 different populations.

Perhaps the most careful study of relative survival under adverse environmental conditions has been carried out by Komai and Emura (1955). They exposed 3 genotypes of *Bradybaena similaris* (brown banded, brown unbanded, and yellow banded) to –5·5 °C for successive 1·5 h periods, and recorded the snails that died. In 5 replicate experiments the brown unbanded morphs showed the highest mortality. All their experimental snails were the offspring of laboratory crosses between brown banded individuals, so that the problem of differential "physiological conditioning" among the morphs is eliminated. The experimental design, however, introduces a problem of its own. The origins of the parental snails are not recorded. If they were hybrids between populations from different localities, artificial linkage disequilibrium may have been generated, and this in turn might well be reflected in differential survival. Of course, if there were natural linkage disequilibrium, Komai and Emura's results would anyway not reflect the action of the colour and pattern genes themselves. There is no obvious physical reason why brown unbanded shells should survive less well at low temperatures.

This illustrates the difficulty of showing that a particular case of selection is directly related to the variation being studied. It can only be done when there is a clear mechanistic connection between the selective agent and the gene product (for a further discussion of this problem, see Clarke, 1975). Unfortunately, although there are strong hints that the thermal properties of the shell-colours and patterns may directly affect their survival, this has not yet been satisfactorily demonstrated. The correct procedure would be to study a series of paired samples of morphs from extremely restricted localities, using incident thermal radiation, and to enquire if the pattern of mortality is more or less exactly related to the thermal properties of the shells themselves.

(b) *Correlation between phenotype-frequencies and environmental factors.* In a series of natural populations, a significant correlation between the frequency of a phenotype and some objectively-defined environmental factor can only be regarded as good evidence for natural selection if 2 conditions are satisfied. First, the phenotype must not have been produced by the environment itself (i.e. it must be genetic, or otherwise of separate origin). Second, either the populations must be statistically independent of each other, or it must be possible to correct for their mutual dependence. As we have seen, the first condition is satisfied by the majority of discontinuous variants. The second condition, however, is more difficult to meet, and the problems involved can be illustrated by an example from the studies of Murray and Clarke (unpublished) on polymorphism in *P. taeniata*.

On the island of Moorea in French Polynesia, *P. taeniata* is polymorphic for shell-colour, and one morph is a deep purplish-brown. Murray and Clarke studied the frequencies of this morph in a series of 50 separate populations within a single valley at the north-west corner of the island. They found a highly significant positive correlation between frequency and altitude. This immediately suggested an interpretation in terms of climatic selection, with the darker morphs being better adapted to the cooler and wetter conditions on the heights. However, when they sampled the adjacent valley, they found an equally impressive negative correlation between frequency and altitude. In fact, there was a high concentration of purple-brown shells in the bottom of the second valley, which gradually decreased in an irregular morph-ratio cline across the mountains into the first valley. The correlations with altitude were entirely misleading because the different populations were not statistically independent of each other. Those with similar frequencies were related by geographical position, even though the distances between them were much greater than the size of the panmictic unit. One explanation of such a pattern (not necessarily the correct one) would be that there were 2 isolated populations, differing in the frequency of purple-browns, which spread across the mountains and hybridized, forming a cline.

If a cline of this sort is found to correspond to a change in some environmental factor, the probability of the correspondence being in a specified direction is exactly a half. Thus, in order to demonstrate a significant association between the frequency and the factor, at least 6 independent clines need to be examined. The total number of populations is almost irrelevant (although of course there must be at least 2 for each cline).

Background. One study that has successfully overcome this problem is the work of Cain and Sheppard (1950, 1954) on the polymorphism of *C. nemoralis*. They found, in the region of Oxford, England, that the frequencies of both colour and banding variants were related to habitat. Brown, pink and unbanded shells were more frequent in woods, whereas yellow and 5-banded shells were more frequent in hedgerows and "rough herbage". The associations were genuine because the habitats were not geographically grouped, but occurred in a series of independent patches so that, in effect, Cain and Sheppard were sampling a large number of independent clines. They interpret the associations as responses to the background colour of the habitats in which the snails live, and suggest that visual predators are selectively removing morphs that stand out against the background. While this is undoubtedly the most convincing interpretation, caution is necessary. First, although Sheppard (1951) clearly demonstrated visual selection, he did not categorically show that it was the consequence of differential crypsis (rather than, for example, colour-preferences on the part of the birds). Second, the distribution of shell colour (although not, perhaps, of banding) is compatible with an explanation in terms of differences in the thermal properties of the shells. Darker shells are expected to be favoured in shaded habitats, lighter shells in places more exposed to the sun.

Similar relations between the frequencies of morphs in *C. nemoralis* and the type of habitat have been reported for other regions of England by Currey *et al.* (1964), Cameron (1969b), and Greenwood (1974), and for regions of France by Lamotte (1959) and Arnold (1970). Cain (1953) has suggested that the frequencies of morphs in 4 exceptional downland colonies, with unusually low frequencies of yellow shells, are related to visual selection by rabbits (which are colour-blind) on the basis of tone rather than colour.

Clarke (1960), studying the closely-related species *C. hortensis*, found that in the Oxford region it responds to habitat in a different way. Yellow shells predominate in all habitats, but colonies in woodlands tend to have a high proportion of banded shells in which the bands are fused together, a condition giving an overall brown appearance that matches the background of brown leaf-litter. Colonies from grasslands, on the other hand, show a relatively high percentage of unbanded forms, which more closely resemble the uniform green background of grass. The different habitats are scattered, and the associations are genuine. Clarke interpreted them in terms of visual selection by predators, and postulated that the responses are different from those of *C. nemoralis* because pinks and browns in *C. hortensis* are at a disadvantage for other (non-visual) reasons. Among mixed colonies of *C. nemoralis* and *C. hortensis* there is a significant positive correlation between the proportions of pale (effectively yellow) shells in the 2 species (Clarke, 1962b). Although selective predation is the most satisfactory explanation of these results, once more the effects of microclimatic factors cannot be excluded.

Associations between morph-frequencies and habitat types in *C. hortensis* have been reported for other regions of England by Carter (1968b), Cameron (1969b), Greenwood (1974) and Pallas-Clark (unpublished).

Carter (1968b) found that on the Purbeck Hills, where pinks and browns are common in *C. hortensis*, it varies with habitat in the manner that is usual for *C. nemoralis*. Richards and Murray (1975) found that in Lynchburg, Virginia, where pinks and browns are absent in *C. nemoralis*, it varies with habitat in the manner that is usual for *C. hortensis*.

Parkin (1971), in his study of *A. arbustorum*, found significant associations between the frequencies of shell-colour morphs and habitats. Dark brown shells predominate in woods, whereas pale browns and yellows are more frequent in open habitats. This is true of 5 separate subdivisions of a major valley in Derbyshire. Jones *et al.* (1974) report that samples of

Hygromia striolata from "darker" habitats have higher proportions of dark-shelled and dark-bodied morphs than "lighter" habitats. The samples were taken in pairs, containing one sample from each type of habitat, and the results from the 16 pairs are highly significant. Both Parkin (1971) and Jones *et al.* (1974) attribute their findings to the results of selective predation, but again microclimatic effects have not been excluded.

Sacchi (1956) examined the variation of *Pseudotachea splendida* in Eastern Spain. He found high frequencies of dark banding in arid habitats (palm clumps, pines on chalk) and lower frequencies in humid habitats (reeds, macchia, pines on dunes). This was true of 4 separate regions, and appears to be a genuine phenomenon, but its interpretation is difficult. Although predation by birds and mammals certainly occurs, Sacchi does not describe the backgrounds produced by the different habitats so that it is not clear whether or not the changes of morph-frequency can be explained in terms of visual selection.

A result that is certainly genuine, but also difficult to interpret, is the finding by Lewis (1968, 1977) that banded morphs of *Cochlicella acuta* are more frequent in the stabilized regions of sand dunes (on short turf) than in less stable (and sandier) parts. Lewis compared pairs of samples from a wide range of European colonies, between the latitudes of 43 ° and 58 °N, and in every case the difference was in the same direction. This finding provides very strong evidence for natural selection, but it is puzzling because although *Cochlicella* is known to be predated by several species of birds, the banded forms appear to be more conspicuous on short turf, and should therefore be less frequent than on the striped background of marram stems in the sandier areas. The observed pattern of variation is also unexpected if selection acts on the (presumed) thermal properties of the morphs.

Climate. In addition to studies comparing morph-frequencies in particular restricted habitats, there have been several larger-scale attempts to relate these frequencies to climatic factors such as temperature and humidity. They have suffered from the difficulty mentioned at the beginning of this section, that the clinal relationship of a frequency with some variable is only good evidence of selection if it can be observed in at least 6 entirely independent situations.

Cain and Currey (1963a), observing the pattern of gene-frequencies in populations of *C. nemoralis* on the Marlborough Downs, Wiltshire, suggested that the frequency of brown shells was associated with areas that were subject to accumulations of cold air, but they did not quantify this postulated relationship. Carter (1968b) found that in 6 subdivisions of the Berkshire Downs there were relatively more samples of *C. nemoralis* containing browns on the escarpment and lower chalk shelf than on the high chalk plateau. This would be significant ($p < 0{\cdot}05$) if the subdivisions were independent, but in fact the grouping of browns is such that the patches overlap the borders of the subdivisions. Carter's finding could be regarded as support for Cain and Currey's proposal, since cold air does accumulate at the bottom of steep slopes. However, there are many scarps on the Marlborough and Berkshire Downs where browns are not found. Cain and Currey's (1963b) samples from the Salisbury Plain do not give any indication of an association between scarps and brown shells (which are very rare), nor do the samples of Arnold (1971) from the Sussex South Downs. In Cameron's (1969c) study of the vicinity of the famous Rickmansworth frost hollow, no browns were found. Cameron reports, however, that there was a significant tendency for yellows to be more common near the bottoms of valleys, and he argues that yellows are more resistant to low temperatures (citing the experiments of Lamotte, 1959—see above). Arnold (1971) found the proportion of yellows to be greatest on the higher parts of the South Downs, as did Carter (1968b) on the Berkshire Downs. Arnold suggests that this is consistent with yellows being more resistant to extreme environments. At intermediate altitudes he found relatively high frequencies of pinks, which he associates with mild climatic conditions. In the extremely mild conditions of Western Ireland, Cameron (1969c) found high frequencies of pinks in one region, but high frequencies of yellows in another. In the almost equally mild climate of Cornwall, Harvey (1971) found a high frequency of yellows which decreased in more easterly (and probably less mild) locations. In the Scilly Isles, however, Murray (1966) found a high frequency of pinks. Finally, to complete the confusion, at the northern limits of the range of *C. nemoralis* in Britain, where it is presumably subject to extreme conditions, Jones (1971) found regions with high frequencies of pinks, others with high frequencies of yellows, but none with high frequencies of browns. The most northerly known population of *C. nemoralis* in Britain (at Foyers on Loch Ness) is monomorphic for pink.

Cain (1968) in a survey of sand-dune populations of *C. nemoralis*, has suggested that the frequency of browns may be related to the presence of a complex topography that allows the accumulation of cold air. He classified 24 separate dune systems into simple and complex topographies (some dunes had both). Those in the complex category, 8 in number, all contained browns. Of the 22 dunes in the simple category, only 2 contained browns. The difference is statistically significant, and seems to provide good evidence of natural selection, although it would be desirable if the simple and complex categories could be defined objectively. However, evidence for natural selection is not necessarily evidence for climatic selection. The low lying areas of complex dunes are often different in their visual background; sometimes they are large blow-outs in which moisture accumulates and makes the sand darker, and sometimes they are filled with dense vegetation under which the background may also be dark. Simple and complex dunes also differ in their flora, and in the distribution of predatory animals.

Local climatic effects on the colour polymorphism of *C. nemoralis* have been claimed for various regions of Europe. Guerrucci-Henrion (1966) suggested that the frequencies of colours in Brittany were associated with rainfall, but this was disputed by Harvey (1971). Arnold (1968) found that in the Pyrenean valley of the Garonne, yellows were common at low altitudes, became rarer at intermediate altitudes, and then became commoner again in the heights. In the Segre-Valira valleys, yellows were common at low altitudes and rarer at intermediate ones, but did not evidently increase at high altitudes. Cameron *et al.* (1973), working in a nearby region, found a positive relation between yellows and altitude in one valley, a negative relation in another, and no relation in a third.

Various workers have postulated the action of climatic selection on the banding polymorphism of *C. nemoralis*. Lamotte (1951) has found that for several regions of France the mean frequency of unbanded is positively related to the mean July temperature. He also reported that, in several Pyrenean valleys the proportion of this morph is higher at low altitudes, lower at intermediate altitudes, and higher again at high altitudes. He interprets this pattern as due to the fact that cloudiness is most frequent at intermediate altitudes (although rainfall and temperature, and other measures of aridity, show no intermediate maximum). Contesse (1963, quoted by Lamotte, 1966) reports a positive association, in the Alps, between the frequency of unbanded and the mean July temperature. There was a negative association with mean annual rainfall. These surveys, however, did not take into account the nature and background of the individual habitats, and it is difficult to know the extent to which habitat and climate were confounded. In a more careful study of two Pyrenean valley-complexes, Arnold (1968) found results similar to Lamotte's, with the lowest frequencies of unbanded at intermediate altitudes. He suggests that the unbanded morph is at an advantage both in very cold and very hot environments. He supported this suggestion with some observations within the Segre Valley (Arnold, 1969). He found that samples on the valley-bottom consistently had lower frequencies of unbanded than samples from the arid hillsides above them, and suggested that this was because the unbanded shells were more resistant to incident heat. It should be pointed out that in his first paper Arnold reports another comparison between valley-bottom and hillside samples (for the Garonne Valley, and the Valle de Aran). Here he found that riverside samples had significantly more unbanded shells than hillside samples, and suggested that the unbanded shells were more resistant to the adverse conditions produced by the ponding of cold air. This nicely illustrates one of the problems of climatic hypotheses; differences in either direction can be explained by postulating one factor or another. Until these factors can be measured objectively, it is necessary to proceed with a healthy scepticism.

Cameron *et al.* (1973), studying 3 other Pyrenean valleys, found that in one valley the frequency of unbanded was positively related to altitude, and in the other two there was no such relation.

For Britain it has been argued by Arnold (1971) that the presence of the 5-banded morph is associated with a mild climate, and Cain (1968) has suggested that on dunes single-banded morphs are associated with warmer and drier exposures, and 5-banded with cooler ones. The evidence for these suggestions has not been quantified.

Jones (1973), in a large-scale survey that includes most of the published samples of *C. nemoralis* from Europe, reports that there is a clear and significant overall relation between the frequency of yellow shells and mean summer temperature. He could find no such association between any of the banding morphs and any of the climatic variables tested. We note that the correlation between colour and summer temperature does not necessarily indicate a causal relation between the two (as Jones, 1973, seems to imply). The survey apparently did not include any information about the habitats of the samples, and latitudinal changes in the proportions of the different kinds of habitats could produce the effects that Jones observed, as could changes in the nature of the habitats themselves.

Jones (1973, 1974) also provides evidence of climatic selection in *Cepaea vindobonensis*. He finds that faint-banded shells are more frequent on the steep sides of limestone basins (poljes) in Yugoslavia than on the flat bottoms of the basins. Not all mountain slopes have faint-banded snails, but where the morph does occur it is generally confined to these slopes. Jones argues that this relation is due to the basin's known tendency to accumulate cold air. Fully-banded individuals will gain an advantage in the early morning because they will warm up more quickly. On the other hand, they may be at a relative disadvantage on the mountain slopes because of the dangers of overheating. This seems a reasonable hypothesis, but it does not explain the rather sporadic appearance of the faint-banded, nor has it yet been supported by survival experiments. Indeed, because crosses have not been done, it is still possible that the faint-banded morphs are directly produced by the environment in which they are found.

Parkin (1972) reports a single clinal association between mean annual rainfall and the frequency of the yellow unbanded morph in *A. arbustorum*. This result was obtained from samples within a valley complex in Derbyshire, England, and it remains to be confirmed for other regions. We note that the association is the reverse of what would be expected from the, presumed, thermal properties of the shells.

Summarizing the work on climatic selection, it is clear that although many patterns of variation have been attributed to climate, in general the degree of repeatability from region to region seems to be small. It has often been argued, both explicitly and implicitly, that combinations of microclimatic and climatic factors can be very different from place to place, so that such repeatability is not to be expected. However, until these factors are objectively defined, so that it is impossible to juggle hypothetical combinations of influences in such a way that any pattern becomes explicable, the evidence for climatic selection must be regarded as weak. Furthermore, even if real and repeatable associations were discovered, their attribution to the direct effects of climate may be incorrect because the distributions of food, competitors and predators are almost certain to be similarly associated. Care is particularly needed because there are no satisfactory experiments showing that the morphs are differentially susceptible to adverse climatic conditions. Despite these uncertainties, experiments and observations on other organisms lead us to expect that differences in the thermal properties of the morphs will indeed be found to have selective importance.

(c) *Changes of phenotype-frequencey with time.* When phenotypes are genetically determined, changes in their frequencies with time can provide powerful evidence of natural selection, but only if certain dangers are avoided. The first danger is the possibility that a population might become extinct or reduced in numbers after the initial observation, and the area be repopulated by migration from another source with a different array of phenotypes. The second, and more serious, danger derives from the fact that morph-frequencies among the pulmonates can change over very short distances indeed (i.e. over a few metres). If earlier and later samples are not taken from exactly the same place, an apparent (but unreal) change of phenotype frequency may be produced. The third danger is that variations in the behaviour of the morphs or the collectors may produce temporally inconsistent biases in sampling. The fourth danger is that biassed sampling by the collectors may itself change the frequencies in the population.

Because of these dangers, a series of observations on a single "population" or "colony" is not normally sufficient, particularly if the area of study is large. Unless population-sizes and migration rates can be closely monitored and continuously estimated, it is necessary to examine a series of well-defined adjacent populations, and to show that there have been consistent changes in all (or most) of them.

Studies of single colonies of *C. nemoralis* have been carried out by van

Heurn (1943, 1945), Schnetter (1950), Lamotte (1951), Goodhart (1956, 1958b), and Wolda (1963), and of *C. hortensis* by Schnetter (1950). All have reported changes of phenotype-frequency, but only the study by Wolda (1963) has avoided the dangers mentioned above. He examined a marked population artificially established on a small island. The population was composed of snails from several different localities which differed in the frequencies of their phenotypes. Wolda found that significant changes of phenotype-frequencies could be ascribed entirely to changes in the proportions of snails from different localities, nicely illustrating potential hazards of sampling from a heterogeneous the population.

Cain and Currey (1968) observed the changes over a 4-year interval in 6 subdivisions of a *C. nemoralis* colony on the Marlborough Downs. The dimensions of the total sampling area were approximately 60 × 70 m. In 3 of the subdivisions they observed significant decreases in the frequencies of brown shells. They attributed these changes (at least in part) to selective predation by thrushes, but did not discuss the possibility that they might have been the consequence of immigration from outside the study area.

Clarke and Murray (1962a, b), working on the dunes at Berrow, Somerset, studied 20 populations of *C. nemoralis* that had previously been sampled by Boycott and Diver in 1926. A comparison of the 2 sets of samples showed that at 18 localities the frequencies of single-banded shells had increased, and at 2 localities they had decreased. Dark brown shells had decreased in frequency at 12 localities, and had increased at 1 locality (at another locality the frequency had not changed, and at the remaining localities dark browns were not found). Extinction and repopulation is extremely unlikely to be a cause of the changes because the localities were scattered over a distance of 2,000 m, and because clines in phenotype-frequency between populations were similar in slope on the 2 occasions. Errors in locating the samples could hardly have been responsible because the changes were highly consistent from population to population. Variations in the behaviour of the morphs are also very unlikely to be an important factor because the later samples were collected under a wide range of weather conditions (and because subsequent collections at different times of the day and year have confirmed them). Disparities between the methods of sampling used by the earlier and later workers are almost certainly not a contributory factor because deliberate variations in sampling technique (some of the collections were made by touch rather than sight) did not affect the results, and because the trends have generally continued in subsequent samples (Murray and Clarke, 1978). Finally, the changes were not a consequence of the sampling itself because repeated samples in the same year did not differ from each other, and because in subsequent years there is no evident relation between the change in a population and the number of times it has been sampled.

On the basis of the samples, and assuming a generation time of 3 years, Clarke and Murray (1962b) estimated that the coefficient of selection in favour of 00300 was about 6%, and against brown about 5%.

Clarke and Murray relate the changes of frequency to the changing ecology of the dunes. Since 1926, they have been overgrown with sea-buckthorn, and have become more stable. The overgrowth of buckthorn may have allowed an increase in the number of song thrushes and consequently have intensified the visual selection on the snails. The stabilization of the dunes has involved the replacement of marram clumps by other grasses and herbs, often with a consequent change of background to a more uniform green. Both changes would be expected to favour 00300 and to disfavour browns, but the effects of microclimatic shifts, or other factors correlated with stabilization, were not excluded.

Clarke *et al.* (1968) carried out a similar study on the dunes at Bundoran, Eire. They found no evidence of consistent changes in the frequencies of phenotypes between 1924 and 1961, neither were there signs of any extensive ecological changes. We must note, however, that in this study the geographical correspondence between the earlier and later sampling localities was inexact.

Several workers have shown that it is possible to score pattern phenotypes (and sometimes colours) in fossil and subfossil samples of pulmonate shells (Diver, 1929; Lamotte, 1951; Owen, 1966; Currey and Cain, 1968; Cain, 1971). Currey and Cain (1968) and Cain (1971) have made use of this possibility to compare phenotype-frequencies in ancient and modern samples from the same regions. For *C. nemoralis*, they report that there are significantly more unbanded shells in pre-Iron Age samples than in corresponding modern ones, but that there is very little change from the early Iron Age to the present day. They explain the apparent differences between the results for the 2 periods in terms of climatic selection. Apart from the fact that changes correlated with climate are not necessarily direct consequences of climatic selection, there is a difficulty because the 2 sets of comparisons (pre-Iron Age to Iron Age, Iron Age to modern) are not statistically heterogeneous. In other words there is no significant evidence that there has been a historical change in the pattern of selection. The firm conclusion is that the subfossil samples as a whole have a higher proportion of unbanded shells than the modern ones. Since the difference occurs in many pairs of samples taken over a wide area of Southern England, it is almost certainly the result of natural selection. Further than that, we cannot go.

C. Selection on polymorphisms other than shell-colour and pattern

Gause and Smaragdova (1940) have claimed that sinistral individuals of *Fruticicola lantzi* survive starvation better than dextrals. However, their report gives no indication that the individuals of different coil came from the same area, so that their conclusion must be treated with reserve.

We know of only one study that claims to have found evidence of selection acting on protein polymorphisms in a pulmonate. Brussard (1974) reports that 13 out of 17 samples of *C. nemoralis* collected in Virginia showed excesses of heterozygotes at a leucine aminopeptidase locus (*LAP*-2). He states that this is statistically significant by the sign test ($p = 0{\cdot}025$). However, the sign test takes as its null hypothesis equal probabilities of deviations in the positive and the negative directions. With random sampling from a population in Hardy-Weinberg equilibrium, such equal probabilities of excesses or deficiencies of heterozygotes can only be assumed if the gene-frequencies of the various alleles are all equal (which in this case they are not). When the gene-frequencies depart from equality, even with random sampling the expectation is that there will be more samples with excesses of heterozygotes that deficiencies (in other words, the distribution of deviations from Hardy-Weinberg expectation will become skewed). When this is taken into account Brussard's result is no longer significant.

For another locus (*phosphoglucomutase*-2), Brussard claims that large populations of *C. nemoralis* show excesses of heterozygotes, whereas small ones show deficiencies. It seems likely that the excesses are again normal consequences of sampling, whereas the deficiencies may have been due to a tendency, when snails were scarce, for Brussard to search a larger area, and to include within one sample several groups with different gene-frequencies.

There are, we believe, no studies of selection in relation to chromosomal polymorphism in the pulmonates.

IV. Linkage disequilibrium, selective interaction and the supergene

When 2 loci are linked, it is sometimes found that among the individuals of a population the alleles at one locus are not randomly distributed in relation to the alleles at the other locus. Some combinations of alleles may therefore be unexpectedly common or rare. This situation has been called linkage disequilibrium.

Linkage disequilibrium can be produced in at least 4 ways. First, the population may go through a bottleneck of numbers so that only a few individuals with particular combinations of alleles can, by chance, give rise to the next generation. These combinations will then be over-represented in subsequent generations. Second, there may be migration from a population in which one combination of alleles predominates into another population in which a different combination predominates. Third, it may result from the hitch-hiker effect if a gene adjacent to the loci concerned mutates to a favourable form and spreads through the population, carrying with it a particular combination of alleles. Fourth, natural selection may directly and disproportionately favour particular combinations of alleles, and act against others.

Once linkage disequilibrium has been established it will decay at a rate proportional to the frequency of recombination between the loci unless selection or migration acts to maintain it. For closely linked genes the rate of decay may be so small that the phenomenon can persist for hundreds, even thousands, of generations even without further selection or migration. Thus, from the theoretical point of view, the existence and persistence of linkage disequilibrium cannot be regarded as evidence of selection unless the population sizes are large, the loci are very loosely linked, and the effects of migration can be excluded. Since in the majority of cases we know nothing of population sizes, recombination rates or migration rates, our arguments are heavily constrained.

Despite these cautionary words, there does seem to be an observable link between strong selection and the presence of linkage disequilibrium. In cases such as the polymorphisms involved in Batesian mimicry, the disequilibrium is so intense that particular combinations of alleles (those giving rise to poor mimics) are virtually absent, and the other combinations behave as if they were alleles at one locus. The combinations in such close association have been termed supergenes, and their occurrence in the animal and plant kingdoms has been reviewed by Ford (1971).

The best case for selectively-maintained linkage disequilibrium among the pulmonates is provided by the data of Cain and Sheppard (1954), who showed for *C. nemoralis* that in the Oxford region the direction of disequilibrium depends on habitat. In woodland populations pink-unbanded and yellow-banded shells tend to be relatively common, pink-banded and yellow-unbanded shells relatively rare. In hedges and grasslands the disequilibrium can go in either direction. This pattern is explicable in terms of visual selection because, against the uniform background of brown leaf litter, the pink-unbanded shells are the most cryptic of the 4 possible types, and the yellow-unbanded shells the least so. For a yellow snail in a wood, the advantage of banding in the improvement of crypsis is great. For a pink snail it is relatively small. For populations in visually more heterogeneous habitats like hedgerows and grasslands, the direction of relative advantages and disadvantages is less predictable, as is the direction of disequilibrium.

Dark brown *C. nemoralis* tend almost invariably to be unbanded, regardless of habitat, but it is still uncertain to what extent this is due to a pleiotropic effect of the brown allele in suppressing the manifestation of banding (see Cain *et al.*, 1960).

Other loci that may be involved in disequilibria with the colour and banding loci are those for white lip (Cook and Peake, 1960; Arnold, 1966, 1968; Clarke *et al.*, 1968) and for spread bands (Carter, 1968b; Johnson, 1976). The directions of the disequilibria may vary, but they have not been associated with particular environmental variables.

Disequilibria between shell-colour and pattern loci occur in every pulmonate species that has been examined from this point of view, including *C. hortensis* (Clarke, 1960), *P. taeniata* (Murray and Clarke, unpublished), *L. flammulata* and *L. aurora* (Barker, 1968, 1969), *B. similaris* (Komai and Emura, 1966), *A. arbustorum* (Parkin, 1971) and *C. acuta* (Lewis, 1968). It is clear that the existence of these disequilibria, and particularly of the linkage associated with them, demands some explanation. Even making very conservative estimates about the numbers of chromosomes in the species concerned, the probability of finding by chance that seven pairs of specified loci are linked is very small.

The linkage is unlikely to be the consequence of the colour and pattern loci having arisen by the duplication of structural genes, because there is no necessary biochemical connection between them (the banding loci in *Cepaea*, for example, determine the structure of the shell, not the presence of band pigment, which is determined by the *hyalozonate* locus). One possibility is that the loci have become linked as a consequence of natural selection. It was Fisher (1930) who first pointed out that if 2 loci interact in such a way that particular combinations of alleles are favoured, selection will act to reduce recombination between them. Inversions or translocations that bring them together will spread through the population. Since it is rather unlikely that an inversion or translocation would only include the genes under selection, such a process might well generate linkage disequilibria with other polymorphic genes. It is worth noting in this context that Johnson (1976) reports a disequilibrium between the *malate dehydrogenase*-1 locus and the colour-banding supergene in *C. nemoralis*, and that Bantock (1972) has found differences in chiasma frequencies between populations of the same species.

An alternative way of generating supergenes has been suggested by McLaren (unpublished). It may be that the colour and pattern loci are not structural but regulatory. An ancestral regulatory gene exercizing control over both colour and pattern could have become duplicated, and the 2 new components could each have come to specialize in controlling one locus. This arrangement would have an advantage over unitary control if there were a need for disequilibrium in one direction in one region, but in the opposite direction in another region.

A final possibility is that in the ancestral pulmonate the colour and pattern loci just happened by chance to be linked, and that this linkage has persisted among the descendants. Although we cannot formally reject this possibility, it nevertheless seems unlikely because of the large differences between species in the range of colours and the manifestation of pattern. In this context it would be instructive to examine the linkage relations of loci controlling ornamentation in pulmonates that are polymorphic for the colour and pattern of the periostracum as well as the shell (some species of *Liguus*, for example).

V. Maintenance of polymorphism

At the level of the gene, a large part of all variation, whether continuous or discontinuous, probably comes into the category of polymorphism (although we know rather little about the frequencies of alleles at the loci responsible for continuous variation). Thus studies of the systems that can act to maintain polymorphisms are fundamental to our understanding of evolutionary processes.

Work on this subject with the pulmonates has been almost entirely confined to the problem of explaining the existence, and persistence, of polymorphisms for colour and pattern. The matter of persistence is an important one, because it suggests that many of these polymorphisms are not merely transitory phenomena. For some species, fossil populations show that the polymorphisms have persisted for thousands, even tens of thousands, of years (Diver, 1929; Lamotte, 1951; Owen, 1966; Currey and Cain, 1968). The existence of apparently homologous polymorphisms in closely related species (for example, of *Cepaea* and *Pseudotachea*, of *Limicolaria*, and of *Partula*) suggests that the polymorphisms may be older than the species themselves, and in some cases older than the genera.

There are 4 obvious ways of explaining these phenomena. The first appeals to random processes and suggests that the polymorphisms are, in the main, selectively neutral. The second attributes them to the joint effects of directional selection and migration. The third proposes that they are the result of heterozygous advantage, and the fourth advocates the importance of frequency-dependent selection. The explanations are not mutually exclusive, although many authors have treated them as if they were.

A. Random processes

The belief that colour and pattern polymorphisms have little effect on the survival or reproduction of the animals that manifest them has been a popular one. Early workers (e.g. Crampton, 1916, 1925, 1932; Diver, 1939) were greatly influenced by the large changes of phenotype-frequency that can take place in apparently uniform habitats.

The most sophisticated development of this view has been the work of Lamotte (1951), who was able to show that the distribution of the unbanded gene in a series of French populations of *C. nemoralis* agrees with a theoretical distribution based upon Sewall Wright's (1948) theory of random genetic drift, if it is assumed that the mutation rate to and from the bandless allele is of the order of 10^{-4}. Lamotte's arguments for the neutrality of the polymorphism were strengthened by 4 additional arguments. First, he found no obvious correlation between the frequency of unbanded shells and any environmental factors. Second, he argued that although thrush-predation occurs in France, most of it happens in the winter and is non-selective. Third, he could find no correlations between the phenotype-frequencies of *C. nemoralis* and *C. hortensis* living in mixed colonies. Fourth, he suggested (for both unbanded and yellow; see Lamotte, 1952) that the variation between small colonies is greater than the variation between large ones.

Subsequent experiments and observations have not supported these arguments. For a detailed criticism of Lamotte's case, the reader is referred to Cain and Sheppard (1954), but the following points are worthy of mention.

In the first place, the approximate correspondence of an observed distribution to a theoretical one merely demonstrates that they are compatible, not that the principles underlying the theoretical distribution represent reality. This is particularly true when the observed distribution is based on a wide range of possibly heterogeneous habitats and the theoretical distribution includes a special assumption (high mutation rate) for which there is no experimental evidence. Second, there are compelling reasons to believe that the polymorphism in *Cepaea* is subject to selection strong enough to override the effects of random drift in all but the very smallest populations (see above). Some of the evidence comes from studies on Lamotte's French localities (Lamotte, 1959; Arnold, 1970). Third, even if selective predation by thrushes is relatively unimportant in France (although this is by no means certain), there are many other predators that hunt by sight and that are potential selective agents. There may also be selectively important variations in microclimate. Fourth, Clarke (1962b) has shown that while in England, as in France, there is no correlation between the phenotype frequencies of *C. nemoralis* and *C. hortensis* in mixed colonies when they are compared by the method of Lamotte (i.e. when the percentages of yellow-shells or unbanded-shells are compared), nevertheless when the visual properties of the phenotypes are taken into consideration, there are positive correlations (in the percentage of effectively pale shells) between the species, as well as some negative ones (which are discussed below). There is good evidence of selection on both species in mixed colonies, and Lamotte's failure to find such evidence is a consequence of the fact that *C. nemoralis* and *C. hortensis* produce visually similar morphs by different genetic routes. Finally, selective as well as neutralist hypotheses predict that the variance of phenotype frequencies between small colonies should be greater than between large ones. As pointed out by Cain and Sheppard (1954) the larger colonies will tend to be more alike because the diverse effects of selection in different parts of the colonies will tend to cancel out when the colony is considered as a whole. Small colonies will usually live in more homogeneous environments, and probably in more extreme ones. Furthermore, if polymorphisms are subject to density-dependent selection, a decrease of population size will tend to reduce heterozygosity, and thus to mimic the effects of random genetic drift (Clarke, 1972).

Goodhart (1962) has advocated the importance of the founder principle (Mayr, 1954) in determining the phenotype-frequencies in a linear series of populations of *C. nemoralis* in Cambridgeshire. The populations, on an artificial river bank, were flooded 5 years before Goodhart's study, and he postulates that the numbers were drastically reduced, so that pockets of different phenotypes occurred at random in different places. This, he argues, can explain the large changes of phenotype-frequency between populations living in an apparently uniform grassy environment.

While Goodhart's explanation may be correct, it is only one of several possibilities. An apparently uniform habitat is not necessarily uniform with respect to factors important to *Cepaea*. The herbaceous flora certainly varies from place to place along the bank, as does the occurrence of other pulmonates—and the incidence of predation by rats. It is dangerous to assume because a relationship between phenotype frequency and environment has not been found that none exists. Furthermore, Murray (1964) has shown how the habit of sperm storage in *C. nemoralis* can act to minimize the effects of both the founder principle and random genetic drift. The mean number of matings contributing to each brood is estimated to be not less than 2, so that in some circumstances the effective population size may be substantially larger than the total number of adults.

B. Selection and migration

If selection favours one genotype in one population, a second genotype in another, and if there is migration between the two, polymorphism may be maintained. The likelihood of polymorphism depends on the relative magnitudes of the selective values and migration rates (for a discussion, see Endler, 1973, 1977). Unfortunately we know too little about these magnitudes in populations of pulmonates, so it is difficult to assess the importance of balances between selection and migration in maintaining polymorphisms. There must be little doubt, however, that they play some part. It is therefore surprising that, as far as we are aware, no single study has explicitly suggested that they may have a significant role.

C. Heterozygous advantage

The advantage of heterozygotes is often suggested as a method of maintaining polymorphism, but it has rarely been demonstrated. Cain and Sheppard (1954) and Ford (1971) have argued that it is important in *Cepaea*, but the only possible evidence of heterozygous advantage that we can find in the literature is the report by Komai and Emura (1955) that double heterozygotes for colour and banding in *Bradybaena similaris* have faster growth rates than homozygotes. Both heterozygotes and homozygotes were the offspring of crosses between brown-banded shells, but Komai and Emura did not record the origin of the parents, and it is quite possible that the apparent heterozygous advantage at the colour and banding loci was due to the fact that the chromosomes carrying the different alleles originated in different populations (i.e. that it was chromosomal heterosis rather than heterozygous advantage at the colour and banding loci themselves).

D. Frequency-dependent selection

It is usual in the theory of population genetics to assume that the selective values of genotypes are constant, but cases are known where they are not (for example, in the polymorphisms associated with Batesian mimicry). If the selective values of genotypes vary with their frequency in such a way that they are advantageous when rare, but disadvantageous when common, then a polymorphism can be maintained.

It was Cain and Sheppard (1954) who first suggested, in relation to the polymorphism of *C. nemoralis*, that predators might concentrate on common varieties of their prey to the extent that rarer varieties would gain an advantage. They concluded, however, that this kind of behaviour is unlikely to be important in maintaining the polymorphism.

Since 1954, a good deal of evidence has accumulated suggesting that this conclusion was premature. Studies on fish (Popham, 1942; reanalysed by Clarke, 1962a), birds (Allen and Clarke, 1968; Croze, 1970; Allen, 1972a, b, 1974, 1976; Manly *et al.*, 1972) and small mammals (Soane and Clarke, 1973) all strongly indicate that predators can indeed concentrate on the commoner varieties of polymorphic prey and disproportionately neglect the rarer ones. The birds that have shown such behaviour include song thrushes, which are major predators of *Cepaea* and other polymorphic snails (Allen, 1972a; Harvey *et al.*, 1975).

Support for the frequency-dependent behaviour of predators comes also from observing the proportions of different species of prey taken by the predators (Tinbergen, 1960; Mook *et al.*, 1960; Gibb, 1962; Murdoch, 1969) although, unlike the studies with polymorphic prey, these observations have not formally excluded the possibility that the predators were seeking areas where the prey were most dense.

Recently, Horsley (personal communication) has demonstrated that the direction of frequency-dependent selection depends upon the density of the prey. At very high densities the direction is reversed, so that rare forms are at a disadvantage (apparently because they stand out against the background of the common forms; see also Allen, 1972b). Clarke (1962a) suggested that frequency-dependence should disappear at very low prey densities, because the predators would concentrate on none of the morphs (see also Greenwood, 1969). Owen (1965, 1969) has claimed that *L. martensiana* is more polymorphic at higher population densities than at lower densities, and has explained the relationship in similar terms. However, it is uncertain that his populations are statistically independent of each other (because of geographical grouping), so that the phenomenon may not be real.

If visual selection on polymorphic variants should act only to disfavour the less cryptic forms without regard to the extent to which they stand out from the norm, it is difficult to understand why the polymorphic varieties are maintained as distinct entities. Why, for example, do clear yellow *C. nemoralis* occur at all in woods? It would seem perfectly simple for the yellow pigment to be obscured. Why do single-banded (00300) shells so frequently have white areas on either side of the dark band, an arrangement that serves to accentuate the band, and to emphasize the distinctness of the morph? The work of Allen (1972a) has provided an answer. He shows, using artificial baits, that predators may select not only in a frequency-dependent manner, favouring rare forms even when they seem relatively conspicuous, but also that they exert disruptive selection when they are presented with a more or less continuously-graded series of colours. Thus they should act to increase the diversity of their prey, as well as to maintain polymorphism. Presumably the habitat will impose some limits to this diversity. The most extremely conspicuous morphs may suffer a disadvantage at all frequencies, and be eliminated. In this context it is worth mentioning that Cain *et al.* (1960) have suggested that some of the morphs in *C. nemoralis* mimic distinct components of their background. This mimicry, by itself, would not be sufficient to maintain the polymorphism. It could, however, help to maintain the distinctness of the morphs. None the less, there are some morphs that do not seem to resemble any component of the background on which they live (bright pink shells, for example) and very few are truly cryptic in the sense that they could escape a close attention to their immediate surroundings.

One consequence of frequency-dependent selection by predators (which has been called apostatic selection by Clarke, 1962a) is that it should produce divergence between sympatric polymorphic species. If a particular morph is common in one species (for reasons not directly related to its visual properties) then selection should act against the same morph in other species. We would therefore expect, in a series of mixed colonies inhabiting visually similar habitats, a negative correlation between the frequencies of particular varieties in one species, and the frequencies of the same varieties in the other. In a series of mixed colonies of *C. nemoralis* and *C. hortensis*, Clarke (1962b) found a significant negative correlation ($p < 0{\cdot}02$) between the frequencies of yellow effectively unbanded shells in the 2 species among samples from open habitats, and a non-significant negative correlation ($p < 0{\cdot}1$) among samples from woodlands. Carter (1967) has objected to Clarke's interpretation on a number of grounds.

In particular Carter compared the proportions of visually similar morphs of *C. nemoralis* and *C. hortensis* in mixed colonies with the proportions of pairs of adjacent unmixed colonies. He argued that if the frequencies of a given type were more dissimilar in mixed colonies than in pairs of adjacent unmixed colonies, this would be evidence for apostatic selection. However, he found just the reverse. There was a significant tendency for the 2 species to be more alike when they occurred together in 1 colony. This result can only be explained by assuming that the backgrounds (or other relevant factors) of the members of the unmixed pairs tended to differ significantly from one another. Clarke (1969) argues that if this is so, the comparison is no longer a valid test of apostatic selection because the differences between the habitats of the members of unmixed pairs could obscure the effects that the observations were designed to detect. The difficulty of searching for character displacement in this way is that habitats allowing the coexistence of 2 species may be very different ecologically from those allowing only single species. The comparison of mixed colonies, as carried out by Clarke (1962b) avoids this difficulty. Nevertheless, Clarke has not absolutely ruled out the possibility that the negative correlations are the consequence of lumping together areas with distinct morph frequencies, and it remains to be seen whether they are a general phenomenon, or particular to only a few districts. In any event, the experimental evidence for apostatic selection itself as a general phenomenon is very strong, even though its direct effects on the polymorphism of Cepaea have yet to be demonstrated experimentally.

E. Other mechanisms for maintaining polymorphism

Theoretically, there are several other ways that polymorphism can be maintained. They include gametic selection and meiotic drive, differential selection in the 2 sexes, and negative assortative mating. The first 2 have not been investigated in the pulmonates and the third does not apply because pulmonates are almost invariably hermaphrodites. Several workers have found that *C. nemoralis* mates at random with respect to shell-colour and banding (Schilder, 1926, 1950; Lamotte, 1951; Schnetter, 1950).

Summarizing what is known about the maintenance of polymorphism in the pulmonates, we conclude that there may be some effects of random genetic drift, and the founder principle, in producing local differences of gene-frequency, but that they are unlikely to be very important in maintaining polymorphisms. There is a *prima facie* case, but no evidence, that balancing forces of selection and migration may play a part. The same applies to heterozygous advantage. The only mechanism for which there is strong experimental evidence about a potential selective agent is frequency-dependent selection by predators.

VI. Variations between populations

Various studies of *C. nemoralis* have shown that the pattern of variation observed by Cain and Sheppard (1954) in the Oxford region, where phenotype frequencies respond to habitat, is by no means universal. Although, as we have seen, it is found elsewhere, there are regions in which another pattern prevails. On sand dunes (Clarke and Murray, 1962a; Clarke *et al.*, 1968; Cain, 1968), downlands (Cain and Currey, 1963a, b; Carter, 1968; Arnold, 1971) and near the northern limits of its range (Jones, 1971), *C. nemoralis* may show the preponderance of a few morphs, apparently regardless of habitat or background over wide areas (very much larger than the panmictic unit). Between such areas the morph frequencies may change violently within distances of 200 m or less, often in apparently uniform environments. These area effects are not confined to *C. nemoralis*. Similar phenomena occur in *C. hortensis* (Clarke, 1961), *B. similaris* (Komai and Emura, 1955), *Cerion* sp (Mayr and Rosen, 1956), *P. taeniata* (Clarke, 1968) and *P. suturalis* (Clarke and Murray, 1971).

Cain and Currey (1963a) suggest that the area effects in *C. nemoralis* may be due to cryptic environmental differences (probably climatic, see above) that produce sharp changes of selective forces over short distances. They argue that random genetic drift cannot be the present cause of the pattern because the areas are too large. Goodhart (1963) argues that the differences between areas are due to the founder principle of Mayr (1954; see also Diver, 1939). He suggests that the first groups of snails that arrived in each area were small populations differing by chance from each other. Initial differences between these populations then persisted because of the evolution of balanced gene-complexes. When the groups expanded and came together, interpenetration was prevented by the relative inviability of hybrids between the balanced systems, and clines became established along the lines of contact. Such clines, he argues, could be maintained indefinitely. Wright (1965) has put forward a similar argument.

Cain and Currey (1963c), replying to Goodhart, suggest that different balanced gene-complexes would not be equally successful and that one group ought to replace the others unless each was best suited to its own particular area. They also argue that the geography of the clines is difficult to reconcile with the view that they are hybrid zones (but see Endler, 1973).

Clarke (1966) has pointed out that the evolution of balanced gene-complexes in different areas does not require founder populations. An algebraic model of a cline showed that area effects could evolve in a continuous series of populations as a result of the accumulation of interacting modifier genes. He argued that there should be an evolutionary trend towards the steepening of morph-ratio clines. This prediction has now been verified experimentally by Endler (1973).

Support for the view that area effects represent different sets of balanced (or coadapted) gene-complexes comes from the work of Clarke and Murray (reported by Clarke, 1968) on *P. taeniata*. In a zone of transition between 2 area effects for shell-colour they found evidence of concomitant changes in the frequency of the banded phenotype, in shell-size, and in the average numbers of intrauterine eggs and young. There were also signs of a developmental breakdown in the expression of banding. These observations suggest that the zone of transition is indeed a zone of hybridization between groups that differ in a wider array of genes than those responsible for colour.

Further support for this view has recently come from the work of Johnson (1976) on an area effect in *C. nemoralis*, which shows a high frequency of the single-banded gene. In addition to this gene, Johnson studied 6 polymorphic enzyme loci, the colour locus, the unbanded locus, and the spread-banded locus. He found that 3 loci (*spread-bands*, *esterase-F*, and 6 *phosphogluconate dehydrogenase*) were clearly associated with the area effect. If these genes represent a random sample of all polymorphic loci, we conclude that a substantial proportion of them must be involved in area effects. The 2 enzyme loci associated with the area effect were in linkage equilibrium with each other, and with the single-banded locus.

Thus it seems that, after all, some or all of the area effects may simply be geographical races or subspecies, with perhaps a limited degree of reproductive isolation between them. This does not deny the importance of the environment in determining their borders, but makes a simple relation between the colour or pattern phenotype and a single environmental factor less likely. In one place, one factor may limit the range of a subspecies, in another place, another factor. In both places, however, this limit will be represented by a sharp change in the frequency of the same observable phenotype.

Clarke (1966) has pointed out that the steepening of a morph-ratio cline may proceed to the point where the balanced gene-complexes on either side of it are no longer reproductively compatible. There would then be selection for reproductive isolation between them. Clarke is thus proposing a model of parapatric (or semisympatric) speciation. Clarke and Murray (1969) argue that the pattern of subspecies and species of *Partula* on the Pacific island of Moorea is explicable in these terms. Two forms may behave as distinct species in one locality, but hybridise freely or intergrade in another (Murray and Clarke, 1968). Such changes can take place within 200 m, or less, and it is possible to find genetic connections, at one place or another, between 8 of the 11 named species on the island. In *P. suturalis*, Clarke and Murray (1969) have observed what seems to be an isolating mechanism in the process of evolving. *P. suturalis* is sinistral in the northern part of the island, but dextral in the south, with a zone of polymorphism between. The dextral forms appear to have evolved as a result of the presence of a sympatric sinistral species (*P. mooreana*). Because the change of coil also involves a change in the position of the genital aperture (from left to right, or vice-versa) it acts as a partial isolating mechanism, reducing the likelihood of crossing between the species.

It is possible that the curious pattern of variation found in the genus *Cerion* by Mayr and Rosen (1956) could also be explained in terms of parapatric differentiation (see also Gould *et al.*, 1974; Woodruff, 1975b).

Thus, the study of pulmonates has generated a case for speciation without geographical barriers. This is not a conventional view, but the conventional view has come from studies of organisms, such as birds, with much greater mobility. The case is certainly not yet proved, but it may provide an impetus to further work on speciation in the pulmonates.

VII. Concluding remarks

Research on the pulmonates has produced some of the strongest evidence that natural selection acts directly on genetically determined visible characters, and it has revealed some very curious patterns of genetic differentiation in space. Although we do not yet fully understand the pattern of selective forces acting on any single character; and although many experiments have foundered because of the problems of sampling organisms whose range of movement is very small, the hazards of work with the pulmonates should not blind us to their very great advantages for the study of natural variation.

VIII. Acknowledgements

We are much indebted to Dr M. A. Carter, Professor J. J. Murray Jnr and to the late Professor P. M. Sheppard, F.R.S., for their very helpful criticisms of the manuscript.

IX. References*

Abeloos, M. (1945). *C. r. Séanc. Soc. Biol. Paris* **139,** 13–14.

Allen, J. A. (1972a). Apostatic selection: the responses of wild passerines to artificial polymorphic prey. Ph.D. Thesis, University of Edinburgh.

Allen, J. A. (1972b). *Nature, Lond.* **237,** 348–349.

Allen, J. A. (1974). *Heredity, Lond.* **33,** 361–372.

Allen, J. A. (1976). *Heredity, Lond.* **36,** 173–180.

Allen, J. A. and Clarke, B. (1968). *Nature, Lond.* **220,** 501–502.

*Arnason, E. and Grant, P. R. (1976). *Evolution, Lancaster, Pa.* **30,** 499–508.

Arnold, R. W. (1966). Factors affecting gene-frequencies in British and continental populations of *Cepaea*. D. Phil. Thesis, University of Oxford.

Arnold, R. W. (1968). *Phil. Trans. R. Soc. Ser. B* **253,** 549–593.

Arnold, R. W. (1969). *Evolution, Lancaster, Pa.* **23,** 370–378.

Arnold, R. W. (1970). *Genetics, Princeton* **64,** 529–604.

Arnold, R. W. (1971). *Heredity, Lond.* **26,** 277–298.

Arthur, W. W. (1973). *Lymnaea stagnalis*; a study of interpopulation variation. B.Sc. Thesis, New University of Ulster.

Aubertin, D. (1927). *Proc. zool. Soc. Lond.* **1927,** 553–582.

Bantock, C. R. (1971). *J. Biol. Educ.* **5,** 25–34.

Bantock, C. R. (1972). *Heredity, Lond.* **29,** 213–221.

Bantock, C. R. (1974). *Heredity, Lond.* **33,** 409–412.

Bantock, C. R. and Bayley, J. A. (1973). *J. anim. Ecol.* **42,** 247–261.

* Due to circumstances beyond our control, publication has been delayed for more than 4 years. References marked with an asterisk are papers that provide important new information about some of the phenomena discussed, but they do not necessitate any serious revision of the text.

*Bantock, C. R. and Noble, K. (1973). *J. Linn. Soc.* (*Zool.*) **53,** 237–252.
*Bantock, C. R. and Price, D. J. (1975). *Evolution, Lancaster, Pa.* **29,** 267–277.
*Bantock, C. R., Noble, K. and Ratsey, M. (1973). *Heredity, Lond.* **30,** 397–398.
*Bantock, C. R., Bayley, J. A. and Harvey, P. H. (1975). *Evolution, Lancaster, Pa.* **29,** 636–649.
Barker, J. F. (1968). *Heredity, Lond.* **23,** 81–98.
Barker, J. F. (1969). *Am. Nat.* **103,** 259–266.
*Bengtson, S. A., Nilsson, A., Nordstrom, S. and Rundgren, S. (1976). *J. Zool. Lond.* **178,** 173–188.
Boettger, C. R. (1954). *Biol. Zbl.* **73,** 312–333.
Boycott, A. E. (1938). *Proc. malac. Soc. Lond.* **23,** 101–108.
Boycott, A. E. and Diver, C. (1923). *Proc. R. Soc. Ser. B* **95,** 207–213.
Boycott, A. E., Diver, C., Garstang, S. and Turner, F. M. (1930). *Phil. Trans. R. Soc. Ser. B* **219,** 51–131.
Brussard, P. F. (1974). *Nature, Lond.* **257,** 713–715.
*Brussard, P. F. (1975). *Evolution, Lancaster, Pa.* **29,** 402–410.
Brussard, P. F. and McCracken, G. F. (1974). *Heredity, Lond.* **33,** 98–101.
Burnet, B. (1972). *Genet. Res.* **20,** 161–173.
Cain, A. J. (1953). *J. Conch., Lond.* **23,** 333–336.
Cain, A. J. (1959). *J. Conch., Lond.* **24,** 352–353.
Cain, A. J. (1962). *Phil. Trans. R. Soc. Ser. B* **253,** 499–517.
Cain, A. J. (1971). *In* "Ecological Genetics and Evolution" (R. Creed, Ed.). Blackwell, Oxford.
*Cain, A. J. (1977). *J. Conch., Lond.* **29,** 129–136.
Cain, A. J. and Currey, J. D. (1963a). *Phil. Trans. R. Soc. Ser. B* **246,** 1–81.
Cain, A. J. and Currey, J. D. (1963b). *J. Linn. Soc.* (*Zool.*) **45,** 1–15.
Cain, A. J. and Currey, J. D. (1963c). *Heredity, Lond.* **18,** 467–471.
Cain, A. J. and Currey, J. D. (1968). *Phil. Trans. R. Soc. Ser. B* **253,** 447–482.
Cain, A. J. and Sheppard, P. M. (1950). *Heredity, Lond.* **4,** 275–294.
Cain, A. J. and Sheppard, P. M. (1952). *Heredity, Lond.* **6,** 217–231.
Cain, A. J. and Sheppard, P. M. (1954). *Genetics, Princeton* **39,** 89–116.
Cain, A. J., King, J. M. B. and Sheppard, P. M. (1960). *Genetics, Princeton* **45,** 393–411.
Cameron, R. A. D. (1969a). *J. Linn. Soc.* (*Zool.*) **48,** 83–111.
Cameron, R. A. D. (1969b). *Proc. malac. Soc. Lond.* **38,** 439–450.
Cameron, R. A. D. (1969c). *J. anim. Ecol.* **38,** 547–553.
*Cameron, R. A. D. and Williamson, P. (1977). *J. Anim. Ecol.* **46,** 173–180.
Cameron, R. A. D., Carter, M. A. and Haynes, F. N. (1973). *Heredity, Lond.* **31,** 43–74.
Carter, M. A. (1967). *Heredity, Lond.* **22,** 117–139.
Carter, M. A. (1968a). *Proc. Linn. Soc. Lond.* **179,** 241–249.
Carter, M. A. (1968b). *Phil. Trans. R. Soc. Ser. B* **253,** 397–446.
Clarke, B. (1960). *Heredity, Lond.* **14,** 423–443.
Clarke, B. (1961). Some factors affecting shell colour polymorphism in *Cepaea*. D. Phil. Thesis, University of Oxford.
Clarke, B. (1962a). *In* "Taxonomy and Geography", Systematics Association publication no. 4 (D. Nichols, Ed.), pp. 47–70.

Clarke, B. (1962b). *Heredity, Lond.* **17,** 319–345.
Clarke, B. (1966). *Am. Nat.* **100,** 329–402.
Clarke, B. (1968). *In* "Evolution and Environment" (E. T. Drake, Ed.). Yale University Press, New Haven and London.
Clarke, B. (1969). *Heredity, Lond.* **24,** 347–352.
Clarke, B. (1972). *Am. Nat.* **106,** 1–13.
Clarke, B. (1975). *Genetics, Princeton* (Suppl.) **79,** 101–113.
Clarke, B. and Murray, J. (1962a). *Heredity, Lond.* **17,** 445–465.
Clarke, B. and Murray, J. (1962b). *Heredity, Lond.* **17,** 467–476.
Clarke, B. and Murray, J. (1969). *Biol. J. Linn. Soc.* **1,** 31–42.
Clarke, B. and Murray, J. (1971). *In* "Ecological Genetics and Evolution" (E. R. Creed, Ed.). Blackwell, Oxford.
Clarke, B., Diver, C. and Murray, J. (1968). *Phil. Trans. R. Soc. Ser. B* **253,** 519–548.
Comfort, A. (1951a). *Biol. Rev.* **26,** 285–301.
Comfort, A. (1951b). *Proc. malac. Soc. Lond.* **29,** 35–43.
Cook, L. M. (1965). *Evolution, Lancaster, Pa.* **19,** 86–94.
Cook, L. M. (1967). *Heredity, Lond.* **22,** 397–410.
Cook, L. M. (1969). *Proc. malac. Soc. Lond.* **38,** 351–358.
Cook, L. M. and King, J. M. B. (1966). *Genetics, Princeton* **53,** 415–425.
Cook, L. M. and O'Donald, P. (1971). *In* "Ecological Genetics and Evolution" (E. R. Creed, Ed.) pp. 93–108. Blackwells, Oxford.
Cook, L. M. and Peake, J. F. (1960). *Proc. malac. Soc. Lond.* **34,** 1–11.
Crampton, H. E. (1916). *Publs Carnegie Instn* **228,** 1–311.
Crampton, H. E. (1925). *Publs Carnegie Instn* **228A,** 1–116.
Crampton, H. E. (1932). *Publs Carnegie Instn* **410,** 1–335.
Croze, H. T. (1970). *Z. Tierpsychol.* **5,** 1–85.
Currey, J. D. and Cain, A. J. (1968). *Phi. Trans. R. Soc. Ser. B* **253,** 483–498.
Currey, J. D., Arnold, R. W. and Carter, M. A. (1964). *Evolution, Lancaster, Pa.* **18,** 111–117.
di Cesnola, A. P. (1907). *Biometrika* **5,** 327–399.
Diver, C. (1929). *Nature, Lond.* **124,** 183.
Diver, C. (1939). *J. Conch. Lond.* **21,** 91–141.
Diver, C. and Andersson-Kottö, I. (1932). *J. Genet.* **35,** 447–525.
Diver, C., Boycott, A. E. and Garstang, S. (1925). *J. Genet.* **15,** 113–200.
Emberton, L. R. B. and Bradbury, S. (1963). *Proc. malac. Soc. Lond.* **35,** 211–219.
Endler, J. A. (1973). A study of morph-ratio clines. Ph.D. Thesis, Edinburgh University.
Endler, J. A. (1977). "Geographic Variation, Speciation and Clines". Princeton University Press.
Evans, H. J. (1960). *Heredity, Lond.* **15,** 129–138.
Falconer, D. S. (1960). "Introduction to Quantitative Genetics". Oliver and Boyd, Edinburgh and London.
Fisher, R. A. (1930). "The Genetical Theory of Natural Selection". Oxford University Press.
Forbes, G. S. and Crampton, H. E. (1948). *Biol. Bull. mar. biol. Lab., Woods Hole* **182,** 26–46.

Ford, E. B. (1945). *Biol. Rev.* **20,** 72–88.
Ford, E. B. (1971). "Ecological Genetics", 3rd edition. Chapman and Hall, London.
Gause, G. F. and Smaragdova, N. P. (1940). *Am. Nat.* **74,** 562–572.
Getz, L. L. (1962). *Nautilus Philad.* **76,** 70–71.
Gibb, J. A. (1962). *Ibis* **104,** 106–111.
Goodhart, C. B. (1956). *Proc. Linn. Soc. Lond.* **167,** 50–67.
Goodhart, C. B. (1958a). *J. anim. Ecol.* **27,** 47–57.
Goodhart, C. B. (1958b). *Proc. Linn. Soc. Lond.* **169,** 163–167.
Goodhart, C. B. (1962). *J. anim. Ecol.* **31,** 207–237.
Goodhart, C. B. (1963). *Heredity Lsnd.* **18,** 459–465.
Gould, S. J. (1969). *Bull. Mus. comp. Zool. Harv.* **138,** 407–532.
Gould, S. J., Woodruff, D. S. and Martin, J. P. (1974). *Syst. Zool.* **23,** 518–535.
Greenwood, J. J. D. (1966). *Proc. malac. Soc. Lond.* **37,** 119–125.
Greenwood, J. J. D. (1969). *Heredity, Lond.* **24,** 157–161.
Greenwood, J. J. D. (1974a). *Heredity, Lond.* **33,** 17–31.
*Greenwood, J. J. D. (1974b). *Evolution, Lancaster, Pa.* **28,** 513–526.
Guerrucci-Henrion, M. A. (1966). *Archs. Zool. exp. gén.* **107,** 369–417.
Harvey, P. H. (1971). *Heredity, Lond.* **26,** 365–372.
*Harvey, P. H. (1976). *Heredity, Lond.* **36,** 1–10.
Harvey, P. H., Birley, N. and Blackstock, T. H. (1975). *Genetica* **45,** 211–216.
Heath, D. J. (1975). *Oecologia, Berl.* **19,** 29–38.
Johnson, M. S. (1976). *Heredity, Lond.* **36,** 105–121.
*Johnson, M. S., Clarke, B. and Murray, J. (1977). *Evolution, Lancaster, Pa.* **31,** 116–126.
Jones, J. S. (1971). Studies on the ecological genetics of *Cepaea.* Ph.D. Thesis, University of Edinburgh.
Jones, J. S. (1973). *Science, N.Y.* **182,** 546–552.
Jones, J. S. (1974). *Heredity, Lond.* **32,** 165–170.
*Jones, J. S. (1975). *Genetica* **45,** 217–225.
*Jones, J. S. and Irving, A. J. (1975). *Biol. J. Linn. Soc.* **7,** 262–271.
*Jones, J. S. and Parkin, D. T. (1977). *Am. Nat.* **111,** 1014–1017.
Jones, J. S., Briscoe, D. A. and Clarke, B. (1974). *Heredity, Lond.* **33,** 102–106.
*Jones, J. S., Leith, B. H. and Rawlings, P. (1977). *Ann. Rev. Ecol. Syst.* **8** 109–143.
Kleiner, A. (1936). *Oiseau, Paris* **2,** 233–251.
Komai, T. and Emura, S. (1955). *Evolution, Lancaster, Pa.* **9,** 400–418.
Lamotte, M. (1950). *J. Conch. Paris* **90,** 120–190.
Lamotte, M. (1951). *Bull. biol. Fr. Belg.* (Suppl.) **35,** 1–239.
Lamotte, M. (1959). *Cold Spring Harb. Symp. quant. Biol.* **24,** 65–84; discussion, 84–86.
Lamotte, M. (1966). *Lav. Soc. Malac. Ital.* **3,** 33–73.
Lanyon, P. G. H. (1966). *Rep. Marlboro. Coll. nat. Hist. Soc.* **105,** 44–49.
Levan, G. and Fregda, K. (1972). *Hereditas* **71,** 245–252.
Lewis, G. (1968). Polymorphism in the shell characters of certain helicid molluscs, particularly the genus *Cochlicella.* D.Phil. Thesis, University of Oxford.

Lewis, G. (1975). *Biol. J. Linn. Soc. Lond.* **7,** 147–160.
Lewis, G. (1977). *Phil. Trans. R. Soc. Ser. B* **276,** 399–451.
Lim Boo Liat (1966). *J. Zool. Lond.* **148,** 554–568.
Lomnicki, A. (1969). *Nature, Lond.* **223,** 1073–1074.
Manly, B. F. J., Miller, P. and Cook, L. M. (1972). *Am. Nat.* **106,** 719–736.
Manwell, C. and Baker, C. M. A. (1968). *Comp. Biochem. Physiol.* **26,** 195–209.
Mayr, E. (1954). *In* "Evolution as a Process" (J. Huxley, A. C. Hardy and E. B. Ford, Eds). Allen and Unwin, London.
Mayr, E. and Rosen, C. E. (1956). *Am. Mus. Novit.* **1806,** 1–48.
Mook, J. H., Mook, L. J. and Heikens, H. S. (1960). *Archs néerl. Zool.* **13,** 442–465.
Morris, D. (1954). *Br. Birds* **47,** 33–49.
Murdoch, W. W. (1969). *Ecol. Monogr.* **39,** 335–354.
Murray, J. (1962). Factors affecting gene-frequencies in some populations of *Cepaea*. D. Phil. Thesis, University of Oxford.
Murray, J. (1964). *Evolution, Lancaster, Pa.* **18,** 283–291.
Murray, J. (1966). *Proc. malac. Soc. Lond.* **37,** 167–181.
Murray, J. (1975). *In* "Handbook of Genetics" (J. C. King, Ed.). Academic Press, New York and London.
Murray, J. and Clarke, B. (1966). *Genetics, Princeton* **54,** 1261–1277.
Murray, J. and Clarke, B. (1968a). *Heredity, Lond.* **23,** 189–198.
Murray, J. and Clarke, B. (1968b). *Evolution, Lancaster, Pa.* **22,** 624–698.
Murray, J. and Clarke, B. (1976a). *Heredity, Lond.* **37,** 253–269.
Murray, J. and Clarke, B. (1976b). *Heredity, Lond.* **37,** 271–282.
*Murray, J. and Clarke, B. (1978). *Malacologia* **17,** 317–330
O'Donald, P. (1968). *Nature, Lond.* **217,** 194.
Oldham, C. (1934). *Proc. malac. Soc. Lond.* **21,** 131–138.
Owen, D. F. (1963). *Science, N.Y.* **140,** 666–667.
Owen, D. F. (1965). *Proc. zool. Soc. Lond.* **144,** 361–322.
Owen, D. F. (1966). *Science, N.Y.* **152,** 71–78.
Owen, D. F. (1969). *J. Zool. Lond.* **159,** 79–96.
Oxford, G. S. (1973a). *Heredity, Lond.* **30,** 127–139.
Oxford, G. S. (1973b). *Biochem. Genet.* **8,** 365–382.
Oxford, G. S. (1973c). *Comp. Biochem. Physiol.* **44B,** 1–10.
*Oxford, G. S. (1975). *Heredity, Lond.* **35,** 361–370.
Parkin, D. T. (1971). *Heredity, Lond.* **26,** 35–47.
Parkin, D. T. (1972). *Heredity, Lond.* **28,** 49–56.
Piaget, J. (1928). *Bull. Soc. zool. Fr.* **53,** 13–18.
Pilsbry, H. A. (1912). *J. Acad. nat. Sci. Philad.* (2nd Ser.) **15,** 427–472.
Pilsbry, H. A. (1945). "Land Mollusca of North America (North of Mexico)". Academy of Natural Sciences, Philadelphia.
Popham, E. J. (1942). *Proc. zool. Soc. Lond.* **112,** 105–117.
*Price, D. J. (1975a). *Heredity, Lond.* **35,** 221–229.
*Price, D. J. (1975b). *Caryologia* **28,** 261–262.
*Price, D. J. and Bantock, C. R. (1975). *Evolution, Lancaster, Pa.* **29,** 278–226.
Richards, C. S. (1968). *Malacologia* **7,** 109–116.

Richards, C. S. (1969a). *Malacologia* **9,** 327–338.
Richards, A. V. and Murray, J. (1975). *Heredity, Lond.* **34,** 188–131.
Richards, C. S. (1969b). *Malacologia* **9,** 339–348.
Richards, C. S. (1970). *Nature, Lond.* **227,** 806–810.
Richards, C. S. (1971). *J. Invert. Pathol.* **17,** 53–58.
Richards, C. S. (1973). *Malac. Rev.* **6,** 199–202.
Richards, C. S. and Merritt, J. W. Jr. (1972). *Am. J. trop. Med. Hyg.* **21,** 425–434.
Richardson, A. M. M. (1972). Ph.D. Thesis, University of Exeter.
Richardson, A. M. M. (1974). *Nature, Lond.* **247,** 572–573.
*Richardson, A. M. M. (1975). *Proc. Malac. Soc. Lond.* **41,** 481–488.
Roszkowski, W. (1914). *Rev. Suisse Zool.* **22,** 457–539.
Sacchi, C. F. (1955a). *Stud. Ghisler. Ser. III* **2,** 43–66.
Sacchi, C. F. (1955b). *Boll. Zool.* **22,** 563–651.
Sacchi, C. F. (1956). *Boll. Zool.* **23,** 461–501.
Sacchi, C. F. (1957). *Atti. 1st Veneto. Sci.* **115,** 285–298.
Schilder, F. A. (1926). *Z. indukt. Abstamm.-u. VererbLehre* **39,** 249–280.
Schilder, F. A. (1950). *Biol. Zbl.* **69,** 79–103.
Schilder, F. A. and Schilder, M. (1957). "Die Bänderschnecken. Eine studie sur Evolution der Tiere". Fisher Verlag, Jena.
Schnetter, M. (1950). *Verh. dt. zool. Marburg* **1950,** 192–206.
Schropp, K. (1931). *Gezurdh Ing.* **1931,** 789–736.
Schwabl, G. and Murray, J. (1970). *Evolution, Lancaster, Pa.* **24,** 424–430.
Sedlmair, H. (1956). *Biol. Zbl.* **75,** 281–313.
Selander, R. K. and Kaufman, D. W. (1973). *Proc. natn. Acad. Sci. U.S.A.* **70,** 1186–1190.
Selander, R. K. and Kaufman, D. W. C. (1975). Genetic population structure and breeding systems. *In* "The Isozymes". Academic Press, London and New York.
Sheppard, P. M. (1951). *Heredity, Lond.* **5,** 125–134.
Soane, I. D. and Clarke, B. (1973). *Nature, Lond.* **241,** 62–64.
Sturtevant, A. H. (1923). *Science, N.Y.* **58,** 269–270.
Tinbergen, L. (1960). *Archs néerl. Zool.* **13,** 265–343.
Torre y Huerta, C. de la. (1950). *Mem. Soc. Cub. Hist. nat.* **20,** 5–20.
*Valovirta, I. and Halkka, O. (1976). *Hereditas* **83,** 123–126.
Van Heurn, W. (1943). *Basteria* **8,** 59–63.
Van Heurn, W. (1945). *Basteria* **9,** 39–43.
Watt, W. B. (1968) *Evolution, Lancaster, Pa.* **22,** 437–458.
Welch, d'A. A. (1938). *Smithson. misc. Collns* **103,** 1–236.
Welch, d'A. A. (1954). *Proc. Acad. nat. Sci. Philad.* **106,** 63–107.
Welch, d'A. A. (1958). *Proc. Acad. nat. Sci. Philad.* **110,** 123–212.
Weldon, W. F. R. (1902). *Biometrika* **1,** 109–124.
Wild, S. V. and Lawson, A. K. (1937). *J. Conch. Paris* **20,** 351–361.
Williamson, M. H. (1957). Competition between molluscs. D.Phil. Thesis, University of Oxford.
Williamson, M. H. (1959). *Proc. R. phys. Soc. Edin.* **27,** 87–93.
*Williamson, P., Cameron, R. A. D. and Carter, M. A. (1976). *Nature, Lond.* **263,** 496–497.

*Williamson, P., Cameron, R. A. D. and Carter, M. A. (1977). *J. Anim. Ecol.* **46,** 181–194.
Wolda, H. (1963). *Archs néerl. Zool.* **15,** 381–471.
Wolda, H. (1965). *Archs néerl. Zool.* **16,** 280–292.
Wolda, H. (1967). *Evolution, Lancaster, Pa.* **21,** 117–129.
Wolda, H. (1969). *J. anim. Ecol.* **38,** 305–327.
Wolda, H. (1970). *Res. Pop. Ecol.* **12,** 185–204.
Wolda, H. (1972). *Neth. J. Zool.* **22,** 428–455.
Wolda, H. (1973). *Neth. J. Zool.* **23,** 162–188.
Woodruff, D. S. (1975a). *Malac. Rev.* **8,** 47–55.
Woodruff, D. S. (1975b). *Malac. Rev.* **8,** 128.
Wright, S. (1948). *Evolution, Lancaster, Pa.* **2,** 279–295.
Wright, S. (1965). *Proc. R. Soc. Ser. B* **162,** 20–104.
Young, F. N. (1960). *Bull. Flo. St. Biol. Sci.* **5,** 259–266.

Chapter 6

Slugs — a study in applied ecology

P. J. HUNTER

Department of Education and Science, Elizabeth House,
York Road, London, UK

I. Introduction

There is no single criterion to define precisely what is or is not a slug; certainly the most obvious features are the reduction or absence of a shell and, perhaps more significantly, the incorporation of the visceral organs into the head-foot. Slugs have a polyphyletic origin and seven families of Mollusca contain terrestrial forms; the Veronicellidae and Rathouisiidae in the order Soleolifera; and the Arionidae, Philomycidae, Limacidae, Testacellidae and Athoracophoridae in the order Sigmurethra, suborder Aulacopoda. The distribution, habits and body form of these animals vary widely from brightly coloured fungus feeders such as the Athoracophoridae to more soberly coloured carnivores, such as the Testacellidae.

In many circumstances slugs can act as pests, but the density at which they achieve pest status varies from place to place and from time to time. A high density of slugs can be acceptable in many crops at many times of the year, but much lower densities are not acceptable when a crop is being grown that is particularly susceptible to attack or when it is in a particularly susceptible phase of growth. Precise information on the extent of slug damage is scant, but they are known to attack a wide range of agricultural and horticultural crops throughout the world (Hunter and Runham, 1972; Stevenson, 1977). Although some tropical forms, like members of the Veronicellidae can act as pests, the major problem is associated with slugs found in temperate regions where crops such as wheat, potatoes, sugar beet and brassicas are particularly vulnerable to damage. Here the most important species are: *Deroceras reticulatum* (Müller), *Arion hortensis* Ferussac* and *Milax budapestensis* (Hazay) (Hunter, 1969). Control of slugs is at present effected by many methods, including modification of the habitat by additional cultivations and modifying planting or harvesting dates; growing non-susceptible crop varieties; or chemically, by using poisoned chemical baits or sprays. For the successful and most effective application of this array of control measures information is required on, at least, certain critical aspects of the ecology of the animal.

II. Life cycles

The variability of slugs has already been noted and this characteristic is as true of their ecology as it is of their body form. Even their life cycles may vary between regions of only slightly different climates and between years of different prevailing weather conditions.

Several studies to ascertain the life cycles of the 3 main pest species in the UK have been completed (e.g. Hunter, 1968). Broadly, these have been conducted in two ways. First, cultures of slugs, confined out of doors in conditions as near to natural as possible, have been closely observed throughout a number of generations; for example, Hunter (1968) kept groups of slugs (usually 4 per group) in terylene net bags or perforated plastic containers (25 cm deep and 5 cm in diameter) which were sunk into the soil. The animals were maintained on a substrate of soil and provided with food of green vegetation and wheat grains. They were weighed periodically and the soil was searched for eggs; these were then counted and kept again out of doors until they hatched. Second, regular sampling of field populations have been conducted to demonstrate whether data from cultures give a reasonable indication of the life cycle of

* Davis (1977) has shown that this species is probably a complex of 3 species.

natural populations. This is laborious and difficult because some slugs live deep in the soil (occasionally down to a depth of 3 m) and are frequently highly aggregated in distribution (Hunter, 1966). South (1964) reviewed various methods of sampling and concluded that the most reliable technique was to dig samples of soil and later extract the slugs from this soil in the laboratory by one of two extraction methods. The first relies on soil-washing, and consists of breaking the soil down over a set of sieves using a high-pressure water jet; each of the sieves is then dipped into a tank containing magnesium sulphate solution so that the slugs and the eggs float to the surface where they are picked off. This method has the advantage that eggs as well as slugs can be found but results in some loss of weight by the slugs during the process. The second technique, flooding, requires setting the soil in sieves or perforated bowls and gradually raising the level of water from below so that the slugs are forced to crawl to the soil surface where they can be collected. By this method slugs are recovered in a relatively natural condition but eggs are not found. Slugs extracted from soil samples can be counted, weighed and dissected to show whether they are mature or immature. The times of appearance of large numbers of mature or very young stages gives a good indication of breeding seasons.

Some studies, using confined cultures and field sampling have demonstrated the life cycles of the three main pest species in various parts of England. For *A. hortensis* information concerning the duration of egg development is shown in Fig. 1a; the data from both cultures and the one field study is included (Hunter, 1968). Eggs took longer to develop in the winter than in the summer so that, even with egg-laying at all times of year, there was a tendency for a peak of hatching to occur in May to June. In cultures (Fig. 1b) the peak was somewhat later in August; most young slugs grew quickly during the autumn and spring to mature and lay eggs at about a year old, although some were not ready to lay their eggs until the following spring. The field samples (Fig. 1c) demonstrated that the proportion of the population unable to mature within a year was probably smaller than indicated by the cultures. A higher proportion of the 1962/3 generation, compared to the 1963/4 generation, was prevented from maturing by a prolonged winter.

In *M. budapestensis*, the hatching period for the eggs (Fig. 2a) was even more concentrated than for *A. hortensis*. Data from cultures (Fig. 2b) indicated that slugs hatching in July to August—again a little later than most of the field population—grew quickly during their first autumn, spring and summer to lay just before and after their second winter. These eggs hatched when the culture slugs would have been about two years old. There was some evidence from different cultures that slugs hatching in April were able to grow fast enough during their first autumn and

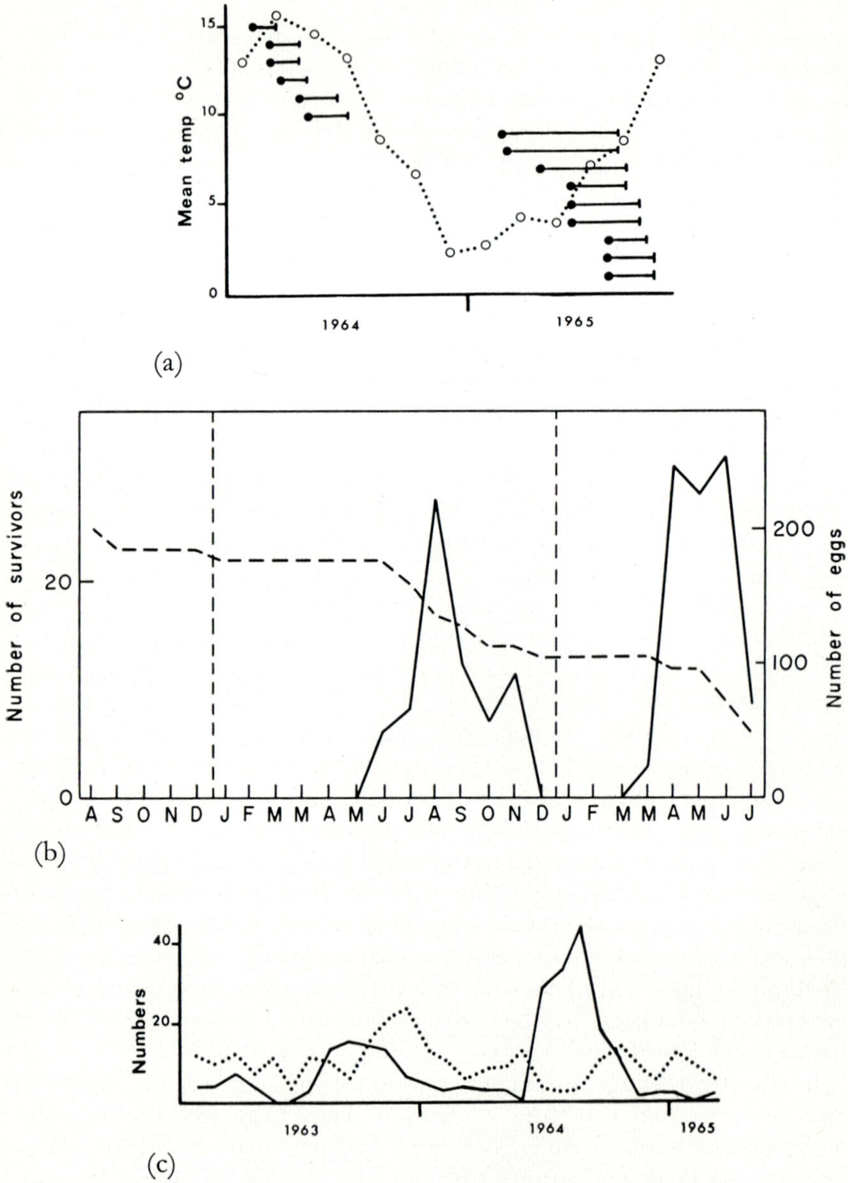

Fig. 1. Life cycle of *A. hortensis*: (a) egg-hatching (● eggs laid, | eggs hatched) and mean temperature (dotted line); (b) solid line, numbers surviving out of 20 (dotted line), numbers of eggs laid (solid line); (c) numbers of mature (dotted line) and newly-hatched (solid line) slugs in samples.

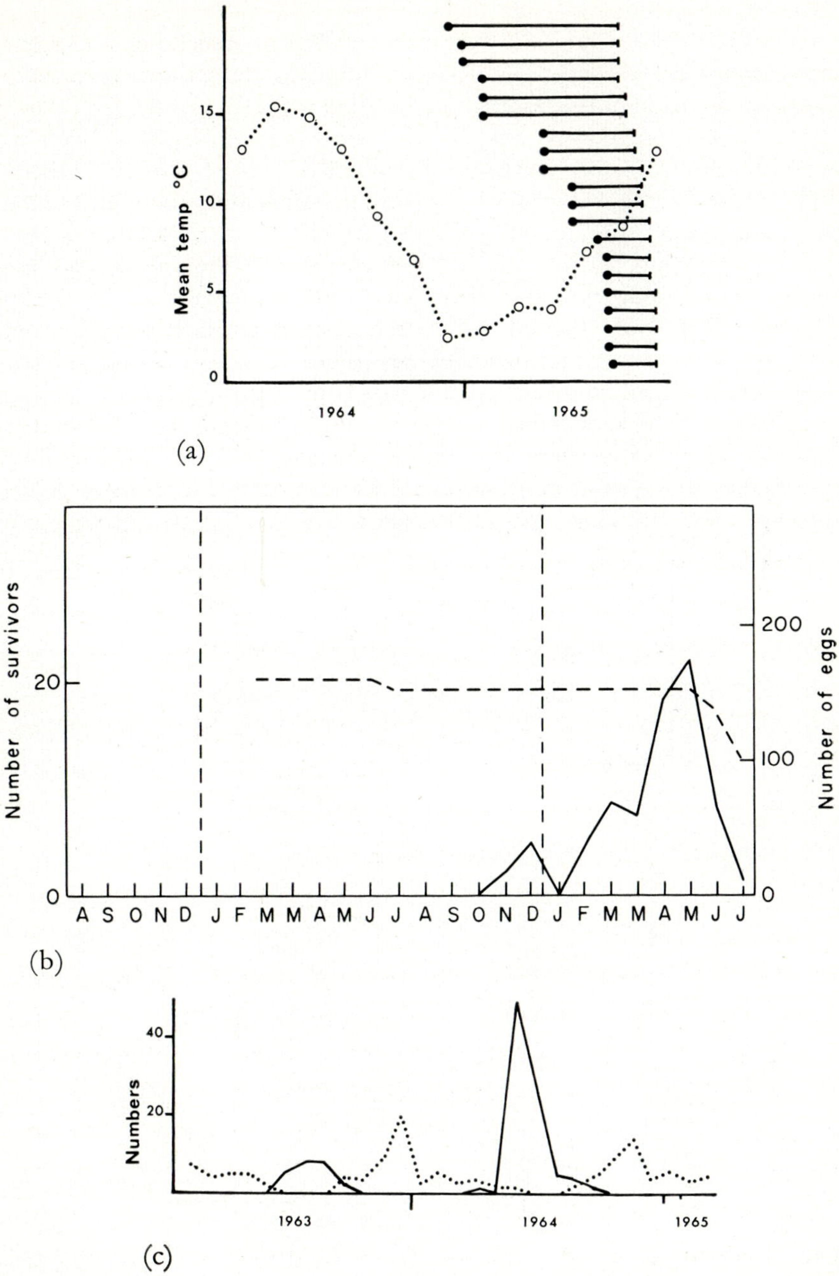

Fig. 2. Life cycles of *M. budapestensis*: (a) egg-hatching; (b) numbers surviving and numbers of eggs laid; (c) numbers of mature and newly-hatched slugs in samples. (Symbols as in Fig. 1.)

spring to mature and lay eggs at about a year old but that those hatching even a short time after this were prevented from breeding until their second autumn. The data from field sampling (Fig. 2c) indicated that only a small proportion of the population departed from the normal life cycle.

The generation interval for *D. reticulatum* is much more difficult to establish and probably varies widely between quite similar climatic regions and possibly between years of different prevailing weather. On the basis of an earlier study in the north of England, Hunter (1968) proposed that the breeding seasons of this species were in the spring and the autum, there being two complete generations each year. However, a later sampling study carried out in the east of England (Hunter and Symonds, 1971) suggested that there are alternate "leapfrogging" generations and this could indicate that the data from the original study should be reassessed. The appearance of very young and mature slugs in samples (Fig. 3a) suggests that there were two different sets of slugs. One of these appeared as young slugs in the spring of 1963 and grew to maturity by the autumn; these laid eggs during the late autumn and early spring, with

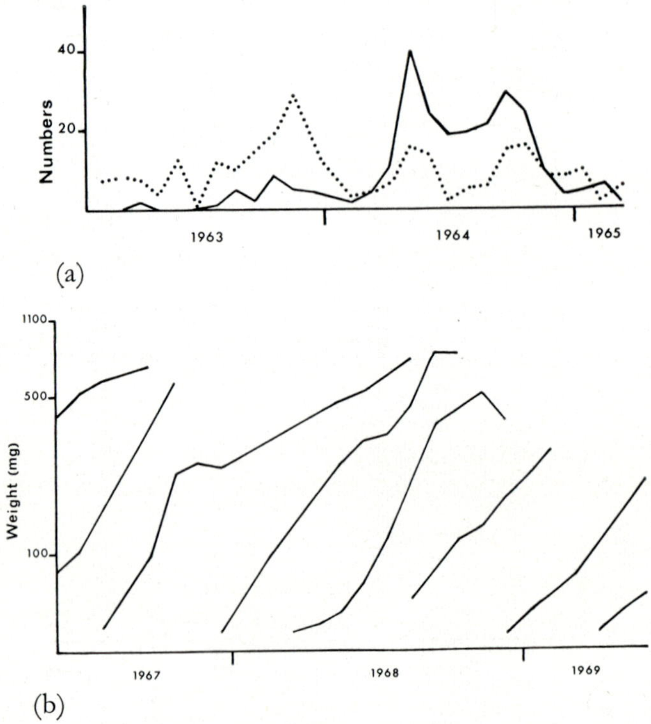

Fig. 3. Life cycles of *D. reticulatum*: (a) growth in weight of generations from field samples; (b) numbers of mature and newly-hatched slugs in samples. (Symbols as in Fig. 1.)

hatching concentrated (in the same way as for *M. budapestensis*) to give the peak of newly hatched slugs in April, 1964. The other set appeared with a peak of young in the autumn of 1963: these grew during the late autumn (development was delayed in the winter) and the spring of 1964 to become mature during the autumn. When the complete data from this study were analysed and the mean growth rate for each generation was plotted, a pattern emerged with intervals between generations of about 9 months duration (Fig. 3b). Unfortunately this investigation covered an unusually wet summer so that the pattern may not have been typical, but at least a high degree of variation in the generation interval of this species was demonstrated.

III. Natural control

Studies on the applied ecology of slugs have not tended to concentrate on those factors which would have proved of greater interest in more academic investigations. There has been less attention given to the way in which a population fits into its ecosystem and more attention given to those environmental factors which can be influenced by man and can affect the population. Generally, ecological studies are concerned with the factors influencing the mortality rate of the various stages of growth. For studies of pests such as slugs the important feature is their ability to increase in numbers from a density which is acceptable to the farmer to one which is not.

Features of a population which can contribute to this capacity to increase in numbers are fecundity, rate of development and mortality of individuals before they have laid their eggs. Environmental factors can affect these parameters either directly (here called primary factors) or indirectly (secondary factors).

A. Primary controlling factors

1. Fecundity

Life cycle studies have shown that *A. hortensis* in confined cultures lays about 50 eggs per individual, *M. budapestensis* about 30 eggs and *D. reticulatum* about 200 eggs. The effect of environmental factors on the fecundity of slugs has not been fully established by field trials, but it is known that they do not lay eggs in cold or dry weather. In experiments in the laboratory it has been shown that the number of gametes produced and the ratio between numbers of sperm and ova appear to be altered by a variety of factors. Low temperature delays the development of the gonad (Lusis, 1966; Smith, 1966; Runham, unpublished), while high temperatures lead to a reduction in numbers or even a total loss of the oocytes (Lusis, 1966; Smith, 1966). Starvation retards development generally, although sperm development is less affected and a good diet promotes differentiation of the oocytes (Richter, 1935).

2. *Rate of development*

The effect of environmental conditions on the developmental rate of slugs has been demonstrated in a range of laboratory experiments (e.g. Carrick, 1938; Abeloos, 1944). It has been shown that *M. budapestensis* eggs hatch in about 3 weeks at 20 °C but take over 4 months to hatch at 7·5 °C. In 12 weeks, *A. hortensis*, when kept at constant temperatures of 20 °C will grow to a weight of 500 mg; but in 12 weeks at 5 °C will only reach a weight of 50 mg. These differences in growth in weight represent difference in growth to maturity and consequently temperature has a marked effect on the generation interval of slugs under field conditions.

Life cycle studies also have shown that considerable variation occurs in the rate of development. Individuals of *A. hortensis* hatching in July/August, instead of the more normal time of May/June, had a lower probability of completing the life cycle within a year and laid their eggs in the autumn: these eggs were then prevented from hatching by the cold weather until the following spring—a generation interval of over 18 months instead of 12 months. Similarly, *M. budapestensis* hatching in April, instead of May, sometimes completed their life cycle in one year instead of two.

A change in the generation interval of a population can have two effects on density. First, it can directly affect the rate of increase in numbers; populations increase in an exponential fashion and the shorter generation interval leads to a greater capacity to increase in numbers. Second, a change of generation interval may expose a population to adverse conditions at a vulnerable stage in its development. The life cycles of slugs have generally evolved so that their breeding is in the autumn or the spring and thus avoid extremes of frost or drought. If these breeding seasons are altered slightly so that they fall when the eggs and young stages are exposed to extreme weather conditions, more will be killed than if they fall at times of milder prevailing weather.

3. *Mortality*

There are virtually no data on the death rate of eggs either in the field or under more controlled laboratory conditions. In the field, eggs are laid underground and it is difficult to simulate their natural habitat so that they can be closely observed. In the laboratory, there is either a 100% successful hatching rate or, if the moisture or temperature regime is allowed to deteriorate, all the eggs are killed. Most species do not survive long exposure to temperatures of less than 0 °C. Similarly eggs cannot withstand extreme dehydration, although *D. reticulatum* embryos survive a loss of water from the egg of up to 80% by weight if immediately rehydrated (Bayne, 1968). At the other extreme eggs cannot withstand immersion in water for over 4 days at 20 °C (Arias and Crowell, 1963). Slug eggs are also killed by predacious soil insects and dead eggs have been noted to be infested with unidentified fungi and nematodes.

The effect of extremes of temperature or humidity on juvenile or adult slugs also requires considerable further study. Experiments on the effect of environmental factors on the mortality of slugs are extremely difficult to interpret, as the mucus of the slug interferes with any artificially established temperature or humidity regime. Nevertheless, exposure to extremes of temperature has been shown in a number of experiments and observations to have a significant effect. For example, experiments by Pinder (1969) demonstrated that *D. reticulatum* and *A. hortensis* cannot live for long when the temperature falls below –3 °C. Observations of field populations (South, 1965; Hunter, 1966) reported occasional decreases in population density after severe frosts. A preliminary series of laboratory experiments by Hunter (1965) indicated that adult slugs cannot survive for long periods when the relative humidity of the atmosphere falls below 90% saturated.

Many parasites are known to infest slugs and some are known to kill them. In certain areas of England large numbers of *D. reticulatum* are infested with the larvae of Sciomyzid flies (Stevenson and Knutson, 1966) and *A. hortensis* in particular is often heavily infested with nematodes (Hunter, 1967a). Predators must also kill a large number of slugs. Starlings, gulls, jackdaw and rook are known to feed on slugs (Collinge, 1921; Holyoak, 1968; Thorpe and Hope, 1908; Blezard, 1967). However, apart from the jaw and, in some species the shell, slugs are completely digested by predators and are not easily recognized in the gut of predatory birds. It is not, therefore, easy to estimate the impact of these predators on populations but, since flocks of up to 700 gulls are not uncommon, these could presumably make a considerable impact. Hedgehogs and other small mammals are known to eat slugs and predacious insects such as carabid, and staphylinid beetle larvae have been observed feeding on them.

In order to evaluate the relative effect of these different factors, a number of attempts have been made to assess the mortality rates of populations in the field. Fig. 4 summarizes data from one of these studies. Samples were taken every month during 1963, 1964 and the first part of 1965, and the slugs from these samples were sorted into generations by reference to their weights—most samples contained more than one generation, but there was a fairly clear distinction between the weights of the slugs in the older and younger generations. The numbers in each generation (5 generations for *D. reticulatum*, 4 for *A. hortensis* and 4 for *M. budapestensis*) are shown for each monthly sample.

Although, like other studies of this kind, it was not possible to draw firm

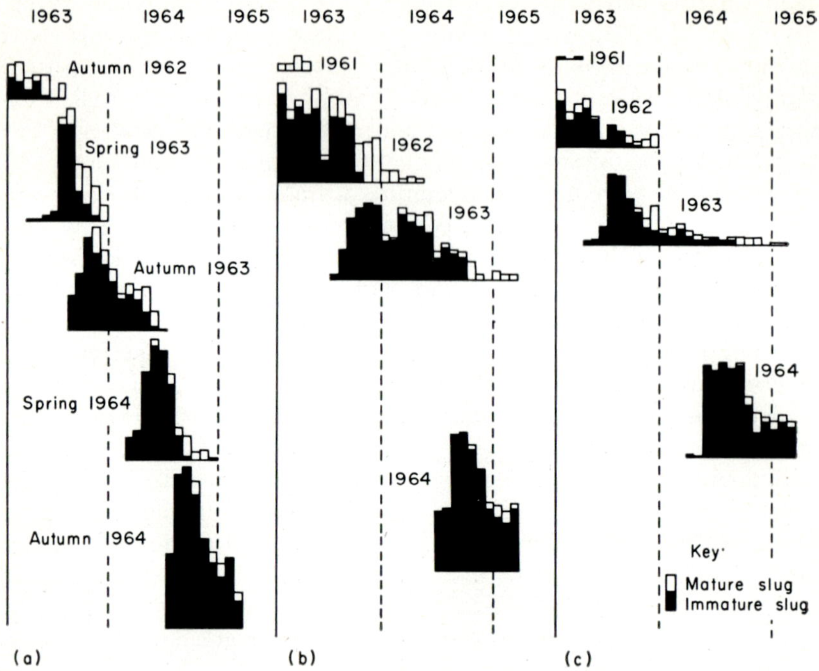

Fig. 4. Numbers of slugs in individual generations during 1963, 1964 and January to March 1965. One sample was taken every 4 weeks and the numbers in each sample, in each generation (see text), for (a) *D. reticulatum*, (b) *A. hortensis* and (c) *M. budapestensis* are shown as vertical bars.

conclusions from this work, it was clear that numbers in each generation were subject to significant reductions only on the very few occasions when there was severe frost or drought (apart, that is, from the natural decline in numbers as the slugs matured and died of old age). But even these reductions were relatively inconsequential when compared with the death rate that has been calculated to have occurred in the egg or very young stages. From estimates of the number of eggs produced per square metre of sampling area, set against the numbers of slugs appearing in the same area in the next generation, the death rate of eggs and young stages must have been well over 90% for all 3 species. There was some evidence that a high proportion of this occurred in the first few days after hatching when the young slugs were unable to move far in search of food or shelter from unfavourable environmental conditions. The weather and the extent of shelter within or on top of the soil during this critical period must, therefore, be the most powerful determinants of the density of the next generation.

B. Secondary controlling factors

The factors outlined above have all been observed to have a direct effect on the density of the slug population, but the intensity of any particular controlling effect and the probability that there is any effect at all depend on a number of secondary circumstances:

(i) The intensity of the controlling effect exerted by extremes of weather or by predators depends to a considerable extent on the amount of shelter available to the slug population. This shelter may be available in a number of forms, for example, there is evidence (Gould, 1961) to suggest that the amount of trash left on the surface after harvesting crops has an influence on the density of populations in the following year. Also experiments (Hunter, 1967b) have suggested that when the number of macrospaces and cracks in the soil are reduced by extensive compaction slug populations are reduced significantly.

(ii) The density of the slug population itself can significantly affect the proportion of population killed. Although the numbers of slugs in the population will probably have no effect on the intensity of control exerted by cold or drought it may have a considerable effect on the proportion of that population killed by certain parasites or predators. A given population of Sciomyzid flies could kill a higher proportion of a sparse *Deroceras* population than of a dense one, because the flies search more diligently in a sparse population than in a dense one, and thus there is an increased probability that a greater proportion of the population is reached. This "inverse density dependent" effect would not operate so strongly with non-specific parasites or predators because a sparse slug population would not necessarily mean that the other available sources of food to the parasite or predator would be similarly sparse.

(iii) The controlling effects by parasites and predators are influenced by a large number of environmental factors, some of which will have no connection with the slug population itself. For example, non-specific parasites and predators will have many sources of food in the community and the availability of these food supplies will affect the density of predators and parasites which in turn will have an effect on the slug. Similarly, the weather conditions could also have a different effect on the parasites and predators than on the slugs, and each parasite or predator species will have its own tolerance levels and availability of shelter.

(iv) The intensity of any controlling effect through shortage of food will be determined by a whole series of environmental factors which relate to the ability of the food itself to grow and flourish and to the presence or absence of competitors. The suitability of the climate for plant growth, the numbers of non-slug species present which feed on the same sources of food, the general state of soil conditions and availability of plant nutrients will all have an effect on plant growth and this will significantly affect the availability of that food to the slug. There seems little evidence however that food is ever a limiting factor for slugs (Duval, 1971).

There is, therefore, a whole series of factors which have a direct or indirect effect on the density of slug populations. It would be possible to further consider the system in more detail by taking account of anything which influences the secondary controlling factors so that a complete network of environmental factors could be seen to affect the population in some way. Nevertheless, for an applied ecological study it is probably sufficient to consider only factors having a primary or secondary effect. Some of these factors are undoubtedly less important to the population while others are critical. For slugs only a broad assessment of population dynamics has been made and the critical controlling factors have only been tentatively identified. The sampling studies described, however, have indicated that the rate of increase in population density depends largely on general weather conditions affecting rate of development and mortality, and on predators and parasites affecting mortality; while the intensity of control exerted by the latter two of these factors is in turn largely determined by the availability of shelter.

IV. Artificial control

Knowledge of the ecology of pests can be used in a variety of ways to lead to more efficient and effective methods of artificial control. First, the study of natural systems of population control will demonstrate whether man can influence or manage those parts of the natural system which are most critical in determining the population's rate of increase. From the information at present available it seems that weather, parasites and predators are the factors which have the greatest direct influence, with the intensity of their effect depending to a large extent on the availability of shelter. There is clearly no way in which man could change the weather, but the availability of shelter can be altered. The surface shelter afforded by vegetation can be kept to a minimum. In a study of slugs in pasture, South (1965) and Pallant (1974) have shown how *D. reticulatum* tend to aggregate in tussocks of grass, demonstrating that shelter is important to this species, and a further study (Gould, 1961) suggested that the removal or burning of surface trash immediately after harvesting might reduce damage to subsequent crops. This surface shelter, however, would be more important to *D. reticulatum* than to *A. hortensis* or *M. budapestensis* which live deeper in the soil (Hunter, 1968). Also, the soil can be compressed to reduce the extent of macrocracks and spaces. This has the added advantage of restricting movement through the soil towards growing wheat seeds or potato tubers but it is rarely possible or desirable in the heavy soils where slugs are most common.

Reducing available shelter not only exposes slugs to adverse weather conditions but also to predators and some parasites (mainly birds, Sciomyzid flies and possibly some of the predacious insect larva). To this extent a limited amount of "biological control" is possible, but so far it has been impossible to manage the predators or parasites of slugs to achieve the measure of biological control that has been possible for some other pests. In general, biological control for control of pests of agricultural crops has only been commercially viable where the pest occurs in high densities, in relatively isolated communities and on crops of high value. Although the last of these is sometimes true for slugs, the first two are usually not. The only serious suggestion that biological control might be possible for the control of slugs (Curtis, 1860) was that ducks should be kept for the purpose.

Knowledge of the ecology of slugs provides additional information that enables predictions to be made concerning when and where damage is likely to occur prior to it actually happening. Certainly many control methods are more effective if applied at predetermined times of year and not on an *ad hoc* basis after the damage has occurred. It can be assumed that damage will tend to be more likely when development has not been held up by long dry or cold periods, when there have been few severe droughts or frosts and where shelter in the soil or on the surface has been kept to a minimum. A number of empirical methods of predicting damage using baits or traps have been developed, for example, counting the number of wheat grains damaged during a given period or determining the damage to a sample of potatoes before planting the crop (Symonds, 1973). These techniques will provide useful specific guidance where necessary.

If prediction is successful, it will then be possible to time the control measures so they will have the greatest effect on population. The life cycle studies have shown when the breeding seasons occur for the main species, and it is clearly more effective to apply control measures just prior to egg-laying. For potatoes in Northern Europe, the main pest species are *A. hortensis*, which breeds in June and July, and *M. budapestensis*, which lays its eggs in late autumn or early spring. However, most chemical control methods are at present based on the use of attractive baits and these cannot be used when the weather is cold or dry; nor are they successful during the growing season of the crops, since the pests are more likely to be attracted to the growing tubers than to the bait. It is not therefore possible to time the treatment ideally to suit all these factors but application in the spring, before the seed potatoes are planted and after the cold winter is finished, would control the *A. hortensis* before their breeding season and the *M. budapestensis* before the spring batch of egg-laying. The main damage to winter wheat is to the grain shortly after drilling, and the important pest species is *D. reticulatum* which breeds in the spring and the autumn. An application of chemical before the autumn breeding season and before the crop is drilled would seem to offer the best combination of timing available. Other crops of which *D. reticulatum* is the major pest, for example, sugar beet, peas and brassicas, are planted in the spring so that the control measures can be applied before the spring breeding season. Care is needed in the choice of molluscide as slug species vary in their susceptibility to the available chemicals and in their susceptibility at different ages (Crowell, 1977). Weather also affects the behaviour of slugs and hence the severity of crop damage. Activity is strongly affected by temperature in association with humidity (Baker, 1973) or perhaps more correctly vapour pressure deficit (Crawford-Sidebotham, 1972). Chemicals are best applied at times of maximum activity.

Field studies have demonstrated that numbers of slugs and activity increase rapidly as the autumn progresses; this opens the way to generally planting crops so that susceptible periods of growth do not coincide with the period of this increased danger. For potatoes, increasing damage is probably also due to a decline in the amount of food on the surface when there is less alternative food available, and to biochemical changes within the tuber as it matures which render it susceptible to damage. This last point about the difference in susceptibility of tubers containing different biochemical compositions is demonstrated by the very fine degrees of choice shown by slugs offered different varieties of potatoes (Gould, 1965; Winfield *et al.*, 1967; Hunter *et al.*, 1968) and by the variation in slug damage with variations in the timing of the biochemical changes in the tuber (Pinder, 1974). For wheat, there are a number of conflicting factors involved in the increase of damage as the autumn progresses. If the wheat is not sown until late autumn the slug population will be at a lower level of activity than during the earlier warmer weather. The numbers of slugs will be high and there is the danger that the temperature may be so low as to delay germination of the wheat, yet not quite low enough to entirely suppress slug activity. A further consideration is that later drilling may render the wheat more susceptible to other pests.

The knowledge of the ecology of pest species of slugs can therefore help towards its effective integrated control, but the detail of knowledge which is required for such control need not go beyond the broad general trends of population increase and decrease, and the main factors significantly affecting the population density. There is little value in detailed studies on those parts of the system that have no significant effect on pest numbers. An approach must be developed which concentrates on those factors having the maximum effect, and those which give a clear indication at an early stage of further avenues of research which should not be considered.

V. Acknowledgements

I am very grateful to Dr N. W. Runham, Dr D. L. Gunn and Dr J. P. Dempster for their constructive comments on this chapter, and to B. V. Symonds for his comments and the use of unpublished data.

VI. References

Abeloos, M. (1944). *Bull. biol. Fr. Belg.* **78,** 215–256.
Arias, R. O. and Crowell, H. C. (1963). *Proc. Calif. Acad. Sci.* **62,** 83–97.
Baker, A. N. (1973). *Proc. malac. Soc. Lond.* **40,** 329–333.
Bayne, C. J. (1968). *J. Zool. Lond.* **155,** 401–411.
Blezard, E. (1967). *Trans. Carlisle nat. Hist. Soc.* **11,** 48–74.
Carrick, R. (1938). *Trans. R. Soc. Edinb.* **59,** 563–597.
Collinge, W. E. (1921). *J. Minist. Agric. Fish.* **27,** 1114–1121.
Crawford-Sidebotham, T. J. (1972). *Oecologia (Berl)* **9,** 141–154.
Crowell, H. H. (1977). *Agric. Exp. Stn, Oregon State Univ. Bull.* **628.**
Curtis, J. (1860). "Farm Insects". Blackie and Son, London.
Davies, S. M. (1977). *J. Conch.* **29,** 173–187.
Duval, D. M. (1971). *J. Conch.* **27,** 249–251.
Gould, H. J. (1961). *Pl. Path.* **10,** 148–146.
Gould, H. J. (1965). *Pl. Path.* **14,** 109–111.
Holyoak, D. (1968). *Bird Study* **15,** 147–153.
Hunter, P. J. (1965). Some aspects of the biology and ecology of *Agriolimax reticulatus* (Müller), *Arion hortensis* Ferussac and *Milax budapestensis* (Hazay). Ph.D. Thesis, University of Newcastle upon Tyne.
Hunter, P. J. (1966). *J. anim. Ecol.* **35,** 543–547.
Hunter, P. J. (1967a). *Proc. malac. Soc. Lond.* **37,** 385–386.
Hunter, P. J. (1967b). *Pl. Path.* **16,** 153–156.
Hunter, P. J. (1968). *Malacologia* **6,** 379–389.
Hunter, P. J. (1969). *NAAS Quart. Rev.* **85,** 31–37.
Hunter, P. J. and Runham, N. W. (1972). *Trop. Sci.* **13,** 191–197.
Hunter, P. J. and Symonds, B. V. (1971). *Nature, Lond.* **229,** 349.
Hunter, P. J., Symonds, B. V. and Newell, P. F. (1968). *Pl. Path.* **17,** 161–164.
Lusis, O. (1966). *Proc. malac. Soc. Lond.* **37,** 19–26.
Pallant, D. (1974). *J. Conch.* **28,** 185–191.
Pinder, L. C. V. (1969). The biology and behaviour of some slugs of economic

importance, *Agriolimax reticulatus*, *Arion hortensis* and *Milax budapestensis*. Ph.D. Thesis, University of Newcastle upon Tyne.

Pinder, L. C. V. (1974). *J. appl. Ecol.* **11,** 439–452.

Richter, E. (1935). *Naturwissenschaften* **69,** 507–544.

Runham, N. W. and Hunter, P. J. (1970). "Terrestrial Slugs". Hutchinson, London.

Smith, B. J. (1966). *Malacologia* **4,** 325–349.

South, A. (1964). *Ann. appl. Biol.* **53,** 251–259.

South, A. (1965). *J. anim. Ecol.* **34,** 403–419.

Stephenson, J. W. and Bardner, R. (1977). Rothamsted Experimental Station, Report for 1976 Part 8, 169–187.

Stephenson, J. W. and Knutson, L. V. (1966). *J. econ. Ent.* **59,** 356–360.

Symonds, B. V. (1973). *Pl. Path.* **22,** 30–34.

Thorpe, D. L. and Hope, L. E. (1908). Report on the food of the blackheaded gull. Cumberland County Council rep. 1907, 3–27.

Winfield, A. L., Wardlow, L. R. and Smith, B. F. (1967). *Pl. Path.* **16,** 136–138.

Chapter 7

Pulmonate molluscs as intermediate hosts for digenetic trematodes

D. S. BROWN*

Department of Zoology, British Museum (Natural History), London, UK

* External Staff, Medical Research Council.

I. Introduction

The first thorough description of the relations between a digenetic trematode and its snail host resulted from investigations into the life cycle of a liver-fluke (*Fasciola hepatica* L.) and its development in *Lymnaea truncatula* (Müll.) by Thomas (1883). Ever since, numerous biologists have been attracted by the seemingly inexhaustible variety of life cycles in which these parasites utilize molluscan hosts. Still many life cycles remain unknown but the importance of the Pulmonata in the evolution of the modern Digenea is clearly evident. There are different opinions among systematists about the position and structure of the group Trematoda; in the conventional system the Trematoda constitutes a class of the phylum Platyhelminthes and contains 3 orders, Monogenea, Aspidogastrea and Digenea. The Digenea are characterized by a complex life cycle, in which usually at least one of the hosts is a mollusc.

Trematode parasites are essentially a component of the fauna associated with aquatic environments and Erasmus (1972) rightly points out that they are generally overlooked until epidemic disease makes the association inescapable. Free-living digenean larvae may be exceedingly abundant in water—an infected snail may discharge clouds of cercariae and in the words of Basch (1975) "snails in their natural habitats may be subjected to a barrage of miracidia derived from numerous species of trematodes parasitic in local vertebrates".

Some of the digeneans that develop in pulmonates are well known for their medical or veterinary importance, which is largely due to the outstanding success of their snail hosts in a variety of freshwater habitats. Schistosomiasis and fascioliasis remain urgent problems, partly because the snail hosts continue to be abundant, despite attempts to control them by chemical and non-chemical means. Turning to the time-scale of evolution, the success of pulmonates in colonizing the land has provided digeneans with the opportunity to move far from their aquatic origins and gain access to terrestrial vertebrates that have little contact with water.

Fortunately, much of the extensive and rapidly increasing literature dealing with relations between pulmonates and the Digenea is gathered into reviews. Recent contributions cover the intermediate hosts of schistosomiasis and fascioliasis (Jordan and Webbe, 1969; Berrie, 1970; Kendall, 1970) and wider fields (Ulmer, 1971; Brooks, 1969), while Wright (1971a) and Erasmus (1972) describe the biology of trematodes with special reference to experimental studies, many of them involving pulmonate snails. I have assembled here some diverse data in an attempt to provide a broad view of pulmonate snails in relation to their digenean parasites. Much of it concerns various aspects of host-specificity, and many of the observations described were made on organisms of medical or veterinary interest, for in recent years financial support for such studies has been forthcoming. However, there is some doubt whether this support will continue to be available at the recent level, and perhaps circumstances will dictate that promising lines of inquiry are pursued in less well-known host parasite combinations. Perhaps this account may be of some value in suggesting to malacologists a variety of possible investigations which could be carried out in their local fauna. In laboratory investigations, it seems that pulmonates must continue to be of major importance because of the relative ease with which they may be maintained.

II. The digenetic life cycle and the pulmonates

Almost all known digeneans have early stages which are parasitic in molluscs (gastropods, scaphopods or lamellibranchs) and the adults are internal parasites of vertebrates. A common life cycle comprises a phase of asexual multiplication in a primary (first intermediate) molluscan host, followed by a period of encystment within a secondary intermediate host, and finally development to maturity with sexual reproduction within a vertebrate (definitive) host. In the most complex life cycles the parasite requires a third intermediate host. Usually the first and second intermediate hosts are molluscs. Penetration into the first molluscan host is achieved either within water by a free-swimming miracidium larva, or the parasite's eggs may be ingested by the host and hatch in its gut. The established miracidium commonly develops into a mother sporocyst which gives rise in one type of life cycle to one or more generations of daughter sporocysts, or in another type of life cycle to one or more generations of rediae. Sporocysts are sac-like organisms which absorb nutriment through their surface, while rediae are equipped with a pharynx and gut and can ingest pieces of host tissue. Daughter sporocysts and rediae produce another stage, the cercaria, which in most life cycles has a brief free-swimming existence, terminated by encystment as a metacercaria. Encystment may occur either on vegetation (as in *Fasciola*) or within a second intermediate host, which may be a different individual of the species which serves as first intermediate host. In some life cycles the second intermediate host is ingested by the definitive host. The schistosomes are unusual in having a cercaria that penetrates directly into the definitive host. An extensive range of information about life cycles may be obtained from Wright (1971a), Erasmus (1972) and Canning and Wright (1972).

There are no fossils to signpost the remote history of the Digenea but it is generally believed that their ancestors were primarily parasites of molluscs and that a vertebrate host was added later to the life cycle. The main evidence cited for this view is that development of a particular digenean usually occurs in a smaller range (in taxonomic terms) of first intermediate hosts than of definitive hosts. Suggestive as may be the narrower host-specificity in the molluscan hosts, caution is necessary in comparing taxonomic categories of animals so diverse as molluscs and vertebrates.

In the view of Llewellyn (1965) the Digenea might have arisen from a marine rhabdocoel-like ancestor that took to feeding on molluscs, probably early bottom-living gastropods and lamellibranchs, by extra-corporeal digestion. By entering the wound and thus gaining access to the inner tissues the predator may have become an endoparasite. Because of the lack of a simple outlet for the eggs, the adult may have acquired the habit of leaving the host, and advantages in disseminating eggs possibly led to the development of a tailed, free-swimming dispersive adult. A second phase in the life cycle may have been initiated through the adult being ingested by vertebrates, avoiding digestion by virtue of the persistence of some of its juvenile endoparasitic adaptations and thus being able to exploit the rich food resources offered by the new kind of host. Wright (1971a) drew attention to another possibility that the ancestors of the Digenea belonged to the Mesozoa, which are parasitic in a variety of marine invertebrates, including cephalopods and lamellibranchs.

Both these evolutionary pathways begin in the sea, in which case the establishment of digenetic life cycles involving a mollusc and a vertebrate could have occurred long before the origin of the Pulmonata. Early gastropods, lamellibranchs and cephalopods are known from the Cambrian period (540 million years ago) and the earliest fishes (including Agnatha) have been found in Ordovician deposits (500 million years ago) (Rensch, 1959), but pulmonates first appear much later just before the Carboniferous period (310 million years ago) (Morton, 1964). If the association between ancestral digeneans and molluscs did begin in the sea it is remarkable that no larval digeneans are known to occur in either the Cephalopoda or the Amphineura, and that the ancient elasmobranch fishes have no characteristic genera of digenetic flukes (Baer, 1952). Llewellyn (1965) suggested that cephalopods might have escaped the attentions of the parasites by being pelagic and fast-moving from an early time, but this consideration does not apply to the amphineurans which are sedentary and apparently vulnerable to attack.

The possibility that the ancestral digeneans first became parasites of molluscs in fresh waters was raised by Baer (1952). He suggested that as a result the parasites could have been already closely adapted to gastropods and lamellibranchs before being introduced into the marine fauna. If this view is correct, pulmonates could have been involved in the evolution of the Digenea from an early stage. Although the adults of about one-third out of more than 60 families of Digenea are confined to fishes (Dawes, 1947) and a number are restricted to amphibians or reptiles, the parasites have also radiated extensively in relation to birds and mammals. The evolutionary history of these higher vertebrates and the recent pulmonates has been largely concurrent and this coincidence has been of great significance for the digenetic flukes, since pulmonate intermediate hosts have become of major importance in allowing the parasites to exploit the resources offered by warm-blooded definitive hosts.

To the parasites, the important attributes of pulmonates are their abundance in a variety of fresh waters and their success in colonizing the land. By serving as food for some mammals and birds that make only occasional contacts with free water, terrestrial pulmonates provide a link enabling the parasites to complete their most adventurous life cycles. Although the characteristic digenetic life cycle is dependent on water at several stages (moist conditions are required for the hatching of eggs, the snail host is often infected by a free-swimming miracidium and the vertebrate host may be penetrated by a free-swimming cercaria), water-dependence has been overcome in a variety of ways. Cercariae may resist desiccation by encysting on temporarily submerged plants which are eaten by grazing animals when the water recedes (e.g. *Fasciola*), or the encystment of cercariae in aquatic larvae of airborne insects may result in the infection of bats and non-aquatic birds. Such life cycles do not depend on terrestrial pulmonates, but it seems unlikely that the flukes would have taken their greatest steps towards independence of water without the existence of land snails.

Adaptation to land snails has occurred independently in several families of Digenea by curtailment of free-living stages and it is possible to mention here only a few examples. A fully terrestrial cycle is demonstrated by *Urotocus tholonetensis* Timon-David which develops in *Helicella arenosa* (Zeigler). Fluke eggs which are ingested by the snail hatch within its intestine and the cercariae do not leave their host but encyst within the sporocyst; development is completed in a magpie (*Pica pica* [L.]) that eats the infected snail (Timon-David, 1957). Some other terrestrial life cycles include 2 intermediate hosts which may both be pulmonates, or a pulmonate and an arthropod. The transference of the parasite from one intermediate host to the other takes place in humid conditions, sometimes by means of a cercaria with a brief free-swimming life, or an infective larva (sporocyst or cercaria) coated with mucus may be expelled from the pneumostome of the snail and be eaten by the second host (Timon-David, 1960; Carney, 1967). Infection of the definitive host in some terrestrial life cycles seems to be aided by modifications in the behaviour of the intermediate host that make it conspicuous to predators; individuals of *Succinea* carrying the conspicuous pulsating brood-sacs of *Leucochloridium* or *Neoleucochloridium* in their tentacles are characteristically found resting in exposed positions on marshy vegetation (literature reviewed by Ulmer, 1971). It is likely that the investigation of life cycles which depend on the predation of infected terrestrial pulmonates will provide further interesting examples of behaviour modified by the larval parasites.

Too many life cycles are unknown to allow more than a tentative estimate of the proportion of digeneans that develop in pulmonate snails. However, an analysis by Ewers (1964) of the snail hosts recorded for 279 species of flukes parasitizing mammals and birds indicates that pulmonates are very important (Table I). When snail families are ranked

Table I. Molluscan families arranged according to the number of digenetic trematode species (using birds or mammals as definitive hosts) reported from them (data from Ewers, 1964).

Family	Number of trematode species and families (in brackets)
Planorbidae	87 (15)
Lymnaeidae	71 (13)
Hydrobiidae	42 (14)
Thiaridae	25 (9)
Physidae	24 (8)
Potamididae	15 (7)
Succineidae	11 (4)
Zonitidae	9 (3)
Helicidae	9 (4)
Viviparidae	8 (6)

according to the number of fluke species recorded from them in Ewers's data pulmonates occupy the first 2 places and 4 other positions among the first 10. Seven out of a total of the 27 trematode families included in the survey were known from pulmonates only, 8 families were restricted to prosobranch hosts, while a further 12 families were recorded from pulmonates and prosobranchs. Among pulmonates the considerably higher incidence of parasitism reported in the Basommatophora than in the Stylommatophora is probably due in part to the closer attention paid to freshwater snails as intermediate hosts.

A compilation by Erasmus (1972) of the numbers of "species" of cercariae recorded from freshwater molluscs in Europe also shows the importance of pulmonates, though the prosobranch *Bithynia tentaculata* (L.) is outstanding as the host of no less than 51 kinds of cercariae (Table II). Many unknown immature stages of adult flukes infecting marine and estuarine birds are likely to develop in marine snails, and consequently the relative importance of prosobranch snails as intermediate hosts could be greater than indicated by Table I, but many freshwater and terrestrial pulmonates are probably involved in undescribed life cycles and little attention has been paid to the estuarine Ellobiidae. It seems likely that few if any prosobranch families will be found to support a richer fauna of digeneans than the Planorbidae and the Lymnaeidae, and these families include some species that seem especially suitable intermediate hosts to judge from the large numbers of different cercariae known to develop in them. Nearly 30 cercarial species have been recorded from *Lymnaea palustris* (Müll.) in Europe (Table II), while Cort *et al.* (1937) mentioned 21 species from the North American *L. emarginata* Say. In South Africa, Porter (1938) found 43 species in *L. natalensis* Krauss, 22 in *Bulinus tropicus* (Krauss) and 18 in *Biomphalaria pfeifferi* (Krauss).

Table II. The number of cercarial "species" recorded from certain molluscs living in some freshwater habitats in Europe From Erasmus (1972) and based on data from a number of authors.

Mollusc	Number of cercarial "species"
Gastropoda	
Pulmonata	
Lymnaea stagnalis	29
L. palustris	17
L. truncatula	3
L. ovata	7
L. auricularia	6
L. peregra	20
Physa fontinalis	1
Planorbis corneus	10
P. carinatus	4
P. vortex	1
P. planorbis	2
Prosobranchia	
Bithynia tentaculata	51
Valvata piscinalis	4
Bivalvia	
Sphaerium corneum	2
Anodonta sp	1

The outstanding variety of larval digeneans supported by some snail species is partly due to their abundance and wide distributions which result in contact with the miracidia or eggs of numerous species of parasite. It may also reflect physiological characters which predispose some snails to be utilized as hosts, and it is a tribute to their biological resilience that they can support this parasite burden. However, although one snail species may support a diverse fauna of digeneans, each parasite species is more or less restricted in its range of snail hosts. This host-specificity is of fundamental biological interest and studies on pulmonate snails have contributed much to our present understanding of the phenomenon. We may never be able to explain fully why a particular digenean develops in a given range of snail hosts, for this depends partly on unknowable evolutionary events, but it is becoming possible to envisage the mechanisms by which specificity is maintained at the present day. Related topics will occupy much of the following 2 sections, commencing with a review of the intermediate host/parasite relations of *Fasciola* and *Schistosoma*.

III. *Fasciola* and *Schistosoma*

Recent research has had 2 different effects on concepts of the specificity shown by larval digeneans in relation to their snail hosts. On the one hand, the synonymies produced in some genera of pulmonates by taxonomic revision have reduced dramatically the earlier reported lists of intermediate hosts for certain parasites. At the same time it has become evident that the limits of host-specificity may be very difficult to define, because the ability of a parasite to develop in a given snail is not necessarily an all or nothing affair. Indeed, compatability between a parasite and its snail hosts may be thought of as a continuous variable, responding to a number of factors including the age of the individual snail and genetic variation in both snail and parasite. In some cases the completely successful development of a parasite has been achieved in the laboratory using a snail host which does not seem to be involved in natural transmission.

A. *Fasciola*

This is a comparatively small genus, containing perhaps no more than 2 valid species (Kendall, 1954) and the taxonomy of the intermediate hosts is relatively simple. Adult liver-flukes are normally parasites in the bile ducts and gall-bladders of herbivorous animals which become infected by ingesting metacercariae encysted on vegetation. Eggs are passed in the faeces and hatch, under suitable conditions of temperature and moisture, to produce free-swimming miracidia which must enter a suitable species of *Lymnaea* to continue development. Within the snail host multiplication takes place through the formation of successive larval stages, finally cercariae leave the host to pass a brief free-swimming life and then encyst, commonly upon vegetation. *F. hepatica* is a common parasite of domestic stock in temperate areas, particularly sheep and cattle, while *F. gigantica* Cobbold infects a similar range of hosts in tropical and sub-tropical regions.

There is little doubt that the snail hosts of *Fasciola* are confined to the Lymnaeidae. Kendal (1954) used the revision of this family by Hubendick (1951) to suggest that the previous lists of intermediate hosts for *F. hepatica*, one of which included over 20 species, should be simplified to comprise only *L. truncatula* (Europe and Asia), *L. bulimoides* Lea (North America) and *L. tomentosa* (Pfeiff.) (Australia). The first 2 species belong to 1 group (genus *Galba* of authors) and are not closely related to *L. tomentosa*, but all 3 snails are ecologically similar and thrive in temporary bodies of water, being capable of surviving drying out of the habitat. To this list must be added *L. columella* Say, which is an intermediate host in Hawaii, Puerto Rico, New Zealand and probably South Africa (Alicata, 1953; Pullan and Whitten, 1972; Boray, 1974). The amphibious habits of all these lymnaeids favour the encystment of metacercariae on vegetation which herbivores are likely to eat when feeding at the water's edge or in damp hollows.

The usual snail hosts of *F. gigantica* all belong to the superspecies of *L. auricularia* (L.), namely *rubiginosa* Michelin (Malaya), *rufescens* Gray (Indian subcontinent) and *natalensis* (Africa). Although these snails can live in very shallow water, they are associated with permanent water-bodies and are not amphibious. Consequently, many cercariae of *F. gigantica* probably encyst on vegetation growing in deeper water and the infection of grazing animals may depend to a considerable extent on their being driven by arid conditions to seek food growing in water.

In West Pakistan where both species of liver-flukes occur, Kendall (1954) found that *F. hepatica* replaced *F. gigantica* above 1220 m in relation to the distributions of their different snail hosts and no evidence was obtained of a common intermediate host. However, *L. truncatula* has been reported to transmit both parasites in Kenya (Dinnik and Dinnik, 1957). Occasional records of unusual intermediate hosts for either of the common *Fasciola* species are understandable in the light of experimental investigations reviewed by Kendall (1965, 1970) and Boray (1969).

Under experimental conditions the complete development of *F. hepatica* (including the production of viable metacercariae) has been obtained in *L. stagnalis* (L.) *L. palustris*, *L. glabra* (Müll.) and *L. peregra* (Müll.), though none of these are normally intermediate hosts. These snails were capable of infection only when freshly hatched or young and there was a high level of mortality (Boray, 1966). Similarly for *F. gigantica*, full development was obtained in newly hatched *L. stagnalis*, *L. peregra* aged up to 4 weeks, and the adults of *L. tomentosa* and *L. truncatula* (though only 5% of the latter produced metacercariae). Using miracidia labelled with radioselenium, Nansen *et al.* (1976) found that *F. hepatica* penetrated a variety of species of *Lymnaea*, but not any of the non-lymnaeid pulmonates or the prosobranchs which were also exposed. When *L. truncatula* and *L. peregra* were exposed together to radioactive miracidia, *L. truncatula* proved to be consistently the more attractive snail (Christensen *et al.*, 1976a).

The results of these experiments suggest that a variety of lymnaeids could be exploited as intermediate hosts by *Fasciola* species, given suitable opportunities for contact in nature. There is evidence that local strains of liver-flukes are adapted to particular snail hosts, and that such adaptation can be rapidly modified, in experiments by Rao (1966) and Boray (1966). Rao observed that a West African strain of *F. gigantica* successfully penetrated *L. rufescens* from India and produced sporocysts, but a significantly smaller number of rediae developed than in the normal host, *L. natalensis*. The transmission of *F. hepatica* in Australia by *L. tomentosa* could be an example of adaptation to a new snail host in recent historical time, since the parasite is believed to have been introduced into Australia during the course of European colonization. Modification of compatibility over an even shorter period has been demonstrated in laboratory experiments (Boray, 1966) in which a German strain of *F. hepatica* was repeatedly passaged through *L. tomentosa*. On the first passage the German strain produced a lower rate of infection and a higher proportion of larvae in abnormal sites than did the Australian strain, but by the third passage the German strain achieved a significantly higher level of infection and was better tolerated. In another of Boray's experiments, the second passage of Australian *F. hepatica* through European *L. peregra* produced infections in older snails, larval development time was shorter and the survival rate of infected snails was higher than in the first passage.

Transport by man of both parasites and snail hosts evidently is an important factor in the development of new host/parasite relationships involving *Fasciola*. Boray (1966) suggested that if *F. gigantica* were introduced into Australia or New Zealand from neighbouring Indonesia or Malaysia, it could achieve compatability with *L. tomentosa* and spread rapidly. Undesirable consequences have arisen, and are likely to continue to do so, from the introduction of snails into new areas. The North American *L. columella* has been introduced in recent years to a number of countries including South Africa, where it has become established in both temperate and tropical regions, providing a potential host for *F.*

hepatica over a far greater area than *L. truncatula*, which is confined to the temperate highveld (Van Eeden and Brown, 1966). Moreover, *L. columella* is probably a more efficient intermediate host of *F. gigantica* than the local host *L. natalensis*, because the more amphibious habits of the introduced snail cause it to frequent marginal vegetation readily grazed by cattle and sheep. *L. columella* has become established also in New Zealand and is responsible for an increased incidence of fascioliasis in that country (Pullan and Whitten, 1972).

Although the transmission of fascioliasis is dependent on a restricted number of lymnaeid species, occasional records of infection of unusual snail hosts are to be expected and the possibility of transmission by such species in limited foci cannot be excluded. If the physiological adaptation of particular strains of liver-fluke to certain snail hosts is genetically based, it could lead, under suitable conditions of isolation, to the emergence of new parasite species. Undoubtedly the snail hosts also show intra-specific genetic variation of characters bearing on their host/parasite relationships. Both infectivity on the part of parasites and susceptibility in their snail hosts have been investigated more extensively in the relations between *Schistosoma* and its intermediate hosts.

B. *Schistosoma*

Comprising about 16 species, *Schistosoma* is a much larger genus than *Fasciola* and its members have followed a number of different evolutionary pathways in relation to snail hosts. *S. japonicum* Katsurada is present in eastern Asia extending westwards to Thailand, occurring as a number of geographical races recognizable by morphological and biological characters, and adapted to different species or infra-specific races of the prosobranch genus *Oncomelania*. A different prosobranch (*Lithoglyphopsis apertus* Ros.) is a host for the Mekong strain of *S. japonicum* (Liang and Van der Schalic, 1975). The other species of *Schistosoma* show an entirely different intermediate host-specificity and develop only in pulmonates belonging to the Planorbidae. Two of them, *S. spindale* Montgomery and *S. nasale* Rao, are widespread parasites of domestic ungulates in south-eastern Asia and are transmitted by *Indoplanorbis exustus* (Desh.) (Bulininae). A comparatively richer fauna of 13 species of *Schistosoma* has been recorded from mammals in Africa (Berrie, 1970) and 1 of them (*S. mansoni* Sambon) has been introduced into the Neotropics.

The African species have eggs with either a lateral (*S. mansoni*) or a terminal spine (*S. haematobium* (Bilharz) and related species); *S. mansoni* develops only in species of *Biomphalaria*, whereas the terminal-spined species are restricted to species of *Bulinus* (apart from possible exceptions mentioned under *S. haematobium*). Although some aspects of the relations between the parasites and their snail hosts are obscured by taxonomic questions (Dinnik and Dinnik, 1965; Pitchford, 1965; Wright *et al.*, 1972; Southgate and Knowles, 1974), the substantial amount of information so far obtained provides the most detailed picture available for a group of digeneans in relation to their intermediate hosts. It is possible to treat in some detail here only the 2 most important parasites of man.

1. *Schistosoma mansoni*

The adult worms are normally found in man in the inferior mesenteric vein and its tributaries. Eggs pass from the body with the faeces and hatch if deposited in water, to produce free-swimming miracidia. The period of intra-molluscan development preceding the first shedding of cercariae can be as short as 3–4 weeks at a suitable temperature. Unlike those of *Fasciola*, the cercariae do not encyst but have a comparatively long free-swimming life.

There are about 10 species of *Biomphalaria* living in Africa, all of them either involved in the transmission of intestinal schistosomiasis or regarded as potential intermediate hosts. However, there is some evidence for the existence of different strains of the parasite and of infra-specific variation in susceptibility in the snails. Parasite strain differences were suspected from geographical variation in the symptoms of disease and the results of snail infection experiments (Wright, 1962), and were confirmed by Saoud (1965a) who distinguished between parasite strains from Puerto Rico, Tanzania and Egypt according to their morphology and biological characters. *B. alexandrina* (Ehrenb.) from 6 localities in Egypt were exposed to *S. mansoni* from Cairo and Alexandria and gave frequencies of infection varying from 0–97% (Cridland, 1968), and marked differences in infection rates in snails from different Egyptian localities were also obtained using a strain of parasite from Tanzania (Cridland, 1970). However, Dawood and Chu (1973) concluded that *B. alexandrina* from 4 provinces of Egypt are equally susceptible to infection with *S. mansoni* from Alexandria.

A gigantic though unplanned experiment, comparable to the introduction of the liver-fluke into Australia, seems to have been carried out with *S. mansoni* in the Neotropics. It is generally agreed that *S. mansoni* was carried from Africa to South America and some Caribbean islands during the course of the slave-trade and the parasite has subsequently become so well established as to constitute a familiar public health problem. Its intermediate hosts in the New World are all species of *Biomphalaria* (Barbosa *et al.*, 1968), though at one time these snails were classified in several different genera (*Australorbis*, *Tropicorbis*, *Taphius*). It is noteworthy that not 1 species of *Biomphalaria* is believed to occur in both Africa and the Neotropics, though some parallel evolution has occurred and the *B. choanomphala* (Martens) group of the African great lakes have their counterparts in the lacustrine species *B. andecola* (d'Orb.) and *B. prona* (Martens). Out of 17 species recognized in the Neotropical region, 8 are known to be intermediate hosts, 4 of these possibly having a recent common origin (Barbosa *et. al.*, 1968). The most important host under natural conditions is *B.* (= *Australorbis*) *glabrata* (Say), which is one of the freshwater snails most commonly maintained in laboratories, and is possibly the most intensively studied pulmonate of all. This snail shows considerable infra-specific variation in compatibility with particular strains of *S. mansoni* and is the subject of extensive genetical studies. Clear differences in infectivity between Neotropical strains of *S. mansoni* are described by Richards (1975a) in relation to *B. glabrata*. Reciprocal cross-ings between parasite strains from Puerto Rico and St Lucia suggested the existence of sex linkage in the inheritance of infectivity to particular strains of the snail.

Despite the fact that the evolution of *Biomphalaria* in Africa and in the Neotropical region has been independent for a long time, the colonization of the Neotropics by *S. mansoni* apparently required only slight adaptation on the part of the parasite. In fact some strains of *B. glabrata* can be directly infected with African strains of *S. mansoni*, and Saoud (1965b) observed that an Egyptian strain of the parasite passaged repeatedly through this snail host became no longer infective to its original host *B. alexandrina*. Taylor (1970) confirmed earlier reports that African *Biom-pharia* are susceptible to Neotropical strains of *S. mansoni*.

2. *Schistosoma haematobium*

Adult worms normally occur in man in the veins of the vesicle complex or in the portal vein and its mesenteric branches. Eggs passed in the urine hatch if discharged into fresh water. If the free-swimming miracidium penetrates a suitable species of *Bulinus*,* cercariae can emerge after 3 weeks at an optimum temperature. *S. haematobium* occurs mainly in the African continent, in Arabia and South West Asia and on some islands in the Indian Ocean; limited foci have been reported from the island of Cyprus, Portugal and India.

Before considering intermediate host-specificity a brief review of the taxonomy of *Bulinus* may be helpful. In 1915 when the life cycle of *S.*

* An unusual snail host has been reported for *S. haematobium* in Portugal, where *Planorbarius metidjensis* (Forbes) has been regarded as responsible for transmission in the Algarve province (review by Azevedo, 1969). However, it seems possible that *P. metidjensis* may be only an occasional host and of secondary importance to *B. truncatus* which also occurs in Portugal. A freshwater limpet belonging to the genus *Ferrissia* is claimed to be the intermediate host for a small focus of *S. haematobium* in western India.

haematobium was first described there were over 100 named species of *Bulinus*, all of them founded on shells. At present the total number of species recognized from morphological and biological features is less than 40 (Table III). Mandahl-Barth (1958) reduced a large number of names to synonymy and established 4 species groups, typified by *B. africanus*, *B. forskali*, *B. tropicus* and *B. truncatus*. These groups were

Table III. The distribution within the genus *Bulinus* of susceptibility to infection by *Schistosoma haematobium*. (X = a natural intermediate host, * = susceptibility demonstrated in the laboratory.

Species of *Bulinus*	Susceptibility to infection
The *B. africanus* group	
B. africanus (Krauss)	X
B. nasutus (Martens)	X
B. abyssinicus (Martens)	X
B. globosus (Morelet)	X
B. ugandae Mandahl-Barth	—
B. jousseaumei (Dautz.)	X
B. obtusus Mandahl-Barth	—
B. umbilicatus M-B.	—
The *B. truncatus/tropicus* complex	
(i) Diploid or chromosome number unknown	
B. angolensis (Morelet)	—
B. coulboisi (Bgt)	?
B. depressus Haas	—
B. liratus (Tristram)	—
B. natalensis (Küster)	*
B. nyassanus (Smith)	—
B. obtusispira (Smith)	X
B. succinoides (Smith)	—
B. tropicus (Krauss)	—
(ii) Tetraploid or with higher polyploid chromosome number	
B. hexaploidus Burch	—
B. guernei (Dautz.)	X
B. mutandaensis (Preston)	—
B. octoploidus Burch	*
B. permembranaceus (Preston)	—
B. rohlfsi (Clessin)	X
B. transversalis (Martens)	—
B. truncatus (Audouin)	X

Table III—continued

Species of *Bulinus*	Susceptibility to infection
The *B. forskali* group	
B. forskali (Ehrenb.)	—
B. mariei (Crosse)	—
B. scalaris (Dunker)	—
B. canescens (Morelet)	—
B. senegalensis Müller	X
B. camerunensis M-B.	*
B. crystallinus (Morelet)	—
B. beccarii (Paladilhe)	X
B. cernicus (Morelet)	X
B. bavayi (Dautz.)	*
The *B. reticulatus* group	
B. reticulatus M-B.	*
B. wrighti M-B.	X

Notes:

(i) Not all the snails have been tested for susceptibility in the laboratory but it seems reasonably certain that all the major intermediate hosts under natural conditions are now known.

(ii) A particular strain of the parasite is not necessarily infective to all the susceptible snails indicated, or even to all populations of a single species.

(iii) *B. obtusispira* stands apart from the other diploid members of the *B. truncatus/tropicus* complex and is related in certain respects to the *B. africanus* group (Wright, 1971b).

(iv) It has been claimed that certain forms of *B. coulboisi* are susceptible to infection (Mandahl-Barth, 1965), but the identity of such snails is far from clear.

distinguished by characters of the shell, radular teeth and soft parts. Subsequently, a *B. reticulatus* group has been segregated from the *B. forskali* group, partly according to immunological characters (Wright, 1971b). The *B. truncatus* and *B. tropicus* groups proved to be less clear-cut than at first supposed and special attention has been devoted to this complex because its members differ markedly in their susceptibilities to infection by *S. haematobium*.

Cytological investigations have helped understanding of the *B. truncatus/tropicus* complex, by showing that some species (including *B. tropicus*) have 18 pairs of chromosomes, whereas others (including *B. truncatus*) have 36 pairs and some populations have 54 or 72 pairs (Burch, 1964; Brown and Wright, 1972). Diploid populations ($2n = 36$)

have been found only to the south of the Sahara, but tetraploid populations have a mainly North African range. Octoploid snails are common in the highland of Ethiopia and are reported from South-West Arabia, while the hexaploid form is known only from Ethiopia. The tetraploids *B. truncatus* and *B. rohlfsi* are important intermediate hosts of *S. haematobium*, but although *B. natalensis* of southern Africa has anatomical characters similar to *B. truncatus* it is more closely related to *B. tropicus* according to the diploid chromosome number, electrophoretic characters of egg proteins and immunological reactions (Burch and Lindsay, 1970; Brown *et al.*, 1971). *B. natalensis* is not known to transmit *S. haematobium* under natural conditions, but successful infection has been reported from the laboratory (Lo *et al.*, 1970; Mandahl-Barth *et al.*, 1976). It seems likely that the tetraploids which are natural intermediate hosts originated in diploid populations having limited susceptibility, and that chromosome doubling somehow led to a greatly increased compatibility with *S. haematobium*.

The distribution within *Bulinus* of compatibility with *S. haematobium* is set out in Table III. Compatibility may be considered at the 3 levels of the species group, the species and the population.

(i) Species group. Some highly effective natural intermediate hosts are found in all the four major groupings of *Bulinus* (Table III). Within the *B. truncatus/tropicus* complex there is an association between compatibility and a tetraploid complement of chromosomes, but compatibility is not necessarily maintained at higher levels of polyploidy, as Lo (1969) has found octoploid snails from Ethiopia to be poor hosts for *S. haematobium* from Egypt. Extensive discussion has centred on the possibility that certain strains of *S. haematobium* are specially adapted to snails belonging either to the *B. africanus* or to the *B. truncatus* group. A case was put forward by Le Roux (1958) for the recognition of a northern parasite (*S. haematobium*) and a southern parasite, *S. capense* (Harley), supposing that they had different intermediate host specificities. Wright (1967) expressed this concept in the form that there exists "a composite species of 2 morphologically similar forms". *S. haematobium* sensu strictu developing only in snails of the *truncatus* and *forskali* species groups in North Africa, the Middle East, parts of West Africa, Mauritius and the Malagasy Republic. *S. capense* was considered to develop in species of the *africanus* complex, generally in Africa south of the Sahara. It appeared that the 2 forms of parasite had a mosaic distribution in West Africa. However, no convincing morphological differences between the northern and southern parasites have been found (Pitchford, 1965) and their host specificities are by no means exclusive. The distinction between the *B. globosus*-borne and *B. rohlfsi*-borne strains of parasite in Ghana is not complete (Paperna, 1968), while *B. africanus* group snails from South Africa have been infected with *truncatus*-borne strains from Iran (Howaldt and Armstrong, 1969) and Egypt (Webbe and James, 1971). It therefore seems that adaptation to snails belonging to the *B. africanus* group on the one hand and the *B. truncatus* group on the other has not involved the parasites in any fundamental dichotomy. It seems satisfactory to regard *S. haematobium* as a widely distributed and complex species with considerable freedom in evolving compatibility with snail hosts, depending on their local availability.

(ii) Species. Each of the major groupings of *Bulinus* also includes one or more species which apparently do not serve as natural intermediate hosts, and this impression has been confirmed by observations in the laboratory in the case of those species so far tested. *B. ugandae* was originally described as a subspecies of *B. globosus*, but its resistance to infection in the laboratory led to recognition as a distinct species and neither natural nor experimental infections have been reported to the present day. *B. forskali* is the most common and widely distributed member of its group, but there is no proof that it is involved in the transmission of *S. haematobium*, even though related species are known to be natural intermediate hosts. These susceptible snails all have restricted distributions, in West Africa (*B. senegalensis*), South-West Arabia (*B. beccarii*), Madagascar (*B. bavayi*) and Mauritius (*B. cernicus*). Wright (1971b) suggested that they might have been formerly more widespread and could have suffered in competition with *B. forskali*, which might derive advantage from its freedom from infection by *S. haematobium*.

Only 2 species are placed in the *B. reticulatus* group, but they show markedly different susceptibilities to *S. haematobium*. Although the typical species has a large though discontinuous range in the African continent, there are no reports of natural infection and only limited success has been obtained with laboratory infections (Wright, 1977). In contrast, *B. wrighti* of southern Arabia is a natural host and is susceptible to parasite strains from a remarkable variety of localities (Wright, 1971b).

High variation in susceptibility to infection within the *B. truncatus/tropicus* complex is well illustrated by recent investigations of Ethiopian snails. Formerly identified as "*B. truncatus sericinus*", the numerous and diverse populations living in the Ethiopian highlands and lowlands were all regarded as potential intermediate hosts for *S. haematobium*, and it was puzzling to parasitologists that transmission of the parasite had rarely been demonstrated. Cytological examination has shown that "*B. truncatus sericinus*" comprises snails having at least 4 different chromosome numbers and therefore capable of behaving as distinct species (Brown and Wright, 1972). The cytological groups are diploid (some populations conforming to *B. tropicus* and others to *B. natalensis*), tetraploid (*B. truncatus*), hexaploid and octoploid.

The proven or suspected transmission sites for *S. haematobium* in

Ethiopia are associated with the Awash river in the eastern lowlands where the presumed intermediate host is *B. abyssinicus* (Martens), a member of the *B. africanus* group. In view of the limited area for transmission it seems probable that this parasite strain is not fully compatible with Ethiopian snails belonging to the *B. truncatus/tropicus* complex. However, so far these snails have been extensively tested only against parasite strains from Egypt (Brown and Wright, 1972; Lo, 1969, 1972). Ethiopian *B. truncatus* proved susceptible to infection in the laboratory and a low rate of infection was achieved in octoploids, but no success has been reported with diploid or hexaploid snails. It therefore seems likely that only *B. truncatus*, in addition to *B. abyssinicus*, is a potential intermediate host for *S. haematobium* in Ethiopia. However, *B. truncatus* is rare in the highlands (where higher polyploid snails are common), and this could be an important factor bearing on the apparent failure of the parasite to become established there. The level of transmission of *S. haematobium* in Ethiopia could be greatly increased by a possible shift to *B. truncatus* of the strain at present carried by *B. abyssinicus*. This process could be encouraged by the current development of irrigation schemes in the lowlands. A number of irrigation systems associated with the Awash river have been colonized by *B. truncatus* and consequently this snail is becoming increasingly abundant in areas inhabited by *B. abyssinicus*, which is essentially an inhabitant of undisturbed marshes. There is thus greater opportunity for miracidia of the strain normally transmitted by *B. abyssinicus* to attempt to infect *B. truncatus*.

(iii) Population. Paperna (1968) observed that strains of *S. haematobium* normally transmitted by *B. globosus* in Ghana did not develop with equal success in all of the stocks of this snail infected in the laboratory. These stocks were derived from different natural populations. It would not be surprising if species of *Bulinus* were found to have a capacity for genetically based variation in susceptibility similar to that shown by *Biomphalaria glabrata* in relation to *S. mansoni*, see p. 318.

3. Local specificity and adaptability

Both *Fasciola* and *Schistosoma* provide evidence of the adaptation of infra-specific strains of parasites to particular species or even populations of snail host. For such adaptation to take place there must be adequate isolation at the population level. Schistosomiasis is often transmitted in isolated sites because of the habits of people and as a result of the discontinuous distribution of the snail hosts. Similarly, the dispersal of the parasites is reduced by the tendency of a rural community to use water at traditional places, while the passive transport of an infected snail to a new transmission site is probably a rare event. Since these isolating factors are neither complete nor historically long-lasting, the accumulation of genetic differences in the interacting populations of snails and parasites rarely proceeds far enough for reproductive barriers to be created before the populations are reunited with their neighbours (if extermination does not occur before that). The partial differentiation and later reabsorption of local gene-pools probably produces a large store of genetically determined variation in widely distributed snails and their parasites. It is thus not surprising that the parasites can readily utilize unfamiliar hosts when the opportunity arises, as has occurred in the colonization of Australia by *F. hepatica* and of the Neotropics by *S. mansoni*. It is noteworthy that long-standing isolation between geographical areas, which presumably led to the evolution of the distinctive morphological characters of *L. tomentosa* and also of the numerous Neotropical *Biomphalaria*, did not result in the accumulation of physiological differences of any great consequence to the larval digeneans. Even so, the restriction of larval *Fasciola* to lymnaeids and the total dependence of *S. mansoni* on *Biomphalaria* reminds us that snail/parasite relationships remain flexible only within certain limits, which have been irrevocably fixed at some remote stage in evolution.

IV. Location of snail host

Limited observations on the behaviour of cercariae which enter snails in order to encyst as metacercariae indicate a response to substances diffusing from their hosts (Ulmer, 1970), but a more substantial body of evidence for chemical attraction comes from studies of miracidia. Wright (1959a) suggested that contact is established through 3 behavioural phases: (i) attraction of the miracidium to the environment in which the host is likely to be found; (ii) random movements until the parasite comes within chemical range of the host; (iii) chemotactic responses to the host.

This behavioural scheme is still applicable in the light of subsequent observations and responses to dissolved substances have been extensively studied in a few species of miracidia, but host-specific substances have so far eluded detection. Host-finding by the miracidium of *S. mansoni* is reviewed by Chernin (1974).

The scanning capacity of the miracidium is impressive and using the simplest experimental system of a single miracidium of *S. mansoni* and *B. glabrata*, Chernin and Dunaven (1962) found that infection took place at distances of 33 cm downwards and 88 cm horizontally in large aquaria. Upatham (1972, 1973) found that miracidia could infect snails at depths twice those of most natural habitats on St Lucia and over a distance of 9 m in natural standing water. Miracidia of *S. haematobium* travelled a distance of at least 2·7 m to infect *B. globosus* (Shiff, 1968).

Remarkable as the achievements of the miracidium are it does not always move unerringly to its objective and can easily be distracted. In long narrow channels in which "target" snails are caged at one end and miracidia are introduced at the other end, most or all of the target snails can be protected from infection by the introduction of resistant "decoy" snails (Chernin, 1968a, b). Miracidia may irreversibly damage themselves by making unsuccessful attempts to penetrate resistant snails (Chernin and Perlstein, 1969). Interference with host-finding by miracidia of *S. mansoni* has been confirmed in outdoor laboratory ponds and semi-natural transmission sites in St Lucia (Upatham, 1972; Upatham and Sturrock, 1973). These authors found that 5 kinds of aquatic animals including fish, tadpoles and an especially effective resistant snail (*Physa marmorata* Guilding) all reduced infection amongst target snails, and they discuss the possibility of harnessing the decoy effect to reduce transmission of *S. mansoni*.

Despite its small size, the miracidium is equipped for complex and subtle behaviour by a relatively large brain, which occupies about one-eighth of the body volume in *F. hepatica* (Wilson and Denison, 1970a), and a variety of what appear to be sense organs are revealed by electron microscopy (Brooker, 1972). Miracidia respond to a variety of stimuli including gravity, temperature, water current and light. White light acts as a complex factor and the wavelengths which elicited the greatest response from miracidia of *Schistosomatium douthitti* (Cort), which develops in *Lymnaea*, were in the blue-green region of the spectrum that penetrates deepest into clear water (Wright *et al.*, 1972). Rigorous experimental techniques are necessary in order to interpret the responses to physical factors that bring a miracidium into the vicinity of a possible snail host, for different factors may interact and the age of the miracidium also has to be taken into consideration.

Laboratory and field observations indicate that both *S. mansoni* miracidia and *B. glabrata* congregate near the edges of shallow waterbodies and consequently the outer margins are the important sites of miracidium–snail contact. Complexities in the location of *B. globosus* by miracidia of *S. haematobium* under natural conditions have been gradually elucidated by Shiff (1969, 1974). Shiff's experiments were carried out on the Rhodesian highveld, which experiences a marked seasonality in climate. Winter days are bright and sunny with air temperature reaching 25 °C, but the nights are cold and temperatures at the bottom of ponds remain low, between 9 and 15 °C, though the surface layer may reach 25 °C during the days. In experiments performed during the summer, when daily temperatures in ponds may vary from 18–27 °C, the miracidia were negatively phototactic and most infections of *B. globosus* occurred at the bottom of the experimental pond in both shaded and unshaded areas and at the surface in the shade. No infections occurred at the unshaded surface. However, a different pattern was obtained when the same experiments were carried out in the winter. Then, many infections occurred at the surface in both shaded and unshaded areas, but few were obtained on the bottom. The suspected interaction between light and temperature was demonstrated in laboratory experiments using a horizontal chamber, in which the negative phototactic response under warm conditions could be reversed by lowering the temperature. This seasonal difference in behaviour pattern seems to be related to the habits of the snail host. In summer *B. globosus* is dispersed throughout its habitat and the tendency of the miracidia to seek shade will bring them to the clumps of vegetation and other surfaces on which snails are browsing, whereas in winter the snails tend to congregate at the surface, apparently basking in the warm upper layer of water, which is attractive also to the miracidia.

Numerous observers have reported conspicuously modified behaviour in miracidia that arrive near a snail. This consists of a great increase in the frequency with which the miracidium turns and alters its course while swimming (Davenport *et al.*, 1962; Wilson and Denison, 1970a). It is thought that this kinetic response increases the chance of random contact with a potential host. Many authors have believed that miracidia in this state are responding to chemical stimuli and snail mucus attracted early attention as a possible source of stimulating substances. Wright (1959b) suggested that the fluorescent substances characteristic for the mucus of certain species of *Lymnaea* might attract miracidia to specific hosts, though in later experiments miracidia of *S. mansoni* were found to respond equally well to water which had contained *B. globosus* and *L. peregra* as to extracts of the normal host (Wright, 1966b). The use of water conditioned by snails has provided a rewarding means for testing the responses of miracidia, and active substances have actually been extracted from conditioned water in independent experimental programmes, by Shiff and Kriel (1970) studying *B. globosus* and by Chernin (1970) using *B. glabrata*. The substance produced by *B. glabrata* and named "miraxone" by Chernin is alcohol-soluble, thermostable, stable in storage and has a low molecular weight. Miraxone (or at least a substance producing similar responses from miracidia) was produced by resistant strains or species of *Biomphalaria* as well as by those susceptible to infection, and also by other planorbids, lymnaeids and *B. truncatus*. Miracidia of *F. hepatica* as well as *S. mansoni* were tested in these experiments and responded in a similar way. Miraxone also stimulates *S. mansoni* miracidia to try penetrating artificial and natural substrata (Chernin, 1972). They avidly attacked agar gel soaked in *B. glabrata*-conditioned water but also attacked agar soaked in water conditioned by helisomid or lymnaeid snails. It thus appears that responses to miraxone are generally non-specific both in relation to the snail that produced the substance and to the species of miracidium responding. Etges *et al.* (1975) suggest that one of the components of miraxone is serotonin, a substance of general distribution amongst animals, which is apparently released through the epithelia of the tentacles and upper head-foot region of *B. glabrata*.

Some years before the isolation of miraxone, MacInnes (1965) had used agar blocks to test the responses of the miracidium of *S. mansoni* to a variety of substances (not of snail origin) and found that a sialic acid, short-chain fatty acids and some amino acids attracted miracidia and stimulated attempts at penetrating the gel. An analysis of water conditioned by *B. glabrata* demonstrated the presence of 18 (possibly 20) amino acids which were released by the snails (MacInnes *et al.*, 1974). Snail conditioned water (SCW) produced a positive response from miracidia when tested in small chambers of the type used by Chernin to test miraxone, but in order to obtain quantitative data MacInnes *et al.* (1974) devised a choice chamber which allowed the numbers of miracidia entering different parts of the chamber to be recorded. That part of the chamber containing SCW attracted significantly more entries than did the control part and a similar result was obtained using an artificial SCW, made by preparing a mixture of amino acids at concentrations equivalent

Table IV. Bioassay of water conditioned by *Biomphalaria glabrata* and of an amino acid mixture equivalent to the amino acids determined to be present in it (from MacInnes *et al.*, 1974). p = probability calculated from Student's t-test, 50 miracidia used per experiment.

Time (min)	Entries of *S. mansoni* miracidia into field of view (mean ± standard error)		Number of experiments	p
	control	snail conditioned water		
10	22 ± 2	38 ± 6	5	0·05
20	55 ± 4	78 ± 5	4	0·05
	control	amino acid mixture		
10	21 ± 2	34 ± 4	66	0·02
20	54 ± 2	89 ± 4	6	0·001

to those found to be present in natural SCW (Table IV). This experiment is especially important (as is that by Wright and Ronald (1972), described below) because of the proof that it is actually the amino acids which modify miracidial behaviour. The authors point out that SCW contains a variety of components other than amino acids, including a peptide fraction and a preliminary analysis by gas-liquid chromatography indicated the presence of at least 6 peaks whose identity and ability to stimulate miracidia have not yet been resolved.

Parallel studies using species of *Lymnaea* have produced results rather similar to those described for *B. glabrata*. A variety of substances thought to be potentially attractive to miracidia were found in the mucus of *L. truncatula* by Wilson (1968), including 16 free amino acids and some fatty acids. Further experiments showed that the miracidium of *F. hepatica* responded to the mucus of *L. truncatula* as to the whole snail, the attractive component seemingly being a small molecule, soluble in water and lipid (Wilson and Denison, 1970a). Accordingly, miracidia were tested with short-chain fatty acids (C^6–C^{10}) and a significant increase in the rate of turning occurred that was proportional to the chain length of the molecule (Wilson and Denison, 1970b). The authors suggested that miracidia might be sensitive to all small organic molecules with one or more carboxyl groups. At higher concentrations short-chain fatty acids stimulated miracidia to attach themselves to glass surfaces (Wilson *et al.*, 1971). Wright and Ronald (1972) identified 17 amino acids in water which had contained *L. palustris* (all but 2 of them were found also by MacInnes *et al.* (1974) in water conditioned by *B. glabrata*) and found that an artificial mixture of these same acids was attractive to miracidia of *S. douthitti* under certain conditions of temperature and illumination. Some single amino acids were attractive but others seemed to be repellent.

The source of amino acids and other substances released by snails could be mucus and faeces, and amino acids may also be lost from the epithelial cells as a means for osmoregulation (Potts, 1967). They could also be released from plant material during feeding by snails. There is clearly a need for more precise identification of miraxone and other substances present in snail conditioned water; some of them, apparently proteins (Levy *et al.*, 1973), inhibit the growth of *L.* (*Fossaria*) *cubensis* (*Pfeiff.*) and *B. glabrata*. Whatever effects such substances have on miracidia they seem unlikely to contribute to the host-specific relationship between a particular species of miracidium and its required intermediate host. However, the possibility of miracidial reponse to host-specific stimuli cannot be entirely dismissed. Ulmer (1971) observed that when a *Helisoma trivolvis* (Say) had been restrained at the edge of a petri dish for a period of hours, a miracidium of *Megalodiscus temperatus* (Stafford) released at the centre of the dish sometimes swam straight to the snail as if following a gradient. Further, when placed equidistant between a specimen of *H. trivolvis* and a resistant snail (*L. reflexa* (Say) about 90% of the miracidia were seen to be attracted to the host species. None the less, the balance of evidence from investigations of host location by miracidia favours the view that host-specificity is determined essentially by events that occur at the molluscan surface (though attachment and attempted penetration can be stimulated by non-specific stimuli) and within the body of the snail. One is surprised at the wastage of miracidia that seems likely to take place during host location, but presumably this is compensated for over the evolutionary time-scale by the value of chance encounters in leading to the exploitation of new intermediate hosts.

V. Establishment and survival in the snail host

The establishment and successful development of a larval digenean in its snail host is the outcome of a combination of favourable circumstances which are not easily disentangled and a bewildering variety of terms are used in discussions. The central concept is "compatibility", a compatible host being one in which the parasite can live, carry out its life processes and perpetuate its species, and these functions are completely or partially prevented in an incompatible host (Cheng, 1968). A snail which cannot be infected with a particular parasite is often described as "refractory", which is a broad term giving no indication of the cause of refractoriness, while a "susceptible" snail is one in which the parasite does develop successfully. "Insusceptibility" and "resistance" are often employed loosely, though they can have different meanings. Insusceptibility can be regarded as the failure of the parasite to establish itself because of passive attributes of the snail, among which have been postulated properties of its surface and the lack of certain biochemical factors necessary for the parasite development. Resistance, on the other hand, is a term commonly applied to active responses to infection, usually cellular, which are innate or possibly can be acquired. The terms used in this account will usually be those of the original authors.

A. Insusceptibility

Reference has already been made to the damage that miracidia may suffer during an unsuccessful attempt at penetration and they may also die because the effort has exhausted their energy reserves. Barriers to penetration possibly lie in the topographical or biochemical properties of the snail surface. Kinoti (1971) suggested that an organ at the anterior extremity of the miracidium of *Schistosoma mattheei* (Veglia and Le Roux) might be used for attachment to the host, the degree of fit between this organ and the snail's epithelium determining the success or failure of penetration. A similar view of the formation of a stable attachment by the miracidium of *F. hepatica* to *L. truncatula* is taken by Wilson *et al.* (1971), who suggest that this may depend on interaction between the corrugated surface of the miracidial apical papilla and microvilli (about 1 μm long) which project from the host's epithelial cells. Schutte (1975) postulates the existence of a penetration barrier at the surface of *Biomphalaria salinarum* (Morel.) in order to explain the small number of miracidia of *S. mansoni* which became established as primary sporocysts. He observed that although snails were exposed to 10 miracidia each, usually only 1 or 2 primary sporocysts developed. No cellular reaction was seen in serial sections around the primary sporocysts, the number of larvae that degenerated after penetration was negligible and there was no evidence suggesting acute reaction with rapid destruction of larvae.

In order to identify causes of insusceptibility arising within the snail's body, a knowledge of the biochemical needs of digenean larvae is essential, but although the satisfaction of the nutritional requirements of the developing larva is an essential condition for a compatible host/parasite relationship, the general nature of the parasite's needs are not at all well known and any host-specific requirements are not understood. Progress on these fronts can be hoped for from the culturing of larval parasites *in vitro* (Voge and Seidel, 1972; Hansen, 1975). The sporocyst larva is perhaps most suitable for this kind of investigation, because lacking a mouth all its biochemical requirements are transported through the body wall and there is not the complication of the direct ingestion of host tissue as in the case of the redia larva. Regrettably one cannot take at face value the report of Muftic (1969) who claimed to have achieved the *in vitro* development of sporocysts and cercariae from miracidia of *S. mansoni*. According to Muftic this took place in snail haemolymph to which he added a substance extracted from tissues of *B. glabrata* and having properties similar to ecdysones (arthropod moulting hormones). The author suggested that the development of trematode larvae might depend on morphogenetic substances, possibly steroids, produced in the ovotestis of the snail host. Bayne (1972) endeavoured to reproduce Muftic's results, using the same methods so far as this was possible, but without success and he drew attention to several aspects of Muftic's account that throw doubt on its accuracy.

Several reports indicate that the production of cercariae is reduced, or even wholly prevented, by the starvation of snails, and observations by various authors on *B. glabrata* and other snails suggest that glucose, either obtained as such or derived from host glycogen, passes through the sporocyst wall and becomes stored as glycogen in the cercarial body (Erasmus, 1972). Southgate (1970) points out that the way in which the parasite uses the host's lipid substances could depend on whether or not an anaerobic environment is produced by the developing larvae, and this would depend in part on the intensity of infection. There is a substantial amount of evidence that trematode larvae utilize free amino acids taken up from the tissues and haemolymph of prosobranch and pulmonate hosts (Targett, 1962; Cheng, 1963; Porter and Gamble, 1971) and it may be that an important element in compatibility is the provision by the snail of an appropriate array of amino acids in suitable concentrations for the build-up of protein in the parasite. However, Lee and Cheng (1972) believe that the concentrations of amino acids in the haemolymph of *B. glabrata* are too low to satisfy the rate of protein synthesis in larval *S. mansoni*. They confirmed previous observations of a reduction of the total protein concentration in the haemolymph and demonstrated that this is largely due to a reduction in haemoglobin, which was observed also by Coles (1971). These observations provide an explanation for the increase in the heartbeat rate of infected *B. glabrata* reported by Lee and Cheng in other papers, and which may be supposed to compensate for the decreased oxygen-carrying capacity of the blood. The selective depletion of a particular form of snail haemoglobin by sporocysts of *S. mansoni* is reported by Anteson and Williams (1975) who suggest that this could have resulted either from selective uptake by the sporocyst of entire haemoglobin molecules having an optimal size and charge, or from exogeneous proteolytic degradation of this particular population of haemoglobin molecules.

B. Resistance

1. *Innate humoral factors*

Although some molluscan substances have an antibody-like effect, no substance comparable to vertebrate immunoglobin has yet been demonstrated (Tripp, 1970). The existence of an "immune-like" substance was suggested by Benex and Lamy (1959) to explain why miracidia of *S. mansoni* were immobilized by aqueous extracts of whole *P. corneus* (L.) and *B. glabrata* from a resistant strain, but not by extracts of *B. glabrata* from a susceptible strain. Immobilization of many miracidia of *S. mansoni* was also obtained in tissue extracts from uninfected *B. truncatus* and *H. caribbaeum* (d'Orb.) (Michelson, 1964). Other evidence for the existence of apparently innate humoral factors stems from the observed death of parasites within snails in the absence of cellular reactions, as in the case of *Trichobilharzia elvae* (Miller) that penetrated 2 species of *Lymnaea* which were not normally hosts (Sudds, 1960) and species of Echinostomatidae transplanted into abnormal hosts (Heyneman, 1966). Heyneman found that larvae transplanted from natural host to natural host (either *L. rubiginosa* or *Indoplanorbis exustus*) continued to develop normally, but larvae transplanted to the abnormal host died, apparently because of "physiological rejection" (though they might have failed to develop because the incompatible host did not satisfy their nutritional requirements). Faulk *et al.* (1973) present evidence for the existence in the haemolymph of *B. glabrata* of proteins capable of binding antigens derived from *Echinostoma lindoense* (Sandground and Bonne), and present in greater quantities in infected snails than in uninfected snails. The mode of action of any humoral factors is unknown, though a technique for the parabiotic joining of snails by means of capillary tubing (Heyneman *et al.*, 1971) could be of potential value in the investigation of this and other aspects of compatibility.

2. *Cellular responses*

Molluscs show a well developed capacity for phagocytosis and encapsulation, which appears to provide the main internal defence mechanism against invading parasites. Cellular responses are produced in *B. glabrata* by a variety of materials (Tripp, 1961), including some that are biologically inert and which stimulate reactions similar to those elicited by parasites.

In a resistant strain of *B. glabrata*, larvae of *S. mansoni* were destroyed within 24 to 48 hours after penetration (Newton, 1952). A cellular infiltration around the miracidium was followed a fibrotic type of encapsulation and degeneration of the parasite, whereas development proceeded without cellular reaction in snails from a compatible strain. In hybrid snails derived from crossing the resistant and the susceptible strains, the parasites developed normally in some snails, but in others there was a tissue reaction of variable intensity and a number of snails contained some parasites surrounded by a tissue reaction and others developing normally (Newton, 1954). These results were attributed to genetic differences between individual snails and also to variation amongst the miracidia, possibly of a genetic nature. Brooks (1953) invoked possible variation among miracidia to account for the development of a few normal sporocysts of *S. mansoni* in *Tropicorbis havanensis* (Pfeiff.), a snail in which the majority of the trematode larvae were encapsulated and destroyed. Kinoti (1971) observed that the few miracidia of *Schistosoma mattheei* successful in penetrating the incompatible snail *B. truncatus* were destroyed by a strong infiltrating reaction of amoebocytes, whereas there was no reaction to the early larval stages of this parasite in *B. africanus*, a natural intermediate host.

Although cellular reactions to larval digeneans may be commoner in incompatible species or strains of snails, they are seen also in satisfactory intermediate hosts. Sudds (1960) recorded the death of sporocysts of *S. douthitti* accompanied by marked tissue reaction in the normal host *Physa gyrina* (Say). Only a small proportion of miracidia of *S. mansoni* survived to produce mother sporocysts even in a "highly susceptible" strain of *B. glabrata* (Pan, 1965). Pan (1965) distinguished 2 types of tissue response:

(i) minimal and focal usually associated with mother sporocysts less than 48 hours old and (ii) generalized and proliferative appearing 6–7 weeks after infection when the cercariae had begun to emerge. The first type of response apparently was not an effective means of resistance as it was associated with only a small number of degenerating sporocysts which might have been killed by humoral factors.

It is interesting to compare the cellular reactions described for *B. glabrata* with those seen in African *Biomphalaria*. Both Meuleman (1972) and Schutte (1974) found that tissue reactions were generally less than in the Neotropical species. Schutte notes an absence of Pan's type (i) response in *B. salinarum*, tissue reactions did not appear until about 13 weeks after infection and the structure of the digestive gland was not obviously modified. He concludes that a more compatible relationship exists between South African *S. mansoni* and *B. salinarum* than between Pan's Puerto Rican parasite strain and *B. glabrata*. However, it should be noted that wide variation in compatibility is known to exist between various strains of snail and parasite in the Neotropical region.

Marked cellular reactions to escaping cercariae seem to be rather common; they have been observed, for example, by Probert and Erasmus (1965) in the case of a normal trematode parasite of *L. stagnalis*, and by Kinoti (1971) in *B. africanus* infected with *S. mattheei*. Kinoti suggested that cercarial antigens were liberated through the rupture of the sporocyst walls and stimulated the host's amoebocytes at a distance.

The amoebocytes known to play an important part in the defences of at least some pulmonates against larval trematodes appear to arise in a variety of sites. Pan (1958) reported several sites of amoebocyte production in *B. glabrata*, including an aggregation of "lymphoid" tissue in the wall of the kidney adjacent to the paricardium. Kinoti (1971) described a similar "amoebocyte organ" in the mantle of several species of *Bulinus*. On the other hand, amoebocytes are present in the blood and in connective tissue throughout the body of *L. stagnalis*, and no definite amoebocyte organ has been found (Sminia, 1977).

The production and behaviour of amoebocytes in *B. glabrata* infected with *Echinostoma lindoense*, a natural parasite of birds, has been described in detail by Lie and his colleagues. Lie *et al.* (1975a, 1976) describe an amoebocyte-producing organ located between the pericardium and the posterior epithelium of the snail's mantle cavity, a position different from that occupied by the lymphoid tissue found by Pan (1958). In uninfected snails this organ was small and thin, but it became distinctly larger 2 or 3 days after a second infection with *E. lindoense*. Some *B. glabrata* are highly susceptible to *E. lindoense*, but laboratory-bred snails show a wide range of natural resistance (Lie and Heyneman, 1976a). In most resistant snails all sporocysts are destroyed in the peripheral tissues soon after miracidial penetration. In less resistant individuals some sporocysts reach the ventricular cavity of the heart, where they are then encapsulated. Aggregations of amoebocytes form in the ventricle in response to the presence of sporocysts, but this does not always result in the destruction of the parasites (Lie and Heyneman, 1976b). When sporocysts do escape encapsulation the host snail becomes highly susceptible to reinfection with the same parasite and its amoebocytes apparently have lost their ability to recognize the parasite as foreign. Snails from normally susceptible strains were enabled to recognize and destroy sporocysts which developed from irradiated miracidia (Lie *et al.*, 1975).

Some pulmonates produce cellular reactions to larval trematodes that do not usually destroy the parasites but serve to contain them and thereby protect the snail from extensive invasion and widespread damage. The sporocysts of certain plagiorchid flukes are characteristically found abutting on their host's alimentary tract, apparently because their migration is confined by proliferation of the basement membrane surround the snail's gut (Schell, 1962). At least 1 larval fluke seems to have taken advantage of cellular reactions by the host, for although *P. gyrina* produces a marked tissue reaction described as an "isolating-protective capsule" (Byrd and Maples, 1969) around larvae of *Dasymetra conferta* Nicoll, these usually complete their development (though in an unusual manner without the differentiation of the familiar sporocyst or redia). Some of the larvae are killed as the result of encapsulation, but the capsule is of benefit to the parasite as well as the host, for the surviving larvae derive sustenance from it and apparently use it as a substitute for their reduced somatic tissues.

3. *Acquired resistance*

One source of evidence to support the view that snails can acquire immunity, is the rarity in nature of individual snails carrying multiple infections with more than 1 species of digenean. Cort *et al.* (1937) carried out an extensive survey, by determining the incidences of 17 species of trematodes in over 7000 specimens of *L. emarginata angulata* (Sow.) from 2 lakes in North America. Multiple infections were detected in 12·3% of 4304 positive snails (529 double, 17 triple, 1 quadruple). Statistical analysis indicated that although the observed frequencies for some combinations of trematodes found in multiple infections were in accordance with random association, other combinations of parasites were significantly rarer than expectation, suggesting the existence of some kind of immunity. On the other hand, in view of the high incidence of infections with 2 species of larval trematodes in the estuarine prosobranch *Velacumantus australis* (Quoy and Gaimard), Ewers (1960) concluded that infection with 1 species of trematode might predispose a snail to infection by another. This possibility was also suggested by Bourns (1963) who investigated the incidence of 6 species of trematodes in 1659 specimens of *L. stagnalis* from 1 locality in Canada. Multiple infections were detected in 13·9% of 801 positive snails (215 double, 15 triple) and the frequencies of the various combinations of parasites found in single snails varied from about half to 27 times the frequencies that would be expected in the event of random infection. For 4 combinations of parasites the divergence of the frequencies from expectation was highly significant at the 0·01 probability levels; in each of these cases a xiphidocercaria was present and the observed number of multiple infections exceeded expectation. Multiple infections in *L. peregra* in Scotland also all included a xiphidocercaria (Wright, 1966b).

The interpretation of multiple infection rates in snails in nature has usually depended on the assumption that the snails in a particular locality are equally exposed to infection by all the trematodes involved, but this may not always be so. Ewers (1960) drew attention to the localized high incidence of multiple infections that could arise as a result of the simultaneous evacuation from a bird of the eggs of 2 or more species of trematodes. On the other hand, a significant reduction from an expected number of multiple infections could be produced by antagonism between different larval trematodes in the same snail, for Wesenberg-Lund (1934) observed echinostome rediae to prey upon the larvae of other trematodes and this behaviour has been investigated extensively by Lie *et al.* (1968), and recently in Egypt by Rysavy *et al.* (1974). Echinostome rediae are facultative predators of other trematodes developing in snails including *Biomphalaria*, *Indoplanorbis* and *Lymnaea*. When 2 different echinostomes occur in the same snail 1 species assumes a dominant role and attacks the other; indirect antagonism has also been observed that possibly involves competition for nutrients or the production of inhibiting substances by the parasite larvae. The outcome of these antagonisms may be the repression or elimination of subordinate trematode species from double infections. It may be concluded that statistically significant deficiencies in observed frequencies of natural multiple infections in snails are unsatisfactory evidence for acquired resistance unless the underlying factors are thoroughly understood.

There is, however, evidence from the laboratory for acquired resistance in a series of extensive investigations summarized by Cort *et al.* (1945) in which *L. stagnalis* and *L. emarginata* were exposed to infection by *Cotylurus flabelliformis* (Faust). This strigeid fluke requires a second intermediate host, normally a different individual of the same species of snail that serves as the first intermediate host, which the cercaria enters to encyst as a metacercaria. It was observed that the presence of sporocysts of *C. flabelliformis* in a snail almost entirely prevented its penetration by cercariae. Cort *et al.* (1945) suggested that this "immunity" might be due to some chemical change in the superficial tissues of the snail. Any such changes that take place seem to be surprisingly rapid, as Anteson (1970) failed to obtain even 1 double infection in *L. catascopium* Say exposed simultaneously to miracidia of *C. flabelliformis* and *Diplostomum flexicauda* (Cort and Brooks).

Further evidence for acquired resistance has been obtained from experiments on *B. glabrata* infected with *S. mansoni*. Aqueous tissue extracts prepared 9 days or more after infection produced immobilization of about 90% of miracidia of *S. mansoni*, while extracts from uninfected snails from the host strain did not produce immobilization (Michelson, 1964). Extracts from some of the resistant snail species tested, including *P. corneus*, *L. palustris* and various forms of *Helisoma* produced no significant immobilization, although *B. truncatus* and 1 strain of *H. caribbaeum* consistently gave moderate levels of immobilization. In order to find out if miracidial immobilizing substance had any specific association with *S. mansoni*, Michelson (1963) tested tissue extracts from snails innoculated or infected with an acid-fast bacillus, bovine albumen, polystyrene spheres, *S. mansoni* eggs, echinostome metacercariae and a nematode. Significantly high levels of immobilization were observed only in extracts from snails innoculated with *S. mansoni* eggs or infected with the nematode, and the percentage of miracidia immobilized was far lower than in extracts from snails infected with *S. mansoni* larvae. Michelson pointed out that the observed effects were not necessarily due to an antigen–antibody type of reaction but might have been caused by toxins produced by the parasites or by products resulting from alterations in the host's metabolism.

Some strains of *B. glabrata* are resistant to *E. lindoense*, and sporocysts developing in such snails are killed after migrating to the heart by amoebocytes (Lie and Heyneman, 1975). When these snails were then reexposed to miracidia of the same species, the sporocysts were destroyed earlier, soon after penetration and mostly in the head-foot region. The response of the snail tissues appeared to be highly specific, for the sensitized hosts remained susceptible to other trematodes. Sensitized snails reexposed to miracidia developed an even higher degree of resistance, expressed by the destruction of a higher proportion of the sporocysts before they were able to migrate from the host's peripheral tissues (Lie and Heyneman, 1976c).

Despite the demonstration of immobilizing substance(s) produced by *B. glabrata*, snails already infected with *S. mansoni* can be reinfected (Pan, 1965; Kagan and Geiger, 1964). File (1975) confirmed these observations and suggested that because schistosomes are dioecious it is an advantage to establish double infections in snails because any bi-sexual infections arising will increase the probability of a successful completion of the life cycle through cercariae of both sexes entering the definitive host. Successful second infections of *S. haematobium* have been obtained also in *B. truncatus* (Wadji, 1966). Using miracidia labelled with radioselenium, Christensen *et al.* (1976b) found that the penetration of *F. hepatica* into *L. truncatula* was not influenced by an already existing infection.

C. Inheritance of resistance and insusceptibility

The patterns of host-specificity in the relations between larval digeneans and their snail hosts suggest that the inter-relationships have evolved on a genetic basis, and direct evidence for the genetic background in the snail is beginning to accumulate. An experimental difficulty lies in the self-fertilizing ability of many freshwater pulmonates that makes necessary a genetic marker (albinism has been commonly used) to ensure that the intended crossings in hybridization experiments have actually occurred.

Newton (1953, 1954, 1955) obtained F^1 hybrid progeny from crossing 2 strains of *B. glabrata*, one (from Puerto Rico) compatible with a Puerto Rican strain of *S. mansoni* and the other (from Brazil) resistant to this parasite. The hybrid snails gave an intermediate frequency of infection and it appeared that tissue reactivity was influenced by multiple genetic factors. However, a considerable proportion of snails from the resistant strain that were exposed when young became infected and viable cercariae were shed (as in the case for unusual lymnaeid hosts of *F. hepatica*). Other investigators (Paraense and Correa, 1963; Kagan and Geiger, 1965) have also obtained progeny of intermediate susceptibility by crossing susceptible and resistant strains of *B. glabrata*.

An extensive and continuing programme of genetical investigations by Richards (1972, 1973, 1975b) has revealed the complicated manner in which insusceptibility to a single Puerto Rican strain of *S. mansoni* is inherited in *B. glabrata*. The snails were derived from the colonies originally established by Newton, and another stock founded with snails from Puerto Rico. By selection and self-fertilization, 3 main lines of snails have been isolated: juvenile and adult susceptible, juvenile-susceptible/adult-refractory and juvenile and adult refractory. Snails of a fourth susceptibility type showed juvenile susceptibility but modification of adult susceptibility, with unpredictable infection frequencies and delayed parasite development. Juvenile susceptibility was apparently controlled by several genetic factors, while juvenile snails became refractory when adult as the result of a single dominant factor. The mechanisms acting in the various forms of insusceptibility studied by Richards have yet to be described in detail, though differences between snail strains in the disposition of aggregations of amoebocytes appear to involve genetic factors (Richards,

1975c). There may be a special epidemiological significance for the juvenile-susceptible/adult-refractory strain of *B. glabrata*, for while infected juvenile snails could ensure the survival and transmission of the parasite, those snails escaping infection until maturity would have several months of refractory reproductive life, allowing the maintenance of a flourishing snail population. Clearly, epidemiological models of the transmission of intestinal bilharziasis should take into account possible differences in the parts played by juvenile and adult snails. Another interesting point is that the single gene responsible for adult insusceptibility was derived by Richards (1975c) from 2 snails, and the snail lines of the fourth type displaying inconsistent adult susceptibility were all traceable to a single snail. The frequencies of these genes in natural populations of *B. glabrata* have yet to be determined.

The existence of heritable factors in *B. nasutus* promoting resistance to infection by *S. haematobium* was suggested by McClelland (1965) to account for a gradual decline in infection rates observed in a colony of snails maintained in the laboratory for several years. Selection appeared to have occurred as a result of snails that had not become infected during experiments being restored to the laboratory stock. Further experiments by Sturrock (1968) demonstrated that even the first generation of laboratory-bred generation of *B. nasutus* was much less susceptible to infection than snails from the natural parental population, and it appears that this short period of laboratory maintenance had significantly increased the proportion of resistant snails.

D. Infectivity of miracidia

The numerous reports of variation in the infectivity of individual miracidia from a single strain of parasite are largely unexplained, though genetic differences between miracidia have been suspected. The selection of infective miracidia of *S. mansoni* by repeatedly passaging the parasite through an incompletely resistant strain of *B. glabrata* produced a doubling in the average infection rate within 12 months (Kagan and Geiger, 1965). As a general principle, it is reasonable to suppose that the establishment of a successful infection depends on the genetic constitutions of both the individual miracidium and an individual snail, and hence the infection rates obtained in experiments reflect the gene frequencies in particular populations of both miracidia and snails. From this point of view, Basch (1975) reinterprets the data derived from experiments with *L. stagnalis* used by Donges (1974) to support his idea that snail populations are homogeneous with respect to infection by larval trematodes.

Fortunately, the separation of the sexes in *Schistosoma* facilitates the production of inter-specific hybrids and interesting observations have been made on the infectivity of hybrid miracidia (Wright and Southgate, 1976). The first step in attempting to hybridize 2 schistosome species is to infect a number of snails with a single miracidium of one or other of the potential parental species. Consequently, the cercariae produced by each snail are of 1 species and 1 sex, which can be determined by the preliminary infection of a mouse or hamster and recovering the adult male or female worms. More experimental mammalian hosts can then be infected with cercariae of both sexes, combining the 2 schistosome species. Pairing between worms of different species and the production of eggs is not conclusive evidence of hybridization by itself, because a female worm carried by the male of a different species can grow and produce eggs by parthenogensis. The criteria used to detect true hybridization include the shape of the eggs laid by the F_1 worms, and the infectivity of the miracidia to snails. That hybrid miracidia combine to a greater or less degree the infectivities of their parent species seems to have been first reported by Le Roux (1954) and is demonstrated in extensive investigations (Taylor, 1970; Wright and Southgate, 1976). A particularly significant cross has been achieved between the species *S. mattheei* and *S. intercalatum* Fisher, which show clearly distinct specificities in regard to snail hosts (Wright, 1974). In these experiments the Lower Guinea strain of *S. intercalatum*, which develops only in members of the *B. forskali* group, was crossed with *S. mattheei*, developing only in some members of the *B. africanus* group. Hybrid miracidia are infective to members of both the snail groups characteristic for the parental schistosomes and this hybrid infectivity persisted through the first 3 filial generations.

E. Evolution of compatibility and host-specifity

Many relationships are compatible in the sense that both the host snail and the larval trematode successfully perpetuate their species, but an individual infected snail may suffer marked pathological effects (Wright, 1966c; Cheng, 1967). These effects may not directly cause death and indeed infected snails may live longer and grow faster than uninfected ones in the laboratory, but under natural conditions it is possible that infected snails have a lowered tolerance of adverse environmental conditions. Even though infected snails can survive extensive damage to their tissues, their reproductive capacity is often completely or partly destroyed. For example, 2 or 3 weeks after infection with *F. hepatica*, egg production by *L. truncatula* ceased entirely and was not resumed (Hodasi, 1972). This was so in snails infected at the ages of 2 weeks, 4 weeks (the age at which oviposition normally began) and 6 weeks (Table V). The total egg production of snails infected at 2 weeks was depressed to a mere 3% of that achieved by uninfected snails during their lifetime. Hodasi (1972) found that the ovotestis of infected snails degenerated and atrophied, even though this organ was rarely invaded by the parasite. Some other examples of a decline in fecundity of infected snails are discussed in Section VI.c.

Table V. Weekly egg production of 6 individual *Lymnaea truncatula* infected with *F. hepatica* and 2 infected snails (data from Hodasi, 1972).

Age (weeks)	Uninfected (mean for 2 snails)	Infected at 2 weeks of age (mean for 4 snails)	Infected at 6 weeks of age (mean for 2 snails)
3	22	15	19
4	85	29	67
5	145	0	104
6	150	0	139
7	106	0	128
8	142	0	146
9	125	0	57
10	102	0	0
11	133	0	0
12	88	0[a]	0
13	83	0[a]	0
14	87	0	0
15	70	0	0
16	43	0	0[a]
17	45	0	0
18	63	0	0
19	35	0[a]	0
20	5[a]		0
21	0[a]		0[a]
TOTAL	1531	44	661

[a] Death of 1 or more snails.

If, as seems likely, cellular reactions and other means by which snails can prevent the establishment of a parasite have a genetic basis, natural selection might be expected constantly to favour a reduction in susceptibility and an improvement in the defences of the host. How then do digeneans maintain compatible relations with their snail hosts despite their vulnerability to defensive reactions? One possible reason is that genetic factors controlling defence mechanisms in the host may affect other kinds of attribute, so that their selective value is not determined solely by the effects of parasitism, but also in relation to the whole genetic background and in relation to varying environmental conditions. Consequently, the balancing of advantages and disadvantages might lead to the establishment of polymorphisms, which could be highly complicated in a species such as *B. glabrata*, known to have multiple genetic factors affecting resistance to *S. mansoni*.

Natural selection acts not only on the snail hosts but also on the larval digeneans and it is likely that these play an active part in creating and maintaining compatible relations by evading the defensive mechanisms of their hosts. This theory has been developed mainly by Wright (1966b, 1971a, 1974). A primary consideration is that while snails are capable of cellular reactions to a variety of foreign bodies including invading digenean larvae, these reactions are elicited far less frequently than might be expected. Particularly striking is the case where one species of parasite triggers a cellular reaction in a snail while another related species develops successfully without any apparent host response (as in the relations between *S. haematobium*, *S. intercalatum* and *B. crystallinus*). It is reasonable to assume that the successful miracidia are not recognized by the snail as being foreign bodies and evidence that the living parasites have an active means of protection is provided by the fact that dead parasites are attacked and encapsulated. Moreover, the release of cercariae can result in sporocysts which have remained immune to attack throughout development being destroyed. Adult schistosomes are believed to avoid provoking the immunological defences of their vertebrate hosts by carrying on their surface a camouflage of "host antigen" (Smithers *et al.*, 1969). It is not known whether the camouflaging substances are absorbed from the host or are synthesized by the parasite, but it is possible that similar devices are employed by the intra-molluscan stages. Wright (1974) points out that the mechanisms employed may not be the same throughout the Digenea and indeed are unlikely to be the same in *Schistosoma* and *Fasciola*, in view of the fundamentally different morphological changes that their miracidia undergo after penetrating into the snail host.

Thus one may envisage a continually shifting balance between the defensive reactions of the snail host and the counter-defence employed by larval digeneans, and hence the preservation of continuous though dynamic states of compatibility. A broadly similar view of the parasite's strategy to avoid destruction is taken by Heyneman *et al.* (1971) who extracted antigens from rediae of *E. lindoense* that had developed in *B. glabrata*. Despite careful washing of the rediae, it was demonstrated that snail antigenic material was responsible for a number of the antibodies produced in rabbit serum. This snail material could have been carried over because it was closely bound to the redial surface, but it might have been present in the redial gut. Further experiments of this kind using sporocysts, which having no mouth are incapable of ingesting host materials, would be of great interest.

In the case of a parasite which is capable of developing in some races of a snail species but not in others (e.g. *S. mansoni* in *B. glabrata*), Wright (1971a) suggests that the restriction is due to an inability to synthesize more than a narrow range of antigens, rather than to limited capacity for absorption. On the other hand, where a species of parasite can utilize hosts belonging to different molluscan families it seems probable that absorption of host antigen is the means of camouflage. Some pulmonates and prosobranchs are outstanding as hosts for a wide variety of digeneans, possibly because their defensive mechanisms are poorly developed or because their antigens can be mimicked by many parasites. Wright speculates that the study of such interractions between pulmonates and their digenean parasites might even throw light on problems of tissue immunity in vertebrates.

VI. The snail as a food source

The nutritional requirements of larval digeneans have already been considered in relation to host-specificity (p. 311). Further attention to the snail host as a food resource raises some interesting and fundamental aspects of the host/parasite relationship that deserve more investigation.

A. Migration within the host and site of development

Many larval digeneans perform a migration within their molluscan host, usually via the blood circulatory system and probably in relation to their requirements for food and space. For example, mother sporocysts of *Schistosoma* species usually develop in the tissues of the snail's foot, whereas daughter sporocysts move to other sites, especially the digestive gland, where development is completed. In general, larval digeneans are found in characteristic sites among a variety of molluscan organs and tissues with a regularity that indicates the operation of factors other than random behaviour (Ulmer, 1971). Daughter sporocysts and rediae frequently inhabit the digestive gland and ovotestis, presumably because these organs are particularly rich in accessible nutritive substances.

A predisposition for certain molluscan organs is also shown by some larval digeneans in their second intermediate host. The cercariae of certain echinostome flukes migrate to the kidney of species of *Lymnaea* and *Indoplanorbis* and there encyst as metacercariae to the exclusion of almost all other tissues (Najarian, 1954; Heyneman, 1966). The cercariae of members of the family Brachylaimatidae that enter the land-snails *Anguispira alternata* (Say) and *Mesodon thyroidus* (Say) to encyst as metacercariae do so in characteristic sites, being found only within the pericardial chamber or in the kidney (Ulmer, 1971).

A remarkable example of site restriction is shown by the cercariae of *Cotylurus flabelliformis* in abnormal snail hosts. These cercariae usually enter a lymnaeid to complete their development into metacercariae, but a planorbid or a physid snail can be utilized providing these are infected with sporocysts of certain other species of trematodes, in which the *Cotylurus* cercariae become hyper-parasites (Cort *et al.*, 1945; Bourns, 1963). Perhaps in this way the hyper-parasites obtain resources which would not otherwise be available to them in an insusceptible snail. Lie *et al.* (1968) made the interesting suggestion that a predatory redia may gain advantage by feeding on another larval trematode which has already expended metabolic energy in converting snail tissue to trematode tissue.

B. Intensity of infection

The proportion of the total body weight of heavily infected snails that is contributed by larval trematodes was estimated to be about 20% by Hurst (1927) and Wesenberg-Lund (1931) in the cases of *Echinostoma revolutum* (Froelich) in *Physa occidentalis* Tryon, and *Leucochloridium paradoxum* Carus in *Succinea putris* (L.). Data given by Cheng (1971) indicate that this proportion can be about 27% for *Physa sayii* (Tappan) infected with *Echinostoma revolutum*.

Further calculations of the biomasses of digenean larvae supported by their snail hosts would be of great interest, but intensity of infection has been more frequently estimated in terms of the numbers of larvae produced. A single miracidium of *F. hepatica* ordinarily gives rise to about 40 rediae in *L. truncatula* but an infection with 2 miracidia yields the increased number of more than 70 rediae (Kendall, 1964). *B. truncatus* exposed to 2 or more miracidia of *S. haematobium* shed more cercariae than snails exposed to a single miracidium (Chu *et al.*, 1966). Such observations suggest that the number of miracidia which establish successful infections has some bearing on the numbers of sporocysts, rediae and eventually of cercariae shed by a snail.

Generally it seems that the numbers of parasite larvae depends also on the size of the snail, and the larvae of some species appear to exploit the resources of the host to the point of destruction. For example, many individuals of *B. tropicus* collected in Africa had so much of their body volume occupied by amphistome larvae that the continued survival of the snail seemed unlikely. However, other factors, at present obscure, may also play a part in determining the population density of digenean larvae within an individual snail, for in some cases the biomass of the parasites seems to reach a plateau well before achieving its full potential (Lie *et al.*, 1968). This indicates the operation of mechanisms that prevent premature exhaustion of the host's resources. File (1975) suggests that schistosomes, being dioecious, could derive advantage from not allowing the progeny of a single miracidium to exhaust the resources of their host. Establishment of a second infection by a different miracidium (possibly of a different sex) would increase the chances of cercariae of both sexes being transmitted to the definitive host.

C. Management of host's resources

In a series of papers dealing mainly with the prosobranch *Hydrobia ulvae* (Pennant), Rothschild (1938) discussed the possible relation between 2 phenomena, the destruction of the gonads in snails infected by some larval digeneans, and the generally large size of individual snails found naturally infected (reviewed by Rothschild and Clay, 1952; Cheng, 1971). Rothschild (1938) suggested a causal relationship in the words "an increase in the soft parts of the body as well as the shell, is thus of great advantage to the parasite. The faculty of producing this increase in size is a character which is presumably most susceptible to selection and its widespread occurrence among the Trematoda is therefore not surprising."

This idea has attracted considerable attention and it has been suggested that naturally infected snails could tend to be large because such individuals produce larger amounts of attractive mucus, and consequently are more likely to attract miracidia (Croll, 1966). It should also be noted that in certain strains of *B. glabrata* it is only adult snails that are susceptible to infection with *S. mansoni*. Controlled experiments comparing the growth rates of infected and uninfected snails have been confined almost entirely to pulmonates, especially *Biomphalaria* species infected with *S. mansoni*. It is important to realize that in this combination the parasite larvae complete their development in the digestive gland and do not damage the ovotestis directly.

A slight increase of the growth rate in *Biomphalaria* is observed in the early stages of infection (Pan, 1962, 1965; Sturrock and Sturrock, 1970; Meuleman, 1972). Hodasi (1972) also observed a temporary acceleration of growth in *L. truncatula* infected with *F. hepatica*. An increase in the thickness of the shell of *H. anceps* (Menke) infected with an echinostome was noted by Etges (1961a, b) and this aspect was further investigated in *P. sayii* by Cheng (1971). Snails bred in the laboratory and aged about 6 weeks were divided into 2 groups, one of which was maintained as a control while each snail in the other group was exposed to 20 miracidia of *Echinostoma revolutum*. At 11 weeks several weights were determined for each snail. It was found that the whole parasitized snails were significantly heavier than the uninfected snails, but their soft tissues weighed about the same and the discrepancy remaining after allowing for the weight of the parasites was accounted for by an increase in the weight of the shell.

From histological and histochemical observations, Cheng (1971) suggested that heavier calcification of the shell in parasitized snails was a consequence of an increase in the number of calcium spherites in the digestive gland, followed by release of ionic calcium when cells containing spherites were destroyed by the parasites.

Thus, work on pulmonates has not provided evidence of any significant and sustained increase in the body size of infected snails that could properly be termed "giganticism". However, interference with the reproductive activities of the host is believed to be of significance to the parasites, even though it may not involve obvious destruction of the gonad. McClelland and Bourns (1969) observed that the ovotestis and accessory reproductive organs of *L. stagnalis* infected with *Trichobilharzia ocellata* (La Valette) remained rudimentary in size, even though the reproductive system was not invaded by the parasite (a similar effect was observed in *L. truncatula* by Hodasi, 1972). The authors suggested that this could be due to an indirect chemical mechanism, perhaps through the parasites producing substances exerting a hormonal effect, and they urged that "the inhibition of molluscan reproduction by trematode parasites must be regarded as a major adaptation which serves on the one hand to spare snails from the double burden of producing both eggs and cercariae, and on the other hand as a resources-management device whereby one portion of a snail population serves its own kind and its parasites by reproducing, while the remainder serves the parasite by incubating cercariae over an extended period of time". McClelland and Bourns (1969) further suggested that "it is more expensive in nutritional terms for a snail to produce eggs than it is to produce cercariae", and support for this view is provided by Bourns (1974) in a comparison of the total protein and carbohydrate expended in the form of eggs and cercariae produced by *L. stagnalis*.

A marked reduction in fecundity is generally evident in snails infected with schistosomes. Oviposition by *Biomphalaria* may cease entirely as early as 2 weeks after infection (Meuleman, 1972; Etges *et al.*, 1975), though it may be resumed in the late stages of infection. The pattern of decline in fecundity does not appear to be due to starvation which could be caused by the parasite withdrawing the nutritional resources of the host. Meuleman observed that oviposition by *B. pfeifferi* was significantly reduced in the period 7–10 days after exposure, before daughter sporocysts had become conspicuous in the digestive gland. Oviposition had ceased entirely before any histological changes were visible in the ovotestis. Meuleman considers the possibility that these effects are due to the interference by the parasite with the endocrine system of the host. This line of enquiry has been stimulated by the work of Joose and colleagues on neurosecretion in freshwater pulmonates, and is being actively pursued in the combination *L. stagnalis*/*T. ocellata* because this comparatively large host allows greater use of experimental techniques (Sluiters, in press).

If, as seems likely, the larvae of a single species of digenean may impose a severe burden on a snail population, then resources management could be an especially acute problem when different species of parasite share the same snail host. The pulmonate species with the greatest variety of recorded digenean parasites are abundant and widely distributed, but the total parasitic fauna carried by a properly defined population of hosts has rarely been studied. Clearly the geographical distribution of definitive hosts will play an important part in governing which parasites are transmitted in a particular locality, and as would be expected the number of digeneans recorded from one snail population is generally considerably less than the variety known from the snail throughout its range. However, it seems that the resources of a single snail population could not support an unlimited number of parasite species, although there are several ways in which the host's resources might be shared. That the larvae of different species occurring in multiple infections may occupy different sites is shown by observations on *L. peregra*, in which sporocysts of one kind were found in the muciparous and oothecal glands, while the larvae of a different species were present in the digestive gland, and in one snail the rediae of a third parasite were confined to an area close to the stomach (Wright, 1966b). Inter-specific competition between parasites might also be reduced by natural selection for mechanisms (such as parasitic castration) which could increase the resources extractable from the individual intermediate host. As a result, a reduction could take place in the number of individuals of the host snail needed by a parasite to produce sufficient larvae for the maintenance of its life cycle. The variation in susceptibility observed in relation to age suggests the further possibility that some pulmonates could support different parasites at different periods in their life, as observed in the prosobranch *Littorina saxatilis* (Olivi), which is infected when juvenile by 1 digenean and by 4 other species when adult (James, 1965).

Due to the discontinuous nature of their habitat, local faunas of freshwater snails may be considered as isolated patches (or islands) of food resources, available for colonization and exploitation by digenean parasites which are dispersed mainly through the distribution and movements of their definitive hosts. The total fauna of Digenea in terms of species and the biomass of each one, which is supported by a defined population of a single snail species, is the outcome of complicated and delicately balanced processes. Freshwater pulmonates, by virtue of the ease with which they may be maintained in the laboratory and the abundance of relatively isolated natural populations, offer highly suitable subjects for the investigation of a fascinating branch of ecology.

VII. References

Alicata, J. E. (1953). *J. Parasit.* **39,** 673–674.

Anteson, R. K. (1970). *Ann. trop. Med. Parasit.* **64,** 101–107.

Anteson, R. K. and Williams, J. F. (1975). *J. Parasit.* **61,** 149–151.

Azevedo, J. F. de (1969). *Bull. Soc. Path. exot.* **62,** 348–375.

Baer, J. G. (1952). "Ecology of Animal Parasites". University of Illinois Press, Urbana.

Barbosa, F. S., Berry, E. G., Harry, W. H., Hubendick, B., Malek, E. A. and Paraense, W. L. (1968). "A Guide to the Identification of the Snail Intermediate Hosts of Schistosomiasis in the Americas". World Health Organisation, Washington, DC.

Basch, P. F. (1975). *Int. J. Parasit.* **5,** 449–452.

Bayne, C. J. (1972). *Parasitology* **64,** 501–509.

Benex, J. and Lamy, L. (1959). *Bull. Soc. Path. exot.* **52,** 188–193.

Berrie, A. D. (1970). *In* "Advances in Parasitology" (B. Dawes, Ed.) Vol. 8, pp. 43–96. Academic Press, London and New York.

Boray, J. C. (1966). *Ann. trop. Med. Parasit.* **50,** 114–124.

Boray, J. G. (1969). *In* "Advances in Parasitology" (B. Dawes, Ed.) Vol. 7, pp. 95–210. Academic Press, London and New York.

Boray, J. C. (1974). *In* "Proceedings of the 3rd International Congress on Parasitology" 506–507.

Bourns, T. K. R. (1963). *Can. J. Zool.* **41,** 937–941.

Bourns, T. K. R. (1974). *J. Parasit.* **60,** 1046–47.

Brooker, B. J. (1972). *In* "Behavioural Aspects of Parasite Transmission" (E. U. Canning and C. A. Wright, Eds) pp. 149–171. Academic Press, London and New York.

Brooks, C. P. (1953). *J. Parasit.* **39,** 159–163.

Brooks, W. M. (1969). *In* "Immunity to Parasitic Animals" (G. J. Jackson *et al.*, Eds) Vol. 1, pp. 149–171. Meredith Corporation, New York.

Brown, D. S. and Wright, C. A. (1972). *J. Zool. Lond.* **167,** 97–132.

Brown, D. S., Oberholzer, G. and Van Eeden, J. A. (1971). *Malacologia* **11,** 171–198.

Burch, J. B. (1964). *Malacologia* **1,** 387–400.

Burch, J. B. and Lindsay, G. K. (1970). *Malac. Rev.* **3,** 1–18.

Byrd, E. E. and Maples, W. P. (1969). *J. Parasit.* **55,** 509–526.

Canning, E. U. and Wright, C. A. (1972). "Behavioural Aspects of Parasite Transmission". Academic Press, London and New York.

Carney, W. P. (1967). *Can. J. Zool.* **45,** 131–134.

Cheng, T. C. (1963). *Ann. N.Y. Acad. Sci.* **113,** 289–321.

Cheng, T. C. (1967). *In* "Advances in Marine Biology" (F. S. Russell, Ed.) Vol. 5, pp. 1–424. Academic Press, London and New York.

Cheng, T. C. (1968). *Pacif. Sci.* **22,** 141–160.

Cheng, T. C. (1971). *In* "Aspects of the Biology of Symbiosis" (T. C. Cheng, Ed.) pp. 103–137. Butterworth, London.

Chernin, E. (1968a). *J. Parasit.* **54,** 509–516.
Chernin, E. (1968b). *Experientia* **24,** 973.
Chernin, E. (1970). *J. Parasit.* **56,** 287–296.
Chernin, E. (1972). *J. Parasit.* **58,** 209–212.
Chernin, E. (1974). *Am. J. trop. Med. Hyg.* **23,** 320–327.
Chernin, E. and Dunaven, C. A. (1962). *Am. J. trop. Med. Hyg.* **11,** 455–471.
Chernin, E. and Perlstein, J. M. (1969). *J. Parasit.* **55,** 500–508.
Christensen, N. O., Nansen, P. and Frandsen, F. (1976a). *Parasitology* **73,** 161–167.
Christensen, N. O., Nansen, P. and Frandsen, F. (1976b). *Z. Parasitenk.* **50,** 67–71.
Chu, K. Y., Sabbaghian, H. and Massoud, J. (1966). *Bull. Wld Hlth Org.* **34,** 121–130.
Coles, G. C. (1971). *Trans. R. Soc. trop. Med. Hyg.* **65,** 686–687.
Cort, W. W., McMullen, D. B. and Brackett, S. (1937). *J. Parasit.* **23,** 504–532.
Cort, W. W., Brackett, S., Olivier, L. and Nolf, L. O. (1945). *J. Parasit.* **31,** 61–78.
Cridland, C. C. (1968). *Bull. Wld Hlth Org.* **39,** 955–961.
Cridland, C. C. (1970). *Bull. Wld Hlth Org.* **43,** 809–815.
Croll, N. A. (1966). "Ecology of Parasites". Harvard University Press, Cambridge.
Davenport, D., Wright, C. A. and Causely, D. (1962). *Science, N.Y.* **135,** 1059–1060.
Dawes, B. (1947). "The Trematoda of the British Fishes". Ray Society, London.
Dawood, I. K. and Chu, K. Y. (1973). *J. trop. Med. Hyg.* **76,** 48–50.
Dinnik, J. A. and Dinnik, N. N. (1957). *Rep. E. Afr. vet. Res. Org.* **1956–1957,** 50.
Dinnik, J. A. and Dinnik, N. N. (1965). *Bull. epizoot. Dis. Afr.* **13,** 341–359.
Dönges, J. (1974). *Int. J. Parasit.* **4,** 403–407.
Erasmus, D. A. (1972). "The Biology of Trematodes". Edward Arnold, London.
Etges, F. J. (1961a). *Trans. Am. Microscop. Soc.* **80,** 221–226.
Etges, F. J. (1961b). *J. Parasit.* **47,** 453–458.
Etges, F. J., Carter, O. S. and Webbe, G. (1975). *In* "Pathobiology of Invertebrate Vectors of Disease" (L. A. Bulla and T. C. Cheng, Eds) pp. 480–496. New York Academy of Sciences.
Ewers, W. H. (1960). *Nature, Lond.* **186,** 990.
Ewers, W. H. (1964). *Parasitology* **54,** 571–578.
Faulk, W. P., Lim, H. K., Jeong, K. H., Heyneman, D. and Price, D. (1973). *In* " 'Non-specific' Factors Influencing Host Resistance" (W. Braun and J. Ungar, Eds) pp. 24–32. Karger, Basel.
File, S. K. (1975). *J. Parasit.* **61,** 75–78.
Fretter, V. and Graham, A. (1962). "British Prosobranch Molluscs". Ray Society, London.
Hansen, E. L. (1975). *In* "Pathobiology of Invertebrate Vectors of Disease" (L. A. Bulla and T. C. Cheng, Eds) pp. 426–436. New York Academy of Sciences.

Heyneman, D. (1966). *Exp. Parasit.* **18,** 220–223.
Heyneman, D., Lim, H. K. and Jeong, K. H. (1971a). *J. Parasit.* **57,** 446–442.
Heyneman, D., Faulk, W. P. and Fudenberg, H. H. (1971b). *Exp. Parasit.* **29,** 480–492.
Hodasi, J. K. M. (1972). *Parasitology* **65,** 359–369.
Howaldt, H. G. and Armstrong, F. I. (1969). *Trans. R. Soc. trop. Med. Hyg.* **63,** 149.
Hubendick, B. (1951). *K. svenska Vetensk. Akad. Handl.* **3,** 1–223.
Hurst, C. T. (1927). *Univ. Calif. Publs Zool.* **29,** 321–404.
James, B. L. (1965). *Parasitology* **55,** 93–115.
Jordan, P. and Webbe, G. (1969). "Human Schistosomiasis". Heinemann, London.
Kagan, I. G. and Geiger, S. (1964). *J. Parasit.* **50,** 474–475.
Kagan, I. G. and Geiger, S. (1965). *J. Parasit.* **51,** 622–627.
Kendall, S. B. (1954). *Ann. trop. Med. Parasit.* **48,** 307–313.
Kendall, S. B. (1964). *In* "Host-Parasite Relationships in Invertebrate Hosts" (A. E. W. Taylor, Ed.) pp. 51–75. Blackwell, Oxford.
Kendall, S. B. (1965). *In* "Advances in Parasitology" (B. Dawes, Ed.) Vol. 3, pp. 59–95. Academic Press, London and New York.
Kendall, S. B. (1970). *In* "Advances in Parasitology" (B. Dawes, Ed.) Vol. 8, pp. 251–258. Academic Press, London and New York.
Kinoti, G. K. (1971). *Parasitology* **62,** 161–170.
Le Roux, P. L. (1954). *Trans. R. Soc. trop. Med. Hyg.* **43,** 3.
Le Roux, P. L. (1958). *Trans. R. Soc. trop. Med. Hyg.* **52,** 12.
Lee, F. O. and Cheng, T. C. (1972). *Exp. Parasit.* **31,** 203–216.
Levy, M. G., Tunis, M. and Isseroff, H. (1973). *Nature, Lond.* **241,** 65–66.
Liang, Y. S. and Van der Schalie, H. (1975). *J. Parasit.* **61,** 915–919.
Lie, K. J., Basch, P. F., Heyneman, D., Beck, A. J. and Audy, J. R. (1968). *Trans. R. Soc. trop. Med. Hyg.* **62,** 299–319.
Lie, K. J. and Heyneman, D. (1975). *Int. J. Parasit.* **5,** 621–625.
Lie, K. J. and Heyneman, D. (1976a). *J. Parasit.* **62,** 292–297.
Lie, K. J. and Heyneman, D. (1976b). *J. Parasit.* **62,** 298–302.
Lie, K. J. and Heyneman, D. (1976c). *J. Parasit.* **62,** 51–58.
Lie, K. J., Heyneman, D. and Yau, P. (1975a). *J. Parasit.* **61,** 574–576.
Lie, K. J., Heyneman, D. and Lim, H. K. (1975b). *Int. J. Parasit.* **5,** 627–631.
Lie, K. J., Heyneman, D. and Jeong, K. H. (1976). *J. Parasit.* **62,** 286–291.
Llewellyn, J. (1965). *In* "Evolution of Parasites" (A. E. R. Taylor, Ed.) pp. 47–78. Blackwell, Oxford.
Lo, C. T. (1969). *Malac. Rev.* **2,** 135–136.
Lo, C. T. (1972). *Malacologia* **11,** 225–280.
Lo, C. T., Burch, J. B. and Schutte, C. H. J. (1970). *Malac. Rev.* **3,** 121–126.
MacInnes, A. J. (1965). *J. Parasit.* **51,** 731–746.
MacInnes, A. J., Bethel, W. M. and Cornford, E. M. (1974). *Nature, Lond.* **248,** 361–363.
Mandahl-Barth, G. (1958). "Intermediate hosts of *Schistosoma*. African *Biomphalaria* and *Bulinus*". World Health Organization, Geneva.
Mandahl-Barth, G. (1965). *Bull. Wld Hlth Org.* **33,** 33–44.

Mandahl-Barth, G., Frandsen, F. and Jelnes, J. E. (1976). *Trans. R. Soc. trop. Med. Hyg.* **70,** 88.
McClelland, W. F. J. (1965). *Rep. E. Afr. Inst. med. Res.* **1963–1964,** 15.
McClelland, G. and Bourns, T. K. R. (1969). *Exp. Parasit.* **24,** 137–146.
Meuleman, E. A. (1972). *Neth. J. Zool.* **22,** 355–427.
Michelson, E. H. (1963). *Ann. N.Y. Acad. Sci.* **113,** 486–491.
Michelson, E. H. (1964). *Am. J. trop. Med. Hyg.* **13,** 36–42.
Morton, J. E. (1964). "Molluscs" 3rd edition. Hutchinson, London.
Muftic, M. (1969). *Parasitology* **59,** 365–371.
Najarian, H. (1954). *J. Morphol.* **94,** 165–197.
Nansen, P., Frandsen, F. and Christensen, N. O. (1976). *Parasitology* **72,** 163–171.
Newton, W. L. (1952). *J. Parasit.* **38,** 362–366.
Newton, W. L. (1953). *Exp. Parasit.* **2,** 242–257.
Newton, W. L. (1954). *J. Parasit.* **40,** 1–4.
Newton, W. L. (1955). *J. Parasit.* **41,** 526–528.
Pan, C. T. (1962). *J. Parasit.* **48,** 20.
Pan, C. T. (1965). *Am. J. trop. Med. Hyg.* **14,** 931–976.
Paperna, I. (1968). *Ann. trop. Med. Parasit.* **62,** 13–26.
Paraense, W. L. and Correa, L. R. (1963). *Rev. Inst. Med. trop. Sao Paulo,* **5,** 15–22.
Pitchford, R. J. (1965). *Bull. Wld Hlth Org.* **32,** 105–120.
Porter, A. (1938). *Publ. South African Inst. med. Res.* **8,** 1–492.
Porter, C. A. and Gamble, W. (1971). *Comp. Biochem. Physiol.* **40B,** 335–340.
Potts, W. T. W. (1967). *Biol. Rev.* **42,** 1–41.
Probert, A. J. and Erasmus, D. A. (1965). *Parasitology* **55,** 77–92.
Pullan, N. B. and Whitten, L. K. (1972). *N. Z. vet. J.* **20,** 69–72.
Rao, C. M. P. (1966). *J. Helminth.* **40,** 131–140.
Rensch, B. (1959). "Evolution Above the Species Level". Methuen, London.
Richards, C. S. (1973). *Am. J. trop. Med. Hyg.* **22,** 748–756.
Richards, C. S. (1975a). *J. Parasit.* **61,** 233–236.
Richards, C. S. (1975b). *Parasitology* **70,** 231–241.
Richards, C. S. (1975c). *In* "Pathobiology of Invertebrate Vectors of Disease" (L. A. Bulla and T. C. Cheng, Eds) pp. 394–410. New York Academy of Sciences.
Richards, C. S. and Merritt, J. W. (1972). *Am. J. trop. Med. Hyg.* **21,** 425–434.
Rothschild, M. (1938). *Novit. Zool.* **41,** 84–102.
Rothschild, M. and Clay, T. (1952). "Fleas, Flukes and Cuckoos". MacMillan, London.
Rysavy, B., Barus, V., Moravec, F. and Yousif, F. (1974). *Folia Parasit.* **21,** 161–168.
Saoud, M. F. A. (1965a). *J. Helminth.* **39,** 101–112.
Saoud, M. F. A. (1965b). *J. Helminth.* **39,** 363–376.
Schell, S. C. (1962). *J. Parasit.* **48,** 387–394.
Schutte, C. H. J. (1975). *S. Afr. J. Science* **71,** 8–20.
Shiff, C. J. (1968). *J. Parasit.* **54,** 1133–1140.
Shiff, C. J. (1969). *J. Parasit.* **55,** 108–110.

Shiff, C. J. (1974). *J. Parasit.* **60,** 578–583.
Shiff, C. J. and Kriel, R. L. (1970). *J. Parasit.* **56,** 281–286.
Sluiters, J. F. (in press). Proceedings of the Sixth European Malacological Congress, Amsterdam, 1977. *Malacologia.*
Sminia, T. (1977). *Malacologia* **16,** 255–256.
Smithers, S. R., Terry, R. J. and Hockley, D. J. (1969). *Proc. R. Soc. B* **171,** 483–494.
Southgate, V. R. (1970). *Parasitology* **61,** 293–299.
Southgate, V. R. and Knowles, R. J. (1974). *J. nat. Hist.* **9,** 273–314.
Sturrock, B. M. (1968). *Ann. trop. Med. Parasit.* **62,** 393–397.
Sturrock, B. M. and Sturrock, R. F. (1970). *Ann. trop. Med. Parasit.* **64,** 357–363.
Sturrock, R. F. and Upatham, E. S. (1973). *Int. J. Parasit.* **3,** 35–41.
Sudds, R. H. (1960). *J. Elisha Mitchell sjient. Soc.* **76,** 121–133.
Targett, G. A. T. (1962). *Ann. trop. Med. Parasit.* **56,** 210.
Taylor, M. G. (1970). *J. Helminth.* **44,** 253–314.
Thomas, A. P. (1883). *Quart. J. micr. Sci.* **23,** 99–133.
Timon-David, J. (1957). *Ann. Parasit. Hum. Comp.* **32,** 219–242.
Timon-David, J. (1960). *Ann. Parasit. Hum. Comp.* **35,** 251–267.
Tripp, M. R. (1961). *J. Parasit.* **47,** 745–751.
Tripp, M. R. (1970). *J. retic. Soc.* **7,** 173–182.
Ulmer, M. J. (1971). *In* "Ecology and Physiology of Parasites" (A. M. Fallis, Ed.) pp. 123–160. Adam Hilges, London.
Upatham, E. S. (1972). *J. Helminth.* **46,** 277–283.
Upatham, E. S. (1973). *Int. J. Parasit.* **3,** 289–297.
Upatham, E. S. and Sturrock, R. F. (1973). *J. Parasit.* **59,** 448–453.
Van Eeden, J. A. and Brown, D. S. (1966). *Nature, Lond.* **210,** 1172–1173.
Voge, M. and Seidel, J. S. (1972). *J. Parasit.* **58,** 699–704.
Wadji, N. (1966). *Trans. R. Soc. trop. Med. Hyg.* **50,** 774–776.
Webbe, G. and James, C. (1971). *J. Helminth.* **45,** 403–413.
Wesenberg-Lund, C. (1931). *K. dansk. Vidensk. Selsk. Skr.* (9) **4,** 89–142.
Wesenberg-Lund, C. (1934). *K. dansk. Vidensk. Selsk. Skr.* (9) **5,** 1–223.
Wilson, R. A. (1968). *Comp. Biochem. Physiol.* **24,** 629–633.
Wilson, R. A. and Denison, J. (1970a). *Comp. Biochem. Physiol.* **32,** 301–313.
Wilson, R. A. and Denison, J. (1970b). *Comp. Biochem. Physiol.* **32,** 511–517.
Wilson, R. A., Pullin, R. and Denison, J. (1971). *Parasitology* **63,** 491–506.
Wright, C. A. (1959a). *Ann. trop. Med. Parasit.* **53,** 288–292.
Wright, C. A. (1959b). *J. Linn. Soc. Zool.* **44,** 222–237.
Wright, C. A. (1962). *In* "Bilharziasis; Ciba Foundation Symposium" (G. E. W. Wolstenholme and M. O'Connor, Eds) pp. 103–120. Churchill, London.
Wright, C. A. (1966a). *In* "Proceedings of the First International Congress on Parasitology" pp. 1058–1059.
Wright, C. A. (1966b). *Helminth. Abst.* **35,** 207–224.
Wright, C. A. (1967). *In* "Bilharziasis" (F. K. Mostoffi, Ed.) pp. 3–7. Springer-Verlag, Berlin and New York.
Wright, C. A. (1971a). "Flukes and Snails". Allen and Unwin, London.
Wright, C. A. (1971b). *Phil. Trans. R. Soc.* **260B,** 299–313.
Wright, C. A. (1974). *J. nat. Hist.* **8,** 545–548.

Wright, C. A. (1977). *In* "Medicine in a Tropical Environment" (J. H. Gear, Ed.) pp. 291–300. Balkema, Cape Town.

Wright, C. A. and Southgate, V. R. (1976). *In* "Genetic Aspects of Host–Parasite Relationships" (A. E. R. Taylor and R. Muller, Eds) pp. 55–86. Blackwell, Oxford.

Wright, C. A., Southgate, V. R. and Knowles, R. J. (1972). *Trans. R. Soc. trop. Med. Hyg.* **66,** 28–64.

Wright, D. G. S. and Ronald, K. (1972). *Can. J. Zool.* **50,** 855–860.

Wright, D. G. S., Lavigne, D. M. and Ronald, K. (1972). *Can. J. Zool.* **50,** 197–200.

Chapter 8

Ecology of freshwater pulmonates

W. D. RUSSELL-HUNTER

Department of Biology, Syracuse University, New York and Marine Biological Laboratory, Woods Hole, Massachusetts, USA

I. Introduction

In modern animal ecology, classic distributional studies are being complemented by work both on physiological ecology and on population dynamics. As is true for most other groups of freshwater invertebrates, earlier ecological studies on molluscs consisted largely of descriptive faunal lists, the best of which incorporated either some correlation of microhabitat to precise environmental conditions, or some serial approach (either temporal or spatial) in surveying different levels of species diversity. More recently, 2 further classes of studies have become important: first, field investigations of physiological aspects of the interactions between molluscs and their environment (particularly in comparative studies of closely-related forms or of interpopulation infraspecific differences in measurable physiological parameters); and secondly, both field and laboratory studies on matters of autecology (particularly investigations of growth, reproduction, and life cycle and their various controls leading either to models for population dynamics or to attempts at actuarial bioenergetics). For the more restricted group of the pulmonate snails, such 3-fold classification of ecological studies is still appropriate, though all 3 areas—distributional, physiological and population—are affected by the paramount significance of certain primary physiological adaptations in this basically air-breathing stock.

The freshwater molluscs of the world include a relatively small number of bivalve genera, and both prosobranch and pulmonate gastropods. The groups of bivalves and of gill-bearing prosobranchs in fresh waters are all most closely related to marine littoral forms and, in certain cases, clearly their immediate descendants. The primary physiological adaptation for life in fresh waters shown by them is capacity for osmoregulation. On the other hand, the larger number of freshwater snails which are pulmonates of the order Basommatophora represent an air-breathing stock with varying degrees of readaptation to aquatic life (Russell Hunter, 1957, 1964).

It is obvious that, in the majority of environments provided by fresh waters, physical, chemical, and trophic-biotic conditions vary more widely than they ever do in the sea. Less obvious but even more significant is the fact that most freshwater environments are transient in time and exhibit a high degree of small-scale, short-term isolation (Hubendick, 1952, 1954, 1960; Russell Hunter, 1952, 1957, 1961b, 1964, 1970). The "ancient" lakes (such as Tanganyika, Baikal, Ochrida, and certain lakes in Celebes, China, and the Philippines) represent exceptions to this pattern (see Boss, Chapter 9 of this volume); but the majority of populations of freshwater animals (including pulmonate snails) live in—and have evolved in—bodies of water more transient than, and considerably smaller than, even medium-sized lakes (Russell-Hunter, 1964, 1970).

In view of these general ecological constraints on pulmonate snails in fresh waters, this chapter is divided into 4 principal sections: 3 on distributional, physiological, and population ecology respectively, and one on certain broader evolutionary considerations.

II. Distributional ecology of freshwater pulmonates

A. Pulmonates and "Aufwuchs"

In their general distribution, freshwater pulmonates are benthic animals living in the shallower waters of lakes, ponds and marshes, and of rivers, streams and ditches. Especially in some genera, but to some extent in the class as a whole, there are many forms which are ubiquitous and cosmopolitan. Narrow zoogeographical ranges are less usual: some genera, and possibly some species, occur in the fresh waters of all 5 of the world's larger land masses. Limited habitat niches are less usual: many forms are euryoecic, occurring in a wide range of environmental conditions wherein their success is dependent on capacity for considerable physiological variation (Russell Hunter, 1957, 1964; see also Section III of this chapter). Despite this widespread distribution, freshwater pulmonates are essentially benthos of the euphotic zones, and thus marginal in most lakes; only in rare instances are extensive populations found at depths of greater than 4 m.

A consequence of this depth preference is that the proportion of the community biomass made up by freshwater pulmonates is markedly less in large lakes and major rivers than it can be in suitable smaller lakes, ponds and marshes or in smaller creeks, streams and ditches. As will be discussed below, pulmonates of lentic waters are most densely concentrated on (though not usually feeding on) vascular macrophytes (emergent or submerged), while pulmonates of lotic conditions are usually associated with rocks or other hard surfaces. Both sets of preferred substrata, in suitably illuminated depths, are coated with *Aufwuchs*. In the bioenergetics of freshwater communities, *Aufwuchs* constitutes a third important sector of primary production in addition to those of phytoplankton and macrophytic vegetation (Odum, 1957; McMahon *et al.*, 1974). *Aufwuchs* can be defined as the scum flora of diatoms, blue-green and other single-celled algae, filamentous algae, bacteria, fungi, protozoans, and other associated microscopic plants and invertebrates, all attached to hard surfaces including macrophytic plants (hence the epiphytic "scum flora"), rocks, wood and miscellaneous debris in suitably lit waters. Freshwater pulmonate snails are basically herbivores, exhibiting both structural and functional adaptations of their radulae and other oral structures, of their digestive systems and of their locomotory mechanisms for continuous slow microphagy. With the exceptions of a few species–populations feeding on detritus in the deeper waters of lakes or in the darkness of cave-systems, and a few forms which can be facultative carnivores, they are "grazing" animals using radular scraping over suitable surfaces. Their preferred pastures are the *Aufwuchs*-bearing rocks and plants of shallower fresh waters. It must be stated that the mere difference between lentic and lotic waters is rarely important in determining the pulmonate fauna, sizes of habitats on the one hand, and microenvironmental conditions on the other, are usually more important. For example the "stream" limpets of Europe and North America, *Ancylus fluviatilis* (Müll.) and *Ferrissia rivularis* (Say) respectively, are pulmonate snails characteristic of, and often abundant in, smaller swiftly-flowing streams. They also occur on the cobblestone bottoms of wave-swept lakes in similar conditions of clear water movement and relatively high oxygenation. The corresponding pair of "pond" limpets, *Acroloxus lacustris* (L.) (Europe) and *Laevapex fuscus* (C. B. Adams) (North America) are found in suitable reed-beds and on other hard substrata over muds of low oxygenation, no matter whether these habitats are in ponds or in slowly flowing rivers.

B. Faunal lists

Historically, knowledge of the distribution of freshwater pulmonates (and of freshwater molluscs in general) began with faunal lists, usually for the larger lakes. The best of these always incorporated some specific habitat descriptions and, by the middle years of this century, comparative studies allowed loose correlations to be drawn between such environmental factors as water hardness and trophic state, and levels of species diversity or levels of "standing crop" abundance of the pulmonate snails. Little or no experimental analysis of species occurrence and diversity had yet been attempted (Macan, 1950), and only in the last decade have the "crop densities" for some freshwater pulmonates been refined into bioenergetic terms as comparative productivity rates (see Section IV below).

It would be inappropriate here to attempt an historical bibliography of the very large number of publications giving distribution data for freshwater pulmonates, and the following references are in no way comprehensive. Early examples of good descriptive ecology are 2 reports by Baker (1916, 1918) on the freshwater molluscs of Oneida Lake, a moderately large and eutrophic lake in upstate New York, and his later 2-volume survey of the freshwater molluscs of Wisconsin, of which 500 pages are devoted to gastropods (Baker, 1928). The weakness of these works arises from overinflated taxonomy (Baker was a "splitter"). In a classic monograph, Boycott (1936) surveyed existing knowledge of the ecology of British freshwater Mollusca, with detailed notes on the habitats of 62 "real" species, of which 26 are pulmonates. Further distributional data and descriptive ecology for freshwater pulmonates are included in Boettger (1944), Berg (1938, 1948), Hubendick (1947), Macan (1950), Russell Hunter (1953b, 1957), Mirolli (1958), van Benthem Jutting (1959), Bovbjerg and Ulmer (1960), Schutte and Frank (1964), Okland (1964, 1969), Harman and Forney (1970), Harman and Berg (1970) and Clarke (1973). The last of this list is a monographic treatment of the molluscs of "the Canadian interior basin" (about a third of the land area of North America), and it combines a full account of geographical distribution both with a thoroughly modern approach to molluscan systematics (by reducing synonymy), and with full biological notes on each species (of a scholarly completeness rarely seen since the days of Baker and Boycott).

The distributional picture which emerges from all this is hardly ever one of physical, chemical or biotic factors uniquely favouring a particular pulmonate species. As Boycott (1936) said, the needs of most freshwater molluscs (including pulmonates) are so similar that habitats can be classed as good or bad for Mollusca as a whole. In general, a greater diversity of pulmonate species is found in bodies of fresh water with higher concentrations of dissolved calcium, richer plant turnover (more eutrophic), and of larger extent. Greater densities of pulmonates are more specifically associated with eutrophic conditions. Other environmental factors, such as temperature range (including broadly altitudinal and latitudinal considerations), and the finer details of hydrography are much less important than the hardness and the trophic level of the waters. But, in these 2 factors (and in spatial and temporal extent) the substantive diversity of fresh waters is immense, when compared to the uniform and constant environment presented by the sea.

C. Dissolved calcium and trophic state

The content of dissolved inorganic salts varies: for example, the content of dissolved calcium can vary 200-fold in fresh waters, and the ratio of calcium to sodium more than 50-fold. The concentration levels and this ratio vary but little in the oceans, calcium as a percentage of sodium remaining about 3·82 for most seawaters. Freshwater pulmonates are found in waters from about 2 mg litre^{-1} calcium to over 120 mg litre^{-1}. Of the 62 species of freshwater molluscs in Britain (Boycott, 1936), 7 can live in extremely soft waters (< 3 mg litre^{-1} calcium), including 2 pulmonates, *Lymnaea peregra* (Müll.) and *A. fluviatilis*. A distribution map of the "purest" mountain and island waters, where these 2 species are the only snails, would also encompass all the most famous distilleries of malt whisky in Scotland. Twenty-six out of 62 species can live in moderately soft waters (< 10 mg litre^{-1} calcium), a further 6 require waters of at least 10 mg litre^{-1} and the remaining 30 are "calciphile" requiring at least 20 mg litre^{-1}. A single British lake with hard water, Llangorge Lake, has 25 molluscan species (Boycott, 1936), and there are 38 species in the eutrophic, hard-water Oneida Lake (Harman and Forney, 1970). Even very alkaline lakes in North Dakota (Tuthill and Laird, 1963, 1964) with mineral contents too high to be potable for man or farm animals (total dissolved solids in terms of 5–15 g litre^{-1}) can support from 1–5 species of freshwater pulmonates.

In fresh waters, there is also considerable variation in the availability of plant nutrient salts, and this has given rise to the most widely accepted biological classification of lakes: that of oligotrophy $\rightarrow$ eutrophy (corresponding to increasing spring levels of dissolved phosphates and nitrates, and hence increased plant productivity). Dystrophic lakes form a third type, in which acidic organic materials in solution exert an "antibiotic" effect which drastically reduces bacterial breakdown and nutrient recycling, thus minimizing productivity (Berg and Petersen, 1956; Salazkin, 1965). Eutrophic waters like Oneida Lake could support not only the greatest diversity of mollusc species, but also the densest and most productive populations of many of them (see also Section IV of this chapter). Fewer species and less productivity are characteristics of the molluscan faunas of oligotrophic and dystrophic waters. In the trophically "poorest" waters for molluscs in North America the only pulmonate species would probably be *F. rivularis*, *Physa gyrina* Say, and/or *Gyraulus parvus* Say; while in similar waters in Europe the only species would be *A. fluviatilis*, *L. peregra*, and/or *Gyraulus acronicus* (Fér.) (Russell Hunter, 1957). As regards truly dystrophic waters, Klimowicz (1962) has pointed out that smaller "impermanent" dystrophic ponds in Poland can support molluscan faunas which include 14 pulmonate species, while larger nondrying dystrophic ponds do not.

Throughout the world—other environmental factors being equal—harder and more eutrophic fresh waters undoubtedly support more molluscs (both in species numbers and in biomass). It should be noted that all those pulmonate species tolerant of softer waters and more oligotrophic conditions (for example, the six species named above) can also survive in, and are found as part of the more diverse molluscan faunas of, harder waters. This is not exactly paralleled by eutrophy—there may be a few pulmonate species characteristic of more oligotrophic conditions (probably by their inability to adjust to the lower oxygen conditions of extreme eutrophy—see Section III below). However, in clean eutrophic waters though not in "hypertrophic" ones, the molluscan faunas will usually include most of the pulmonate species of oligotrophy.

D. Space and other factors

Size of habitat also influences pulmonate diversity. Russell Hunter (1957) records Loch Lomond as supporting 13 species (now 14) out of the 25 species of gastropods in that area of Scotland, and "smaller highland lochs" (not very dissimilar in water quality) as supporting 7 out of 25 species. Similarly, Harman and Berg (1970) surveyed total molluscan species living in the relatively oligotrophic Finger Lakes of upstate New York, and report 23 species in the larger Seneca Lake and 6 and 7 species respectively in the smaller Owasco and Otisco Lakes (the eutrophic Oneida Lake having 38).

As already noted, freshwater pulmonates are essentially animals of the marginal benthos of shallow waters, but a few species do occur in limited populations at greater depths in certain lakes. The classic case is of populations of *Lymnaea* spp living at depths of 40–200 m in the Lake of Geneva (Roskowski, 1914). The respiratory implications of this and other "continuously submerged" pulmonates are discussed below (Section III, see also Russell Hunter, 1953c; Henderson, 1963). In most Danish and British lakes the only snails penetrating to great depths are prosobranch species of the genera *Valvata* and *Bithynia* (Berg, 1938; Russell Hunter, 1953c), but pulmonate populations living submerged include *L. peregra* to below 9 m in Windermere (Macan, 1950), and that species and *Physa fontinalis* (L.) at about 5·5 m in Loch Lomond (Russell Hunter, 1953c). An interesting recent report (Thut, 1969) concerns Lake Washington, near Seattle, wherein a planorbid, *Gyraulus* sp has its population concentrated between 40 and 55 m, and apparently reproduces all year round. Conditions affecting the penetration of pulmonate species into greater depths in lakes remain obscure, but certain aspects of the biology of any that do so can be confidently predicted. They will always have water-filled mantle cavities (no gas bubbles), they will all be detritus feeders exploiting the fine "rain" of seston from the waters above, they will not have limited breeding seasons, and they will all be relatively small as adults.

A few less comprehensive environmental factors can now be considered. Substrate preferences seem dominated by *Aufwuchs* availability. Even distribution patterns on broad-leaved vascular plants (Pettee and Gifford, 1964) or in reed-beds (Rudescu and Popescu-Marinescu, 1970) for certain pulmonate species probably are based on the provision of appropriate supporting surfaces for the epiphytic *Aufwuchs*. Implicit in several recent papers on the food-supply of freshwater pulmonates (see, for example, Bovbjerg, 1965; Calow, 1970; McMahon *et al.*, 1974) is the fact that only very rarely do they ingest living tissues from vascular macrophytes. A small set of experiments on *L. peregra* (Storey, 1970), which can ingest and process dicotyledon leaves, suggests that this diet can only be utilized if considerable quantities of mineral particles are simultaneously ingested for trituration within the snail's gut. Another feature of higher plant ecology which affects the distribution of freshwater pulmonates is their seasonal die-off. This may result in seasonal migrations for some species (Cheatum, 1934; Clampitt, 1974), but Bickel (1965) has reported on a population of *Physa integra* Hald. where the end of macrophytic growth is accompanied by great mortality of snails, the population being maintained from year to year by the smaller numbers of individuals living overwinter on dock pilings and larger debris. Several species of *Lymnaea*, *Physa*, and *Planorbis* can live temporarily as members of the "neuston", floating upside down under the surface film and using the ciliated foot-sole to collect microscopic organisms and chains on free protein molecules from it. The food available in such surface films in fresh waters is discussed by Goldacre (1949). The mechanics of such pedal ciliary feeding has been elucidated (Cheeseman, 1956; Summers, 1966), and Summers reports respiration data which suggest that film-ingestion could represent a significant food intake under certain circumstances.

The diversity of pulmonate faunas and their productivities in fresh waters are also reduced by environmental temperature extremes (though species of *Lymnaea*, *Physa* and *Australorbis* can live at temperatures from below 0 °C up to around 44 °C: see records and references in McNeil, 1963; Russell Hunter, 1964; McMahon, 1975c); by the risk of environmental desiccation (see Boycott, 1936; Olivier, 1956; Chernin and Adler, 1967); by irregular flash floods (Thorup, 1970); and by brackish-water conditions near the sea or on small islands (Russell Hunter, 1953d; McMillan, 1966).

E. Anthropogenic conditions

There is a large literature on the effects of human modification of natural fresh waters by damming, draining, polluting and so on. As regards pulmonate species diversity, all of man's activities seem to be deleterious to varying degrees, though in some instances productivity of individual species may be increased.

The effects of mild organic pollution by man (for example, with small amounts of sewage) are often indistinguishable from the naturally increasing eutrophy (locally and temporarily, or more generally over periods of time) of many inland water bodies. As stressed in many general accounts of freshwater ecology (e.g. Odum, 1959; Russell-Hunter, 1970), no lake can be permanently classified at a specific trophic level; this and other characteristics of its waters are changing continually. Several lakes whose faunas were first studied 50 years or more ago provide proven cases of reduction in faunal diversity (including extinction of freshwater pulmonate populations). Baker's Oneida Lake, restudied after 50 years by Harman and Forney (1970), had lost 11 species of molluscs and considerable molluscan productivity. Environmental changes, including human pollution, had been intensified in this lake as regards the other molluscs by the explosive population increases of an introduced prosobranch *Bithynia tentaculata* (L.) (Harman, 1968a, b; Harman and Forney, 1970). Lake Okoboji in Iowa lost 11 species of gastropods (8 pulmonates) in about 40 years as a result largely of human activities (Bovbjerg and Ulmer, 1960; see also Clampitt *et al.*, 1960). Loch Leven was a famous trout lake in Scotland, which by nutrient enrichment is becoming more eutrophic (Morgan, 1970). It has been closely studied for more than 80 years of progressive eutrophication. In the last 30 years populations of 6 gastropod species (5 of them planorbid pulmonates) have disappeared from the lake, 2 of these within the 9 years, 1958–1967. Accounts have been given of similar faunal reductions in smaller lakes in Illinois (Paloumpis and Starrett, 1960); increasing biomass of oligochaetes accompanying reductions of molluscan diversity and biomass. Mild amounts of sewage pollution can selectively increase or decrease turnover in certain mollusc populations (see Patrick *et al.*, 1966, for an extensive ecological survey of the Savannah River; and Maitland, 1965, 1966, for local biomass decrease and recovery in the pulmonate limpet, *A. fluviatilis*). Mild amounts of heat pollution produce enhanced generation growth rates and compensating modified fecundity in populations of *Physa virgata* Gould in Texas (McMahon, 1975c). The result in this case was a pattern of life cycle and a level of productivity not much changed by the thermal effluent.

Human interference with water levels is always deleterious. Variable storage dams rarely support a diversity of freshwater pulmonates. Early work on respiration and migrations of *L. peregra* in Loch Lomond demonstrated that rise of lake level can sometimes create starvation conditions for marginal populations of that pulmonate. Grimås (1962) has documented the reduced faunal diversity and productivity after level regulation of Lake Blåsjön in northern Sweden, and Harman (1970) discussed how a reduction of lake level by 2 m in the recent past (c. 150 years) for a meromictic marl lake (which has subfossil shell deposits) caused the extinction of once very abundant populations of 4 larger species of pulmonate snails. Smaller pulmonate species and prosobranch species, not so dependent on marginal shallow water, have survived.

F. Passive dispersal

It remains true that the distribution of freshwater molluscs has largely been accomplished by passive dispersal. Man's activities in modifying drainage, and particularly his early nineteenth century (in Europe and North America) enthusiasm for canals uniting drainage systems, have affected distribution patterns of certain molluscs (Boycott, 1936; Russell Hunter and Warwick, 1957; Harman and Berg, 1970). As Boycott (1936) points out, the "canal basin" of England and Wales at its peak extent united the drainage systems of just under 50% of the area of the country or over 50 of the 215 river catchment systems (including all the larger ones). In considerable areas of north-western Ohio, central and south-western France, and midland England, riverine freshwater pulmonates had, or still have, the opportunity to migrate through habitats exceptional in their continuity. However, the majority of lakes and ponds and a high proportion of the world's streams remain discrete, and passive transfer of snails must account for their being populated. An obvious example which has been well documented (Boycott, 1930, 1936; Shrubsole, 1934, 1937), concerns the many hundreds of cattle ponds of the chalk downs of southern England. Most of these closed ponds have molluscan faunas, and the species diversity (but not the abundance) of each fauna relates to the age of the ponds. Boycott (1930, 1936) was able to conclude that each pond got on average 1 new species every 9 years. There is hardly any doubt that birds are responsible for much molluscan dispersal. Extensive records of snails in transit had been collected by the end of the nineteenth century by H. W. Kew, and a short review of a huge literature of reports is that of Rees (1965). The distribution patterns of freshwater molluscs reflect routes of bird migrations (Nicol, 1936; Bondesen and Kaiser, 1949; Russell Hunter and Warwick, 1957; Russell Hunter, Maitland and Yeoh, 1964; other records in Rees, 1965) and the birds circumstantially implicated include migratory geese, gulls and such reedbirds and waders as bobolink and plover. In addition, there is an intriguing series of records on insect transport. From Europe, North America and India, there are cases of live freshwater pulmonate limpets (mostly species of *Ancylus* or *Laevapex*) being found "in flight" on the elytra of larger water beetles such as *Dytiscus* spp (Johnson, 1904; Annandale *et al.*, 1921; Buttner, 1953; Théodoridès, 1961; Rosewater, 1970; see also Rees, 1965). Rates and extent of passive dispersal and their effects on discontinuities in breeding populations have certain consequences for evolutionary processes in freshwater pulmonates (see Section V below). It is worth noting that such bird and insect transport has random elements in its operation, and the resultant dispersal may have to be modelled in stochastic rather than deterministic forms (Skellam, 1955, 1973).

III. Physiological ecology, acclimation and infraspecific variation

A. Aquatic readaptation

As the distributional ecology of freshwater pulmonates is basically that of marginal shallow-water animals, so their physiological ecology is primarily that of lung-snails. They form the "higher" families of the order Basommatophora and are primarily air breathers showing varying degrees of readaptation to aquatic life. The most important feature of the Pulmonata (including all the freshwater families) is the absence of ctenidia—the characteristic gills, present and structurally homologous in all other major groups of the Mollusca (see Yonge, 1947; Morton, 1958; Morton and Yonge, 1964). The gilless mantle cavity has a roof richly vascularized with thin-walled vessels, its opening to the exterior is narrow and muscular, and the rectum and kidney duct do not open into it; the mantle cavity is a lung. The most primitive pulmonates, the families Ellobiidae and Otinidae (Morton, 1955a, b), are clearly amphibious air-breathers as adults, despite the fact that marine eggs hatching to free-swimming planktonic veligers are retained by the typical salt-marsh ellobiid, *Melampus bidentatus* (Mort.) (Apley, 1970; Russell-Hunter, Apley and Hunter, 1972). Evolutionary trends in the freshwater pulmonates thus contrast with the more usual patterns in other groups (cf. Carter, 1931), wherein freshwater life leads to preadaptations subsequently allowing transition to the land.

Pulmonates have moved from sea to land to fresh water, and the freshwater species can be ranked as a series showing progressively greater degrees of readaptation to aquatic life (Russell Hunter, 1952, 1953c, 1957, 1964). At one end of the series in temperate latitudes are such marsh-dwelling snails as *Lymnaea truncatula* (Müll.) (Basommatophora) and *Succinea* spp (primitive Stylommatophora), species primarily air-breathing and nonaquatic, whose habits result in their occasional submergence in water. At the other end are the most truly aquatic species; for example, some ancylid limpets, where the pulmonate mantle cavity is completely absent and neomorphic gill-lobes are developed. "Intermediate" species are those which are manifestly aquatic in habit, but retain the mantle cavity of characteristic "lung" form, and these include the Physidae and most Lymnaeidae. Many of the larger species in this group retain aerial respiration and are essentially "diving" animals, with their behavioural patterns as dominated by the rhythm of visits to the surface as are those of turtles or whales.

Some of the earliest work on interpopulation physiological variation in freshwater pulmonates (Russell Hunter, 1953c) was concerned with respiration in this intermediate group. Since this variation is extensive and since pulmonate respiratory behaviour is extremely plastic, it is now necessary to describe certain respiratory mechanisms and patterns as properties of all individuals within certain undisturbed populations, rather than as patterns utilized by all populations within a particular species.

B. Pulmonary respiration

Surface breathing may occur in populations living in large lakes and involve them in seasonal (Cheatum, 1934; Clampitt, 1974) or in short-term (Russell Hunter, 1953b) migrations. The latter occur at high water temperatures in certain high-level littoral populations of *L. peregra* in Loch Lomond where the respiratory needs of adult snails may result in their temporary starvation (by forcing migration on to near-sterile substrata), or may cause them to migrate away from young they have produced (Russell Hunter, 1953b, 1957). In relation to this degree of starvation, Bovbjerg (1965) noted the aggregation of *Lymnaea reflexa* Say on to algal patches, but his careful experiments did not reveal any distant chemoreception. Subsequently, work by Calow (1970) and McMahon *et al.* (1974) emphasized the importance of filamentous green algae and similar high quality of *Aufwuchs* to *Lymnaea* spp.

The physiology of surface breathing in freshwater pulmonates has long been studied in the laboratory, culminating in such detailed investigations as those of Precht (1939), Füsser and Krüger (1951), Henderson (1961, 1963), and Jones (1961, 1964a, b). It should be noted that all "diving" pulmonates take up some dissolved oxygen cutaneously from the water as well as utilizing oxygen from the air bubble taken up at the surface. Accordingly in large adults, dives can be longer when dissolved oxygen tensions in the water are higher (Henderson, 1961, 1963; Jones, 1961, 1964a). Similarly surfacing can be less frequent at lower water temperatures (Precht, 1939; Russell Hunter, 1953b, c; Henderson, 1961), and younger (smaller) spat of species-populations surfacing regularly as adults may be able to remain submerged until their growth takes them past some threshold value in their surface–volume relationship (Russell Hunter, 1953b, c).

Certain populations of pulmonate species normally considered to be diving air-breathers are known to occur in large lakes at depths from which periodic excursions to the surface are impossible. The classic case of the Lake of Geneva with *Lymnaea* spp at depths of 40–200 m is extreme, but far from unique. A survey of the condition of the mantle cavity in *L. peregra* and *P. fontinalis* living in Loch Lomond has revealed several distinct physiological states in different populations (Russell Hunter, 1953c). Some populations of each species living near the margin of the lake surface regularly, but only when the water temperature is relatively high, and then much less frequently than in laboratory experiments. Other populations of the 2 species live up to 1 km offshore, although they have not been found in depths of over 6 m. In many populations, including some in relatively shallow water, the mantle cavity remains water-filled throughout life. In young spat from all populations, the mantle cavity is water-filled. In still other populations of the two species the cavity contains a gas bubble and micro-gas-analysis has revealed that in some cases the gas composition is such that the bubble could be used as an exposed physical gill, while in others this is unlikely and a hydrostatic function as an internal bubble more probable (Russell Hunter, 1953c). The respiratory behaviour of these and other species shows considerable adaptive plasticity, and snails from all the above populations took up a pattern of surface breathing after being brought into the laboratory (Russell Hunter, 1953c). As regards the use of the pallial bubble as a physical gill, crucial experiments to prove this function (including surfacing into high nitrogen and high oxygen atmospheres) were subsequently carried out with *Lymnaea stagnalis* (L.) (Henderson, 1961, 1963).

As regards the possible hydrostatic function of the pallial bubble, earlier evidence was contradictory (Jacobs, 1941; Precht and Otto, 1948), field evidence was merely circumstantial (Russell Hunter, 1953c), as were experiments on pulmonary gas content during diving respiration (Jones, 1961). However, Henderson (1961, 1963) demonstrated that, in such large species as *L. stagnalis* and *Planorbarius corneus* (L.), the hydrostatic function may determine the amount of air taken in at surfacing. Henderson showed in laboratory experiments that the normal duration of a dive in such species is of the same order as the time taken to use up oxygen from the bubble until the snail loses its buoyancy, but that much longer dives are possible. It is clear that the snail's detection of its changing buoyancy level plays an important part in the regulation of diving.

In both *P. corneus* and *L. stagnalis* there is always some cutaneous uptake and, under the best possible conditions of oxygen tensions in the water, this can just exceed the contemporary pulmonary uptake, amounting to 60% of the oxygen consumed (Jones, 1961). In most natural environments, pulmonary uptake (and surface breathing) remains more important. In *Planorbarius*, Jones (1961, 1964a, b) claims that the haemoglobin serves to allow greater exploitation of the pulmonary oxygen store during each dive, and that *L. stagnalis*, lacking such a respiratory pigment, makes shorter dives under the same conditions. This contrasts markedly with data on the same 2 species discussed by Henderson (1961, 1963) who found that *L. stagnalis* will increase its air store to a much greater extent than *Planorbarius* under the same conditions, allowing it to adjust for dives lasting twice as long. Despite this disagreement over dive-frequencies, Jones (1961, 1964a) has clearly demonstrated that the haemoglobin in *Planorbarius* is used as a respiratory transport pigment, and is adaptively more important in relation to pulmonary uptake than to cutaneous respiration. Earlier work (see, for example Borden, 1931; Wolvekamp, 1932; Fox, 1945; Zaaijer and Wolvekamp, 1958) had left this matter obscure, and there were unsubstantiated claims for oxygen storage and variations in pigment-synthesis with environmental conditions. There still is a need for comparative field studies to establish any clear relationship between specific ecological conditions and the occurrence of haemoglobin in particular species and races of planorbid snails. Haemoglobins in different freshwater pulmonates may well have different functions (Read, 1966), since Gratzer and Allison (1960), Manwell (1964) and Read (1966) note that molluscan haemoglobins vary considerably in molecular weight and isoelectric properties. It is also possible that the cytochrome systems of freshwater pulmonates are unusual (Mattisson, 1960).

C. Cutaneous respiration

Obviously all populations of freshwater pulmonates with water-filled mantle cavities, and all those species (mostly of freshwater limpet genera) which have no mantle cavity, must derive all their oxygen by cutaneous uptake through exposed surfaces. Growth changes in the surfaces available for such respiration differ in *L. peregra* and *P. fontinalis* (Russell Hunter, 1953c), and these changes can modify the simplest possible relationship between tissue mass (or volume) and available respiratory surface (Russell Hunter, 1953c, 1964). Despite such modification, size does matter and, in several species decrease in the surface:volume ratio consequent on growth clearly limits the size of an individual that can live submerged (Russell Hunter, 1953b, 1957). Cutaneous respiration under water replaces surface breathing most completely in 2 basommatophoran groups, the smaller Planorbidae and the various freshwater limpets, in both of which secondary gills have been developed (Russell Hunter, 1957, 1964). These gill-lobes are neomorphic—clearly not homologous with any part of the ctenidium of other molluscs—and lie outside the original mantle cavity. Continuous bands of ciliated epithelia are almost totally absent from the pallial organs of most freshwater pulmonates (Russell Hunter, 1953c; and unpublished) though they are present and functionally important on the corresponding structures in freshwater prosobranch snails. In the pulmonates, a few scattered ciliated cells occur on pallial structures and can move mucous sheets over their surfaces, but there are no continuous ciliated epithelia capable of water propulsion for enhanced respiratory exchange. The larger pulmonate snails (*Siphonaria* spp) of the marine littoral similarly lack ciliated surfaces outside the mantle cavity (Yonge, 1952, 1960). This absence of cilia may reflect incomplete readaptation from terrestrial life. In freshwater pulmonates, well-ciliated surfaces could increase the respiratory efficiency of the neomorphic gills in *Ancylus*, *Ferrissia*, *Acroloxus*, and *Laevapex*, and of general cutaneous respiration in forms like *Physa*.

Studies of oxygen consumption in freshwater limpets (Berg, 1952, 1953; Berg, Lumbye and Ockelmann, 1958; Burky, 1969, 1971; McMahon, 1972, 1973), and other freshwater snails (Lumbye, 1958; Berg and Ockelmann, 1959; Winberg and Belyatskaya, 1959, 1963; Jones, 1964a; Harrison, 1968; Daniels and Armitage, 1969), have provided valuable data on the ecological factors which can cause changes in purely aquatic respiration. The respiratory rates of *A. lacustris*, usually found in standing water, and *A. fluviatilis*, usually in streams, are of the same order, but the former survives better under anaerobic conditions (Berg, 1952). This seems ecologically significant, but surprisingly *Acroloxus* cannot maintain its normal oxygen consumption at lowered oxygen tension, while *Ancylus* shows considerable respiratory independence, maintaining its normal oxygen consumption at lowered tensions (even at one-third saturation values). Perhaps surprisingly, the North American pond limpet, *Laevapex fuscus*, shows a seasonal shift in its respiratory independence of lowered oxygen tensions (McMahon, 1973). Winter-conditioned limpets of this species have an exceptional oxygen-independent respiratory pattern, while summer-conditioned limpets are oxygen dependent in their respiratory rates. In a survey of respiration rates in 9 species of freshwater snails conducted by Berg and Ockelmann (1959), decrease in available oxygen causes lowered oxygen consumption in all species; this reduction is immediate but gradual in some species, and in others falls sharply but only at a critical level of oxygen tension. Only the prosobranch, *Bithynia leachi* showed a degree of oxygen independence similar to that of winter-conditioned *Laevapex*.

All species studied (Berg, 1952, 1953; Berg *et al.*, 1958; Berg and Ockelmann, 1959; Burky, 1969, 1971; Daniels and Armitage, 1969; McMahon, 1972, 1973) show increase in oxygen consumption with increasing temperature, but there are major differences between species and at different seasons within a species. For example, seasonal variations are found in the oxygen consumption of *A. fluviatilis* (Berg *et al.*, 1958), and can be correlated with reproduction. Berg *et al.* (1958) point out a further correlation with the changing growth rates of this species, using data from Russell Hunter (1953a). A degree of starvation such as must occur seasonally has a marked effect on respiratory rate in *Ancylus*, some effect in *Lymnaea palustris* (Müll.) and in the prosobranch *B. leachi*, and little or no effect in 6 other species (Berg and Ockelmann, 1959). In 3 different stocks of *Australorbis pfeifferi* Krauss Harrison (1968) found oxygen consumption rates to differ in response to calcium bicarbonate concentration. If we disregard the variations introduced by different nutritional and reproductive states, summer respiration rates are normally higher than those in winter. At ambient temperatures, the summer oxygen consumption rates are 2·8–6·0 times higher than the winter ones in *A. fluviatilis* (Berg *et al.*, 1958), 10·5 times higher in *F. rivularis* (Burky, 1969, 1971), and from 9·4–36·0 times higher in different populations of *L. fuscus* (McMahon, 1972, 1973). Temperature acclimation is known to take place in the respiration of a number of freshwater pulmonate species, and may well occur to some extent in the majority of species living in temperate and boreal latitudes.

D. Acclimation and low oxygen stress

In most poikilothermic animals (see Bullock, 1955; Prosser, 1955, 1967; and references therein) acclimation of respiratory rates to sustained temperature changes is compensatory, that is metabolic rates (including oxygen consumption) measured at standard temperatures against controls are raised in animals which have been conditioned for an appropriate period at a lower temperature. Thus cold-acclimated animals have greater respiration rates than comparable warm-acclimated animals when respiration is measured at some common intermediate temperature. Apart from seasonal changes, similar compensatory adjustments are found as latitudinal differences in many aquatic invertebrates and fishes (Berg, 1953; Scholander *et al.*, 1953; Fry, 1958). Three freshwater pulmonates have now been found to show "reverse acclimation", a modification of response (found also in one fish, *Carassius carassius*, 2 barnacle species, a freshwater amphipod, an intertidal limpet, and the land pulmonate, *Helix pomatia* (L.)) termed unusual by Fry (1958) and paradoxical by Prosser (1967). All 3 are freshwater limpets, and reverse acclimation of respiration has been demonstrated in stocks of *A. fluviatilis* by Berg (1953; Berg *et al.*, 1958), of *F. rivularis* by Burky (1969, 1970, 1971), and of *L. fuscus* by McMahon (1972, 1973). Both the modifications of this which occur at low oxygen tensions, and a possible explanation of the adaptive significance in *Laevapex* are discussed more fully by McMahon (1973).

Measured at 100% oxygen tension at 10 ° and 20 °C, respiration rates are higher in the summer-conditioned stocks than in the winter-conditioned stocks for all 3 populations of *Laevapex* studied. Such a respiratory response to temperature acclimation is that which has been termed reverse acclimation. However, *Laevapex* migrates down rocks into conditions of reducing mud just before the onset of winter and thus, during the period of lower temperatures is also living in conditions of low oxygen tension.

McMahon's (1973) respiration experiments with decreasing oxygen tension show that at low oxygen tensions winter-conditioned specimens have a higher respiratory rate than summer-conditioned individuals. Such a response to temperature acclimation at low oxygen tensions may be interpreted as positive or compensatory rather than reverse acclimation. As a result of this, McMahon (1973) has advanced the hypothesis that, in his populations of *Laevapex* at least, "reverse" temperature acclimation at higher oxygen tensions is merely one manifestation of a more general winter-conditioning. These limpets, when winter-conditioned, have a higher respiratory rate at low oxygen tension than do summer-conditioned limpets. Although this acclimatory process appears to be triggered by decreasing temperature, its adaptational significance involves the survival of the inactive overwintering limpet in contact with reducing mud.

A number of freshwater pulmonates, including *L. fuscus*, have considerable capacity for anaerobic respiration. The biochemistry of this has been reviewed (von Brand, 1946; Goddard and Martin, 1966), and most snails have been thought to pay an oxygen debt after experiencing periods of low oxygen stress. There are differences in the metabolites excreted during and after such stress; differences in anaerobic capacity between species (von Brand *et al.*, 1950, 1955) and differences within species which are related to temperature and varying degrees of tissue hydration (von Brand, 1955). Correlation of these differences with known ecology is not yet possible, but species differences in physiological resistance have a discernible biochemical basis. In 11 species of freshwater snails, those with low resistance to anaerobiosis were shown to accumulate lactic acid in the blood during low oxygen stress, while those more resistant to low oxygen stress appear to accumulate less toxic higher fatty acids (Mehlman and von Brand, 1951). In *L. fuscus* McMahon (1973) found that after treatment with low oxygen stress there is no overt payment of an oxygen debt, instead there is a general increase in respiratory rate at all oxygen tensions. These rates are maintained for up to 135 hours after return to fully saturated conditions and appear to be in part a short-term acclimation of respiratory rates to low oxygen tensions. Although there is no direct evidence, it seems likely that metabolic end-products are continuously leached into the environmental water by *Laevapex* during low oxygen stress. Metabolite accumulation is probably more extensive in some of the larger species of such genera as *Lymnaea*, *Planorbarius* and

Australorbis studied by von Brand and his co-workers. It must be remembered that *L. fuscus* (and the other smaller pulmonate limpets such as *Ferrissia* and *Acroloxus*) are relatively tiny snails and physiologically "thin" organisms with no part of their tissues more than a millimeter from the environmental water.

In sum, these respiratory studies attest to the extent of physiological plasticity which is a characteristic of freshwater pulmonates. As will be discussed more fully below (Section V), this characteristic is almost certainly related to the small-scale and transience of most freshwater habitats. This transience is probably also reflected in the co-existence of several degrees of readaptation of an initially air-breathing stock to life in fresh waters (Russell Hunter, 1952, 1953c, 1957). In this respect it is important to note that, in the series of pulmonates considered above, the snails which are air-breathing and amphibious are structurally unspecialized and often more primitive, while those most completely adapted for aquatic life (e.g., the various freshwater limpets) are most specialized structurally (e.g., in pallial morphology, in renal and genital organs, and even in nervous system, see Hubendick, Chapter 1 of this volume). The opposite would be the case if the group had followed the more usual pattern like vertebrates and annelids (Carter, 1931) of preadaptation in fresh waters leading to terrestrial life.

E. Other interpopulation variation

Populations of freshwater pulmonates show great physiological plasticity, and this is most evident in modes of respiration and in patterns of life cycle and reproduction. It is convenient to consider reproductive and demographic variation along with bioenergetic ecology in the next section (IV), but there remain studies concerned with other kinds of interpopulation variation in freshwater pulmonates (shell components, shell growth patterns, radula proportions, and enzyme systems) which can somewhat arbitrarily be discussed here.

In freshwater pulmonates—as in the majority of molluscs—the secreted shells have 2 principal components: a meshwork of hardened protein and crystalline calcium carbonate (Jones, 1969; Russell-Hunter *et al.*, 1968, 1970). The latter is secreted after active uptake directly from environmental water. In the euryoecic species, *L. peregra*, the thickness (and mass) of the calcareous shell varies with the calcium available in the waters (Hubendick, 1947, Russell Hunter, unpublished data from Scottish waters). Thus it appears that *L. peregra* expends about the same energy on shell-making no matter what the environmental hardness. The case of the stream limpet, *Ferrissia rivularis*, in natural creek populations of upstate New York is strikingly different. Although these creek waters vary over

7-fold in dissolved calcium, the highly significant differences in shell calcium found to exist between populations are not correlated (Russell Hunter *et al.*, 1967, 1970). Anabolic concentration ratios ranged from 1609:1–10,615:1 and there was other circumstantial evidence of physiological races. These striking differences between populations are highly significant in relation to the possible use of such molluscs as indicators of environmental radiocontamination with strontium-90 (Nelson, 1964; Rosenthal *et al.*, 1965) which they found to be accumulated indiscriminately with calcium. Likins *et al.* (1963) claimed relative discrimination against strontium in *Australorbis glabratus*, but Van Der Borght and Van Puymbroeck (1964, 1966) demonstrated active uptake direct from environmental water of both calcium and strontium in pulmonates such as *L. stagnalis*, *L. auricularia* (L.) and *P. corneus*. Thus it becomes important that some freshwater mollusc species have populations which differ significantly from each other in the extent to which they do concentrate calcium from the environment. A further survey of total organic carbon and total nitrogen in the shells of the same creek populations of *Ferrissia* was subsequently carried out (Russell-Hunter *et al.*, 1970). Carbon:nitrogen (C:N) ratios are consistently about 3·25:1 indicating that the non-calcareous component is mostly protein, but the total organic material in the shell (related to standard sized limpets) varies 3-fold between populations. When the shell protein figures are considered along with those for shell calcium, there proves to be neither an inverse relation between them (as expected by an "adaptive value" hypothesis) nor any direct relationship (as expected if both relate to levels of energy turnover). It should be noted that, in content of shell protein and of shell calcium (as even more clearly in other measurable characters), the variation within the majority of single populations is very much less than the range of variation for the species as a whole. It seems that genetic controls of shell secretion for the 2 major components are independent, and that the chances of genetic dispersal among the isolated creek populations of this limpet have resulted in some rather inappropriate shells in certain habitats.

The shell component differences in the pond limpet, *L. fuscus* (McMahon, 1975a) are markedly less than those discussed in *F. rivularis*. However, shell calcium in *Laevapex* increases with calcium concentration in the environment just as it does in populations of *L. peregra*. In populations of *L. palustris* Hunter (1975b) reported a third relationship between shell and environmental conditions. His survey covered 14 population-sites, and he found that the ratio of shell calcium to whole animal dry weight changes little throughout growth, and does not vary greatly between populations of *L. palustris*. This proved true over a wide range of calcium concentrations of environmental water, and represents a "regulation" unusual in a species which shows great interpopulation variation in other respects. Patterns of interpopulation variation in shell components obviously differ among closely related freshwater pulmonate species.

In many pulmonates, shell shape is another infraspecific variable. Extensive data have been collected on shell biometrics in *F. rivularis* from a set of 54 populations in a variety of localities in upstate New York of varying trophic conditions and water hardness, and with different degrees of isolation from each other (Russell-Hunter and Nickerson, unpublished). Two kinds of variation in shell indices (ratios) occur. Where H is the height of the limpet shell and L and W are the length and width of its aperture respectively, the steepness index ($3H/L + W$) is allometric and reflects local growth patterns as modified by trophic conditions. On the other hand the roundness index (W/L) is basically isometric during growth, highly specific to each population, and results from the fixed growth gradients of the mantle-edge (which seem to be rather rigidly genetically determined). The steepness index, like other allometric growth relationships (best assessed by computational methods developed by palaeontologists; see Gould, 1966; Vermeij, 1971) directly reflects trophic conditions, and further studies may allow it to be used in assessments of changes in the nutritional history of the individual snail or of the population. For the roundness index—as with other genetic variation—the range of variation is much less within each isolated population than the range of variation for the species as a whole. Relative levels of variation between populations may reflect differences in temporal extent of isolation. Somewhat similar results have been reported for biometrics of other freshwater pulmonates. Hubendick (1951) carried out a thorough survey of shell form in *L. peregra*, and Mozley (1935) discussed interpopulation variation in *L. emarginata* Say and *L. palustris*. A more recent survey of shell shape in *L. palustris* is reported by Hunter (1975b). Covering 14 populations in a variety of trophic conditions in upstate New York, it shows that 2 biometric ratios involving growth along the spiral axis (in this turbinate shell) are allometric and reflect trophic conditions at different sites, while a ratio defining apertural proportions was isometric and surprisingly constant not only within populations but between them. The aperture shape in *L. palustris* seems to be genetically controlled to an unusually fine degree, with little or no variation occurring in its expression.

Interpopulation variation in shell-form and in shell components would seem in some cases to be broadly ecophenotypic and in other cases to be more rigidly genetically determined. It is unlikely that the fine structure of teeth in the radulae of pulmonate snails show much environmentally controlled variation. In the majority of gastropods, radular dimensions seem to be under rigid genetic control and are seemingly non-adaptive, yet there are no observable environmental effects on their genetic expression. Studies of radulae from different populations of *L. peregra* in northwestern Europe (Berrie, 1959) and of *L. palustris* in upstate New York (Hunter, 1972, 1975b) yielded similar results. In ratios reflecting the proportions of certain lateral teeth in the radula, variation within all populations was low, but variation between populations was very high. Berrie (1959) notes that the level of significance for the variation between certain populations of *L. peregra* would have justified the setting up of subspecies. Hunter (1972), with *L. palustris*, used comparisons of differences between pairs of population means by Tukey's test which allows establishment of significance at a series of probability levels. Approximately 43% of all pair-wise comparisons for 1 ratio (for 16 populations) were significantly different at the 0·80 probability level or higher (20% at the 0·99 level). In both investigations, the interpopulation variation proved geographically random: the biggest differences often occurring between adjacent populations, and some widely separated stocks having the closest similarities. In the survey of *L. palustris* in upstate New York, radular differences were never related to shell biometric differences or any environmental conditions. The lack of geographic clines in these genetically controlled ratios can be taken as reflecting the stochastic aspects of passive dispersal.

More direct assessment of genetic isolation involves using the electrophoretic techniques developed by Hubby and Lewontin (1966) to survey allele frequencies of allozymes in a series of populations of a species. Nickerson (1972) surveyed variation in 7 enzyme systems (9 loci) in 16 natural populations of *F. rivularis* living in a wide range of environmental conditions. Cellulose polyacetate strip electrophoreses showed that 5 loci were uniformly dimorphic while 2 were dimorphic at several sites. The proportion of polymorphic loci in individual populations (ranging between 0·56 and 0·78) and the levels of heterozygosity per average individual (ranging between 0·20 and 0·32) were as high or higher than any yet reported for field populations of any species. Geographically, and in relation to all biotic and abiotic environmental factors which were measured, these high levels of enzyme variability were randomly distributed. Once again, there were no clines. Successive generations were surveyed at 3 population sites and showed little alteration in allelic frequencies. Experimental transfers were carried out of young spat limpets between 2 population sites. After growth to adults the allelic frequencies (and shell roundness ratios) were essentially those of their parental stocks. Environmental factors do not seem to have modified phenotypic expression of these allelic variants.

Even without the direct evidence of breeding experiments, it is reasonably certain that such interpopulation differences are genetically determined. The implications of the extent and pattern of these infraspecific variations, and of physiological variability, in freshwater pulmonates will be discussed in relation to certain broader evolutionary considerations in Section V below.

IV. Life cycles, population dynamics and comparative bioenergetics

A. Patterns in life cycles

As has been known for a number of years, considerable infraspecific interpopulation variation occurs in growth rates, reproduction, and the course of the life cycles found in natural populations of freshwater pulmonates in temperate regions (Russell Hunter, 1961b, 1964). In some species such as *Gyraulus albus*, infraspecific differences between populations involve only growth rates and the time and intensity of the breeding season. Three natural populations of *G. albus* studied by Russell Hunter (1961b) in Scotland could differ by as much as 3 months in breeding season (late-April versus late July), and by as much as 3·5 times in their mean carbon biomass at time of breeding (recalculated from data in Russell Hunter, 1961b). Despite such variations, all 3 populations show a simple annual life cycle with complete replacement of generations (a univoltine pattern).

In many other pulmonate species which have been studied, interpopulation differences also involve the seasonal course of the reproductive cycle and, in particular, the number of generations per year. Among the more comprehensive studies on natural populations are those of W. F. De Wit (1955, Holland); R. M. De Witt (1955, 1967, U.S.A.); Geldiay (1956); Duncan (1959); Russell Hunter (1953a, 1957, 1961a, b, 1964); McCraw (1961, 1970); Berrie (1963, 1965); Maitland (1965); Lamberet (1966); Clampitt (1963, 1970); Gillespie (1966, 1969); Burky (1969, 1971); Nickerson (1972); Hunter (1972, 1975a); McMahon (1972, 1975a); Eckblad (1973); Eversole (1974); and McMahon *et al.* (1974). The most common pattern of life cycle seems to be a simple univoltine one with breeding in late spring or early summer. However, considerable variation occurs between populations. Using a modified version of a classification adopted earlier (Russell Hunter, 1961b, 1964) we have 7 principal patterns of life cycle in temperate populations. Two are biennial patterns: 1 with each generation capable of reproducing in 2 successive years, and the other with a prereproductive life of about 22 months. Two are simple annual patterns: one with spring breeding, and the other with late summer breeding. In a further 2, a second generation is involved annually: 1 shows complete replacement of generations, in the other both "early" and "late" generations overwinter and contribute to the next year's "early" breeding season. Lastly, there can be 3 generations successively replacing each other in the course of a year (1 spring, 1 summer, and 1 over-wintering).

The biennial patterns involve larger pulmonates such as *P. corneus* and *L. stagnalis* (Baker, 1911; Berrie, 1965, 1966) in higher latitudes or more oligotrophic waters (i.e. poorer temperature or trophic conditions). Annual cycles with late summer reproduction occur in such genera as *Planorbis*, *Physa* and *Lymnaea*, again typically in harsher environments. Annual cycles with late spring or early summer reproduction have been reported for some populations of almost all the more common species of freshwater pulmonates (for example, see lists in Russell Hunter, 1957, 1961b; Burky, 1969, 1971; Hunter, 1972, 1975a). The pattern of 2 generations per year with incomplete replacement is known in *P. fontinalis* (Duncan, 1959; Russell Hunter, 1961b), *L. palustris* (Hunter, 1972, 1975a), and *F. rivularis* (Russell Hunter, unpublished; Nickerson, 1972). The true bivoltine pattern of 2 generations with complete replacement is known in *P. fontinalis* (De Wit, 1955), *L. peregra* (Russell Hunter, 1961b), *F. rivularis* (Burky, 1968, 1971), and *L. fuscus* (McMahon, 1972, 1975a). A 3-generation pattern is known for *L. truncatula* (Walton and Jones, 1926) and for *P. virgata* and *L. fuscus* (McMahon, 1975b, c).

In different natural populations of *L. peregra* we find 5 out of these 7 patterns (biennial twice breeding, both univoltine patterns and both 2-generation patterns), while in different populations of both *P. fontinalis* and *F. rivularis* all of these last 4 patterns have been found. Implicit in such interpopulation variation in life cycles are considerable interpopulation differences in growth rates, in survivorship, in time of breeding, and in gross fecundity, and—while much of these interpopulation differences seem to be environmentally evoked—at least some infraspecific variation seems to be genetically determined (Russell Hunter, 1961b; McMahon, 1973, 1975a; Hunter, 1975b). Variations in growth rates can often be clearly related to trophic differences in habitats (Russell Hunter, 1953a, 1961b; Burky, 1971; Hunter, 1975a; McMahon, 1975a), and recent work on variation in *Aufwuchs* (McMahon *et al.*, 1974) shows a direct relationship between growth in 2 pulmonate species and both the quantity of *Aufwuchs* available and its nutritional quality. Kept in overwinter starvation conditions, tissue "degrowth" may occur (Russell-Hunter and Eversole, 1976). Variations in mortality may involve set environmental limitations (Russell Hunter, 1953a, 1961b). Variations in time of onset and intensity of breeding are dependent both on environmental factors (water temperature being most important), and on endogenous causes involving the growth and generation prehistory of the snails (Duncan,

1959; Russell Hunter, 1961a, b; McMahon, 1975a) but, again in a few cases involving an inappropriate relation to environmental conditions, some genetically determined differences can be deduced (Russell Hunter, 1961b).

Environmental signals relative to the onset of reproduction require further investigation. No clear-cut evidence has been published for the occurrence in any freshwater pulmonate of the sort of seasonal control by changing day-length which has been described for the primitive salt-marsh pulmonate, *Melampus* (Apley, 1970; Russell-Hunter *et al.*, 1972). In an unpublished thesis (1951), Jenner reported that stocks of *L. palustris* were brought into reproduction by lengthening days at an *L*:*D* of 13·5:10·5 (personal communication from C. E. Jenner, see also discussion in De Witt, 1967). After further work on *Physa pomilia* and *Lymnaea columella* Say, including induction of reproduction during decreasing day-length, De Witt (1967) concluded that initiation of egg-production is a function of temperature rather than day-length. In various pulmonates, Bondesen (1950), De Witt (1954), Geldiay (1956), Apley (1967), Burky (1969), Eversole (1974), and McMahon (1975a, c) have reported "threshold" temperatures for the onset of egg-laying. Burky (1968, 1971) reports on reproduction in *Ferrissia* being temporarily "switched off" by an unseasonable drop in water temperatures.

In the case of variations in fecundity between populations clearly both ecophenotypic and genetically fixed differences occur. Bred in the laboratory, "races" of *Lymnaea* spp can show genetically determined variations in fecundity and in growth rates (Boycott *et al.*, 1930; Boycott, 1936; Forbes and Crampton, 1942a), and "demes" of *A. pfeifferi* show differences in fecundity and survivorship (Harrison *et al.*, 1970). Recently, the genetic basis of similar differences in *A. glabratus* (Say) has been elucidated by careful breeding experiments (Richards, 1970; Richards and Merritt, 1972). On the other hand, McMahon *et al.* (1974) found that not merely the quantity of *Aufwuchs* available, but its nutritional quality assessed both as percentage organic carbon and as carbon:nitrogen (C:N) ratio, had direct effects on gross fecundity in natural populations of *L. fuscus* and *L. palustris*.

Boreal populations of freshwater pulmonates will differ less in life cycle patterns from the temperate ones already described than tropical ones do. Colder latitudes will tend to support either biennial or the more impoverished annual patterns of life cycles. Under those tropical conditions with an annual dry season, natural populations of pulmonates can have an average life span of under 1 year (maximum *c.* 15 months), and a life cycle extending over 1 dry and 2 rainy seasons, or through one rainy season only (Olivier and Barbosa, 1955a, b). In general, tropical breeding seems to be synchronized with rainy seasons, though Webbe (1960) reports field evidence for Tanganyika which shows that "longer" rainy periods may affect snail populations adversely. In more stable fresh waters in the tropics—as in the waters of limestone caves—reproduction in pulmonates may occur all the year round, as a result of high and relatively constant water temperatures.

B. Transfer and density experiments

The above descriptive ecology of life cycles, which resulted from much careful sampling of relatively undisturbed natural populations of pulmonate snails, is now being supplemented by some experimental work and by certain important studies on population bioenergetics (the latter in terms of organic carbon or nitrogen turnover rather than in directly measured calorific values). The beginnings of experimental ecology with pulmonates have involved either field "transfer" experiments between different populations of the same species, or the controlled manipulation of 1 environmental parameter (abiotic or trophic) in an otherwise undisturbed natural population. The first kind have attempted to elucidate the extent to which certain interpopulation differences are genetically based, while the second kind of experiments have largely been directed to questions of population regulation, particularly as regards density-dependent feedbacks with negative effects on growth and fecundity.

Interpopulation transfer experiments with pulmonate snails have been carried out by Geldiay (1956) on *A. fluviatilis*, by Burky (1969, 1971) and by Nickerson (1972) on *F. rivularis*, by McMahon (1972, 1975a) on *L. fuscus*, and by Hunter (1972, 1975a) on *L. palustris*. The results of these transfer experiments have been mixed, and conclusions to be drawn from them are very tentative. Geldiay (1956), with *Ancylus*, studied shell growth rates only and found no differences in these rates between native and transferred stocks. In all the other cases, reciprocal transfer experiments resulted in marked differences in the parameters assessed. In reciprocal experiments involving one population of *L. palustris* from a eutrophic environment and another from relatively oligotrophic conditions, Hunter (1972, 1975a) measured considerable differences both in growth rates and in egg-production but concluded that all these differences could have resulted from the imposed changes in environmental conditions. At the other extreme, Nickerson (1972) using shell biometrics to assess the results of transfer of specimens of *F. rivularis* between oligotrophic and eutrophic environments found that certain shell growth ratios characteristic of stocks were retained by them after transfer. Such shell ratios can be regarded as "racial", i.e. genetically determined, characteristics. This was even more clearly true for the allelic frequencies for nine loci which were assessed by electrophoretic techniques (Nickerson, 1972). With *L. fuscus*, McMahon (1972, 1975a) assessing growth and fecundity in transfer experiments between 2 populations (one from a mesotrophic and the other from a eutrophic environment) found circumstantial evidence for the existence of physiological races. The *Laevapex* population from the mesotrophic environment seems to have either higher assimilation efficiency, or higher anabolic conversion efficiency, or both. Burky (1969, 1971), assessing the same parameters in *F. rivularis* transfers, had rather more equivocal evidence for genetic controls but this was subsequently somewhat strengthened by Nickerson's (1972) results on biometrics and allozymes in the same species. Taken as a whole up to this time, these transfer experiments on pulmonates merely confirm that interpopulation variation in fecundity can involve both ecophenotypic and genetically-fixed differences.

When we turn to feedback aspects of population regulation, we find extensive work on density-dependent effects on freshwater pulmonates in laboratory culture and relatively little on similar effects in the field. Because of the plasticity already noted in growth and reproduction in these snails, culture results are usually difficult to correlate with observations on natural populations. For example, it has long been known that, other conditions being equal, more dense laboratory populations show reduction of growth rates and fecundity (e.g., references in Taylor, 1900; Boycott *et al.*, 1930; Forbes and Crampton, 1942b; Noland and Carriker, 1946). Laboratory work on possible biological control of schistosome hosts has included extensive density experiments, the better designed ones again proving these "crowding" effects in *A. glabratus* (Chernin and Michelson, 1957a, b) and in *Bulinus forskalli* (Ehrnb.) (Wright, 1960), see also Brown, Chapter 7 of this volume). Natural populations of *A. fluviatilis*, *P. fontinalis*, and *L. peregra* in Scotland did not show the expected inverse relationship between adult size and population density, though density varied greatly (up to 5-fold), and adult size considerably (up to ± 17·5%), when surveyed over 9 years (Russell Hunter, 1961a). Field data on fecundity (for *Ancylus* only) suggest a more complex relationship with population density:number of egg capsules per limpet is lower both at very high adult densities and at low densities (Russell Hunter, 1961a, 1964).

Two sets of field experiments on density-dependent regulation in pulmonates merit discussion. Eisenberg (1966, 1970) worked with stocks of *Lymnaea elodes* Say (which may be a subspecies of *L. palustris*) in large wire-mesh enclosures (pens) in a Michigan pond. At 3 population densities (2 altered experimentally) there was an inverse relationship between densities of adults and young. Within 1 reproductive period, a 10-fold initial difference in pen densities had been compensated for by modified fecundity, and the numbers of the next breeding generation were nearly equivalent. In his later study, Eisenberg (1970) showed that food additions (frozen leaf spinach) could cancel the inverse relationship between adult snail density in pens and their egg production, the population response thus being dependent on resource limitation. Eisenberg's demonstrations of density-dependent fecundity and of resource limitations in near-natural populations were surprisingly clearcut, and have perhaps more significance in the literature of population dynamics than in that of pulmonate physiology. Unanswered questions concern the respiratory schesis of the penned snails, egg viability and spat mortality, the quality of the natural food (*Aufwuchs*) available in the pens, and—as is common in much work on density-dependence—the possible concurrence of more than 1 mechanism of density regulation.

Eversole (1974) carried out parallel laboratory and field experiments on the extent of, and the controls for, differential fecundity in *Helisoma trivolvis* (Say). There are undoubtedly 2 (and perhaps 3) different density-dependent mechanisms which modify rates and patterns of egg-laying. In *Helisoma*, an 8-fold increase in population density (in field cages) produces a mean decrease per snail from 1516–385 eggs per breeding season. Such a decrease corresponds to a reduction from 59% of the mean annual N-R.A. (non-respired assimilation, see further explanation below) to 33% directed to egg output. Undoubtedly many other molluscan populations exhibit density-dependent control of fecundity, but it is unlikely that any single causality is universal. For *L. elodes* and *L. stagnalis* trophic limitation can be important (Van Der Steen, 1967; Eisenberg, 1970; Mooij-Vogelaar and Van Der Steen, 1973; Mooij-Vogelaar, Jager and Van Der Steen, 1973; Van Der Steen *et al.*, 1973); while for *Australorbis sudanica* (Martens) specific chemoreception has been claimed (Berrie and Visser, 1963; Berrie, 1968, 1969); and for *Lymnaea pervia* Martens general accumulation of contaminating metabolites has been implicated (Hirata and Kawabata, 1954). In the case of these more recent experiments on *Helisoma* (Eversole, 1974), the single most important influence seems to be the frequency of tactile contacts between adult snails. However, fecundity could also be reduced by sustained egg-laying and the accumulation of spat. Further experiments and analyses revealed that spat numbers had a greater impact on fecundity than did eggs, but both had less than did adult numbers. In all of these experiments on *Helisoma* (Eversole, 1974), the density-dependent reductions of fecundity recorded never resulted from limitation of available food resources.

C. Actuarial bioenergetics

Within the last decade, natural populations of snails have begun to be investigated in terms of the bioenergetic turnover of total organic carbon and of total nitrogen of tissues and shells. (These two kinds of analyses have proved more readily applicable to repeated sampling of natural populations than direct microcalometric determinations of energy content values for tissues.) Investigations in such terms of standing crop biomass and of productivity are reported in Russell-Hunter *et al.*, 1968, 1972; Russell-Hunter, 1970; Apley, 1967, 1970; Apley *et al.*, 1967; Burky, 1969, 1971, 1974; Avolizi, 1970, 1975; Mattice, 1970, 1972; Nickerson, 1972; Hunter, 1972, 1975a; McMahon, 1972, 1973, 1975a; McMahon *et al.*, 1974; Eversole, 1974; and several other communications as yet unpublished. Except for those of Avolizi (1970, 1975, on *Sphaerium*), Burky (1974 only, on *Pomacea*), and Mattice (1970, 1972, on *Bithynia*), all these reports concern basommatophoran pulmonates. All were based on regular collecting from natural freshwater populations except for a series on the primitive salt-marsh pulmonate *Melampus* (Apley, 1967, 1970; Apley *et al.*, 1967; Russell-Hunter *et al.*, 1972). Analyses of total organic carbon (which are usually directly related to calorific values) were made using the "wet-oxidation" method developed by Russell-Hunter *et al.* (1968), and refined to give reproducible results with molluscan growth stages over 6 orders of magnitude (Russell Hunter and Apley, 1966; Russell-Hunter *et al.*, 1972). Analyses of total nitrogen were made on a Coleman model-29 semiautomatic nitrogen analyser which employs a modified micro-Dumas method as described by Gustin (1960). In addition, dry weights, ash-free dry weights, and assessments of shell calcium and shell protein have usually been recorded. Analyses were normally made on selected samples of size-groups chosen from the age structure of natural snail populations collected at times picked to correspond to the major changes in biomass with growth; with state in the reproductive cycle (e.g. prebreeding, midbreeding, postbreeding, adolescent growth, overwintering); and with the seasons of their natural environment.

Values of standing crop production (usually in total organic carbon terms) were normally calculated by first computing appropriate values per 100 snails for each sample. This was done by taking each known size frequency distribution for the population at that sampling time (expressed as per cent distribution), and then summing the products of these frequency values and the appropriate carbon values (in turn derived from sets of regressions of carbon to snail size constructed from the raw data of carbon analyses). Then, for each sampling time, the values per 100 snails could be combined with appropriate field assessments of population densities to give values in milligrams of carbon per square metre of stream or pond bottom (that is, standing crop biomass figures). Subsequently, as discussed below, series of standing crop production figures (along with survivorship and fecundity data) could be transformed into productivity rates.

For the stream limpet, *F. rivularis*, Burky (1971) gives peak (late August) values of standing crop biomass at 192 and 99 mg C m^{-2} for eutrophic and mesotrophic creeks respectively, and Nickerson (1972) computes values for 16 *Ferrissia* populations (all assessed at the peak biomass of late-spring breeding) ranging from 14–160 mg C m^{-2}. For the "pond" limpet, *L. fuscus*, McMahon (1975a) assessed peak biomass values (at different seasons) for 3 populations (7 generations) ranging from $<$ 2–5 mg C m^{-2}. There are comparable figures for the larger pulmonates of the genera *Helisoma* and *Lymnaea*. For 1 population of *H. trivolvis*, Eversole (1974) gives a peak biomass value (late September) of 4·2 g C m^{-2}. For *L. palustris*, Hunter (1975a) computes biomass values for 3 populations (in environments of increasing eutrophy) of 0·3, 0·9, and 0·7 g C m^{-2} in August and 0·4, 1·6, and 1·5 g C m^{-2} in October respectively. It is worth noting that the rank order of different populations of a species (such as those of *L. fuscus* or *L. palustris*) in terms of standing crop biomass (mg C m^{-2}) does not necessarily reflect the rank order of productivity rates (mg C m^{-2} day^{-1}) for such populations (Russell-Hunter, 1970).

Similar values for biomass can be computed from the total nitrogen analyses of snails sampled from populations. Computation of both carbon and nitrogen biomass values for "standard" snails under different biological conditions and at different times of the year allows the assessment of carbon:nitrogen (C:N) ratios of the snail tissues. In an admittedly oversimplified summary, Russell-Hunter (1970) claimed that decreasing C:N ratios (corresponding to build-up of protein fraction) characterized the prereproductive period, while increasing C:N ratios (probably build-up of glycogen or galactogen storage, much less likely fats in pulmonates) characterized the period of "preparation" for overwintering, in natural snail stocks. Molluscan eggs vary greatly in their specific C:N ratios, and a tentative classification by ratio and size into 5 categories is given by Russell-Hunter (1970). With regard to the snails discussed here, all freshwater pulmonates seem to have "large eggs with low C:N ratios" (from which spat emerge as "miniature adults", immediately capable of most adult behaviour, including appropriate patterns of feeding); while the primitive pulmonate *Melampus* has "small eggs of intermediate C:N ratio" (which hatch as free-swimming planktonic veligers, and whose further development involves both anatomical and trophic metamorphoses).

All investigations of the bioenergetic ecology of populations (including attempts at actuarial "balance-sheets") must move beyond biomass and ratio assessments to computations of productivities as rates. All rate measures (of feeding, of growth, of respiration, of reproduction, and of mortality) are essentially measures of biomass per time and can be reduced or converted to values such as milligrams carbon per 100 snails per day, or milligrams carbon per square metre per day. In overall productivity rates the components are similar no matter whether one is considering energy flow in a modal individual snail or for each snail generation-cohort or population as a whole (Russell-Hunter, 1970). In any carbon budget of rates, only a fraction from the productivity of the previous trophic level (in the case of most freshwater pulmonates, from the productivity of the *Aufwuchs*) makes up the total ingested (T.I.). Some part of that is not assimilated (N.A.) and egested as faeces, while the rest as the total assimilated carbon (T.A.) enters the organism (or population). Of the T.A. a considerable part is broken down as respired assimilation (R.A.) for maintenance and other activities, and only the fraction of non-respired assimilation (N-R.A.) remains available for increase in the carbon biomass of the snails by growth and reproduction. (Ecologically, only a fraction of this N-R.A. is in turn available to the next trophic level, that is, to form the T.I. of snail predators.) There are several simple relationships: T.A. = T.I. − N.A.; N-R.A. = T.A. − R.A.; and N-R.A. = T.I. − (N.A. + R.A.); and from these, several ratios such as T.A./T.I.; N-R.A./T.A.; and N-R.A./T.I. can be derived and, if expressed as percentages, can be regarded as "efficiencies" (for example, the first is gut-efficiency). In addition, another important ratio concerns the partitioning of N-R.A. between growth and reproduction.

There are several ways of computing growth in bioenergetic terms. Perhaps the best use of the data from carbon analyses of selected population samples involves the computation of incremental growth in carbon terms per unit population taking into account the appropriate size-frequency (age) structure of the population at each time, and then summing and averaging these increments over appropriate periods of time as rates. Egg-production represents a considerable energetic output from a fertile snail or from a population, and it is important to quantify it in the same terms as growth, assimilation, respiration and the rest. In numerical terms of mean gross annual egg output, available snail data range from some populations of *Ferrissia* at 8 eggs per year per adult (Burky, 1971), of *Laevapex* at 27–70 (McMahon, 1975a), of *Ancylus* at 47 (Russell Hunter, 1953a, 1961a), of *Bithynia* at 201 (Mattice, 1972), of *Physa* at 550, of *Helisoma* at 1900 (Eversole, 1974), of *Lymnaea* at 2200, to *Melampus* at 33,150 (Apley, 1970; Russell-Hunter *et al.*, 1972). In general, lower numerical fecundity corresponds to larger egg-size when adult snails are of comparable biomass (Russell-Hunter, 1970). Burky (1973, 1974) notes that the large amphibious snail, *Pomacea urceus* (Brug.), has relatively "gigantic" eggs (comparable to those of small birds), but still can produce 50–200 eggs per clutch. In carbon terms for *Melampus*, it can be calculated that 87% of the N-R.A. during the reproductive season is directed to egg output (corresponding to 46% of the total annual N-R.A.; Apley *et al.*, 1967; Russell-Hunter *et al.*, 1972). The corresponding seasonal fraction for *Bithynia* is 55% (Mattice, 1972), for *Helisoma* 83% (Eversole, 1974), and for *Ferrissia* 45% (Burky, 1971). It is possible to compute changes in organic carbon and nitrogen contents (and thus C:N ratios) for whole snails, for excised gonads, and for egg-masses produced over the time course of the breeding season. This has been done for *Melampus bidentatus* (Apley *et al.*, 1967; Russell-Hunter *et al.*, 1972) and for *H. trivolvis* (Eversole, 1974) and, in both species, data suggest that gonadial protein turnover does not keep up with (and thus limits) egg output. In a population of *Melampus* with 3 cycles of egg-laying, the C:N ratio of excised gonads rose from a value of 6·3:1 during the first egg-laying cycle to one of 15:1 at the end of the third cycle. One of the most significant aspects of the recent experiments on density-dependent reduction of fecundity, conducted by Eversole (1974) on *Helisoma*, is the demonstration that experimental changes in population density affect the bioenergetics of fecundity both in total organic output and in its relative partitioning. In carbon terms, more crowded snails direct a smaller fraction of N-R.A. to fecundity (63%) than do less crowded snails (74%). These more crowded snails direct less total organic carbon (but a larger fraction, 37%) to growth.

Implicit in all this work on reproduction is the greater bioenergetic cost of "femaleness", and this is highly significant in relation to the adaptive value of protandric hermaphroditism in some molluscs and of sexual dimorphism in others. This is discussed elsewhere in relation to evidence for functional protandry in the pond limpet *Laevapex* (Russell-Hunter and McMahon, 1976) and to an anomalous sex-ratio in *Lacuna* (Russell-Hunter and McMahon, 1975).

D. Productivities and efficiencies

Growth and reproduction rates (in biomass per unit environment per time) can be combined with mortality rates (expressed in the same terms) to give overall productivity rates. Mortality values derived from density measurements (or counts of shells from dead snails) can be combined with appropriate carbon values to provide assessments of successive (and cumulative) carbon values per square metre lost from the populations by death during the time intervals between successive samplings. Using a method of computation and presentation originally set out by Russell-Hunter (1970), at a series of successive census of the population, the appropriate cumulative biomass value for the snails that have died can be added to each standing-crop biomass value for the snail population to give a series of values for cumulative net biomass production. These last values for any generation or cohort of snails can be taken at any point in time and divided by the elapsed time to that point to provide an assessment of productivity rate. Calculated in this fashion, for the pond limpet *L. fuscus* McMahon (1975a) reports productivity rates varying from 0·07 mg C m^{-2} day^{-1}–1·2 mg C m^{-2} day^{-1}, and for 2 populations of *L. palustris* Hunter (1975a) reports rates of 3·7 mg C m^{-2} day^{-1} and 22·8 mg C m^{-2} day^{-1}. In comparison to these 2 freshwater pulmonates (1 small, 1 moderately large), Mattice (1972) computes a productivity rate of 1·4 mg C m^{-2} day^{-1} for a lake population of the prosobranch snail *Bithynia tentaculata*, and Avolizi (1970) computes rates of 4·6 mg C m^{-2} day^{-1} and 13·2 mg C m^{-2} day^{-1} for 2 nearly sympatric species of the freshwater bivalve, *Sphaerium*. Data of other authors, if assessed even as ash-free dry weights, can be converted into similar rates for comparison. Eckblad (1973) presents data on 3 freshwater pulmonates living in an intermittent backwater in upstate New York which are convertible to productivity rates of 3·1 mg C m^{-2} day^{-1} for *L. palustris*, 2·3 mg C m^{-2} day^{-1} for *P. integra*, and approximately 0·03 mg C m^{-2} day^{-1} for *G. parvus* (a relatively minute planorbid). Gillespie (1966, 1969) presents data on molluscan populations in the Madison River at Yellowstone which can be converted to productivity rates of 1·6 mg C m^{-2} day^{-1} for *P. gyrina*; 4·8 mg C m^{-2} day^{-1} for *Gyraulus deflectus* (Say); 1·3 mg C m^{-2} day^{-1} for the prosobranch snail, *Valvata humeralis*; and 4·7 mg C m^{-2} day^{-1} for the bivalve, *Pisidium compressum*. Tilly (1968) gives a calorific estimate (from wet weight biomass values) of energy flow for a population of *P. integra* in Cone Spring, Iowa, which can be converted to a productivity rate of 42 mg C m^{-2} day^{-1}.

In relation to the discussion earlier in this chapter, of the extent of interpopulation variation in freshwater snail species, it is highly significant that differences in productivity rates between populations within species can equal or exceed the differences between species. As noted earlier, the interpopulation differences probably reflect considerable phenotypic plasticity in a euryoecic species like *L. palustris*, contrasting with more genetic variation between populations in the relatively stenoecic pulmonate limpets, *Ancylus*, *Ferrissia* and *Laevapex* (see further discussion in Russell Hunter, 1957; Russell Hunter *et al.*, 1967; Nickerson, 1972; Hunter, 1975a).

A current weakness of actuarial bioenergetics for freshwater pulmonates is that, because of relative technical difficulties, the "best" values for certain budget components are often derived from one particular species for which the other components have been less extensively investigated. For example, the most extensive data on R.A./N-R.A. are those for *Ferrissia* (Burky, 1971) and for *Laevapex* (McMahon, 1972, 1973, 1975a); on the T.A./T.I. relationship those for *L. palustris* (Hunter, 1972, 1975a) and for *Bithynia* (Mattice, 1972); and in the partitioning of N-R.A. into reproduction those for *H. trivolvis* (Eversole, 1974) and for *Melampus* (Apley *et al.*, 1967; Russell-Hunter *et al.*, 1972). Despite these non-complementary data, a number of fruitful comparisons can be made on certain budget ratios. For example, although there is a more than 10-fold difference between the 2 species in their productivity rates, the range of N-R.A./T.A. percentage values for populations of *L. fuscus* is 22–52% (McMahon, 1975a), and N-R.A./T.A. values for 2 populations of *L. palustris* are 22 and 41% (Hunter, 1975a). These values for the 2 species of pulmonates grazing on *Aufwuchs* can be compared to the more sedentary filter-feeding snail, *B. tentaculata* which has an N-R.A./T.A. value of 65% for a lake population (Mattice, 1970). For 2 populations of *L. palustris*, Hunter (1975a) was able to demonstrate that the population in the less eutrophic environment, although having a markedly lower overall productivity rate, compensates to some extent by raising its T.A./T.I. ratio (or gut-efficiency) to 68 against 57% for the population in the more eutrophic site. In comparing his 3 populations of *L. fuscus*, McMahon (1973, 1975a) notes that they all have similar levels of R.A., and therefore the differing levels of T.A. (and presumably of T.I.) for the 3 populations correspond to a parallel series of values for N-R.A. and hence for productivity rates. Indeed it becomes possible to relate fecundity levels directly to the quantity and quality of the *Aufwuchs* as a food source (McMahon *et al.*, 1974).

Perhaps the most meaningful way to make productivity comparisons between dissimilar species and populations is to use some ratio between productivity rate and mean standing crop biomass for the same time period (Russell-Hunter, 1970). Direct comparisons can be made in terms of turnover times (in days) or of the reciprocal ratios in inverse time (i.e. as rate constants). Turnover times for 2 populations of *L. palustris* (Hunter, 1975a) are both about 75 days. For a univoltine population of *Laevapex* (McMahon, 1975a) turnover is about 93 days, and for the 2 generations of a bivoltine population 54 days for the overwintering and 14 days for the summer cohort. There is obviously a direct relationship between average length of life cycle and any ratio of mean standing biomass to productivity rate. With species such as these freshwater pulmonates, where life cycle length is an interpopulation variable (see above, and Russell Hunter, 1961b, 1964), the consequences as regards differential rates of selection are highly significant.

Since the classic review of the consequences of life cycle pattern by Cole (1954), refinements from deterministic to stochastic models of population growth and survival have been proposed (e.g. Skellam, 1955; Holgate, 1967), palaeontologists and others have attempted simple computer simulations of successive changes in size-frequency distributions in populations (e.g. Craig and Oertel, 1966a, b), and, more recently, the value of systems modelling has been advanced (e.g. by Hubbell, 1971 for isopods; and by Calow, 1973 for snails). In particular, Calow's proposed systems models provide for loops of feed-back control at various levels (such as those we know exist controlling fecundity in snails; Eversole, 1974 and see discussion above).

Both more field tests of life cycle patterns and their bioenergetics, and further theoretical work on the implications of the relationship between mean life cycle lengths and productivity: biomass ratios are needed. A minor example (being investigated by the present writer) involves differentiating, in cases where natural populations display 2 breeding seasons each year, between annuals with 2 distinct breeding stocks, and bivoltine forms with incomplete replacement (bivoltine forms with complete replacement can usually be distinguished more easily). It can be theorized that, in the latter (bivoltine) case, both the minimum size at reproduction and the gross fecundity will be higher for the late-born (overwintering) generation than for the early-born one. On the other hand, for that peculiar double annual pattern, differences between sympatric stocks in minimum breeding size need not be associated with parallel differences in gross fecundity. Eventually, other bioenergetic consequences of shifts in life cycle pattern may be elucidated. As noted below (Section V below), such comparative studies of bioenergetics of natural populations of pulmonates are also likely to contribute to our future understanding of evolutionary processes.

V. Evolutionary aspects of ecology

A. Space and time in fresh waters

As for so many important aspects of the "ecological theater" as a setting for the "evolutionary play", G. Evelyn Hutchinson has published an elegant essay (Hutchinson, 1965) on "The lacustrine microcosm reconsidered", in which he comments upon certain scale effects of freshwater environments. In this present short excursus on the evolution of freshwater pulmonates, emphasis is placed on other features of fresh waters involving scale: the twin limits to the extent of freshwater microcosms in space and in time. Much small-scale and short-term isolation of populations can occur.

A general hypothesis about the processes of evolution in molluscs living in fresh waters was put forward some years ago independently by Hubendick (1952, 1954, 1960 on the basis of systematic and taxonomic studies) and by the present writer (Russell-Hunter, 1952, 1953b, c, 1957, 1961b, 1964, 1970, on the basis of surveys of physiological variation). It is that the transience of most freshwater environments, and the consequent high degree of short-term, small-scale isolation of populations which can occur within them, result in rates and modes of evolutionary change which are markedly different from evolution in similar animals living in the sea or on land (or, for that matter, in the few "ancient" lakes of unusual permanence). The theoretical consequences of this environmental mode were summarized by Russell Hunter (1964, p. 103):

The transience of most fresh-water habitats limits the number of generations for which any separate population of fresh-water animals can exist. But, much short-term, small-scale isolation can occur, as fresh-water habitats are geographically discontinuous and mostly small. Genetic isolation can be sufficient to produce some interpopulation diversity, but even limited transfer of individuals between populations by passive dispersal usually results in sufficient gene exchange to prevent full speciation . . . In general, much interpopulation diversity occurs but little full speciation. Further, fresh-water snails show much adaptive plasticity . . . in aspects of their physiology . . . [that] is of fundamental selective value. Particularly in many pulmonates, selection has produced genotypes which can show phenotypic flexibility.

This hypothesis has since been endorsed by other workers including McMichael (1967) in the course of his survey of Australasian molluscan faunas, and Fretter and Graham (1963) in the course of a comparison with processes of speciation in marine littoral molluscs. It is worth recapitulating the limits of fresh waters.

The world extent of lakes and rivers is not great: the total area of inland waters is approximately five million square kilometres, about one-seventieth of the oceans' area. Except for a few major river systems, there is little continuity of habitat for freshwater animals such as pulmonate snails, since all coastal rivers and the majority of mountain streams are as isolated from each other as are closed ponds. These separate units are relatively small, and very numerous. Lake Superior has the largest surface area of any body of fresh water (83,300 km^2); and there are probably only 14 other freshwater lakes with superficial areas $> 15,000$ km^2. Further tabulation (Russell-Hunter and Belanger, unpublished) reveals that there are only about 610 lakes (which can be listed in some tens of book pages) with areas > 100 km^2 (or 38·6 $miles^2$). The majority of lakes are smaller, but there are very many of them, and probably their combined surfaces amount to just over 80% of the world's freshwater area. An estimate of those over 0·1 km^2 (25 acres) for Sweden alone would be $> 10^5$ and for Canada at least 10^6. "Ponds" are even more numerous and, through most of the English-speaking world, the term would imply areas of less than half an acre (0·002 km^2 or 2000 m^2). Ponds only rarely have depths without rooted vegetation.

Limits in time are similarly low. Most bodies of fresh water are relatively transitory in a geological time scale. Most of the larger lakes have existed as such for only a few thousand years (e.g. Lake Ontario, 8000; Loch Lomond, 11,000), and the likely future "life expectancy" of the majority of the world's larger lakes is $< 20,000$ years. Smaller lakes and large ponds are even more transitory, with maximum lifetimes of hundreds of years (and gradual disappearances recorded in the time-scales of recent human history). For this reason, in certain areas of the world there are many more "fossil lakebeds" than there are extant lakes. The great bulk of the world's freshwater faunas, including pulmonate snails, have always lived in lakes with durations of the order of 10^3–10^4 years, or in ponds of perhaps 10^2 years. In contrast, certain relatively constant environments may persist on areas of the continental land surfaces for up to 10^6 years, and in the oceans for many times this period.

Only the very small number of "ancient" lakes with ages of 10^5 years and over (including Baikal, Tanganyika, Ochrida, and Posso) can provide for animal populations a temporal and spatial continuity comparable to terrestrial and marine environments. Only in them do we find faunas both more diverse and more bizarre with a high percentage of endemic forms (Dartevelle and Schwetz, 1948; Schwetz and Dartevelle, 1948; Hubendick, 1952, 1960, 1969; Stankovic, 1960; Kozhov, 1963; see also Boss, Chapter 9 of this volume). As noted before (Russell Hunter, 1957, 1961b, 1964), the majority of populations of pulmonate snails live in—and have evolved in—bodies of water considerably smaller than, and more transient than, even medium-sized lakes such as Loch Lomond (70 km^2) or Lake Oneida (233 km^2).

B. Systematics and biometric variation

Modern systematic revision of groups of freshwater pulmonates (e.g. of the world's Lymnaeidae by Hubendick, 1951; of the freshwater limpets of North America [Basch, 1963], and of elsewhere in the world [Hubendick, 1960, 1964, 1967, 1969, 1970]; and of Canadian Physidae by Clarke, 1973; see also Hubendick, Chapter 1 of this volume) reveals a characteristically small number of genera and species, almost world-wide in distribution, but with a high degree of infraspecific interpopulation variation. This is broadly true of both freshwater prosobranch snails and freshwater bivalves although individual species in these 2 groups are more rarely cosmopolitan. Many genera are still widely distributed. Earlier, the extent of interpopulation differences in shell characteristics in all 3 groups of freshwater molluscs had led to the description of great numbers of museum and geographical "species": the validity of many of these being doubtful. Boss (1971) comments that synonymy ratios tend to be highest among freshwater groups, and records examples of modern revisions involving reductions ranging from 27:1 (lymnaeids) to 6:1 (physids). Excepting only the molluscan faunas of the "ancient" lakes, the systematics of freshwater pulmonates thus reflects the limits of freshwater habitats in space and in time in such features as: the relative paucity of genera and species, their widespread distribution (both zoogeographically and ecologically), and the nearly incredible extent to which each species is divisible into many recognized "varieties", each of these infraspecific forms being relatively uniform in the populations where it occurs.

The extent of physiological as well as biometric variation between different populations of freshwater pulmonate species has been described in earlier sections of this chapter. In all cases where detailed biometry has been done there is 1 consistent feature. The amount of variation within any single population is always very much less than the range of variation for the species as a whole. Many kinds of interpopulation variation appear to be geographically random, with populations having the closest similarities often being far removed from each other. For shell shape and for shell components, the interpopulation variation (and the relative uniformity within each population) could either be determined genetically or be ecophenotypic. Some apparent population uniformity, for example in shell dimensions, is clearly environmentally evoked. Examples of the occurrence within a species of freshwater mollusc of a particular character, in many populations as an ecophenotypic modification, but in occasional populations genetically fixed, are discussed by Russell Hunter (1961b). In other cases with considerable adaptive significance, such as the levels of calcareous and of proteinaceous components in limpet shells (Russell-Hunter *et al.*, 1967, 1970), it seems that the chances of genetic dispersal among the isolated creek populations have resulted in some rather inappropriate shells in certain habitats. Of the 2 principal shell indices assessed in shell biometrics of populations of *F. rivularis*, the steepness index is allometric and reflects local trophic conditions, while the roundness index is basically isometric, highly specific to each population and probably more rigidly genetically determined. In the cases concerning variation in the shapes of radular teeth, the biometric ratios seem to be under rigid genetic control, are relatively non-adaptive, and there are no observable environmental effects on their expression. Both Berrie (1959) with populations of *L. peregra* in north-western Europe, and Hunter (1975b) with populations of *L. palustris* in the north-eastern United States found that interpopulation variation reached highly significant levels but was geographically random with the biggest differences in some cases occurring between adjacent populations. Electrophoretic data on 7 enzyme systems in 16 populations of *F. rivularis* (Nickerson, 1972) showed very high proportions of polymorphic loci and levels of heterozygosity but no evidence of geographic clines.

C. Physiological variation

The broader examples of physiological variation already discussed probably represent the expression of considerably larger portions of the genome. Some of the earliest work on infraspecific physiological variation in snails (Russell Hunter, 1953c) was concerned with respiration and, in general, pulmonate respiratory behaviour is extremely plastic (Berg and Ockelmann, 1959; Burky, 1971; Henderson, 1963; McMahon, 1973; Russell Hunter, 1953c, 1964; see Section III of this chapter). In different populations within a species of *Lymnaea* or *Physa*, one can find rhythmic diving with aerial gas-exchange, or cutaneous aquatic respiration with a water-filled lung, or the use of an exposed gas-bubble as a physical gill. A major point to be emphasized is that the associated behaviour patterns are elaborate and different but, under natural conditions, specific to each population. Brought into the laboratory, some stocks (from the offshore bottom for example) could move to an elaborate behaviour pattern of surfacing not used for many generations. In a crude generalization, the genetic basis for the elaborate behaviour must have been retained unused (and of no local selective value) over perhaps hundreds of generations. Another kind of respiratory variation concerns capacity for temperature acclimation (including the superficially non-adaptive "reverse acclimation" discussed above, Section III). This is probably interrelated with capacity to resist partial anaerobiosis. Once again we find evidence for the existence of some "physiological races", but a great deal more data on stocks with exceptionally high levels of adaptive plasticity.

There is similar but less extensive evidence for physiological variation in the feeding mechanisms and calcium metabolism of some freshwater pulmonates, but the most extensive field data concern reproduction. As already noted (see Section IV above, and references therein) in natural populations of freshwater pulmonates in temperate regions, considerable interpopulation variation occurs in growth rates, in fecundity, and in patterns of life cycle. While much of these infraspecific differences (again usually uniform within each population) in reproductive physiology seem to be environmentally evoked, at least some interpopulation variation in fecundity and life cycle pattern seems to be genetically determined. The latter part of the evolutionary hypothesis set out above concerns such physiological variation more exclusively. It proposes that the adaptive plasticity found in respiration, in reproduction, and in other aspects of the physiology of freshwater snails, is itself of fundamental selective value. Particularly in certain species of freshwater pulmonates, selection has produced genotypes which can show phenotypic flexibility. What appears to have been favoured in their "discontinuous" evolution has been capacity to vary—the possession of adaptive plasticity (see also Russell-Hunter, 1961b, 1970).

As already noted for the infraspecific variation between populations, the remarkable uniformity within populations can be either ecophenotypic or genetically determined. The capacity of pulmonates, especially Lymnaeidae, for self-fertilization has been regarded (Hubendick, 1951) as contributing to any apparent genetic uniformity occurring within populations. Hubendick further suggested that this capacity could have had a retarding effect on lymnaeid evolution. However, existing field data suggest that self-fertilization is rare in natural populations; and it is important to consider certain functional aspects of the one-sided copulatory pattern which is characteristic of the higher limnic Basommatophora. By careful use of genetic markers in the planorbid *A. glabratus*, Richards (1970, and personal communication; see also Richards and Merritt, 1972) has shown that self-fertilization will occur in isolated individuals, but that the presence of foreign sperm in the genital system (from a single copulation) will prevent its occurrence. Markers prove that the male gametes passed in a single copulation will be those responsible for fertilizing the eggs laid over many subsequent weeks. Circumstantial evidence suggests that this continued potency of a single cross-fertilization is true for many other freshwater pulmonates (though not for the primitive saltmarsh genus, *Melampus*), though it has only been proven experimentally in *Australorbis*. In relation to interpopulation variation and intrapopulation uniformity, this copulatory pattern could contribute to and enhance a "founder effect" when new populations are first established, and the effectiveness of any subsequent transfers would be limited by the restricted duration of the copulatory period in that "new" population's life cycle. Perhaps as much as capacity for self-fertilization, this mechanism will intensify the effects of short-term small-scale isolation in freshwater populations. However, there are suggestions in the literature (e.g. De Witt and Sloan, 1959; Hubendick, 1951; Russell-Hunter and McMahon, 1976) that copulatory patterns (and the extent of self-fertilization) may differ among the families of the freshwater Basommatophora.

D. Future evolutionary studies

The most important raw materials for natural selection in animals are not so much genetic differences in obvious characters (such as number of shell-spines, or even shade of shell colour), but rather they are genetically controlled differences in bioenergetic partitioning (Russell-Hunter *et al.*, 1968; Russell-Hunter, 1970). These differences (discussed in Section IV, and elsewhere) should not simply be assessed as altering some "overall conversion efficiency" of the individual organisms or of the populations (though both of these are not without significance in ecological studies). The real efficiencies (those whose variation provides for natural selection) are those based on capacity for such bioenergetic shifts as are required to maintain a "steady-state" of the organism or the population. All established (or perhaps "Darwinian") species-populations become more "efficient" as their feed-back systems allow them closer homeostasis in the dynamic equilibrium of a steady (at least in longer time-scales) conversion of their trophic input into fertile adults of their species. It is mildly satisfying to the writer that, in many cases, this can be assessed in the older evolutionist's terms (which are at present rather unfashionable) of "the adaptive significance" of the modifiable process or structure to the organism. This leads to 2 further rather trite concepts. First, processes of animal evolution by natural selection have been involved with whole animals and their efficiencies as machines to make more of their own kind in immediately succeeding generations, while retaining sufficient flexibility of response (or, in a sense, machine inefficiency) to ensure long-term fitness (or the probability of leaving descendants after many generations). Thus, rather than proportional changes in single alleles, patterns of functional interdependence (largish bundles of integrated genetic material) are what should be regarded as more important in making hypotheses about evolutionary matters (including phylogeny). Evolutionary "success" over, say, 10^7 years implies more than relative degrees of adaptedness and almost certainly involves the maintenance of a genetic repertoire of great flexibility in its expression, but with it an increasing capacity for homeostatic responses. Secondly, modes and rates of evolutionary change are largely determined by the temporal and spatial extent of steady-state populations. Continuity of any environment in space determines the size of a panmictic population; its continuity in time determines the number of generations that a population exists with its gene pool subjected to the same continued selection pressures. As noted elsewhere (Brooks, 1957; Russell-Hunter, 1957, 1970), for future evolutionary studies (as for earlier ones), bodies of fresh water provide natural "experimental" populations which have experienced varying degrees of spatial and temporal limitation. Some such future studies need not be directly concerned with evolutionary questions.

A strict dichotomy between applied and "pure" theoretical science is unrealistic in many aspects of ecology (see further discussion in Russell-Hunter, 1970, wherein this is noted as an exception to the general problems first made explicit by Snow, 1959). As regards the physiology, bioenergetics and evolution of freshwater pulmonate snails, one can point to many reciprocal benefits from the interaction of basic research studies on the one hand, and investigations directed to applied purposes such as pollution and disease control on the other. The immediate applied significance of the research on freshwater pulmonates discussed above (and elsewhere in this volume) can be considered under 3 categories: snails as disease vectors, the detection and assessment of environmental radiocontamination with strontium-90, and general water management including control of pollution. First, all data on variations in productivity and turnover efficiency obtained from natural populations of snails are important in relation to the turnover potential of these populations as vectors for schistosomes and other trematode parasites (see also Brown, Chapter 7 of this volume). Secondly, studies on the field rates of shell metabolism (both for calcium and for protein) can be important in assessing radionuclide contamination of aquatic environments (Nelson, 1964; Rosenthal *et al.*, 1965; Russell-Hunter *et al.*, 1967, 1970). Thirdly, all estimates of C:N ratios, and of turnover rates in natural populations of molluscs, are of value in setting up "biological assay" methods for water utilization planning and for general pollution control (Russell-Hunter, 1961a, 1970; Maitland, 1965, 1966; McMahon, 1975c). These and other matters of immediate value to current human concerns have all been paralleled by slowly maturing contributions from the ecology of freshwater pulmonates to evolutionary studies and theory (Russell-Hunter, 1970). It is to be hoped that future developments in such ecological investigations will continue and expand this symbiosis between basic and applied research. Any future field studies on natural populations of freshwater pulmonates must yield both more sophisticated data on the extent of physiological variation and more bioenergetic measures of differential survivorship and differential fecundity in population life cycles. Both kinds of data, concerning the twin bases on which evolutionary change depends, are needed if we are to move to a more predictive theory of evolutionary mechanisms. As an obvious example, among the major interpopulation variables in many of the species of freshwater pulmonates is life cycle length, which in turn is uniquely important in relation to rate-constants for productivity and hence to selection rates.

As noted elsewhere (Russell-Hunter, 1970), not solely studies in molecular biology, but also investigations on the physiological ecology of whole animals and into the comparative bioenergetics of populations, must contribute to our future understanding of evolutionary processes.

VI. Acknowledgements

This chapter was prepared while my work on freshwater and littoral snails was supported by a research grant from the U.S. National Science Foundation, grant number GB-36757 continued as BMS72-02511-AO1. I am indebted to Christopher H. Price, Robert A. Browne and Perry

Russell-Hunter for help with much reference material, and to Catherine M. Herrity both for similar aid, and for much help in the preparation of the manuscript. I must also thank 5 of my former research associates, Dr Albert J. Burky (now of the University of Dayton), Dr Arnold G. Eversole (Clemson University), Dr R. Douglas Hunter (Oakland University), Dr Robert F. McMahon (University of Texas), and Dr Richard P. Nickerson (Simmons College), and my wife, Myra Russell-Hunter for reading the chapter in manuscript. All errors, misinterpretations, and the culpable eclecticism of certain sections remain my own.

VII. References*

Annandale, N., Prashad, B. and Din, A. (1921). *Rec. Ind. Mus.* **22,** 529–631.

Apley, M. L. (1967). Field and experimental studies on pattern and control of reproduction in *Melampus bidentatus* (Say). Ph.D. Thesis, Syracuse University, 154 pp. (*Diss. Abstr.* **68,** 1527-B, order no. 68–13, 810).

Apley, M. L. (1970). *Malacologia* **10,** 381–397.

Apley, M. L., Russell-Hunter, W. D. and Avolizi, R. J. (1967). *Biol. Bull mar. biol. Lab. Woods Hole* **133,** 455–456.

Avolizi, R. J. (1970). Biomass turnover in natural populations of viviparous sphaeriid clams: interspecific and intraspecific comparisons of growth fecundity, mortality, and biomass production. Ph.D. Thesis, Syracuse University, 150 pp. (*Diss. Abstr.* **32,** 1919-B, order no. 71–23, 429).

Avolizi, R. J. (1975). *Hydrobiologia* **46,** 100–121.

Baker, F. C. (1911). *Chicago Acad. Sci. Spec. Publ.* **3,** 1–539.

Baker, F. C. (1916). *N.Y. State Coll. For., Tech. Publ.* **4,** 1–366.

Baker, F. C. (1918). *N.Y. State Coll. For., Tech. Publ.* **9,** 1–264.

Baker, F. C. (1928). *Wisconsin Acad. Sci. Arts Letts,* and *Wisconsin Geol. Nat. Hist. Surv. Bull.* **70,** 1–507.

Basch, P. F. (1963). *Bull. Mus. Comp. Zool. Harvard Univ.* **129,** 399–461.

Berg, K. (1938). *K. danske vidensk. Selsk.* **8,** 1–255.

Berg, K. (1948). *Folia Limnol. Scand.* **4,** 1–318.

Berg, K. (1952). *Hydrobiologia* **4,** 225–267.

Berg, K. (1953). *Hydrobiologia* **5,** 331–350.

Berg, K. and Ockelmann, K. W. (1959). *J. Exp. Biol.* **36,** 690–708.

Berg, K. and Petersen, I. C. (1956). *Folia Limnol. Scand.* **8,** 1–273.

Berg, K., Lumbye, J. and Ockelmann, K. W. (1958). *J. Exp. Biol.* **35,** 43–73.

Berrie, A. D. (1959). *Ark. Zool.* **12,** 391–404.

Berrie, A. D. (1963). *Nature, Lond.* **198,** 805–806.

Berrie, A. D. (1965). *Proc. malac. Soc. Lond.* **36,** 283–295.

Berrie, A. D. (1966). *Proc. malac. Soc. Lond.* **37,** 83–92.

Berrie, A. D. (1968). *Ann. trop. Med. Parasit.* **62,** 45–51.

Berrie, A. D. (1969). *Malacologia* **9,** 35–36.

Berrie, A. D. and Visser, S. A. (1963). *Physiol. Zool.* **36,** 167–173.

* This manuscript was submitted in 1975. No attempt has been made to update references.

Bickel, D. (1965). *Sterkiana* **18,** 17–20.
Boettger, C. R. (1944). *Tierwelt Nord- und Ostsee* **9,** 241–478.
Bondesen, P. (1950). *Nat. Jutlandica* **3,** 1–208.
Bondesen, P. and Kaiser, E. W. (1949). *Oikos* **1,** 252–281.
Borden, M. A. (1931). *J. Mar. Biol. Ass. U.K.*, N.S. **17,** 709–738.
Boss, K. J. (1971). *Occ. Pap. Mollusks, Harv.* **3,** 81–135.
Bovbjerg, R. V. (1965). *Malacologia* **2,** 199–207.
Bovbjerg, R. V. and Ulmer, M. J. (1960). *Proc. Iowa Acad. Sci.* **67,** 569–577.
Boycott, A. E. (1930). *Trans. Herts. Nat. Hist. Soc.* **19,** 1–25.
Boycott, A. D. (1936). *J. Anim. Ecol.* **5,** 116–186.
Boycott, A. E., Diver, C. Garstang, S. and Turner, F. M. (1930). *Phil. Trans. R. Soc. (B)* **219,** 51–131.
Brooks, J. L. (1957). *In* "The Species Problem" (E. Mayr, Ed.) pp. 81–123. American Association for the Advancement of Science, Washington, D.C.
Bullock, T. H. (1955). *Biol. Rev.* **30,** 311–342.
Burky, A. J. (1969). Biomass turnover, energy balance, and interpopulation variation in the stream limpet, *Ferrissia rivularis* (Say), with special reference to respiration, growth, and fecundity. Ph.D. Thesis, Syracuse University, 104 pp. (*Diss. Abstr.* **30,** 3917-B, order no. 70–1942).
Burky, A. J. (1970). *Acta Cien. Venezolana* **21,** 28.
Burky, A. J. (1971). *Ecol. Monogr.* **41,** 235–251.
Burky, A. J. (1973). *Malac. Rev.* **6,** 59.
Burky, A. J. (1974). *Proc. malac. Soc. Lond.* **43,** 127–143.
Buttner, A. (1953). *Ann. Parasit. Hum. Comp. Paris* **28,** 452–453.
Calow, P. (1970). *Proc. malac. Soc. Lond.* **39,** 203–215.
Calow, P. (1973). *Am. Nat.* **107,** 559–574.
Carter, G. S. (1931). *Biol. Rev.* **6,** 1–35.
Cheatum, E. P. (1934). *Trans. Am. Micros. Soc.* **53,** 348–407.
Cheeseman, D. F. (1956). *Nature, Lond.* **178,** 987–988.
Chernin, E. and Adler, V. L. (1967). *Ann. Trop. Med. Parasit.* **61,** 11–14.
Chernin, E. and Michelson, E. H. (1957a). *Am. J. Hyg.* **65,** 57–70.
Chernin, E. and Michelson, E. H. (1957b). *Am. J. Hyg.* **65,** 71–80.
Clampitt, P. T. (1963). The comparative ecology of the snails *Physa gyrina* and *Physa integra*. Ph.D. Thesis, State University of Iowa, 87 pp. (*Diss. Abstr.* **24,** 2624–2625, order no. 63–7994).
Clampitt, P. T. (1970). *Malacologia* **10,** 113–151.
Clampitt, P. T. (1974). *Am. Midland Nat.* **92,** 275–300.
Clampitt, P. T., Waffle, E. L. and Bovbjerg, R. V. (1960). *Proc. Iowa Acad. Sci.* **67,** 553–568.
Clarke, A. H. (1973). *Malacologia* **13,** 1–509.
Cole, L. C. (1954). *Rev. Biol.* **29,** 103–137.
Craig, G. Y. and Oertel, G. (1966a). *Nature, Lond.* **210,** 438–439.
Craig, G. Y. and Oertel, G. (1966b). *Q. J. Geol. Soc. Lond.* **122,** 315–355.
Daniels, J. M. and Armitage, K. B. (1969). *Hydrobiologia* **33,** 1–13.
Dartevelle, E. and Schwetz, J. (1948). *Mém. Inst. r. colon. belge Sect. Sci. nat. méd.* 8° **14,** 1–118.
De Wit, W. F. (1955). *Basteria* **19,** 35–73.
De Witt, R. M. (1954). *Am. Nat.* **88,** 159–164.

De Witt, R. M. (1955). *Ecology* **36,** 40–44.
De Witt, R. M. (1967). *Malacologia* **5,** 445–453.
De Witt, R. M. and Sloan, W. C. (1958). *Trans. Am. Micros. Soc.* **77,** 290–294.
De Witt, R. M. and Sloan, W. C. (1959). *Anim. Behav.* **7,** 81–84.
Duncan, C. J. (1959). *J. Anim. Ecol.* **28,** 97–117.
Eckblad, J. W. (1973). *Hydrobiologia* **41,** 199–219.
Eisenberg, R. M. (1966). *Ecolcgy* **47,** 889–906.
Eisenberg, R. M. (1970). *Ecology* **51,** 680–684.
Eversole, A. G. (1974). Fecundity in the snail *Helisoma trivolvis*: experimental, bioenergetic and field studies. Ph.D. Thesis, Syracuse University, 139 pp. (*Diss. Abstr.*, 35: 5716-B, order no. 75–10,538).
Forbes, G. S. and Crampton, H. E. (1942a). *Biol. Bull. Mar. biol. Lab. Woods Hole* **82,** 26–46.
Forbes, G. S. and Crampton, H. E. (1942b). *Biol. Bull. mar. biol. Lab. Woods Hole* **83,** 283–289.
Fox, H. M. (1945). *J. Exp. Biol.* **21,** 161–165.
Fretter, V. and Graham, A. (1963). *System. Assoc. Publ.* **5,** 99–107.
Füsser, H. and Krüger, F. (1951). *Z. vergl. Physiol.* **33,** 14–52.
Geldiay, R. (1956). *J. anim. Ecol.* **25,** 389–402.
Gillespie, D. M. (1966). Population studies of four species of molluscs in the Madison River, Yellowstone National Park. Ph.D. Thesis, Montana State University, 50 pp. (*Diss. Abstr.* **27B,** 2184–2185, order no. 66–10,842).
Gillespie, D. M. (1969). *Limnol. Oceanogr.* **14,** 101–114.
Goddard, C. K. and Martin, A. W. (1966). *In* "Physiology of Mollusca" (K. M. Wilbur and C. M. Yonge, Eds) Vol. 2, pp. 275–308. Academic Press, New York and London.
Goldacre, R. J. (1949). *J. anim. Ecol.* **18,** 36–39.
Gould, S. J. (1966). *Biol. Rev.* **41,** 587–640.
Gratzer, W. B. and Allison, A. C. (1960). *Biol. Rev.* **35,** 459–506.
Grimas, U. (1962). *Rep. Inst. Freshwat. Res. Drottningholm* **44,** 14–41.
Gustin, G. M. (1960). *Microchem. J.* **4,** 43–54.
Harman, W. N. (1968a). *Nautilus, Philad.* **81,** 77–83.
Harman, W. N. (1968b). *Nautilus, Philad.* **82,** 72–73.
Harman, W. N. (1970). *Nautilus, Philad.* **84,** 21–30.
Harman, W. N. and Berg, C. O. (1970). *Ohio J. Sci.* **70,** 146–150.
Harman, W. N. and Forney, J. L. (1970). *Limnol. Oceanogr.* **15,** 454–460.
Harrison, A. D. (1968). *Arch. Hydrobiol.* **65,** 63–73.
Harrison, A. D., Williams, N. V. and Greig, G. (1970). *Hydrobiologia* **36,** 317–327.
Henderson, A. E. (1961). Studies on the respiratory and hydrostatic functions of the mantle cavity in two freshwater pulmonate snails. Ph.D. Thesis, University of Glasgow, 81 pp.
Henderson, A. E. (1963). *Z. vergl. Physiol.* **46,** 467–490.
Hirata, K. and Kawabata, T. (1954). *Sci. Rep. Kagoshima Univ.* **3,** 93–99.
Holgate, P. (1967). *J. Theor. Biol.* **14,** 1–10.
Hubby, J. L. and Lewontin, R. C. (1966). *Genetics* **54,** 577–594.

Hubell, S. P. (1971). *In* "Systems Analysis and Simulation in Ecology" (C. Patten, Ed.) pp. 269–324. Academic Press, New York and London.
Hubendick, B. (1947). *Zool. Bidr. Upps.* **24,** 419–559.
Hubendick, B. (1951). *K. svenska. Vetensk Akad. Handl.* (4) **3,** 1–223.
Hubendick, B. (1952). *Ark. Zool.* (2) **3,** 319–323.
Hubendick, B. (1954). *Proc. malac. Soc. Lond.* **31,** 6–11.
Hubendick, B. (1960). *Proc. zool. Soc. Lond.* **133,** 497–529.
Hubendick, B. (1964). *Medd. Göteborgs Mus. Zool. Avd.* **137,** *In Göteborgs K. Vetensk-o. vitther. Samh. Handl.* **9,** 1–72.
Hubendick, B. (1967). *Acta R. Soc. sci. litt. gothoburg,* (*Zool.*) **1,** 5–52.
Hubendick, B. (1969). *Arch. Molluskenk* **99,** 55–67.
Hubendick, B. (1970). *Acta R. Soc. sci. litt. gothoburg,* (*Zool.*) **5,** 4–52.
Hunter, R. D. (1972). Energy budgets and physiological variation in natural populations of the freshwater pulmonate, *Lymnaea palustris*. Ph.D. Thesis, Syracuse University, 115 pp. (*Diss. Abstr.* **33,** 5066-B, order no. 73–9534).
Hunter, R. D. (1975a). *Ecology* **56,** 50–63.
Hunter, R. D. (1975b). *Am. Midl. Nat.* **94,** 401–420.
Hutchinson, G. E. (1965). "The Ecological Theater and the Evolutionary Play". Yale University Press, New Haven.
Jacobs, W. (1941). *Sber. Ges. Morph. Physiol. Münch.* **50,** 45–52.
Johnson, C. W. (1904). *Nautilus, Philad.* **17,** 120.
Jones, J. D. (1961). *Comp. Biochem. Physiol.* **4,** 1–29.
Jones, J. D. (1964a). *Comp. Biochem. Physiol.* **12,** 283–295.
Jones, J. D. (1964b). *Comp. Biochem. Physiol.* **12,** 297–310.
Jones, T. (1969). *Proc. malacol. Soc. Lond.* **38,** 549–550.
Klimowicz, H. (1962). *Polskie Archwm Hydrobiol.* **10,** 271–285.
Kozhov, M. M. (1963). Lake Baikal and its life. *Monographiae Biologicae* **11,** Junk, Hague.
Lambéret, E. (1966). *Bull. Soc. zool. Fr.* **91,** 753–772.
Likins, R. C., Berry, E. G. and Posner, A. S. (1963). *Ann. N.Y. Acad. Sci.* **109,** 269–277.
Lumbye, J. (1958). *Hydrobiologia* **10,** 245–262.
Macan, T. T. (1950). *J. anim. Ecol.* **19,** 124–146.
McCraw, B. M. (1961). *Trans. Am. Micros. Soc.* **80,** 16–27.
McCraw, B. M. (1970). *Malacologia* **10,** 399–413.
McNeil, C. W. (1963). *Ecology* **44,** 187–191.
McMahon, R. F. (1972). Interpopulation variation and respiratory acclimation in the bioenergetics of *Laevapex fuscus*. Ph.D. Thesis, Syracuse University, 130 pp. (*Diss. Abstr.* **33,** 5068-B, order no. 73–9549).
McMahon, R. F. (1973). *Biol. Bull. mar. biol. Lab. Woods Hole* **145,** 492–508.
McMahon, R. F. (1975a). *Proc. malac. Soc. Lond.* **41,** 331–351.
McMahon, R. F. (1975b). *Am. Midl. Nat.* **95,** 174–185.
McMahon, R. F. (1975c). *Ecology* **56,** 1167–1175.
McMahon, R. F., Hunter, R. D. and Russell-Hunter, W. D. (1974). *Hydrobiologia* **45,** 391–404.
McMichael, D. F. (1967). *In* "Australian Inland Waters" (A. H. Weatherly, Ed.) pp. 123–149. Australian National University Press, Canberra.

McMillan, N. F. (1966). *Glasg. Nat.* **18,** 411–417.

Maitland, P. S. (1965). *Proc. malac. Soc. Lond.* **36,** 339–347.

Maitland, P. S. (1966). *Glasg. Univ. Publ. Stud. Loch Lomond* **2,** 1–194.

Manwell, C. (1964). *In* "Oxygen in the Animal Organism" (F. Dickens and E. Neil, Eds) pp. 49–119. Macmillan, New York.

Mattice, J. S. (1970). Trophic biology of a natural population of *Bithynia tentaculata* in terms of ecological energetics. Ph.D. Thesis, Syracuse University, 122 pp. (*Diss. Abstr.* **32,** 1927-B, order no. 71–23,456).

Mattice, J. S. (1972). *Ekol. pol.* **20,** 525–539.

Mattisson, A. G. M. (1959). *Ark. Zool.* **12,** 143–163.

Mehlman, B. and von Brand, T. (1951). *Biol. Bull. mar. biol. Lab. Woods Hole* **100,** 199–205.

Mirolli, M. (1958). *Mem. Ist. ital. Idrobiol.* **10,** 209–316.

Mooij-Vogelaar, J. W. and Van der Steen, W. J. (1973). *Proc. K. ned. Akad. Wet.* (*C*) **76,** 61–68.

Mooij-Vogelaar, J. W., Jager, J. C. and Van der Steen, W. J. (1973). *Proc. K. ned. Akad. Wet.* (*C*) **76,** 245–256.

Morgan, N. C. (1970). *Hydrobiologia* **35,** 545–553.

Morton, J. E. (1955a). *Phil. Trans. R. Soc.* (*B*) **239,** 89–160.

Morton, J. E. (1955b). *Proc. zool. Soc. Lond.* **125,** 127–168.

Morton, J. E. (1958). "Molluscs". Hutchinson, London.

Morton, J. E. and Yonge, C. M. (1964). *In* "Physiology of Mollusca" (K. M. Wilbur and C. M. Yonge, Eds) Vol. 1, pp. 1–58. Academic Press, New York and London.

Mozley, A. (1935). *Genetics, Princeton* **20,** 452–465.

Nelson, D. J. (1964). *Nature, Lond.* **203,** 420.

Nickerson, R. P. (1972). A survey of enzyme and shell variation in 16 populations of the stream limpet, *Ferrissia rivularis* (Say). Ph.D. Thesis, Syracuse University, 149 pp. (*Diss. Abstr.* **33,** 4588-B, order no. 73–7753).

Nicol, E. A. T. (1936). *Proc. R. Soc. Edinb.* **56,** 169–195.

Noland, L. E. and Carriker, M. R. (1946). *Am. Midl. Nat.* **36,** 467–493.

Odum, E. P. (1959). "Fundamentals of Ecology" 2nd edition. Saunders, Philadelphia.

Odum, H. T. (1957). *Ecol. Monogr.* **27,** 55–112.

Ökland, J. (1964). *Folia limnol. scand.* **13,** 1–337.

Ökland, J. (1969). *Malacologia* **9,** 143–151.

Olivier, L. (1956). *J. Parasitol.* **42,** 137–146.

Olivier, L. and Barbosa, F. S. (1955a). *Publçães avuls. Inst. Aggeu Magalhães* **4,** 79–103.

Olivier, L. and Barbosa, F. S. (1955b). *Publçães avuls. Inst. Aggeu Maghalhães* **4,** 105–115.

Paloumpis, A. A. and Starrett, W. C. (1960). *Am. Midl. Nat.* **64,** 406–435.

Patrick, R., Cairns, J., Jr. and Roback, S. S. (1967). *Proc. Acad. nat. Sci. Philad.* **118,** 109–407.

Pettee, R. E. and Gifford, C. E. (1964). *Proc. Indiana Acad. Sci.* **74,** 393–397.

Precht, H. (1939). *Z. vergl. Physiol.* **26,** 696–739.

Precht, H. and Otto, E. (1948). *Verh. dtsch. zool. Keil* **60,** 381–387.

Prosser, C. L. (1955). *Biol. Rev.* **30,** 229–262.
Prosser, C. L. (1967). "Molecular Mechanisms of Temperature Adaptation". Symposium, American Association Advancement of Science, Washington D.C., 398 pp.
Read, K. R. H. (1966). *In* "Physiology of Mollusca" (K. M. Wilbur and C. M. Yonge, Eds) Vol. 2, pp. 209–232. Academic Press, New York and London.
Rees, W. J. (1965). *Proc. malac. Soc. Lond.* **36,** 269–282.
Richards, C. S. (1970). *Nature, Lond.* **227,** 806–810.
Richards, C. S. and Merritt, J. W., Jr. (1972). *Am. J. Trop. Med. Hyg.* **21,** 425–434.
Rosenthal, G. M., Jr., Nelson, D. J. and Gardiner, D. A. (1965). *Nature, Lond.* **207,** 51–54.
Rosewater, J. (1970). *Nautilus, Philad.* **83,** 144–145.
Roskowski, W. (1914). *Rev. suisse Zool.* **22,** 457–539.
Rudescu, L. and Popescu-Marinescu, V. (1970). *Arch. Hydrobiol.* Suppl. **36,** 279–292.
Russell Hunter, W. (1952). *Glasg. Nat.* **16,** 84–85.
Russell Hunter, W. (1953a). *Proc. zool. Soc. Lond.* **123,** 623–636.
Russell Hunter, W. (1953b). *Proc. R. Soc. Edinb.* (*B*) **65,** 84–105.
Russell Hunter, W. (1953c). *Proc. R. Soc. Edinb.* (*B*) **65,** 143–165.
Russell Hunter, W. (1957). *Glasg. Univ. Pub., Stud. Loch Lomond* **1,** 56–95.
Russell Hunter, W. (1961a). *Proc. zool. Soc. Lond.* **136,** 219–253.
Russell Hunter, W. (1961b). *Proc. zool. Soc. Lond.* **137,** 135–171.
Russell Hunter, W. (1964). *In* "Physiology of Mollusca" (K. M. Wilbur and C. M. Yonge, Eds) Vol. 1, pp. 83–126. Academic Press, New York and London.
Russell-Hunter, W. D. (1970). "Aquatic Productivity". Macmillan, New York.
Russell Hunter, W. and Apley, M. L. (1966). *Biol. Bull. mar. biol. Lab. Woods* **131,** 392–393.
Russell-Hunter, W. D. and Eversole, A. G. (1976). *Comp. Biochem. Physiol.* **54A,** 447–453.
Russell-Hunter, W. D. and McMahon, R. F. (1975). *Nautilus, Philad.* **89,** 14–16.
Russell-Hunter, W. D. and McMahon, R. F. (1976). *Trans. Am. Micros. Soc.*, **95,** 174–182.
Russell Hunter, W. and Warwick, T. (1957). *Proc. R. Soc. Edinb.* (*B*) **66,** 360–373.
Russell Hunter, W., Maitland, P. S. and Yeoh, P. K. H. (1964). *Proc. malac. Soc. Lond.* **36,** 27–32.
Russell-Hunter, W. D., Meadows, R. T., Apley, M. L. and Burky, A. J. (1968). *Proc. malac. Soc. Lond.* **38,** 1–11.
Russell Hunter, W., Apley, M. L., Burky, A. J. and Meadows, R. T. (1967). *Science, N.Y.* **155,** 338–340.
Russell-Hunter, W. D., Burky, A. J. and Hunter, R. D. (1970). *Biol. Bull. mar. biol. Lab. Woods Hole* **139,** 402.
Russell-Hunter, W. D., Apley, M. L. and Hunter, R. D. (1972). *Biol. Bull. mar. biol. Lab. Woods Hole* **143,** 623–656.
Salazkin, A. A. (1965). *Zool. Zh.* **44,** 1602–1609.

Scholander, P. F., Flagg, W., Walters, V. and Irving, L. (1953). *Physiol. zool.* **26,** 67–92.

Schutte, C. H. J. and Frank, G. H. (1965). *Bull. Wld Hlth Org.* **30,** 389–400.

Schwetz, J. and Dartevelle, E. (1948). *Mém. Inst. r. colon. belge Sect. Sci. nat. méd* **8° 16,** 1–58.

Shrubsole, G. (1934). *J. Conch. Lond.* **20,** 130–141.

Shrubsole, G. (1937). *J. Conch. Lond.* **20,** 308–310.

Skellam, J. G. (1955). *In* "The Numbers of Man and Animals". (J. B. Cragg and N. W. Pirie, Eds) pp. 31–46 Oliver and Boyd, Edinburgh.

Skellam, J. G. (1973). *In* "The Mathematical Theory of the Dynamics of Biological Populations" (M. S. Bartlett and R. W. Hirons, Eds) pp. 63–85. Academic Press, London and New York.

Snow, C. P. (1959). "The Two Cultures and the Scientific Revolution". Cambridge University Press, New York.

Stankovic, S. (1960). The Balkan Lake Ohrid and its Living World. *Monographiae Biologicae* **9,** 1–357.

Storey, R. (1970). *J. Conch. Lond.* **27,** 191–195.

Summers, W. C. (1966). Quantitative measurements of snail feeding at the water surface. Ph.D. Thesis. University of Minnesota, 80 pp. (*Diss. Abstr.* **28,** 391–B, order no. 67–7794).

Taylor, J. W. (1900). "A Monograph of the Land and Freshwater Mollusca of the British Isles. Vol. 1. Structural and general". Taylor Brothers, Leeds.

Théodoridès, J. (1961). *Vie Milieu* **12,** 195.

Thorup, J. (1970). *Arch. Hydrobiol.* **66,** 447–457.

Thut, R. N. (1969). *Ecol. Monogr.* **39,** 79–100.

Tilly, L. J. (1968). *Ecol. Monogr.* **38,** 169–197.

Tuthill, S. J. and Laird, W. M. (1963). *Nautilus, Philad.* **77,** 47–55.

Tuthill, S. J. and Laird, W. M. (1964). *Nautilus, Philad.* **77,** 81–89.

van Benthem Jutting, W. S. S. (1959). *Basteria* **23,** 106–131.

Van der Borght, O. and Van Puymbroeck, S. (1964). *Nature, Lond.* **204,** 533–534.

Van der Borght, O. and Van Puymbroeck, S. (1966). *Nature, Lond.* **210,** 791–793.

Van der Steen, W. J. (1967). *Archs néerl. Zool.* **17,** 403–468.

Van der Steen, W. J., Jager, J. C. and Tiemersma, D. (1973). *Proc. K. ned. Akad. Wet. (C)* **76,** 47–60.

Vermeij, G. J. (1971). *J. Zool. Lond.* **163,** 15–23.

von Brand, T. (1946). *Biodynamica Monographs* **4,** 1–328.

von Brand, T. (1955). *J. Wash. Acad. Sci.* **45,** 373–377.

von Brand, T., Baernstein, H. D. and Mehlman, B. (1950). *Biol. Bull. mar. biol. Lab. Woods. Hole* **98,** 66–276.

von Brand, T., McMahon, P. and Nolan, M. O. (1955). *Physiol. Zool.* **28,** 35–40.

Walton, C. L. and Jones, W. N. (1926). *Parasitology* **18,** 144–147.

Webbe, G. (1960). *Ann. trop. Med. Parasit.* **54,** 54–59.

Winberg, G. G. and Belyatskaya, J. S. (1959). *Zool. Zh.* **38,** 1146–1151.

Winberg, G. G. and Belyatskaya-Potayenko, J. S. (1963). *Zool. Zh.* **42,** 1593–1600.

Wolvekamp, H. P. (1932). *Z. vergl. Physiol.* **16,** 1–38.
Wright, C. A. (1960). *Ann. trop. Med. Parasit.* **54,** 224–232.
Yonge, C. M. (1947). *Phil. Trans. R. Soc. (B)* **232,** 443–518.
Yonge, C. M. (1952). *Proc. malac. Soc. Lond.* **29,** 190–199.
Yonge, C. M. (1960). *Proc. Ca. Acad. Sci.* (4) **31,** 111–119.
Zaaijer, J. J. P. and Wolvekamp, H. P. (1958). *Acta physiol. pharm. néerl.* **7,** 56–77.

Chapter 9

On the evolution of gastropods in ancient lakes

K. J. BOSS

Museum of Comparative Zoology, Harvard University, USA

I. Introduction

In widely separated portions of the world, certain bodies of fresh water contrast markedly with other comparatively simple temperate or tropical lakes. Distinguished by unique faunas and floras, usually with high degrees of endemism, these lakes differ from other fresh waters in their age. They are, indeed, ancient lakes, the source of considerable interest because of their biotas, and old in the sense of dating from the Pleistocene or even earlier Tertiary times. Brooks (1950) has admirably reviewed the surprisingly rich arrays of species and discussed the probable evolutionary mechanisms behind their elaboration. In extensively documenting the species-flocks of cichlid fishes in the Great African lakes, Fryer and Iles (1972) lend credence to the argument in support of allopatric intralacustrine evolution of species. Appropriately, the whole fascinating process falls within the confines of the stability-time hypothesis of Sanders (1968) which can explain the unexpected diversity in the biota in the deep-sea as well as that of the fauna of the ancient lakes.

The principal features characterizing ancient lakes include these: great age and an exceptional biota, particularly one showing high endemicity. The uniqueness of their molluscan faunas was greeted with surprise by malacologists. Both Woodward (1859), who first received shells collected by Speke on the shores of Lake Tanganyika in East Africa, and Gerstfeldt (1859), who recognized the unusual endemic planorbids in Lake Baikal in Siberia, expressed amazement in describing such remarkable freshwater molluscs.

II. The lakes

Six bodies of fresh water are suitable for analysis since their age has some basis in paleontologic fact and their gastropod faunas are sufficiently studied. Map 1 shows their locations: Lake Baikal in Western Siberia (Map 2), Lake Tanganyika of the rift valley of Eastern Africa (Map 3), Lake Ohrid of the karst district in the South-Western Balkans (Map 4), Lake Titicaca of the Altiplano of Western South America (Map 5), Lake Biwa in South-Central Honshu (Map 6), Japan, and Lake Inlé in the Shan Plateau of Burma (Map 7). Other candidates for consideration as ancient lakes, such as the Malili System in central Celebes (Kruimel, 1917–1918), the Quatro Ciénegas complex of Coahuila, Mexico (Taylor, 1966), and Lake Lanao on Mindanao in the Philippines (Myers, 1960) are either not truly ancient or have been too imperfectly studied malacologically.

Table I lists these lakes along with certain of their physical features. With the exception of Lake Inlé, the age of which is not known with any precision, all the other ancient lakes are unique by their age, variously estimated back to Pliocene or near Miocene times, and unusual in their depths. Located in similarly shaped, narrowly elongate graben troughs, Baikal and Tanganyika are unrivalled as the world's deepest lakes (Halbfass, 1928) while the smallish Lake Ohrid is remarkable for its depth in comparison to its size. The Titicaca basin is of considerable geologic complexity and its present day aspects were formed by epeirogenic elevation during the Pliocene. The Japanese Lake Biwa was formed by

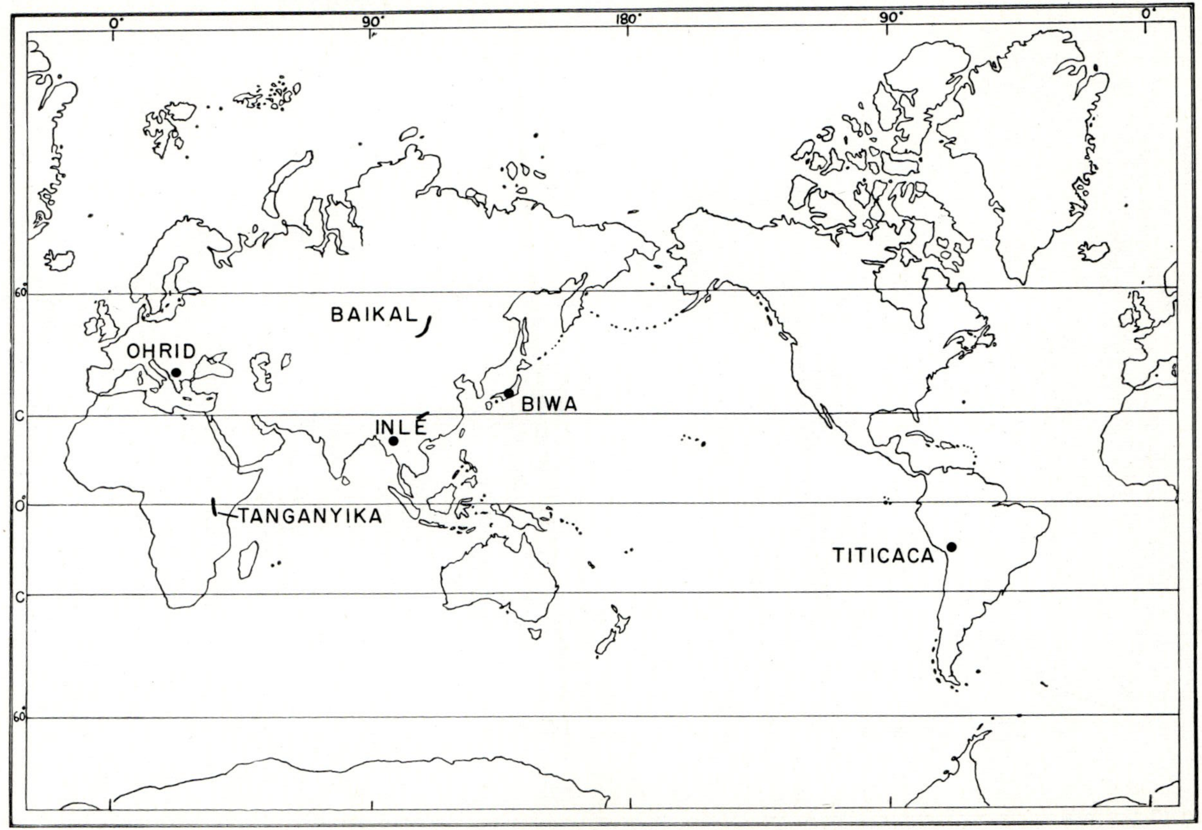

Map 1. The locations of 6 ancient lakes.

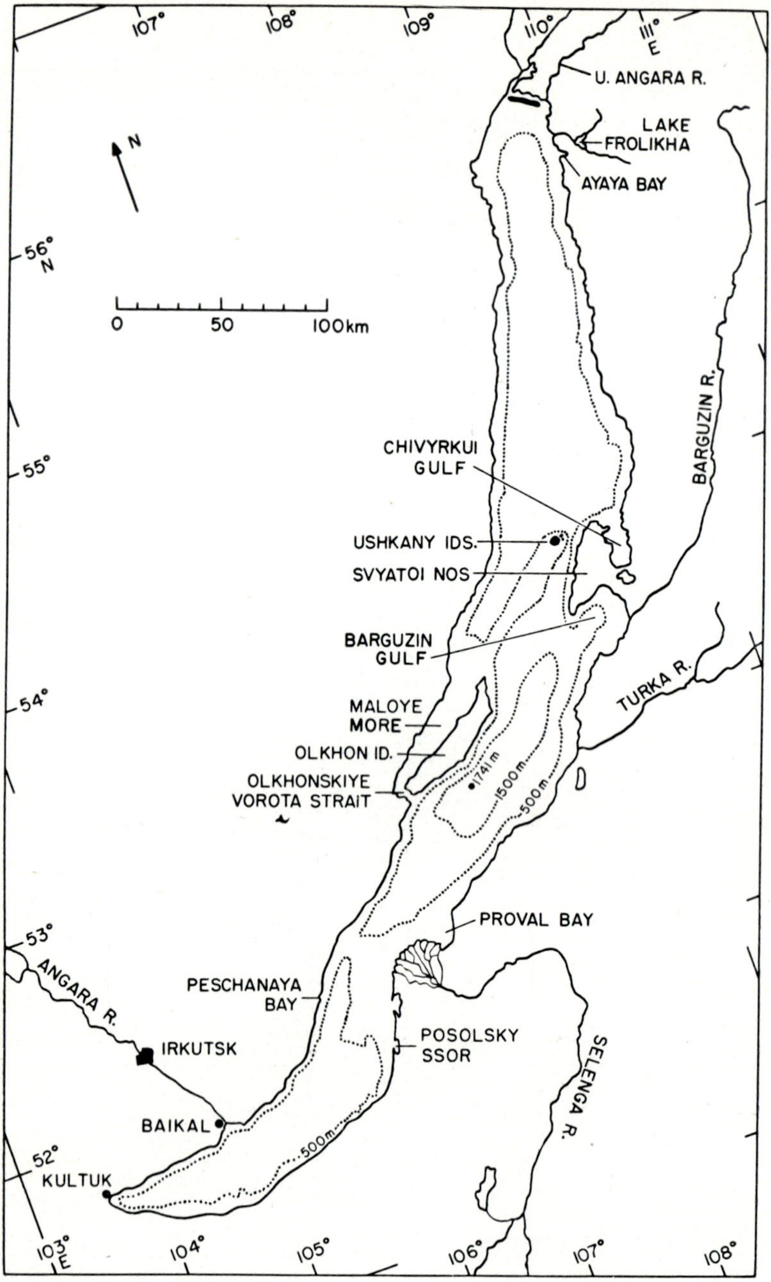

Map 2. Lake Baikal (modified after Kozhov, 1963).

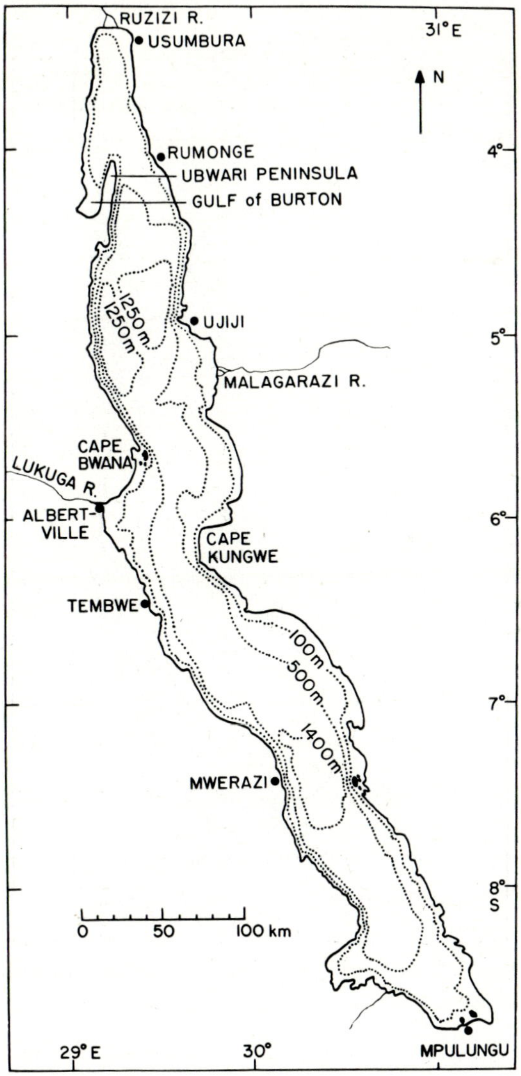

Map 3. Lake Tanganyika (modified after Capart, 1949).

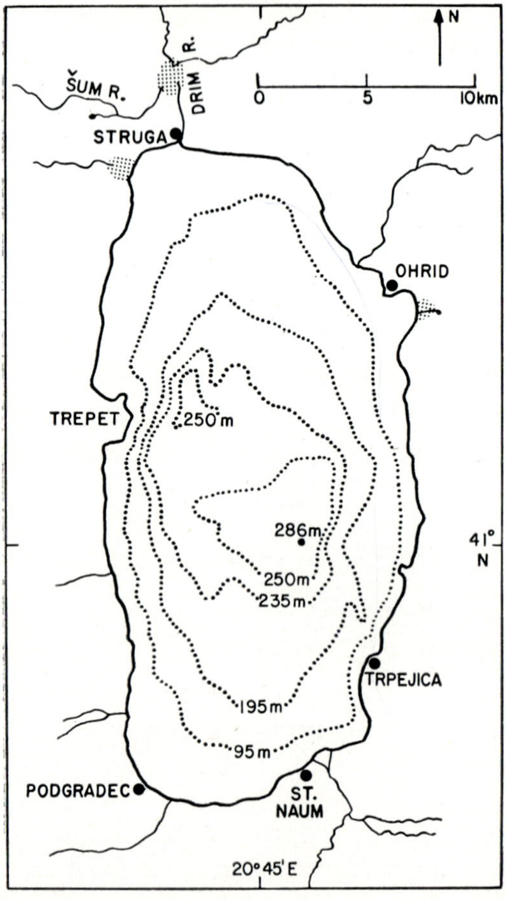

Map 4. Lake Ohrid (modified after Stankovič, 1960).

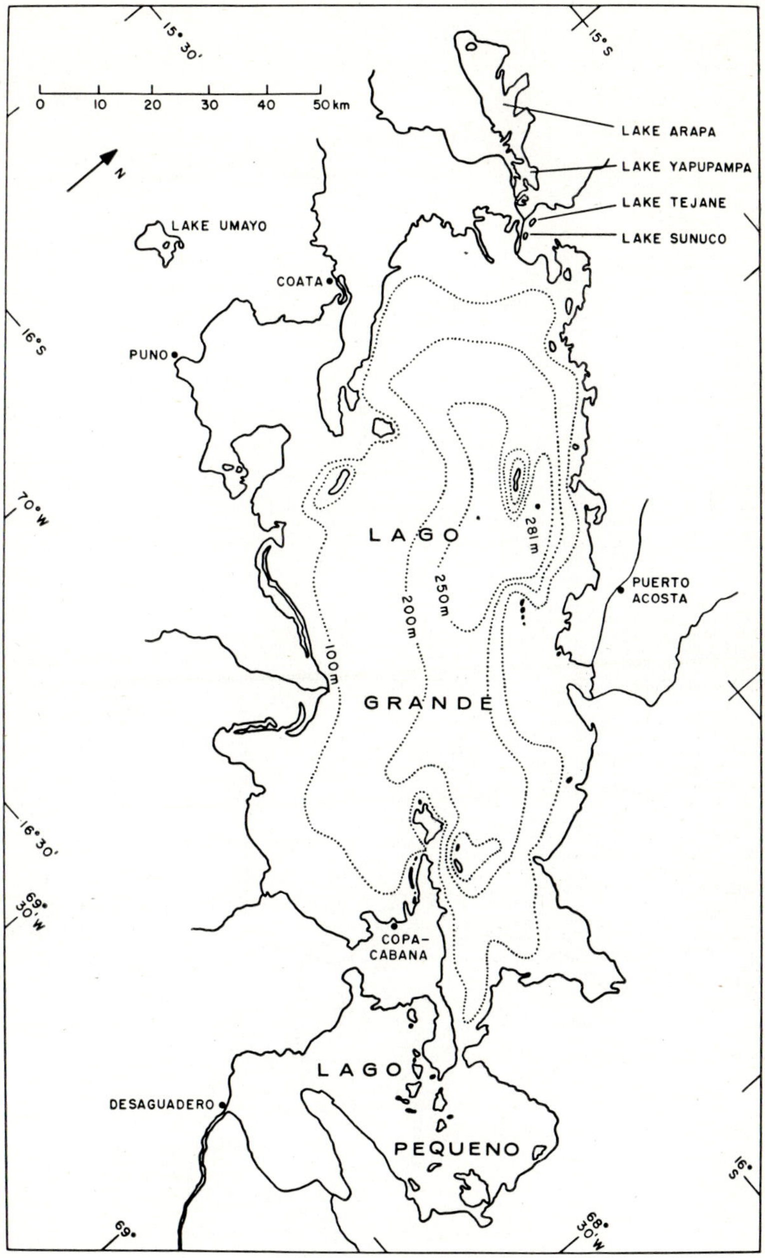

Map 5. Lake Titicaca (modified after Gilson, 1939).

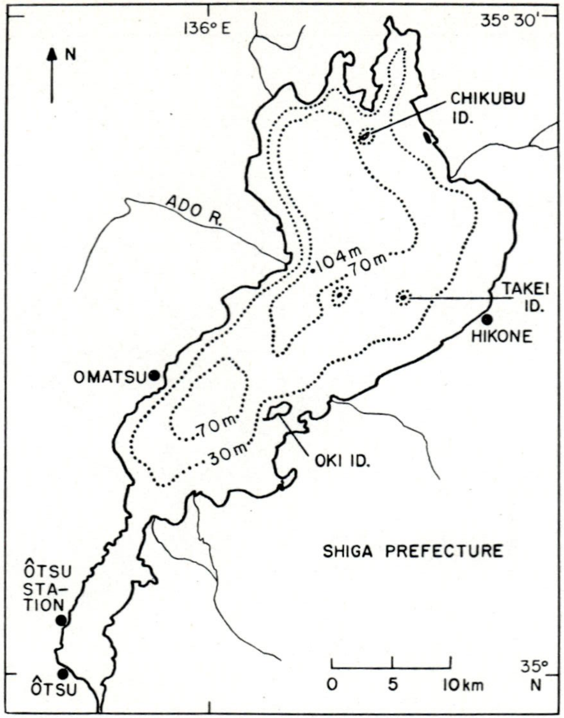

Map 6. Lake Biwa (modified after Mori *et al.*, 1967).

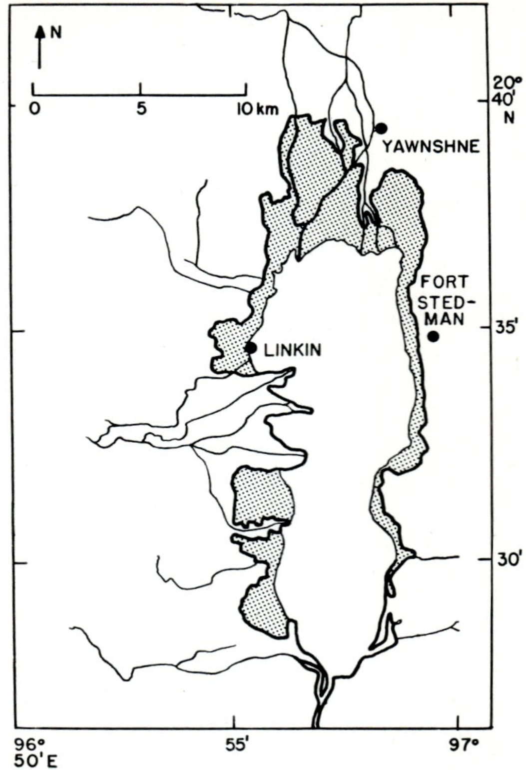

Map 7. Lake Inlé (modified after Annandale, 1918, shaded areas represent zone of floating vegetation).

Table I. Features of 6 ancient lakes.

	Baikal (Siberia)	Tanganyika (E. Africa)	Ohrid (Balkans)	Titicaca (Peru–Bolivia)	Biwa (Japan)	Inlé (Burma)
Location	51– 55 °N 103–110 °E	03–09 °S 29–31 °E	40–41 °N 20–21 °E	15–16 °S 68–70 °W	35 °N 136 °E	20 °N 97 °E
Area (km²)	31,500	34,000	340	7600	675	20
Depth (m)	1741	1470	286	281	104	7
Shape	long, narrow	long, narrow	rounded	irregularly rounded	irregularly elongate	oblong
Age	Mio-Pliocene	Pliocene	late Pliocene	post Miocene	pre Pliocene	? Pliocene
Formation	graben trough	graben trough	graben trough	elevated plateau	upwarped basin	elevated plateau
Contour system (number of basins)	2–3	2	1	complex	1	1
Drainage	Arctic Ocean via Yenisei	Atlantic Ocean via Congo	Mediterranean Sea via Drim	enclosed	Pacific Ocean via Setagawa	Indian Ocean via Salween

pre-Pliocene upwarping around its basin and has been relatively isolated without marked geological disturbance for about one million years. The most authoritative works on the geological, climatologic and hydrographic characteristics of each lake include: Baikal (Kozhov, 1963), Tanganyika (Beauchamp, 1939, 1940; Capart, 1949; Leloup, 1953), Ohrid (Stankovič, 1960), Titicaca (Gilson, 1939; Moon, 1939), Biwa (Horie, 1961, 1962; Mori, 1967), Inlé (Annandale, 1918).

III. The gastropods of the lakes

As is well known, freshwater snail faunas consist, with extremely rare exceptions, of prosobranch and pulmonate taxa. Although the latter are the subject of this essay, generalizations concerning their evolution in the ancient lakes can best be made in contrasting them with the prosobranch fauna. In addition to the systematic or taxonomic breakdown of the gastropods, the snail fauna shows 2 ecological moieties, especially in Lakes Baikal, Ohrid and Tanganyika. One group of species, largely pulmonate, inhabits the shallow water regions around the periphery of each lake while another lives in the lake itself, or the so-called lake proper, at greater depths and under a markedly different ecological regime. The latter element consists principally of prosobranchs and endemic pulmonates.

Table II. The Pulmonata of Lake Baikal and their distribution (modified after Kozhov, 1936; bathymetric ranges within the lake included).

Family	Distribution	
	Baikal	General distribution
Physidae		
Physa fontinalis (L.)	shore zone (0–10 m)	Holarctic
Aplexa hypnoruum (L.)	shore zone (0–10 m)	Holarctic
Lymnaeidae		
Lymnaea truncatula (Müll.)	shore zone (ssors)	Palearctic
Lymnaea palustris (Müll.)	shore zone (0–10 m)	Holarctic
Lymnaea auricularia (L.)	shore zone (+ Baikal proper to 20 m)	Holarctic
Lymnaea ovata (Lind.)	shore zone (0–10 m)	Palearctic

Table II—continued

Family	Distribution: Baikal	General distribution
Lymnaea peregra Müll.	shore zone (0–10 m)	Palearctic
Lymnaea stagnalis (L.)	shore zone (0–10 m)	Holarctic
Planorbidae		
Anisus leucostoma (Miller)	shore zone (0–10 m)	Holarctic
Anisus contortus (L.)	shore zone (0–10 m)	Palearctic
Gyraulus albus (Müll.)	shore zone (0–10 m)	Holarctic
Gyraulus gredleri (Gredler)	shore zone (+ in North province, 0–10 m)	Palearctic
Gyraulus umbiliciferus Kozhov	north Baikal (0–10 m)	endemic
Segmentina nitida (Müll.)	shore zone	Holarctic
Armiger crista (L.)	shore zone	Holarctic
Choanomphalus maacki (Gerstf.)	Baikal to 50 m + Angara River	endemic
Choanomphalus amauronius (Bgt)	Baikal to 100 m + Angara River for 600 km	endemic
Choanomphalus patulaeformis Ldh	Baikal (discontinuous) (0–10 m)	endemic
Choanomphalus dybowskianus (Ldh)	Baikal + Angara River (0–50 m)	endemic
Choanomphalus gerstfeldtianus (Ldh)	Baikal (0–50 m)	endemic
Choanomphalus schrencki Dybowski	Baikal (south) (0–100 m)	endemic
Choanomphalus annuliformis Kozhov	Baikal (north) (0–100 m)	endemic
Acroloxidae		
Acroloxus sibiricus (Gerstf.)	Baikal + Angara River (0–50 m)	endemic
Acroloxus troscheli (Dybowski)	Baikal (0–50 m)	endemic
Acroloxus kobelti (Dybowski)	Baikal + Angara River (0–20 m)	endemic
Acroloxus boettgerianus (Ldh)	Baikal (shallow)	endemic

Table III. The Prosobranchs of Lake Baikal and their distribution (modified from Kozhov, 1936).

	Endemic species	General Siberian species
Valvatidae		
Megalovalvata	4	
Valvata		2
Hydrobiidae		
Kobeltocochlea	2	
Benedictia	4	
Bythniidae		
Bithynia		1
Baicaliidae		
Baicalia	32	
Eubaicalia (6 sp)		
Parabaicalia (4 sp)		
Korotnewia (3 sp)		
Pseudobaicalia (4 sp)		
Microbaicalia (1 sp)		
Baicaliella (1 sp)		
Trichobaicalia (1 sp)		
Dybowskiola (1 sp)		
Teratobaicalia (1 sp)		
Maackia (1 sp)		
Gerstfeldtia (5 sp)		
Baicalia (3 sp)		
Godlewskia (1 sp)		
Liobaicalia	1	

Table IV. The non-thalassoid gastropods of Lake Tanganyika and their distribution (modified after Leloup, 1953).

	Distribution	
	Tanganyika[a]	General distribution
PULMONATA		
Lymnaeidae		
Lymnaea natalensis (Krauss)	shore	Ethiopian
Planorbidae		
Bulinus angolensis (Morel.)	tributaries	Ethiopian
Bulinus africana (Krauss)	shore	Ethiopian
Bulinus forskalii (Ehrenb.)	shore	Ethiopian
Biomphalaria alexandrina (Ehrenb.)	shore	Ethiopian
Gyraulus apertus (Mart.)	shore	Ethiopian
Gyraulus chevalieri (Germ.)	shore	Ethiopian

Table IV—continued

	Distribution	
	Tanganyika[a]	General distribution
Gyraulus coretus (De Blainv.)	shore	Ethiopian
Gyraulus costulatus (Krauss)	shore	Ethiopian
Gyraulus lamyi (Germ.)	shore	Ethiopian
Segmentina kanisaensis Prest.	tributaries	Ethiopian
Segmentina planodiscus Melv. and Pons.	shore	Ethiopian
Ancylidae		
Burnupia caffra (Krauss)	shore–100 m	Ethiopian
Ferrissia tanganyicensis (Smith)	shore–70 m	endemic
PROSOBRANCHIA (NON-THALASSOID)		
Ampullariidae		
Lanistes farleri Graven	shore	Ethiopian
Lanistes graueri Thiele	? shore	Ethiopian
Lanistes olivaceus (Sow.)	shore	Ethiopian
Pila ovata (Oliv.)	shore	Ethiopian
Viviparidae		
Neothauma tanganyicense Smith	shore–65 m	endemic
Viviparus brincatianus (Bourg.)	shore	? endemic
Viviparus costulatus (Mart.)	? shore	Ethiopian
Viviparus unicolor (Oliv.)	? shore	Ethiopian
Hydrobiidae		
Bithynia alberti Smith	shore	Ethiopian
Mysorelloides multiculcata (Bourg.)	shore–50 m	endemic
Tomichia guillemei Leloup	tributaries	endemic
Thiaridae		
"*Melania*" *tanganyicensis* Smith	?	endemic
Melanoides admirabilis (Smith)	shore	endemic
Melanoides tuberculata (Müll.)	tributaries	Ethiopian
Cleopatra bulimoides (Oliv.)	tributaries	Ethiopian
Cleopatra cyclostomoides (Kust.)	tributaries	Ethiopian
Cleopatra guillemei Bourg.	shore	endemic
Cleopatra jouberti Bourg.	shore	endemic
Cleopatra trisulcata Germ.	shore	? endemic
Potadoma ignobilis (Thiele)	shore	Ethiopian
Potadomoides pelseneeri Leloup	shore	? endemic

[a] The term "shore" embraces those habitats separated by Leloup (1953) from the so-called lake proper and includes, e.g. marais, marecage, etang, anse, cours d'eau, ruisseau, rive and lagune-crique. The generalized term Ethiopian means the species occurs outside the confines of Tanganyika, either as a widely distributed species or a more localized distribution in other parts of Africa.

Table V. Geographic, bathymetric and ecologic data on the endemic thalassoid prosobranchs of Lake Tanganyika (after Leloup, 1953; substrate preferences are generalizations, while depths are ranges in which living specimens have been obtained; further ecological restrictions are mentioned in the text, ecl = entire circumference of the lake proper).

Distribution		Depth range (m)	Substrate preference
Syrnolopsidae			
Anceya giraudi Bourg.	ecl	130	sandy mud
Anceya terebriformis (Smith)	?		
Martelia tanganyicensis Dautz.	ecl	9	sandy mud
Syrnolopsis gracilis Pilsbry and Bequaert	ecl	116	sandy mud
Syrnolopsis lacustris Smith	ecl	100	sandy mud
Thiaridae			
Bathanalia howesi Moore	southern	200	rocky
Bathanalia straeleni Leloup	ecl	80	rocky
Bridouxia giraudi Bourg.	ecl	80	solid
Chytra kirki (Smith)	ecl	80	sandy mud
Edgaria nassa (Wood.)	ecl	littoral rocks	rocky
Hithia littorina Ancey	SE coast	beach	sandy
Limnotrochus thomsoni Smith	ecl	60	sandy mud
Paramelania iridescens (Moore)	ecl	150	muddy
Paramelania damoni (Smith)	ecl	65	rocky
Reymondia horei (Smith)	ecl	10	littoral vegetation sand–mud–rock
Spekia zonata (Wood.)	ecl	70	rocky
Stanleya neritinoides (Smith)	ecl	100	rocky
Stormsia minima (Smith)	ecl	15	rocky
Tanganyicia rufofilosa (Smith)	ecl	20	rocky
Tiphobia horei Smith	ecl	125	muddy

Table VI. The Pulmonata of Lake Ohrid and their distribution (modified after Radoman, 1964).

Family	Distribution	
	Ohrid	General distribution
Physidae		
Aplexa hypnorum L.	ponds	Holarctic
Lymnaeidae		
Lymnaea truncatula Müll.	streams	Palearctic
Lymnaea palustris Müll.	ponds and springs	Holarctic
Lymnaea auricularia L.	ponds	Holarctic
Lymnaea peregra Müll.	springs and streams	Palearctic
Lymnaea relicta Pol.	lake proper, springs and streams	endemic
Lymnaea stagnalis L.	ponds	Holarctic
Planorbidae		
Planorbis planorbis L.	springs and ponds	Palearctic
Planorbis macedonicus Stur.	lake proper	endemic
Planorbis corneus L.	lake proper and ponds	Palearctic
Anisus contortus (L.)	springs and ponds	Palearctic
Hippeutis complanatus (L.)	ponds	Palearctic
Armiger crista (L.)	lake proper	Palearctic
Gyraulus trapezoides Pol.	lake proper	endemic
Gyraulus lychnidicus Pol.	lake proper	endemic
Gyraulus crenophilus Hub. and Rad.	springs	endemic
Gyraulus fontinalis Hub. and Rad.	springs	endemic
Gyraulus albidus Rad.	lake proper and streams	endemic
Ancylidae		
Ancylus fluviatilis Müll.	springs and streams	Palearctic
Ancylus tapirulus Pol.	lake proper	endemic
Ancylus scalariformis Stank. and Rad.	lake proper	endemic
Ancylus lapicidus Hub.	lake proper	endemic
Acroloxidae		
Acroloxus lacustris (L.)	lake proper	Holarctic
Acroloxus improvisus Pol.	lake proper	endemic
Acroloxus macedonicus Hadž.	lake proper	endemic

Table VII. The Prosobranchia of Lake Ohrid and immediate vicinity (modified after Radoman, 1964, and Starobogatov, 1970).

Prosobranchia	Habitat[a]
Neritidae	
Theodoxus fluviatilis L.	littoral
Valvatidae	
[b]*Valvata stenotrema* Pol.	littoral
[b]*Valvata hirsutecostata* Pol.	sublittoral
[b]*Valvata rhabdota* Stur.	littoral
[b]*Valvata relicta* Pol.	littoral, sublittoral + profundal
[b]*Gocea ohridana* Hadž.	littoral
Viviparidae	
Viviparus viviparus	littoral
Thiaridae	
Amphimelania holandri C. Pfeiff.	streams, Drim River
Hydrobiidae (1 non-endemic)	
[b]*Pyrgula* (9 sp)	2 littoral
	3 sublittoral
	2 profundal
	2 springs/streams
[b]*Pyrgohydrobia* (3 sp)	1 littoral
	1 springs/streams
	1 extinct
[b]*Ginaia* (1 sp)	littoral
[b]*Pseudamnicola* (22 sp)	10 littoral
	3 sublittoral
	1 profundal
	8 springs/streams
[b]*Horatia* (4 sp)	2 littoral
	1 sublittoral
	1 springs/streams

[a] Littoral = 0–20 m; sublittoral = 20–50 m; profundal > 50 m.
[b] Endemic.

Table VIII. The gastropods of Lake Titicaca and their distribution (modified from Haas, 1955).

Gastropod	Distribution
Pulmonata	
Planorbidae	
Tropicorbis canonicus (Cousin)	Titicaca Basin, western S. America, and Amazonian
[a]*Taphius montanus* (D'Orb.)	Titicaca Basin
[a]*Platytaphius heteropleurus* (Pilsbry and Vanetta)	Titicaca Basin
Ancylidae	
Gundlachia crequii (Bavay)	Titicaca Basin, Lake Langui, and Amazonian
Gundlachia moricandi (D'Orb.)	Titicaca Basin and eastern S. America
Prosobranchia	
Hydrobiidae	
[a]*Littoridina stiphra* Haas	Lake Titicaca
Littoridina cumingii (D'Orb.)	Lake Titicaca and Amazonian
Littoridina cuzcoensis Pilsbry	Río Huatanay, Amazonian
Littoridina lacustris Haas	Lake Titicaca
[a]*Littoridina languiensis* Haas	Lake Langui, Amazonian
[a]*Littoridina popoensis* (Bavay)	Lake Poopó
[a]*Littoridina profunda* Haas	Lake Titicaca
[a]*Littoridina saracochae* Haas	Lagunilla Sarachocha and Langunilla Lagunilla
[a]*Littoridina vestita* Haas	Lagunilla Saracocha
[a]*Littoridina berryi* Pilsbry	Lake Titicaca
[a]*Littoridina andecola* (D'Orb.)	Lake Titicaca
[a]*Littoridina aperta* Haas	Lake Titicaca
[a]*Littoridina forsteri* Haas	Lake Titicaca
[a]*Strombopoma ortoni* (Pilsbry)	Lake Titicaca
[a]*Strombopoma gracile* Haas	Lagunilla Lagunilla
[a]*Rhamphopoma magnum* Haas	Lake Titicaca
[a]*Rhamphopoma parvum* Haas	Lake Titicaca
[a]*Heligmopoma umbilicatum* Haas	Lake Titicaca
[a]*Brachypyrgulina carinifera* Haas	Lake Titicaca
[a]*Limnothauma crawfordi* Haas	Lake Titicaca
[a]*Ecpomastrum mirum* Haas	Lake Titicaca

[a] Endemic.

Table IX. Gastropods of Lake Biwa, Japan (modified after Mori, 1970).

Pulmonata		
	Lymnaeidae	
		Fossaria ollula (Gould)
		Radix japonica (Jay)
		[a]*Radix onychia* (West.)
	Planorbidae	
		Camptoceras hirasei Walker
		Camptoceras prashadi Clench
		Segmentina nitidella Mart.
		Gyraulus hiemantium (West.)
		[a]*Gyraulus biwaensis* (Preston)
		[a]*Gyraulus amplificatus* (Mori)
Ancylidae		
		Ferrissia nipponica (Kuroda)
		Ferrissia japonica Habe and Burch
Prosobranchia		
	Viviparidae	
		Cipangopaludina japonica (Mart.)
		Cipangopaludina malleata (Reeve)
		[a]*Heterogen longispira* (Smith)
		Sinotaia histrica (Gould)
	Valvatidae	
		Valvata japonica Mart.
	Bithyniidae	
		Parafossarulus manchouricus japonicus (Pilsbry)
	Pleuroceridae[b]	
		Semisulcospira reiniana (Brot)
		Semisulcospira decipiens (West.)
		[a]*Semisulcospira multigranosa* Boettger
		[a]*Semisulcospira niponica* (Smith)
		[a]*Semisulcospira reticulata* Kajiyama and Habe
		[a]*Semisulcospira habei yamaguchi* Davis

[a] Endemic.

[b] Davis (1969) provided this list of pleurocerids and also pointed out that *S. habei habei* and *S. nakasekoae* are endemic at Uji in the Seta River, which may be construed as a portion of Lake Biwa.

Table X. Gastropods of Lake Inlé, Burma (modified after Annandale, 1918).

- Pulmonata
  - Succineidae
    - *Succinea indica* Pfeiff.
  - Lymnaeidae
    - *Lymnaea andersoniana* Nevill
    - [a]*Lymnaea shanensis* Annand.
    - [a]*Lymnaea mimetica* Annand.
  - Planorbidae
    - *Planorbis exustus* Desh.
    - [a]*Planorbis velifer* Annand.
    - *Planorbis trochoideus* Benson
    - *Planorbis calathus* Benson
    - *Planorbis caenosus* Benson
- Prosobranchia
  - Hydrobiidae
    - [a]*Hydrobioides nassa* (Theobald)
    - [a]*Hydrobioides physeus* Annand.
    - [a]*Hydrobioides avarix* Annand.
    - [a]*Hydrobioides nana* Annand.
    - *Amnicola alticola* Annand.
  - Viviparidae
    - *Ampullaria winkleyi* Pilsbry
    - *Vivipara lecythis* (Benson)
    - *Taia naticoides* (Theobald) (Upper Burma + Lashio)
    - *Taia theobaldi* (Kobelt) (also near Kalau)
    - [a]*Taia shanensis* (Kobelt)
    - [a]*Taia elitoralis* Annand.
    - [a]*Taia intha* Annand.
  - Thiaridae
    - "*Melania*" *tuberculata* (Müll.)
    - "*Melania*" *terebra* Benson
    - "*Melania*" *baccata* (Gould)
    - *Paludomus ornata* Benson

[a] Endemic.

Table XI. Summary of the diversity and endemicity of the pulmonate and prosobranch gastropods of ancient lakes. For explanation of the calculation of the endemicity ratio, see text.

Lake	Total number of Gastropods	Total Pulmonata	Endemic Pulmonata	Non-endemic Pulmonata	Total Prosobranchia	Endemic Prosobranchia	Non-endemic prosobranchia	Endemicity total endemic / total gastropods	Endemicity ratio
Baikal	72	26	12	14	46	43	3	0·76	2·0
Tanganyika	55	14	1	13	41	30	11	0·56	10·4
Ohrid	71	25	12	13	46	42	4	0·76	1·8
Titicaca	26	5	2	3	21	18	3	0·76	2·1
Biwa	24	11	3	8	13	6	7	0·37	1·7
Inlé	25	9	3	6	16	7	9	0·40	1·5

Table XII. Data on chromosome numbers in families of freshwater gastropods (from Burch, 1965; Burch and Natarajan, 1966; Davis, 1969; Patterson, 1967).

Gastropod	Haploid number	Remarks
Pulmonata		
Acroloxidae	18	
Ancylidae	15–60	polyploids
Planorbidae	16–36	polyploids
Lymnaeidae	17–19	
Physidae	18	
Prosobranchia		
Viviparidae	7–14	
Valvatidae	10	
Hydrobiidae	16–17 (+ ?)	polyploids
Thiaridae	16–19 (to 47)	polyploids
Pleuroceridae	7–20	

The remarkable diversities and endemicities of the pulmonates and prosobranchs of ancient lakes show similar patterns in each lake, providing essential clues about the evolutionary biology of each group. Comparison of these faunas and their radiations permit the elaboration of an explanatory paradigm possessing heuristic value and indicating avenues for further research endeavours.

Tables II–X list the known species of gastropods of 6 ancient lakes based on the more recent and reliable authorities. As far as available, information on the ecological preferences or zoogeographic distributions of each species is provided, and Table XI summarizes these data. In all the ancient lakes, the pulmonates are outnumbered by the prosobranchs. Further, there are clearly fewer endemic pulmonates than endemic prosobranchs. These facts are rendered more startling when the ratio of endemicity* of pulmonates is contrasted with that of prosobranchs. Depending on the accuracy of the data, endemicity in prosobranchs is always greater than that among pulmonates. As shown in Table XIII, prosobranchs are from 1·5–10 times more endemic. Even though the systematic treatments of the various faunas are not comparable, the data illustrated in Table XIII shows a phenomenon which is remarkable in its similarity for each lake. All the molluscan faunas are highly endemic, being as great as 76% in Baikal, Ohrid and Titicaca. Putatively older and

* The endemicity ratio is calculated as:
(number of endemic prosobranchs/total number of prosobranchs)/(number of endemic pulmonates/total number of pulmonates).

Table XIII. Endemicity and the endemicity ratio (see text) for the gastropods of 6 ancient lakes.

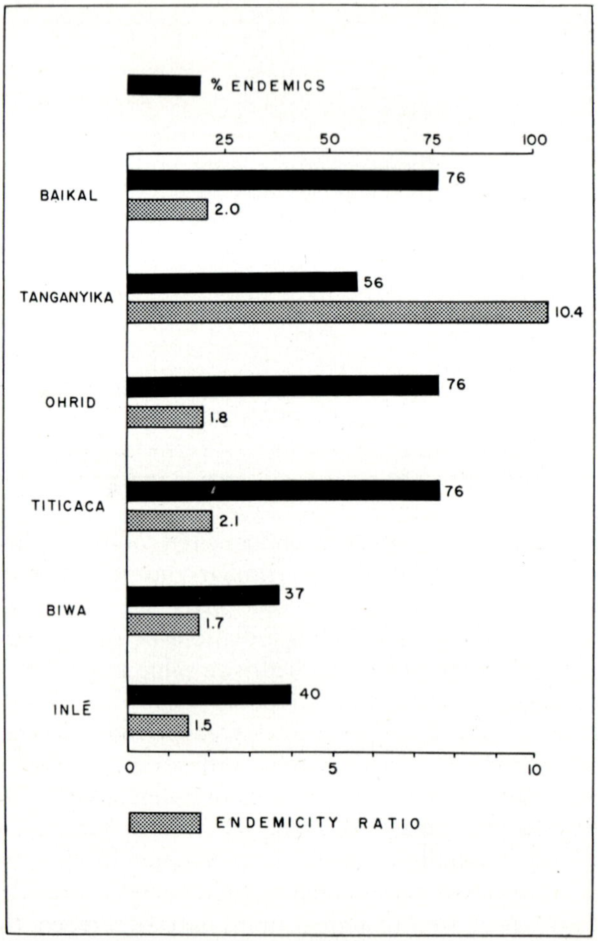

obviously larger, deeper lakes such as Baikal and Tanganyika support both richer faunas and greater endemicities; however, the most important generalization, as shown by the endemicity ratios, is that the pulmonates are less prone to speciation in the lakes than are the prosobranchs.

Briefly, 2 major evolutionary patterns emerge in regard to gastropod faunas of the ancient lakes: the pulmonate moiety shows markedly lower endemicities than does that of the prosobranchs, and the totality of the fauna is segregated into 2 main ecological components. Before considering these most important phenomena, let us review the process of speciation and the evolution of the gastropod faunas in the lakes.

IV. Origin and evolution of lake gastropods

The synthetic theory of evolution proposes that the complex process of speciation in sexually reproducing animals is initiated by the separation of populations of a species by geographic barriers. Such spatial isolation facilitates the gradual divergence of different populations until intrinsic genetic isolating mechanisms have evolved which will prevent introgression if the populations should ever again become sympatric. Essentially this evolutionary phenomenon consists of the division of the gene pool of a species into separate entities which eventually become genetically incompatible, forming discrete, unique units of the natural world—species. To be successful, a new species must possess ecological uniqueness, that is, it must occupy an ecological niche which allows it to survive sympatrically with other species, including closely related ones. The immense complexities behind species formation and their unique properties have been the objectives of the majority of investigations in organismic biology; the subject has been recently reviewed by Mayr (1970).

Arguments concerning the origin of the faunas of the ancient lakes are legion. Although all freshwater gastropods arose from marine ancestors, the theories (Moore, 1903; Dybowski, 1900) purporting to show that ancient lakes represented isolated portions of seas or oceans which became fresh water, so-called *Reliktenseen*, have been shown to be incorrect (Brooks, 1950; Boss, 1973). The great resemblance between the melaniid prosobranchs of Tanganyika and certain marine snails is, clearly, a matter of convergence (Yonge, 1938), and one which is not nearly as exact as the enthusiasm of early naturalists portrayed (Pilsbry and Bequaert, 1927).

From ancient indigenous stocks, the gastropods of ancient lakes evolved within the lakes themselves. Among the Pulmonata, the Planorbidae is the main group to have radiated in Baikal (Table II; Figs 1 and 2), Ohrid (Table VI; Fig. 6), Titicaca (Table VIII) and Biwa (Table IX). Little or no radiation is detectable among other pulmonates such as the lymnaeids and physids though the ancylid and acroloxid limpets show some speciation in Baikal and Ohrid (Tables II and VI, respectively, and Fig. 7). Among prosobranchs, the small cosmopolitan hydrobioid snails show the greatest diversity while melaniid or thiarid stocks exhibit a spectacular radiation, especially in Tanganyika (Table V; Figs 3 and 4). The viviparids boast a considerable diversity in Lake Inlé (Table X, Fig. 4) and a unique form in Lake Biwa (Table IX, Fig. 4). Such intralacustrine speciation, at one time a subject of considerable controversy, has been well-documented in many groups by Brooks (1950).

Fig. 1. The endemic pulmonate *Choanomphalus* from Lake Baikal. 1–3, *Choanomphalus maacki* Gerstfeldt (diameter = 7 mm, MCZ 58646); 4–6, *C. amauronius* Bourguignat (diameter = 7 mm, MCZ 58650).

Opportunity for the speciation process has been provided by the long periods of time each lake basin has been isolated and by the complicated geological changes in the sizes of the lakes, their drainage systems and connections with adjacent bodies of water, allowing geographic isolation of populations and subsequent genetic differentiation. Confirming data are available for Baikal (Pavlovsky, 1941), Tanganyika (Capart, 1949), Ohrid (Stankovič, 1960), Titicaca (Moon, 1939), Biwa (Davis, 1969) and Inlé (Annandale, 1918).

To account for the diversity of the fauna in Lake Tanganyika, Livingstone (1965) calculated that if the 2 main basins of the lake had been separated during periods of low water and that during each period, allopatric populations diverged enough genetically to prevent subsequent inbreeding, only 8 falls of lake level would be capable of producing 256 species from a single parental stock. Although the history of the lake may have allowed such opportunities (Capart, 1949), such extreme success in speciation does not have to be invoked to account for the diversity of the molluscs.

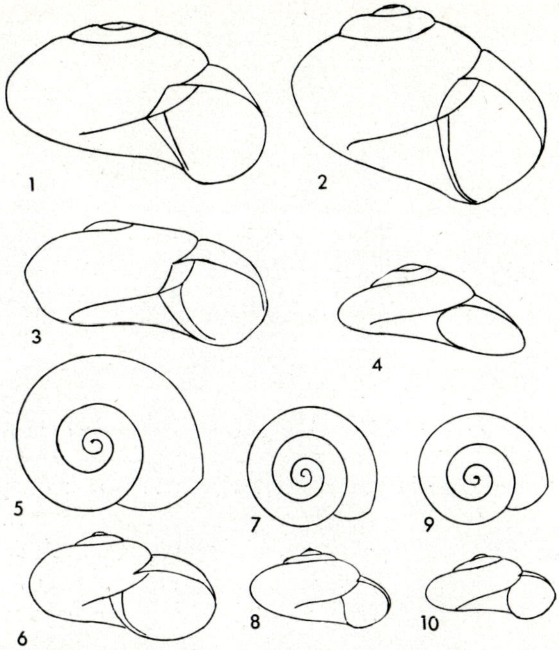

Fig. 2. Representative species of the radiation of the endemic pulmonate *Choanomphalus* in Lake Baikal. 1–3, *C. maacki* Gerstfeldt, showing the range of variation in shape (diameter of shell to 8 mm); 4, *C. schrenki* Dybowski, a representative of the subgenus *Sulcifer* (diameter of shell to 3·5 mm); 5 and 6, *C. amauronius* Bourguignat (diameter of shell to 8 mm); 7 and 8, *C. patulaeformis* Lindholm (diameter of shell to 4 mm); 9 and 10, *C. gerstfeldtianus* Lindholm (diameter of shell to 2·3 mm) (after Zhadin, 1952).

Fig. 3. One of the most unusual thalassoid prosobranchs from Lake Tanganyika. *Tiphobia horei* E. A. Smith (height = 44 mm, MCZ 243714).

Fig. 4. Some representative prosobranchs endemic in ancient lakes. 1–2, a thalassoid melaniid, *Chytra kirki* (Smith), Lake Tanganyika (diameter = 14·2 mm, MCZ 29745); 3, A thalassoid melaniid, *Limnotrochus thomsoni* Smith, Lake Tanganyika (height = 13·8 mm, MCZ 41118); 4, The solute baicaliid, *Liobaicalia stiedae* (Dybowski), Lake Baikal (height = 8·5 mm, MCZ 58668); 5, A viviparid, *Taia intha* Annandale, Lake Inlé (height = 24 mm, MCZ 143696); 6, The monotypic viviparid, *Heterogen longispira* (Smith), Lake Biwa (Height = 28 mm, MCZ 193349); 7, A thalassoid melaniid, *Paramelania* (*Bythoceras*) *iridescens* (Moore), Lake Tanganyika (height = 30 mm, MCZ 29748).

It has been pointed out that linear continuity, or shore line distribution, allows more differentiation or fractionation of populations than areal, lake basin, distribution (Fryer and Iles, 1972). Examples of such splitting of populations is evident for gastropods of ancient lakes and lends credence to the oft cited generalization that a lake is to aquatic organisms as an island is to terrestrial ones: opportunities for the initial phase of speciation, namely geographic isolation, are present. Frequently the isolated populations in the ancient lakes have been accorded different taxonomic treatment by individual authors. Thus, for what some might interpret as a polytypic species of *Pyrgohydrobia*, Radoman and Marinkovič (1959) considered 3 separate species in Lake Ohrid: *P. saintnaumi* Radoman is isolated at a spring at St Naum, *P. jablanicensis* Radoman in a spring on the Śum River, and *P. grochmalickii* (Polinski) in Ohrid from the shore to 3 m. The authors suggest that *saintnaumi* Radoman and *jablanicensis* Radoman were isolated when the lake level dropped and were derived from a common lake ancestor. Leloup (1953) recognized that most of the endemic thalassoid gastropods were distributed around the entire circumference of Tanganyika, but he also noted that some species had specific substrate preferences and formed isolated populations. Better documented examples of such isolation occur among the fishes where Fryer and Iles (1972) show that certain rock-associated species are rigorously limited to substrate types so that gene flow is reduced even between adjacent populations.

Essentially, each species consists of a series of isolates of varying degrees of genetic miscibility. The splitting up of a species into many small, separate populations is of fundamental importance in the speciation process in ancient lakes, for given the extremely long coast lines and discontinuous distribution of specific substrate habitats, intralacustrine barriers become established. A snail in Tanganyika at Usumbura and another in Mpulungu are separated by the distance from London to Vienna; possibilities for intralacustrine barriers and isolation are apparent. Further, important genetic phenomena attend isolated, small populations. Both genetic drift and inbreeding with rapid genetic change may profoundly affect the population. Most frequently such alterations are deleterious but it is only occasionally that success accompanies such changes and a genetically isolated population might establish itself as an incipient species.

Many authors have referred to "explosive speciation" or "explosive radiations" in the ancient lakes. In truth, there is nothing explosive about the speciation of the gastropods in these basins. The radiation of melaniid prosobranchs in Tanganyika has taken millions of years (Leloup, 1953), and the development of the diverse hydrobiid-like baicaliids in Baikal may date from earlier than the mid-Tertiary (Martinson, 1948).

In summary, during the history of the development of the faunas of the ancient lakes, several factors may facilitate intralacustrine speciation. Alterations in the physical conformity of the lakes have allowed adjacent lakes and drainage basins to act as effective geographical isolating barriers. Additionally the lakes themselves have been *refugia* during times of physical stress. Faunas from surrounding bodies of water or adjacent river systems sometimes survived the exigencies of the Pleistocene in the lakes. The viviparid, *Neothauma*, now restricted to Tanganyika, lived in Lake Edward during the Pleistocene (Fuchs, 1936) but survived the arid portions of the period in the basin of Tanganyika. Certain elements of the Baikalian fauna sought refuge in the lake during the glaciation of the Pleistocene. Similarly, the unique viviparids and possibly hydrobiids of Lake Inlé carry on as remnants of a once extensive plateau lake system which otherwise has dried up (Annandale, 1918). Thus, it appears that geographic barriers associated often with ecological specificities have been the principal extrinsic factors causing localization of populations within the lakes. In this regard, the radiations of both the pulmonate and prosobranch gastropods in the ancient lakes are comparable to the more thoroughly investigated radiations of Darwin's finches in the Galapagos (Lack, 1961) or the drosophilid flies of the Hawaiian islands (Mayr, 1970).

V. Lake communities

In the 3 better known ancient lakes, namely, Tanganyika, Baikal and Ohrid, the best malacological studies show that in regard to the gastropods, 2 ecological moieties are similarly present in each lake. Leloup (1953), Kozhov (1963) and Stankovič (1960) all stress that the endemic elements are predominantly limited to the lake proper while others, generally those molluscs of wider distribution, are adapted to the peripheral habitats of the lake. As Tables II–VII indicate, the surrounding shallows of the ancient lakes, including tributaries, springs, shallow bays, ssors, associated ponds, and swamps harbour a molluscan fauna which sharply contrasts with the indigenous and endemic faunas which penetrate greater depths in each lake. These 2 faunal elements have markedly different ecological preferences and ranges of tolerance. The former consists of eurytopic species with broad physiological tolerances and with considerable geographic ranges, the latter of stenotopic forms with narrower tolerances and smaller geographic ranges, being usually largely endemic. The total number of eurytopic species is less than those that are stenotopic. That is to say, the inshore, more cosmopolitan elements are less diverse while the fauna of the lake proper, in addition to being largely endemic, shows greater diversity. The inshore eurytopic moiety consists largely of pulmonates while the lake proper component is made up mostly of prosobranchs and endemic pulmonates.

Thus, along the shallows and shore lines of such lakes as Baikal and Tanganyika, a less diverse, largely non-endemic, eurytopic fauna holds sway; species in this community are adapted to wide fluctuations of physical, environmental parameters. In the Palearctic lakes Ohrid (Table VII) and Baikal (Tables II and III), this fauna has principally Palearctic or

Holarctic representatives. In Tanganyika such an assemblage is overwhelmingly Ethiopian (Tables IV and V). That such division of the faunas of at least 1 of the lakes, namely Baikal, is of great antiquity is shown by outcrops of probably Pliocene or Miocene age near Kultuk where isochronous faunas were similarly separated. Already at that time an endemic Baikalian fauna was segregated from the then more widely distributed inshore fauna which consisted of viviparids, lithoglyphine hydrobiids and unionids, presently of more southerly climes (Martinson, 1951).

Occasionally an element of the inshore fauna invades the lake proper. Such incursions include *Lymnaea auricularia* (L.) to depths of 20 m in Lake Baikal (Table II) and the Ethiopian ancylid, *Burnupia caffra* (Krauss), to 100 m in Lake Tanganyika (Table IV).

Fig. 5. The endemic pulmonate *Gyraulus* (*Carinogyraulus*) *lychnidicus* Hesse from Lake Ohrid (diameter of shell = 4·5 mm, MCZ 168320).

In the lake-proper portions of Lakes Baikal, Ohrid and Tanganyika, a largely endemic element predominates, consisting principally of prosobranchs, but occasionally showing a radiation of pulmonates, as with *Choanomphalus* (Figs 1 and 2) and *Acroloxus* in Baikal (Table II), and *Ancylus* (Fig. 7), *Acroloxus* and *Carinogyraulus* (Figs 5 and 6) in Ohrid (Table VI). This largely endemic malacofauna is highly diverse, consisting of predominantly stenotopic species.

This pattern of ecological segregation of eurytopic and stenotopic species is comparable to that shown in communities studied in the marine environment. Sanders (1968) analysed the interacting factors affecting the diversity of species which make up benthic communities. He noted that Lake Baikal was essentially similar in its possession of 2 distinct faunal regimes with the boreal marine environment off New England. The shallows of the continental shelf harbour an impoverished eurytopic fauna

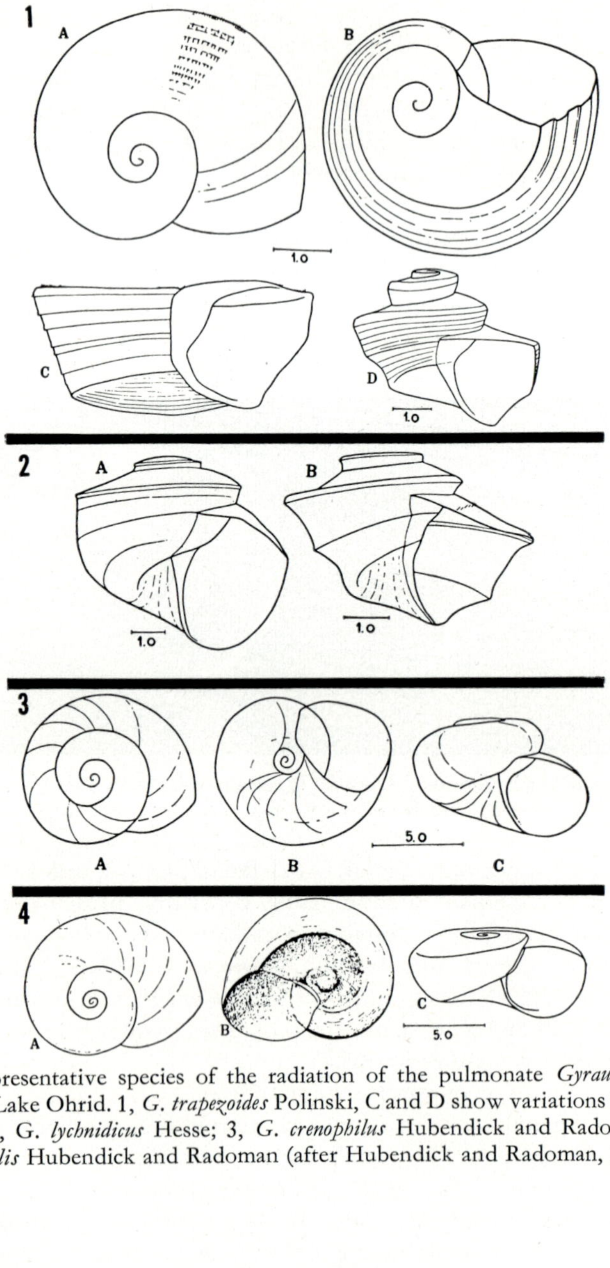

Fig. 6. Representative species of the radiation of the pulmonate *Gyraulus* (*Carinogyraulus*) in Lake Ohrid. 1, *G. trapezoides* Polinski, C and D show variations in height of the spire; 2, G. *lychnidicus* Hesse; 3, *G. crenophilus* Hubendick and Radoman; 4, *G. fontinalis* Hubendick and Radoman (after Hubendick and Radoman, 1959).

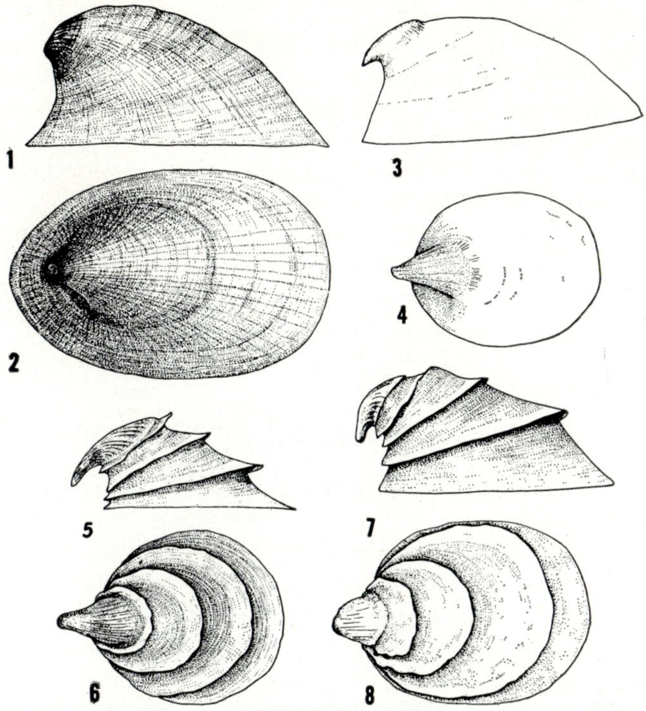

Fig. 7. Representative species of the radiation of ancylid pulmonates in Lake Ohrid. 1–2, *Ancylus fluviatilis* Müller (length to 11 mm, Palearctic); 3–4, *A. tapirulus* Polinski (length to 6 mm, endemic); 5–6, *A. lapicidus* Hubendick (length to 2·5 mm, endemic); 7–8, *A. scalariformis* Stankovič and Radoman (length to 6 mm, endemic) (after Hubendick, 1972).

while the greater depths of the continental slope support a more highly diverse group of stenotopic species. In elaborating reasons to explain this pattern, he contrasted, in abstractions, two kinds of communities, represented in a sense in nature and existing along a continuum. A physically controlled community is one in which physical conditions fluctuate widely and the organisms are thus subjected to severe physiological stress. Such communities, for example, high arctic environments, tropical deserts and the inshore faunas of ancient lakes, are characterized by low diversity or small numbers of eurytopic species. In contrast, the biologically accommodated community exists where physical conditions are relatively constant and uniform for long periods of time. Since little physical alteration occurs in this environment, the organisms evolved principally under selective forces which are biological in nature. Competition, predator prey relationships, and other modalities of biological stress more strongly mediate evolution than do physical forces. The communities, evolved in uniform physical environmental parameters through long periods of time, result in highly diverse assemblages of predominantly stenotopic species. Biotopes which most nearly approach the purely biologically accommodated communities include tropical shallow water regions, the deep sea, tropical rain forests, and the so-called lake-proper portions of the ancient lakes.

Thus it is that the stability-time hypothesis elaborated by Sanders (1968) to account for the differential diversities in on-shore and deep-sea habitats also provides an explanation for the existence of similar differences in the gastropod faunas of the ancient lakes.

VI. The differential speciation of pulmonates and prosobranchs

In all the ancient lakes the pulmonates are numerically fewer than the prosobranchs, but as highlighted previously the pulmonates show a greatly reduced tendency to speciate. The endemicity ratios illustrated in Table XIII are uniformly greater for prosobranchs. Further, among the pulmonates, only the Planorbidae show marked degrees of endemicity while the limpet-like Ancylidae and Acroloxidae exhibit a slight tendency towards endemicity. In what way can we account for such noticeable differences? How is it that pulmonates are less prone to speciation, less likely to form geographically isolated populations which become genetically and ecologically distinct, and how is it that most pulmonate species maintain the cohesion of their respective genetic backgrounds? Finally, how do those pulmonates which have speciated and are endemic in ancient lakes differ from other pulmonates?

In order to be able to develop an explanatory paradigm for the differential endemicities of the 2 principal subclasses of gastropods in ancient lakes, one must re-examine certain of the basic biological features of these 2 groups of molluscs. Several fundamental traits distinguish the prosobranchs and pulmonates, but the nature of the reproductive and respiratory systems demand special consideration. It is suggested that, among other factors, the pulmonates are simply less inclined to speciation in ancient lakes because they are reproductively, genetically and physiologically different from prosobranchs. Further, those pulmonates having formed endemics have physiological adaptations which allow them to invade ecological regimes denied other pulmonates.

A. The role of respiration

Since it is probably of secondary importance and of lesser complexity, the nature of the respiratory system will be considered first. Prosobranchs are characterized by the presence of a gill or ctenidium which lies anteriad of the heart in a mantle cavity. This structure is responsible for respiration, with the exchange of gases taking place in the aqueous environment of the mantle cavity. In contrast, the pulmonates are air-breathing snails which have developed a lung-like sac or pulmonary cavity where respiration transpires. Of course, each group of snails has made advances and alterations in its respective *Bauplan*. The most important of these, especially among the Pulmonata, include the elaboration of respiratory pigments and secondary gill-like structures.

How do respiratory mechanisms—the presence or absence of a gill or lung, the development of oxygen-carrying pigments, the evolution of secondary respiratory organs—affect the process of speciation? Such adaptations provide animals with a means to exploit greater depths in the lake, and thus, encounter new environmental horizons inaccessible to other species. Frequently, these habitats allow the snails to enter new niches, either increasing their ecological diversity or shifting their ecological requirements. Additionally, the new niches may be at greater depths, a situation which in these animals simultaneously reduces dispersal and promotes isolation.

The presence of a gill facilitates maximum penetration of habitable depths in the ancient lakes. An examination of bathymetric distributions in the lakes, especially Baikal (Kozhov, 1936) shows that the gill-bearing prosobranchs, especially endemics, penetrate to the profundal depths with species of *Benedictia* having been dredged below 500 m. Similarly, in Ohrid (Stankovič, 1960), only prosobranchs reach great depths as do, in Tanganyika, such prosobranchs as *Tiphobia*, *Bathanalia* and *Paramelania*, having bathymetric ranges beyond 100 m (Leloup, 1953). No pulmonate has been found living in depths below 100 m in any of these lakes and most species are restricted to much shallower habitats.

That respiratory parameters are instrumental in establishing limits for depth distribution in pulmonates is further substantiated by closer examination of the radiation among the pulmonates themselves. Only those which have special respiratory adaptations succeed in exploiting deeper habitats in the lakes and show a marked incidence of endemicity. In virtually every lake (Tables II–X) the planorbids outnumber other pulmonates and constitute the principal group to have undergone intralacustrine speciation. Of the pulmonates, they almost always have the greatest depth penetration in each lake. Most exemplary are the radiations of the endemic groups *Choanomphalus* (Table II; Figs 1 and 2) and *Carinogyraulus* (Table VI; Figs 5 and 6) in Baikal and Ohrid respectively. According to Kozhov (1936), of the 7 species of *Choanomphalus*, one lives to depths of 10 m, the other to 20 m, 5 species penetrate to depths of 50 m, and one occurs as deep as 100 m. In lesser depth zones, *Carinogyraulus* is variously distributed to 50 m in Ohrid. Unfortunately, there is insufficient information on the occurrence and distribution of the 5 species of *Gyraulus* in Lake Tanganyika.

An important aspect of the respiratory system of planorbids is biochemical. Unlike most other pulmonates, planorbids possess the respiratory pigment hemoglobin (Read, 1966). The physiological and biochemical features of planorbid hemoglobin are variable, and though conflicting results impute different facultative responses to oxygen tension and other parameters (Ghiretti, 1966; Ghiretti and Ghiretti-Magaldi, 1972), apparent correlation of the effective oxygen-binding capabilities of the protein are related to environmental conditions encountered by the snail (Jones, 1972). Perhaps the successful radiation of the planorbids in ancient lakes has, in part, been facilitated by the presence of a more effective hemoglobin, allowing these snails to exploit deeper habitats where conditions of reduced oxygen supply or even near anaerobic situations might obtain. Such a conjecture is open to actual testing, the results of which might well disclaim the statement of Ghiretti and Ghiretti-Magaldi (1972): "It is impossible to draw any conclusion on the evolutionary aspects of hemoglobin from the available data on the presence and distribution of the pigment in molluscs."

Additionally, other respiratory adaptations have been evolved in some pulmonates which may have been instrumental in the radiation of endemics. The acroloxid limpets, small freshwater cap-like snails of Holarctic distribution, have developed a small gill-like structure, the pseudobranch, in lieu of a true ctenidium (Franc, 1960; Hubendick, 1970). A modest radiation of acroloxids has occurred both in Baikal (Table II) and Ohrid (Table VI) with species advancing in depths to 50 m. Further, simple cutaneous respiration, a physiological adaptation of pulmonates, and one especially effective in animals of small body size, allows the exchange of gases to take place through the skin (Hyman, 1967). The elaboration of several species of another freshwater limpet, *Ancylus* (Table VI; Fig. 7) which lacks a pseudobranch, in Lake Ohrid (Hubendick, 1970) may have been facilitated by cutaneous respiration, permitting some species access to habitats in the 50 m contour. Finally, the planorbids also have a secondary respiratory faculty in the development of respiratory folds and projections within the pulmonary cavity (Hyman, 1967).

B. The role of reproduction

The nature of the reproductive system in the different subclasses of freshwater gastropods is of fundamental importance in its complicated effect on the process and rate of speciation. Prosobranchs are gonochoristic or dioecious, having the sexes in separate individuals while pulmonates maintain both sexes in the same individual, which is to say, they are hermaphroditic or monoecious. Although both groups lack free swimming larval stages, differences in their reproductive strategies, including the occurrence or absence of viviparity, dispersal ability, the nature of egg masses, and the fundamental genetic background, profoundly affect the tendency to speciate.

What sort of features, associated with reproduction, promote isolation, the initial step in the speciation process? Several behavioural and developmental aspects of the reproductive biology of these snails may contribute to the marked differences in their endemicities in the ancient lakes. Viviparity or ovoviviparity may be accountable for some reduction of geographic distribution and, concomitantly, gene flow. In several independent lineages of freshwater prosobranchs, females have become live-bearers or viviparous by holding the developing embryos *in utero* or in an incubatory pouch, e.g. hydrobiids, viviparids and melaniids (Lamy, 1928). All these families show marked degrees of endemicity in the ancient lakes. Radiations of hydrobiid-like snails are seen in the Syrnolopsidae of Tanganyika (Table V), the littoridine complex in Titicaca (Table VIII), the pyrguline taxa of Ohrid (Table VII) and the genus *Hydrobioides* in Lake Inlé (Table X). Such genera as *Kobeltocochlea* and *Benedictia* and the related endemic family Baicaliidae in Lake Baikal (Table III) also represent a diversification of hydrobiid-like forms. Among viviparids, Lake Biwa has the endemic genus *Heterogen* (Table IX and Fig. 4), Lake Tanganyika the endemic *Neothauma* (Table IV) and Lake Inlé (Table X, Fig. 4) an intralacustrine radiation of *Taia*. Similarly, melaniid stocks evolved the *Semisulcospira* species complex in Biwa (Table IX) and the spectacular thalassoid gastropods of Lake Tanganyika (Table V, Figs 3 and 4). In contrast, there are no known cases of viviparity in the Basommatophora (Hyman, 1967) and the freshwater pulmonates of the ancient lakes show little of the remarkable radiations seen in the prosobranchs.

Additionally, the nature of the egg itself may play an important facultative role. Bondeson (1950) in an exhaustive study of egg capsules in pulmonates has indicated that among basommatophorans, the physids and lymnaeids have a more primitive structure with less solid external membranes than the apparently stronger, somewhat harder encapsulations found in planorbids, ancylids and acroloxids. The latter groups are the only pulmonates to have shown significant radiation in any of the ancient lakes, notably Baikal and Ohrid (Tables II and VI). Further, in regard to egg masses, physiological tolerances or ontogenetic requirements may be instrumental. For certain Baikalian prosobranchs, Kozhov (1963) showed that the temperature regimes typical of the lake proper rather than the surrounding shallow waters or ssors are responsible for successful gametogenesis and egg deposition.

Certainly both viviparity and the tolerance or physiological requirements of eggs play key roles in the dispersal ability of these gastropods. Although little information is available, passive dispersal of immature stages and, possibly sub-adults, is effected by birds, insects or other external agents (Kew, 1893). Since some prosobranchs are viviparous brooders and since there is some indication that their developmental stages have more strict physiological requirements, species of prosobranchs may have a greater tendency to remain in their home areas, that is, to be philopatric. It is known, for example, that rock-dwelling pleurocerids stay in the same position over long periods of time (Morrison, personal communication). Surely this phenomenon is reflected in the more restricted geographic ranges of many prosobranchs. In contrast, pulmonates are usually not philopatric. Many species have exceptionally extensive geographic ranges, indicative of a successful ability to disperse over great distances. Eggs and immature stages of pulmonates have been collected on the feet of birds (Rees, 1965) and in some experimental work, specimens of physids were actually viable after having passed through an avian digestive tract (Malone, 1965).

One obvious, but frequently overlooked, element causing a drastic reduction in dispersal outside a lacustrine system is the distribution of the species within the body of water itself. In the case of ancient lakes, passive dispersal by avian or aerial agents is sharply curtailed in those species in deeper water, simultaneously contributing to their philopatry. Pulmonates tend to inhabit shallower water where the possibilities of dispersal are greater while prosobranchs and the endemic pulmonates tend to live in deeper water where passive transport is less.

Any factor that reduces dispersal may facilitate speciation since reduced dispersal may mean a diminution of gene flow between populations and the concomitant establishment of intrinsic isolating mechanisms, early steps in the incipience of new species. Features of the biology of an organism, such as a habitat restricted to deep, offshore waters, or parental care, including viviparity and brooding, tend to increase philopatry, decrease dispersal markedly, and dispose populations to greater, more effective isolation and consequent speciation.

C. Genetic strategies*

The nature of the genetic system, its expression in gonochoristic versus hermaphroditic species, requires comment, since ultimately it is most instrumental in the establishment of genetic isolating mechanisms, formation of species, and the maintenance of phyletic cohesion. Unfortunately, the literature of population genetics does not adequately contrast the effects of gonochoristic versus hermaphroditic systems (Kimura and Ohta, 1971). The hermaphroditism exhibited by freshwater basommatophoran pulmonates may be recognized as fitting the low density model as elaborated by Ghiselin (1969). The advantage of the hermaphroditic strategy allows mating with any other member of the species where the probability of encountering a suitable mate is low. However, pulmonates occasionally go a step further and employ self-fertilization, and evidently even parthenogenesis, to establish and maintain populations (Larambergue, 1939). Such isolated populations of pulmonates have long been the bane of taxonomists, because genetic inbreeding and random drift in these isolates frequently produce morphologically distinct animals which have all too often been considered separate species (Hubendick, 1951). That real speciation with the establishment of genetic isolation and ecological compatibility has been effected seldom has been documented. Nevertheless, one of the reasons behind the difficulties in the systematics of pulmonate taxa is the inherent flexibility occasioned by hermaphroditism. Most assuredly, the basic differences in the reproductive biology of pulmonates and prosobranchs have contributed to the different patterns of speciation in the ancient lakes.

How does a genetic system facilitate speciation and what evidence involving genetics, if any, suggests that pulmonates are less prone to speciate than prosobranchs? In addition to the complications introduced by the occurrence of polyploidy, the smaller chromosome numbers and

* The following suggestions concerning the proposed genetic systems and strategies of the gonochoristic prosobranchs and hermaphroditic pulmonates are speculative. I merely suggest that the term open genetic system be applied to such a group as the limnic basommatophoran pulmonates and that of closed system to the freshwater prosobranchs discussed herein.

By an open system, I infer that the genetic background of the organism provides greater heterogeneity and, ultimately, greater phenotypic variability. With a less cohesive or loose genotype, speciation would be less frequent, geographic ranges much larger and physiological tolerances broader. In contrast, the closed system would indicate a more cohesive, tighter genetic system wherein organisms are more prone to speciate. Carson (1975) has discussed the open and closed genetic systems and their evolutionary consequences, and Mayr (1975) has stressed the cohesiveness of the genotype.

the possibility of chromosomal rearrangements indicate that genetic phenomena are instrumental in the speciation of the snails of ancient lakes.

Richards (1970) has summarized much of what is known about the genetics of the medically important planorbid *Biomphalaria*. Although the ecological genetics of polymorphism in terrestrial gastropods, especially *Cepaea*, have received considerable attention (Ford, 1971), the specifics of inheritance in freshwater gastropods are very poorly known. However, considerable inferential data are available as well as some empirical information in the literature of cytogenetics.

As a tentative hypothesis, it is suggested that an open genetic system or background characterizes the Pulmonata while a more restrictive, closed genetic system is a feature of the Prosobranchia. Intuitive as this assumption is, certain evidence lends credence to its support. Since our knowledge of molluscan genetics is limited and rudimentary, attention must be turned towards one of the sources of empirical data which may reflect something about the operative genetic systems in freshwater gastropods. The karyotypes of various taxa have been delineated in an almost purely descriptive literature. This body of information provides extensive lists of chromosome numbers and documents the occurrence of certain cytological phenomena, principally incidences of polyploidy.

Table XII cites the known range of haploid numbers for several families of freshwater gastropods and has been compiled from several review papers. Many of these data are fragmentary and extremely limited for certain families. For example, only single observations were made on the Valvatidae and Acroloxidae. Further, it must be stressed that with the exception of information on *Semisulcospira* in Lake Biwa, none of the data has been amassed on the species living in ancient lakes. Although this information is far from satisfactory, the genetic systems of these snails exhibit complications, particularly in regard to polyploidy, and show that haploid numbers among prosobranchs may be smaller than those of pulmonates. A reduction in chromosome number, as noted subsequently, may mark an increase in the formation of supergenes, a restricting of genetic flexibility or heterozygosity, and, concomitantly, a narrowing of the niche of a species. Such phenomena are indicative of increased isolation along the path to speciation.

Among the Pulmonata, the families Physidae ($n = 18$) and Lymnaeidae ($n = 17$–19) are extremely conservative, showing little variation in haploid number, no tendency to be polyploids, and little or no radiation or endemicity in any of the ancient lakes. Since only a single observation was cited by Burch (1965) for the Acroloxidae, attention must be cast towards the Ancylidae and the Planorbidae both of which have smaller basic haploid numbers than the other pulmonates, show considerable occurrence of polyploidy and have exhibited the most noteworthy radiation among the pulmonates of the ancient lakes, particularly in Baikal and Ohrid (Tables II and VI).

On the whole it appears, except for cases of polyploidy, that the Prosobranchia have a lower mean haploid number. At least a tendency for smaller haploid numbers is apparent in the Pleuroceridae, Valvatidae (only a single observation) and the Viviparidae (Patterson, 1967). The trend is discernible even among the Hydrobiidae and the Thiaridae (or Melaniidae) when contrasted with the pulmonate Physidae and Lymnaeidae, although it is admittedly slight. The prosobranch groups which have speciated most appreciably in the ancient lakes, namely the Hydrobiidae and Thiaridae (or Melaniidae), show, as did their counterparts among the pulmonates, a markedly increased frequency of polyploidy.

The importance of polyploidy in these gastropods can not truly be assessed since it is not known whether auto- or allopolyploidy is involved. However, there is little doubt that its results have posed considerable problems for the taxonomist since the families in which the phenomenon seems to be widespread are notoriously difficult systematically (Burch and Huber, 1966). Geographical and ecological correlations with polyploidy have recently been documented for species of the planorbid *Bulinus* in Africa (Brown and Wright, 1972).

The importance of chromosomal arrangement in the process of speciation has received considerable attention in the exemplary studies of White and his students in Australia. Reduction in chromosome numbers has been correlated with geographical isolation and the tendency towards speciation. The complicated chromosomal system in stick insects or phasmatids in New South Wales is both useful and instructive. Craddock (1970, 1972) elegantly showed that individual populations exhibited their own distinct chromosomal patterns, including complicated changes in sex chromosomes and reduction in haploid number of autosomes. Such alterations might allow for the establishment of genetic isolating mechanisms and thus contribute to the formation of incipient species. Since total DNA content of different populations with these different chromosomal numbers and arrangements remained essentially constant, fusion phenomena of a Robertsonian type were singled out to account for increased numbers of metacentric and, concomitantly, decreased numbers of acrocentric autosomes.

Although the evidence is slight, similar chromosomal phenomena may be implicated in the radiation encountered among the snails in ancient lakes. Burch's study (1968) of the cytotaxonomy of Japanese *Semisulcospira*, including species in the endemic complex of Lake Biwa, revealed strikingly different karyotypes for certain species and showed that imbalance between acrocentric and metacentric chromosomes existed. That something complicated, involving chromosomal alterations, is taking place is indicated by the instability of chromosome numbers and the scattered occurrences of uni-, bi-, tri- and quadrivalents. At least 3 of the endemic species of *Semisulcospira* (e.g. *decipiens* (Westerlund), *niponica* (Smith), and *reticulata* Kajiyama and Habe) show variation with different populations of each having haploids of 12 or 13. Further, although there is not a clearcut pattern of proportionately increased metacentrics with a decrease in acrocentric chromosomes, there is 1 marked phenomenon, all the endemic species of *Semisulcospira* in Lake Biwa show reduced haploid numbers (e.g. *multigranosa* (Boettger), 14; *nakasekoae* Kuroda, 13; *decipiens* (Westerlund), 12 or 13; *niponica* (Smith), 12 or 13; *reticulata* Kajiyama and Habe, 12 or 13; and *habei* Davis, 7) in contrast to the more widely distributed species (e.g. *reiniana* (Brot), 20; *kurodai* Kajiyama and Habe, 18; and *libertina* (Gould), 18).

One of the meanings behind reduced chromosome numbers, particularly when it can be correlated with concomitant reduction in acrocentric chromosomes and increase in metacentric chromosomes is the linkage of genes on to individual chromosomes to form complex supergenes or interacting hereditary units in the genetic system. These in turn may be related to delicate phenotypic adaptations to the environment and further may indicate a narrowing of specific genetic expression, leading even to a reduction in heterogeneity. Indeed, such is the case with the most intensely studied prosobranch in any of the ancient lakes, the species of *Semisulcospira* in Lake Biwa, where it has been shown, as reflected by certain serological features, that macromolecular heterogeneity is comparatively low (Davis, 1972; Davis and Susuzi, 1972), particularly when contrasted with a eurytopic, more widely distributed prosobranch such as the hydrobiid *Oncomelania hupensis* (Davis, 1968). Thus the tendency towards lower haploid numbers in prosobranchs may be significant as an indication that speciation may be more easily effected among such taxa.

Additional evidence supporting the generalization that pulmonates possess an open genetic system while prosobranchs exhibit a more restrictive, closed system comes from ecological theory and population genetics. The possession of an open genetic system implies that the organisms have greater heterogeneity, a wider expression of phenotypes, are more variable phenotypically, more adjustable physiologically, more ecologically tolerant, and less prone to genetic isolation and speciation. Conversely, a closed genetic system implies tighter genetic control, narrower ranges of phenotypic expression, less variability, less physiological adjustability and less ecological tolerance. Organisms with such a genetic background tend to become genetically isolated more easily and speciate more readily.

Further evidence bears inferentially to support the suggested notion that the prosobranch faunas of the ancient lakes have a greater potential for speciation than do the pulmonates. As has been discussed, the latter tend to be numbered among the inshore, eurytopic moiety of the gastropods, while the former have shown the greater radiation with most of the species extending to greater depths in the lakes and exhibiting a characteristic stenotopy. The slight data from karyological studies summarized above lend credence to the idea that prosobranchs tend to have lower chromosome numbers and, thus to elaborate a tighter, more closed genetic system with less heterogeneity than the eurytopic pulmonates.

Such conjectures have received empirical documentation in marine animals. Grassle (1967, 1972) provided a valuable genetic interpretation in regard to complex relationships obtaining between diversity and environmental stability. He indicated that species, where physical parameters are the most important selective agents, maintain higher degrees of genetic variability, the better to cope with the unpredictability of the environment. Such species, usually pulmonates as we have seen, tend to be widely distributed and eurytopic, indicative of their genetic flexibility, comparatively open genetic system, and reduced tendency towards speciation and endemicity. Conversely, stenotopic species, most frequently prosobranchs and endemic pulmonates, in a biologically accommodated community, showing restriction of zoogeographic range and narrowing of physiological tolerances, exhibit reduced genetic variability, with concomitant reduction in heterogeneity and a greater inclination towards speciation. Only recently, Levinton (1973), measuring variability with selected isozyme systems, has shown that among several species of marine bivalves, effective number and absolute number of alleles decreased with increased depth. Although such information is fragmentary, it is suggestive of the regulation of genetic variability by environmental variability and of the increase of heterozygosity in more physically fluctuating regimes. However, numerous other studies have not been able to show a correlation between genetic heterogeneity, as measured by isozyme variation, and the purported stability or instability of the habitat (see Murphy *et al.*, 1976).

In general, the pulmonate gastropods have not shown the spectacular radiation of the prosobranchs. Those groups of pulmonates, particularly the Planorbidae, Ancylidae and Acroloxidae, which do show some intralacustrine speciation, have anatomical, physiological and genetic characteristics which tend to promote the speciation process. The Pulmonata as a group usually have broader physiological and ecological tolerances than do the Prosobranchia. Occupying environments which might be characterized as physically unstable, the pulmonates have more extensive geographical ranges, greater genetic flexibility and what might be called an open genetic system. Prosobranchs, in contrast, tend to be more stenotopic with less genetic plasticity and a more restrictive or closed genetic system, factors which may have greatly facilitated their intralacustrine radiation in the ancient lakes.

VII. Summary

As a result of complex evolutionary processes, which have proceeded during lengths of time variously estimated to be as much as 30 million years, a highly diverse gastropod assemblage has formed in 6 ancient lakes (Baikal, Tanganyika, Ohrid, Titicaca, Biwa and Inlé). Similar patterns of radiation are shown in each lake with each malacofauna consisting of considerable endemic elements. Pulmonates exhibit lesser radiations and are less prone to speciation. In all the lakes they are outnumbered by prosobranchs which have greater diversities and more endemics. Additionally, the fauna is divisible into 2 ecological moieties: an inshore, shallow water eurytopic component, principally of pulmonates, and a lake-proper, deeper dwelling element, predominantly composed of prosobranchs and endemic pulmonates. Possible factors subtending the tendency towards slower speciation in pulmonates are suggested. It is shown how lower diversity may be correlated with anatomical, genetic, geographic, and ecological aspects of the biology of these snails.

VIII. Acknowledgements

I am indebted to the following for assistance during the course of this study: J. Peake, W. Baranowski, A. H. Clarke, G. M. Davis, M. Dent, E. Fischer-Piette, T. Habe, W. Heard, B. Hubendick, R. I. Johnson, V. Jovanovič, E. Mayr, P. Radoman, A. Riedel, O. A. Skarlato, K. Tsunoda, R. D. Turner and H. Waite.

The Milton and Foster Funds of Harvard University provided financial support for important translations.

IX. References

Annandale, N. (1918). *Rec. Ind. Mus.* **14,** 1–214.

Beauchamp, R. S. A. (1939). *Rev. Inst. hydrobiol.* **39,** 316–353.

Beauchamp, R. S. A. (1940). *Nature, Lond.* **146,** 253.

Bondesen, P. (1950). "A Comparative Morphological–Biological Analysis of the Egg Capsules of Freshwater Pulmonate Gastropods". Naturhistorisk Museum, Aarhus.

Boss, K. J. (1973). *Bull. Am. malac. Union*, **1972,** 11–13.

Brooks, J. L. (1950). *Q. Rev. biol.* **25,** 30–60; 131–176.

Brown, D. S. and Wright, C. A. (1972). *J. Zool., Lond.* **167,** 97–132.

Burch, J. B. (1965). *Proc. 1st Eur. malac. Congr.* **1962,** 215–241.
Burch, J. B. (1968). *J. Conch., Paris* **107,** 1–51.
Burch, J. B. and Huber, J. M. (1966). *Malacologia* **5,** 41–43.
Burch, J. B. and Natarajan, R. (1966). *Cytologia* **31,** 109–116.
Capart, A. (1949). *In* "Explor. hydrobiol. du Lac Tanganika" Vol. 2, fasc. 2, pp. 1–16. Inst. Roy. Sci. nat. Belgique.
Carson, H. L. (1975). *Am. Nat.* **109,** 83–92.
Craddock, E. (1970). *Science, N.Y.* **167,** 1380–1382.
Craddock, E. (1972). *Aust. J. Zool.* **20,** 445–462.
Davis, G. M. (1968). *Malacologia* **7,** 17–70.
Davis, G. M. (1969). *Malacologia* **7,** 211–294.
Davis, G. M. (1972). *Comp. Biochem. Physiol.* **41A,** 339–347.
Davis, G. M. and S. Suzuki. (1971). *Veliger* **13,** 207–225.
Dybowski, W. (1900). *NachrBl. dt. malac. Ges.* **32,** 143–152.
Ford, E. B. (1971). "Ecological Genetics". Chapman and Hall, London.
Franc, A. (1960). *In* "Traité de Zoologie V, Classe des Bivalves" (P.-P. Grassé, Ed.). Masson et Cie, Paris.
Fryer, G. and Iles, T. D. (1972). "The Cichlid Fishes of the Great Lakes of Africa. Their Biology and Evolution". Oliver and Boyd, Edinburgh.
Fuchs, V. E. (1936). *J. Linn. Soc.* **40,** 93–106.
Gerstfeldt, G. (1859). *Mem. Acad. imp. Sci. St. Petersburg* **9,** 507–548.
Ghiretti, F. (1966). *In* "Physiology of Mollusca" (K. M. Wilbur and C. M. Yonge, Eds) Vol. 2, pp. 175–208. Academic Press, New York and London.
Ghiretti, F. and Ghiretti-Magaldi, A. (1972). *In* "Chemical Zoology" (M. Florkin and B. T. Scheer, Eds) Vol. 7, pp. 201–217. Academic Press, New York and London.
Ghiselin, M. T. (1969). *Q. Rev. Biol.* **44,** 189–208.
Gilson, H. C. (1939). *Trans. Linn. Soc. Lond.* (3) **1,** 1–20.
Grassle, J. F. (1967). Influence of environmental variation on species diversity in benthic communities on the continental shelf and slope. Ph. D. Dissertation, Duke University, Durham, N.C.
Grassle, J. F. (1972). *In* "5th Eur. Mar. Biol. Symp" (B. Battaglia, Ed.) pp. 19–26. Piccin Editore, Padua.
Haas, F. (1955). *Trans. Linn. Soc. Lond.* (3), **1,** 275–308.
Halbfass, W. (1928). *Geogr. Zeits.* **34,** 516–530.
Horie, S. (1961). *Mem. Coll. Sci. Kyoto Univ.* (B) **28,** 52–71.
Horie, S. (1962). *Mem. Coll. Sci. Kyoto Univ.* (B) **29,** 191–262.
Hubendick, B. (1951). *K. Svenska VetenskAkad. Handl.* (**4**) **3,** 1–223.
Hubendick, B. (1970). *Acta R. Soc. Sci. litt. Goteburg. Zool.* **5,** 1–52.
Hubendick, B. (1972). *Inf. Soc. belge de Malacol.* (1) **8–9,** 109–126.
Hubendick, B. and Radoman, P. (1959). *Ark. Zool.* (2) **12,** 223–243.
Hyman, L. H. (1967). "The Invertebrates Vol. 6 Mollusca I". McGraw-Hill, New York.
Jones, J. D. (1972). "Comparative Physiology of Respiration". Arnold, London.
Kew, H. W. (1893). "The Dispersal of Shells". Kegan Paul, London.
Kimura, M. and Ohta, T. (1971). "Theoretical Aspects of Population Genetics". Monographs in Population Biology No. 4, Princeton University Press.

Kozhov, M. (1936). *Trudў baĭkal. limnol. Sta.* **8,** 1–352.
Kozhov, M. (1963). "Lake Baikal and its life" Vol. XI, Monographiae Biologicae, Junk, Hague.
Kruimel, J. H. (1917–1918). *In* "Geologische en Geographische Doorkruisingen van Midden-Celebes (1909–1910)" (E. C. Abendanon, Ed.) Vol. 4, pp. 1626–1650. Brill, Leiden.
Lack, D. (1961). "Darwin's Finches". Harper and Brothers, New York.
Lamy, E. (1928). *J. Conch., Paris* **72,** 163–197.
Larambergue, M. (1939). *Bull. Biol. mar. biol. Lab. Woods Hole* **73,** 19–231.
Leloup, E. (1953). *Inst. R. Sci. nat. Belgique* **3,** 1–273.
Levinton, J. (1973). *Science, N.Y.* **180,** 75–76.
Livingstone, D. A. (1965). *Limnol. Oceanogr.* **10,** 607–610.
Malone, C. R. (1965). *Nautilus, Philad.* **78,** 135–139.
Martinson, G. G. (1948). *Trudў baĭkal. limnol. Sta.* **12,** 82–106.
Martinson, G. G. (1951). *Trudў baĭkal. limnol. Sta.* **13,** 5–92.
Mayr, E. (1970). "Populations, Species, and Evolution". Belknap Press of Harvard University Press, Cambridge.
Mayr, E. (1975). *Biol. Zbl.* **94,** 377–388.
Moon, H. P. (1939). *Trans. Linn. Soc. Lond.* (3) **1,** 27–43.
Moore, J. E. S. (1903). "The Tanganyika Problem, an Account of the Researches Undertaken Concerning the existence of Marine Animals in Central Africa". Hurst and Blackett, London.
Mori, S. (1970). *Mem. Fac. Sci. Kyoto Univ.* (B) **3,** 22–46.
Mori, S., Yamamoto, K., Negoro, K., Horie, S. and Suzuki, N. (1967). *Mem. Fac. Sci. Kyoto Univ.* (B) **1,** 36–40.
Murphy, L. S., Rowe G. T. and Haedrich, R. L. (1976). *Deep-Sea Res.* **23,** 339–348.
Myers, G. S. (1960). *Evolution, Lancaster, Pa.* **14,** 323–333.
Patterson, C. M. (1967). *Malacologia* **5,** 111–125.
Pavlovsky, E. V. (1941). *Priroda* **3,** 19–31.
Pilsbry, H. and Bequaert, J. (1927). *Bull. Am. Mus. nat. Hist.* **53,** 69–602.
Radoman, P. (1964). *Arh. Biol. Nauka* **15,** (3–4), 101–119.
Radoman, P. and Marinkovič, D. (1959). *Arh. Biol. Nauka* **11,** 73–75.
Read, K. R. H. (1966). *In* "Physiology of Mollusca" (K. M. Wilbur and C. M. Yonge, Eds) Vol. 2, pp. 209–232. Academic Press, New York and London.
Rees, W. J. (1965). *Proc. malac. Soc. Lond.* **36,** 269–282.
Richards, C. S. (1970). *Nature, Lond.* **227,** 806–810.
Sanders, H. L. (1968). *Am. Nat.* **102,** 243–282.
Stankovič, S. (1960). "The Balkan Lake Ohrid and its Living World". *In* Monographiae Biologicae, **9.** Junk, Hague.
Starobogatov, Ya. I. (1970). "The Molluscan Fauna and the Zoogeographical Zonation of the Continental Water Reservoirs of the Earth". Nauka, Leningrad.
Taylor, D. W. (1966). *Veliger* **9,** 152–228.
Woodward, S. P. (1859). *Proc. zool. Soc. Lond.* **1859,** 348–350.
Yonge, C. M. (1938). *Nature, Lond.* **142,** 464.
Zhadin, V. I. (1952). "Mollusks of Fresh and Brackish Waters of the USSR". Keys to the Fauna of the USSR, Zool. Inst. Acad. Sci., **46,** Moscow.

Chapter 10

Distribution and ecology of the Stylommatophora

J. PEAKE

Department of Zoology, British Museum (Natural History), London, UK

I. Introduction

The initial attempts at classifying distribution patterns of terrestrial molluscs were made during the nineteenth century and were developed extensively from proposals formulated for other taxonomic groups. They were concerned almost exclusively with recent spatial relationships and tended to ignore any temporal contribution. For example, one of the first schemes, which was proposed by Woodward (1856), recognized 27 regions corresponding to the botanical divisions suggested earlier by Schouw (1823); in later editions of the same work these regions were homologized with the 6 divisions recognized by Sclater (1858) in his study of avian biogeography. Such a unified approach would have appealed to Wallace (1876) who exhorted "we therefore venture to hope that conchologists will give us the advantage of their more full and accurate knowledge both of the classification and distribution of this interesting group of animals, not to map out new sets of regions for themselves, but to show what kind of barriers have been most efficient in limiting the range of species, and how their distribution is actually effected . . .". Laudable objectives, but linked to an over optimistic view of the state of taxonomic knowledge concerning the group. Nevertheless, the questions posed are those that are still exercizing the minds of both biogeographers and conchologists or malacologists today.

In his 2 monumental works on zoogeography, Wallace (1876, 1881) extolled the virtues of an all enveloping approach; although the patterns he described were derived extensively from a study of vertebrate taxa, other groups, including molluscs, were considered. The scheme which he proposed, with 6 regions and 24 subregions, has been utilized in nearly all subsequent analyses that have attempted a comprehensive approach, even though many workers have proposed their own modifications, e.g. Fischer (1887), Keferstein (1866), Cooke (1895) and more recently Solem (1959) and Te (1972, 1976).

As a corollary to these comprehensive studies a considerable number of local faunistic investigations were, and still are being undertaken. These have increased the resolution of information concerning distribution patterns besides stimulating ecological research. They include the works of Kobelt (1897, 1898) for Europe; Økland (1930) and Waldén (1955) for areas of Scandinavia; Ant (1963, 1969) for parts of Germany, and Ellis (1951) and Kerney (1976) for the British Isles. This close link between biogeography and ecology has been epitomized in the volumes by Hesse *et al.* (1951) and more recently by MacArthur (1972). The development of the revolutionary concepts of continental drift during the past decade have had a profound effect on the development and evolution of concepts regarding biogeography as seen, for example, in the analyses by Raven and Axelrod (1974) for phanerogams and Cracraft (1973) for birds. Yet the concept of a united Gondwanaland had been evoked by both Germain (1925) and Boettger (1936) when considering the recent distribution of the stylommatophoran family Acavidae. They recognized the paradox imposed by organisms with a low vagility yet possessing wide and frequently disjunct distributions extending over more than a single continent. Nevertheless, the majority of workers have based their accounts on the concept of continental permanence (e.g. Solem, 1959), if only for reasons of parsimony, or an ability to evoke extensive land bridges or sunken continents (e.g. Hedley, 1899, Pilsbry, 1900). Today, however, with equal panache many disjunct distributions are attributed to the influence of continental drift! Earlier there had been a tendency to consider oceans as a final obstacle to any interchange between terrestrial faunas, and consequently the probability for long distance dispersal negligible, but if long distance dispersal is feasible, even though its probability is low, then the problem of interpreting recent patterns in terms of reconstructions of land masses cannot be entirely deductive. Moreover, where a number of groups exhibit similar, but strongly disjunct distributions it might be feasible to utilize such reconstructions to make tentative explanations as to the origin of these patterns.

It is important, however, when using information concerning continental drift, to distinguish different levels of "hardness of evidence". For example, the dated arrangements of the continents since the Late Cretaceous based on sea-floor and other corroborative information can be considered to constitute hard evidence, while earlier reconstructions based on palaeomagnetic data alone are often far less reliable (Jardine, 1973). Biological information presents similar problems in terms of the classification of the group and knowledge of the fossil record; certainly data for the Stylommatophora are not of a uniform quality. Indeed, the importance of a phylogenetic classification, as opposed to a phenetic or synthetic, in any biogeographical analysis has been extolled by, for example, Nelson (1974) and Cracraft (1973). Unfortunately considerable taxonomic confusion reigns in the Stylommatophora and the only conclusion that can be reached regarding the classification of the group is that the results readily adapt themselves to a discussion of evolution, even though it is largely an act of faith that the units are considered monophyletic. Intuitively there is a desire to equate taxonomic distinctiveness with the length of historical separation of the groups—a pursuit that can only be indulged with enormous care.

The rather traditional approach to biogeography emphasizes the descriptive aspects of faunas and analyses in terms of the physical features of the environment. Indeed, the type and form of information for terrestrial molluscs has ensured that it is this avenue which has been almost exclusively explored. Alternative approaches consider the biological or ecological properties of biotas and use these to characterize or identify features of populations or communities occupying different areas. That is, elucidating the different adaptive strategies, an approach which has been exploited so successfully, for example, by Hutchinson (1965) and his school. In this area the studies available for land molluscs are very limited and rather scattered.

II. History

Although the fossil record for pulmonates is impoverished compared to that available for marine taxa, it is still possible to extract some information regarding distributional patterns and the persistence of taxa through time. Fortunately many of the available fossils can be equated with modern taxa in spite of the problems inherent in using conchological characters alone.

The earliest faunas rich in Stylommatophora are recorded from Cretaceous and Palaeocene deposits in both the Americas and Europe (Wenz, 1923–1930; Zilch, 1959–1960; Parodiz, 1969). These provide evidence for the existence of highly differentiated taxa in which the major evolutionary trends were already clearly established. Indeed all 5 (or 4) extant orders and suborders have been recorded, and even though new families continue to be added throughout the Caenozoic, the only other period when additions of such magnitude are made is in the recent. Nevertheless, this distribution could be fortuitous reflecting the interests of collectors and bias in the fossil record and in many aspects such a pattern can be equated with the early Cretaceous evolution of the angiosperms (Hughes, 1973; Raven and Axelrod, 1974) and the extensive radiation of insects associated with these plants (Ehrlich and Raven, 1964). Whether there was a contemporaneous reorganization of the evolutionary patterns of terrestrial molluscs is difficult to determine on the basis of present information, even though such a hypothesis is inviting (see pp. 474–476). There are earlier records of fossil land snails from the Palaeozoic, particularly the Carboniferous and Permian of North America, but the taxonomic affinities of these have frequently been disputed. They have been interpreted as either prosobranchs or pulmonates, and, if the latter, as Basommatophora or Stylommatophora (Zilch, 1959–1960; Cox, 1960). Recent research, however, has demonstrated that some of this material can be referred to the family Tornatellinidae and can only be separated at the generic level from extant taxa (in Solem, 1976). It is tempting to speculate on the form of the habitat or the availability of food sources for these early land molluscs; certainly modern vegetation dominated by pteridophytes or gymnosperms is not noted for rich associations of terrestrial molluscs, although on some Pacific Islands small species of pulmonates are frequently found associated with ferns, and some of these are members of the Tornatellinidae. In these instances, the small size of the snails can be correlated with a propensity for successful dispersal and an ability to utilize small spaces as refuges, while the fine structure of the litter or the fern fronds on which the animal has to crawl could favour small size particularly of the foot sole.

Contemporaneous with the development of a highly differentiated stylommatophoran fauna was the evolution of other non-marine forms. Fresh or brackish-water taxa are comparatively numerous in the fossil record, for example, deposits from the Carboniferous and Permian periods are notable for the abundance of bivalve taxa, whose affinities with other groups is unknown, for the true Unionidae did not appear until the Triassic or Jurassic (Cox *et al.*, 1969). While a diverse fauna of freshwater gastropods is recorded from the Jurassic onwards, with both prosobranch and basommatophoran families represented, it is surprising that earlier records for these taxa are either rare or absent considering the habitats which they occupy and the potential for successful preservation. Amongst groups of terrestrial prosobranchs the richest record is for the Tertiary, but there are a number of taxa recorded from the Palaeozoic. There is a dubious record from the Devonian and well-preserved material from the Carboniferous and Permian, but in many instances these have been attributed to extinct families or subfamilies (Wenz, 1923–1930).

The origin and the radiation of non-marine molluscs appears to have occurred during those periods when the fossil record is deficient, that is in the Palaeozoic and Mesozoic. Nevertheless, the recognition by Yochelson and Solem (in Solem, 1976) of extant taxa amongst fossil material from the Palaeozoic demonstrates the enormous potential for future studies.

III. Palaeobiogeography

A. Introduction

The principles of historical biogeography were firmly established by Simpson (e.g. 1940) in many publications analysing the distribution of mammals, and the concepts of sweepstakes, filters and corridors are now a recognized part of the vocabulary for the subject. To this suite must be added those rather specialized terms that reflect the peculiar nature of dispersal on drifting continents, the "Noah's Arks" and the "Viking Funeral Ships" (McKenna, 1973). Yet such notions would appear to have made comparatively little impact on the literature dealing with the distribution of non-marine molluscs: it is problematical whether this reflects the limitations in knowledge of either the distributions or the nature of the factors which have influenced the establishment of these patterns. Speculations on the development and origin of continental faunas of Stylommatophora have frequently been based on presumptions regarding evolutionary rates in different taxonomic groups, followed by estimates of the ages of the faunas; yet both can be regarded as highly tenuous activities.

Evidence accumulated from the fossil record in both the Americas and Europe would indicate that the critical period for the establishment of modern distribution patterns of terrestrial molluscs occurred during the Mesozoic in the Americas and later, during the Tertiary,* in Europe. For

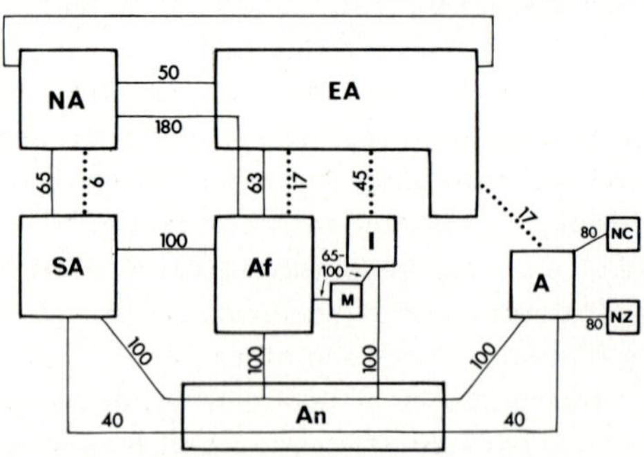

Fig. 1. Isolation of continents: time in million years before present when migration across dry land could have last occurred between continental areas (solid lines), or time when such migration routes were established (dotted lines) (data from Raven and Axelrod, 1974; McKenna, 1973). NA = North America, SA = South America, EA = Eurasia, Af = Africa, I = India, M = Madagascar, A = Australia, NC = New Caledonia, NZ = New Zealand, An = Antarctica.

* N.B. Caenozoic = Tertiary plus Quaternary; Quaternary = Pleistocene plus Holocene (Recent).

at the onset of the Tertiary, and indeed in the Late Cretaceous, the stylommatophoran faunas of these 3 continents were already distinct, with the taxa which characterize the modern faunas of the Americas clearly recognizable *in situ* (Fig. 1). Prior to this time there is a gap in the record and further speculation must be based on the evidence available for the distribution of other taxonomic groups, for example, plants. During the Mesozoic the climates over the whole globe were warm, and although temperature gradients existed, these would probably not have been so marked as during the Tertiary; at this time, a broad equatorial "floristic" belt can be recognized separating 2 polar regions in Siberia and southern Gondwanaland (Barnard, 1973). Within these 3 zones the floristic elements were presumably widely dispersed, although in the equatorial belt it is possible to recognize the establishment of provinces at the species level. For example, in the Early Cretaceous 3 floras can be delimited in this belt, a North American, a European-Central Asian and an East Asian or Western Pacific. Tertiary and modern floras, and indeed faunas, provide a marked contrast to this pattern with the development of a high degree of provinciality that can be associated with increasing latitudinal thermal gradients and the fragmentation of the large continental masses (Valentine, 1973). Information concerning contemporaneous molluscan faunas is lacking, but the limited data from the Palaeozoic would suggest that many of the early terrestrial taxa were small and possibly had a wide distribution. Extrapolating, it would not be unreasonable to suggest that the origin of provinciality, as found in modern faunas of Stylommatophora, must have been established during the Mesozoic and could have coincided with the origin of the angiosperms in the early Cretaceous. Moreover, the rise of this group of plants must have presented a range of new ecological, and consequently evolutionary, opportunities.

The evolution of the molluscan faunas in North America and Europe were markedly different. The facies of the present American fauna were established at the onset of the Caenozoic, while those for Europe were evolving throughout the Tertiary from one dominated by taxa that now have a relict tropical distribution, largely in South-East Asia, to one characterized by temperate or xerophilic elements. While these changes reflected the gradual and probably worldwide deterioration of the climate throughout the Tertiary, why should the effects have been so different in the 2 areas? Perhaps an answer can be found in the geographical positions and the orography of the 2 regions.

Europe has always remained an integral part of a larger Eurasian land mass, even though it has been subdivided into islands by epi-continental oceans. Links with North America must have been diminishing throughout the Caenozoic and though overland pathways could have existed initially across the present North Atlantic, for, at least, warm temperate species these would have been severed about 50 million years B.P. Nevertheless, stepping-stones could have played a role in the continuing interchange of biota between the 2 regions well into the Tertiary. Since the Eocene, however, the major route for many organisms was probably via the Bering Straits even though there is little evidence of its extensive use by terrestrial molluscs; this would appear to reflect the nature of the climatic conditions prevailing along it. Yet the presence of a Holarctic element in both Europe and North America testifies to the interchange of faunas even though the number of species involved is comparatively low. The fossil record and the degree of taxonomic differentiation, would suggest a long and continuous history for this element in both continents. Many of the taxa represented are small species that have a high propensity for passive dispersal. An exception is provided by *Cepaea hortensis* (Müll.), a comparatively large species found on both continents, but whose occurrence in America was the subject of considerable debate: was it ndigenous or introduced by man? (Lindroth, 1957). Indeed, Wegner 1929) regarded its distribution as strong evidence for his theory of continental drift, but it is now widely accepted that man was involved in its post-Columbian introduction into North America (Lindroth, 1957; Waldén, 1963).

The isolation of North America from other continents, except for the northern links with Eurasia, was maintained throughout the Tertiary, until direct connections with South America were established about 6 million years B.P. It has been estimated that during the Cretaceous, the distance separating the continental margins of the 2 American continents was about 2500 km, but this gradually decreased and stepping stones became increasingly important for faunal interchange (Raven and Axelrod, 1974); even though some reconstructions for continental arrangements would not indicate such wide separation, all agree on the isolation of the two land masses.

The climatic variations and possible overall deterioration that occurred during the Caenozoic must have affected both northern land masses and initiated major changes or extinctions in the biotas. In North America it would appear that the molluscan fauna was subsequently re-established from elements which already existed on the continent, and it is these that gave rise to the considerable autochthonous radiation seen in some families. It is remarkable that in this continent there would appear to have been a remarkably low level of extinction at the generic level throughout the Tertiary; the fauna must have slowly evolved and adapted to new climatic regimes in comparative isolation. The record for Europe provides an enormous contrast, massive extinctions occurred during the same period affecting a wide range of taxonomic groups. Subsequently the region was colonized by a fauna dominated by taxa that were associated with cooler or temperate environments. There has been considerable speculation regarding the origin of this biota; Likharev and Rammelmeier (1952) have emphasized the potential of the faunas of northern and eastern regions of Eurasia as sources for colonizing species during the Miocene. Such a hypothesis would agree with the botanical data, for example, Cheney (1940) demonstrated a north-west to south-east orientation of the vegetation zones in Eurasia during the Eocene, with a tropical flora extending north along the western margin of the continent, and a temperate and cool-temperate flora existing in northern and eastern Asia—the Arcto-Tertiary flora of some authors (e.g. Němefc, 1970). Montane environments in this and other regions could also have acted as sources of cold-adapted forms (Lindroth, 1957). Further if such centres existed in north-eastern Eurasia they could also have served as sources for propagules which colonized North America. Yet the position of these areas in many recent continental reconstructions would suggest that, at least, during part of the Mesozoic they would have endured continuous darkness during the winter—possibly not a problem for a snail!

The interpretation of the spatial and temporal patterns would indicate that in the northern land masses the major factors in determining modern distribution patterns have been the isolation of the land masses, the proximity of sources for colonizing propagules and the propensity of the taxa in those areas for dispersal, together with ability of taxa to evolve and survive in new temperature regimes. Indeed, the importance of barriers to dispersal have been evoked to explain differences in the distribution patterns of post-Pleistocene floras in northern North America and Europe. The extension of the floras northwards, after the compression southwards during the Pleistocene glaciations, has been blocked in Europe by the massive Alpine mountain belts arranged along an east–west axis. In America the belts are along a south–north axis and consequently dispersal has not been impeded in these directions. A similar arrangement is found in the terrestrial molluscs in the 2 regions and it is inviting to accept a similar explanation, but a wide variety of other environmental parameters, for example climatic, are associated with such orographic features and these would tend to reinforce the establishment of the observed distribution patterns.

At present, the limited nature of the fossil record for Stylommatophora and prosobranch groups places a major restraint on the recognition of the broad patterns of temporal and spatial distribution other than for North America and Europe. The only other continent for which adequate information is available is South America, and here the record is again confined to the Late Cretaceous and Caenozoic. The data for other areas is either meagre or absent except for records from the Pliocene or

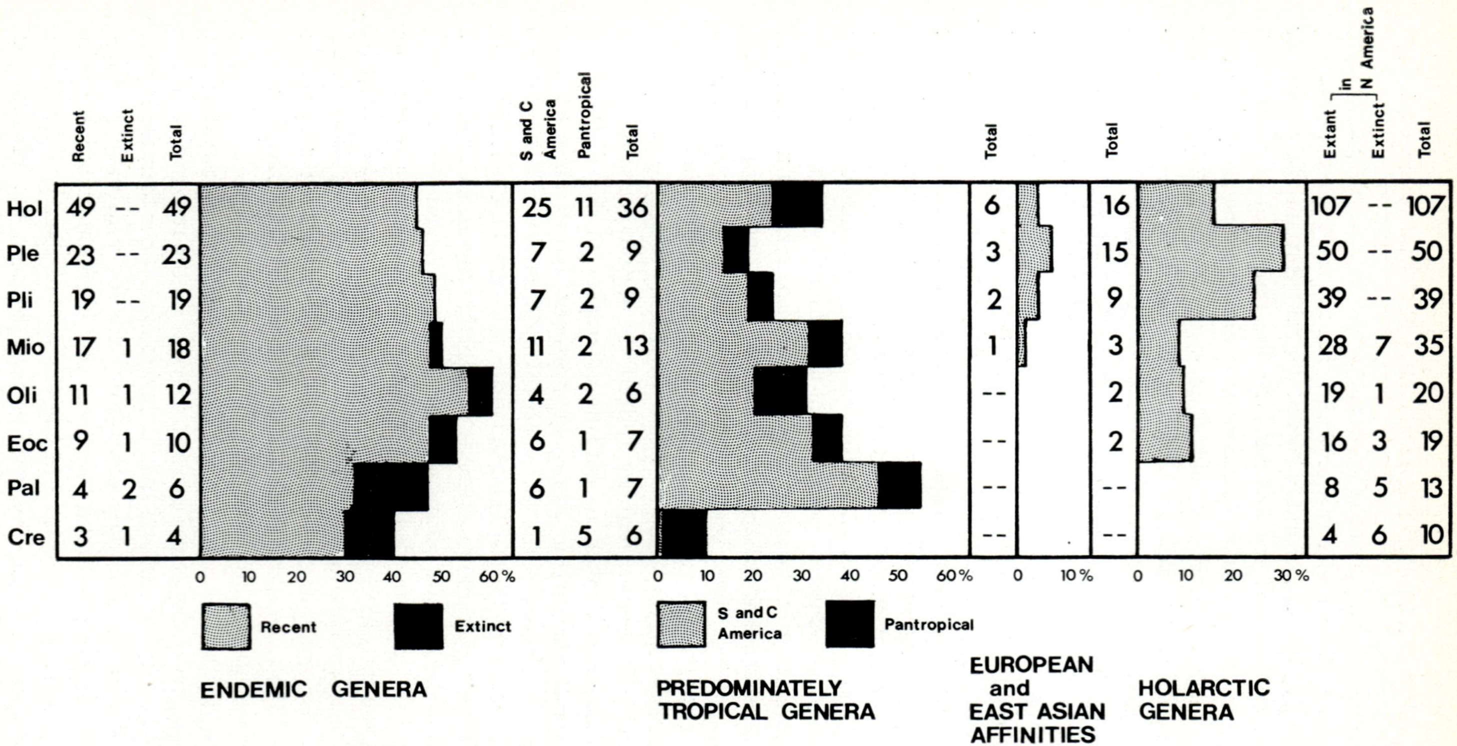

Fig. 2. Development of the North American terrestrial molluscan fauna since the late Mesozoic, plotted as number of genera recorded for each geological period (after Waldén, 1963). A genus has been regarded as present in the area during all periods from the earliest to the latest record, even when not known from all intervening strata. It is important to realize that the data for the Holocene period is not entirely comparable with those for earlier periods. The Pantropical group includes many marine littoral genera and a few taxa with very ancient and it is presumed tropical distributions.

Pleistocene. For example, those of Yen (1943) for China, McMichael (1968) and McMichael and Iredale (1959) for Australia, and Fleming (1975) and Climo (1975) for New Zealand. A few records for the Miocene present exciting discoveries which serve to illustrate the rich potential that is awaiting discovery. From the Lake Victoria area in East Africa, Verdcourt (1963) has described a rich Miocene fauna with many terrestrial elements having affinities with the present fauna existing in the area. Similarly, records for the Pacific Islands suggest that the Miocene faunas of endodontid snails were closely related to those extant on high islands today (Ladd, 1957, 1958; Solem, 1976). While in the Caribbean region, faunas from the Miocene are also closely related to those extant today (Simpson, 1894; Pilsbry and Olsson, 1954; Morrison, 1955), indicating that endemism on these islands was clearly established by this period.

The following account has been drawn extensively from a number of excellent reviews covering various aspects of the palaeobiology of terrestrial mollusca in North America and the Western Palaearctic, these include Henderson (1931, 1935), Licharev and Rammelmeier (1952), Pilsbry (1939–1948), Wenz (1923–1930) and Waldén (1963), and for South America, Parodiz (1969) and Pilsbry (1911) have been the important sources. Figures 2 and 3 were derived largely from Waldén (1963); although these include data relating to all non-marine groups, the exclusion of information for taxa other than Stylommatophora would not materially alter the picture, instead it would remove corroboratory evidence.

B. North America

The North American fauna is dominated by 2 distinct elements, an eastern and a western, characterized at the family levels, and with the antecedents of each being recognizable at the onset of the Tertiary. In the north these groups merge with a Holarctic boreal fauna, composed largely of small species, and in the south with a Mid-American association. The West American fauna extends from the Rocky Mountains to the Pacific Ocean and includes a number of families or subfamilies with either endemic or distinctive elements in that region. Many have an extensive fossil history during the Tertiary, for example, the Helminthoglyptidae, Ammonitellidae, Thysanophorinae and Holospirinae, and in the case of the Oreohelicidae even extending back into the Cretaceous. The eastern group extends from the mountains to the Atlantic Ocean and is dominated by a single family, the Polygyridae, with a long fossil history commencing in the Cretaceous. The persistence of these 2 faunas can be attributed to the initial separation of 2 land areas by a major marine incursion that persisted at least during the Late Cretaceous and early Tertiary and sub-

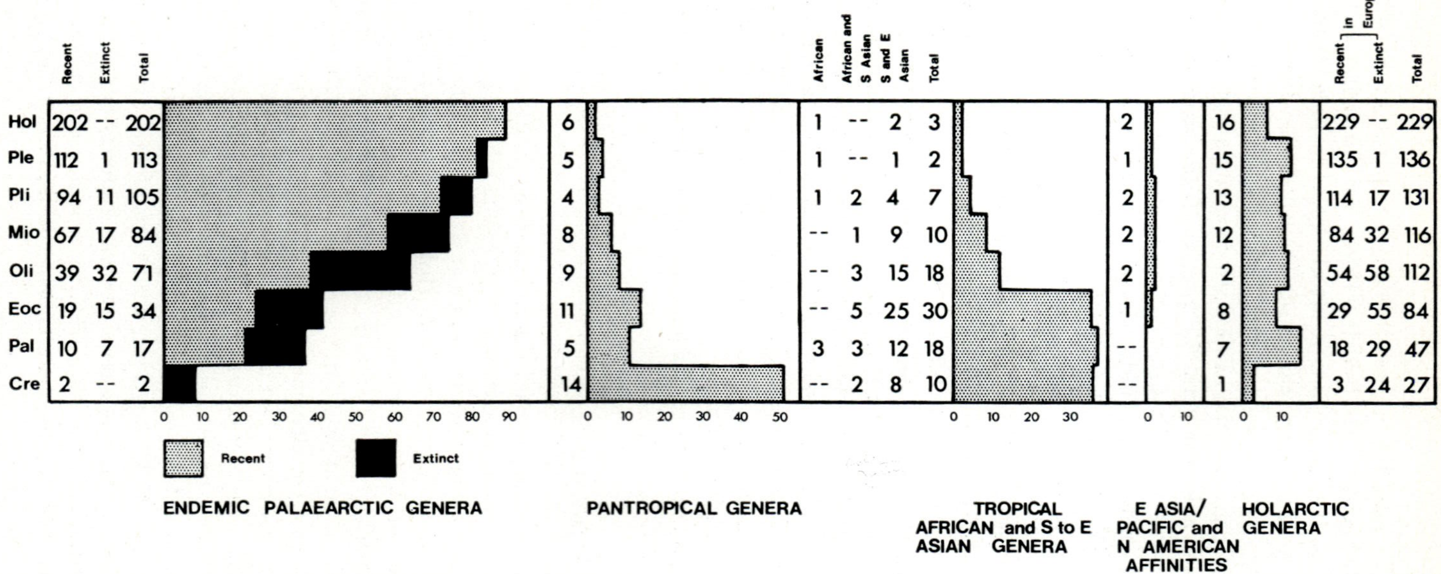

Fig. 3. Development of the European land molluscan fauna since the late Mesozoic (after Waldén, 1963). See notes for Fig. 2.

sequently by unfavourable environmental conditions in the Central Plain. Remarkably few taxa have successfully bridged this gap.

During the Tertiary there was a slight reduction in the proportion of tropical taxa represented in the North American fauna, even though the actual number of genera recorded increased. This element was composed of taxa with either a predominantly wide tropical distribution or one centred on the Middle American region. The latter is defined as the areas bordering the Caribbean Sea, and includes the islands, Central America, northern South America, the southern tip of Florida and Lower California. The fauna of this region is characterized by families that have undergone extensive radiation on many of the islands, for example, the Cerionidae, Oleacinidae, Spiraxidae, Urocoptidae and some prosobranch groups. Many of these have a fossil history extending at least to the Eocene, and in the case of the Spiraxidae to the Palaeocene in Europe. The penetration of this fauna into North America has been remarkably limited; indeed the low level of taxonomic differentiation of the immigrants suggests a comparatively recent colonization, possibly Pliocene–Pleistocene.

A Holarctic element has been present in North America throughout the Tertiary, but with a marked increase in representation during the Pliocene and Pleistocene. The earliest fossil records for the majority of the genera included in this group are from Eurasia, suggesting that dispersal was strongly directional towards North America; this is in agreement with the data for a number of taxonomic groups including, for example, mammals (Simpson, 1947). The molluscan representatives are small forms and the recent dominance in northern areas of such species with fitness for passive dispersal is hardly surprising in view of the markedly fluctuating climates and the high probability of local extinction.

C. Europe

The evolution of the molluscan fauna in the Western Palaearctic has been markedly different; the distribution patterns are more complex with the level of taxonomic differentiation between regions typically at the species or generic levels and not the family. For example, the fauna of the circum-Mediterranean area is isolated from that of northern Europe by the extensive Alpine range, it is characterized by extensive radiation at the species level of such families as Chondrinidae, Clausilidae and Helicidae, even though closely related or congeneric forms also exist in adjacent areas. The complexity of the patterns, in contrast to those in North America, can be attributed to the more diverse orographic history of Europe, the closer juxtaposition of other large land masses and changes in the climate throughout the Caenozoic.

At the end of the Mesozoic the molluscan fauna of Europe was dominated by tropical genera with either circumtropical distributions or affinities to taxa now confined to South-east Asia and to a far lesser extent Africa. Associations with contemporaneous North American or recent European faunas are not recognizable. The affinity with Asia was also reflected in the flora and there is evidence to suggest that in the Eocene, at least, a fairly uniform tropical flora stretched along the Tethys Sea from Britain to Borneo (Hughes, 1973); presumably the fauna exhibited a similar distribution. The general decrease in Europe of the tropical element from the Oligocene onwards is associated with the overall cooling of the climate and a corresponding increase in genera with a temperate distribution. Many of the early representatives of the Palaearctic element that colonized the area prior to the Miocene subsequently became extinct, although by the Pliocene the facies of the recent fauna were clearly established.

The scarcity in the fossil record of taxa with an exclusively southern African affinity suggests an isolated development for Africa; at least, the widespread distribution of some of the smaller species at present common to the 2 areas can be attributed to recent passive dispersal. These include a small number of species found in montane habitats in eastern and southern Africa (Forcart, 1956; Germain, 1928; Hubendick, 1953; van Bruggen, 1977). Such a pattern is consistent with the data for continental drift and the increasing isolation of Africa from Europe during the early Palaeocene to Upper Eocene followed by convergence and direct connection about 17 million years B.P. (Raven and Axelrod, 1974; Smith *et al.*, 1973). Moreover, it has been suggested that massive extinctions of the terrestrial biota could have occurred over large areas of Africa during the Tertiary, thereby reducing the probability of recognizing earlier connections. The fauna of North Africa has had a different history and is included with the circum-Mediterranean and is thus within the Palaearctic region (Sacchi, 1965).

While the distinctions between the North American and European faunas have been emphasized, the isolation of the 2 areas is not absolute. Besides the northern Holarctic element, there are a few genera, also with predominantly small species, common to the 2 regions. Some of these have a more extensive extra-territorial distribution and include, for example, *Gastrocepta*, *Pupoides*, *Cionella* (= *Cochlicopa*), *Ceciliodes* and *Strobilops*; the latter illustrates a problem—a genus with a wide but markedly disjunct distribution, a long fossil history, at least, since the Eocene, and distinct forms recognized as subgenera. Difficulties in interpreting this type of distribution pattern frequently arises; the choice is between accepting either a pre-drift origin or a trans-oceanic dispersal (or a combination of these), albeit at a time when the distances involved may not have been as great as at the present day. For *Strobilops* both America

(Pilsbry, 1927–1935) and Europe (Lindroth, 1957) have been suggested as centres of dispersal. While the distribution of some of the larger species belonging to the family Spiraxidae raises similar, if not more complex problems; in Europe there is an extant genus, *Poiretia* (Miocene–Recent), which together with the fossil genera *Palaeoglandina* (Palaeocene–Pliocene) and *Pseudoleacina* (Palaeocene–Oligocene) are related to the American *Euglandina* (Pliocene–Recent), a genus that appeared as a fossil much later than the European representatives.

D. South America

Information concerning South America has recently been reviewed by Parodiz (1969) and the degree to which the conclusions of this work support the much earlier speculations of Pilsbry (1911) is extremely interesting. They both recognized periods of colonization and radiation, but these were based extensively on information concerning the comparatively rich freshwater faunas; data for terrestrial taxa are far more limited, as readily demonstrated by the numbers of genera recorded during the Tertiary (scored on the same basis as data for Fig. 2): Palaeocene = 1; Eocene = 3; Oligocene = 2; Miocene = 7; Pliocene = 7. The majority of these fossils are referable to those genera or families that are today restricted to South America.

Utilizing the complete body of information, 4 patterns of association have been recognized, largely on the order of appearance in the fossil record. The oldest group contains a number of families of freshwater molluscs which have a very ancient and worldwide distribution; there is no evidence on which to base speculation as to the original distribution of this element. It includes such basommatophoran families as Lymnaeidae, Planorbidae, Physidae and Ancylidae. The second group consists of taxa which are first recorded towards the close of the Cretaceous or in the Palaeocene; these are again freshwater forms, but the majority belong to prosobranch families like Viviparidae, Valvatidae, Pleuroceridae, Thiaridae and Potamididae. For these taxa Parodiz postulated a northern origin, however, a recent study of the freshwater prosobranch genus *Glacidorbis* has provided evidence of faunal links between South America and Australasia (Meier-Brook and Smith, 1975); thus indicating that a southern distribution or route must also be considered. A later group, for which a northern origin has also been suggested, appeared at the end of the Miocene and during the Pliocene; this included the stylommatophoran families Streptaxidae, Helminthoglyptidae and it has been suggested the Clausiliidae; although amazingly neither the Streptaxidae nor Clausiliidae have been recorded in North America! A reciprocal extension of southern taxa northwards into North America occurred at the same time, it included the Bulimulidae and other families with a predominantly Central American distribution. It is important to note that while representatives of earlier colonizations are recorded from as far south as Patagonia, those from the last group have a much more restricted northern distribution. The final group consists of those families with a long history in the continent and which are today either restricted to the region or have undergone considerable autochtonous radiation in it; these include the Chilinidae, Strophocheilidae and the Bulimulidae, even though the latter has a wider and strongly disjunct distribution.

The high level of endemism, visible at all taxonomic levels, testifies to the great isolation of the area, yet the origin and relationships of the fauna are somewhat difficult to equate with the available geological information. The reciprocal interchange of faunas between the 2 Americas can be associated with the plate-tectonic model for the Caribbean (Rosen, 1975) and the gradual fusion of the 2 continents from the Miocene onwards, culminating in final union in the Pliocene about 6 million years B.P. During the Cretaceous when the continents were widely separated it has been suggested that scattered, but perhaps large islands existed in the region of Central America; towards the end of this period the importance of oceanic barriers could have been further reduced by uplift of land areas and the development of volcanic islands. Nevertheless, faunal interchange across this route appears to have been precarious until the Miocene. An Antarctic route to other parts of Gonwanaland, existed during the Cretaceous and this persisted until, at least, 100 million years B.P. and possibly later (Raven and Axelrod, 1974; McKenna, 1973); consequently there appears to be no overriding argument for postulating a northern dispersal route for those few taxa with a southern disjunct distribution.

There is a noticeable absence in the fossil record of a clear affinity between the terrestrial molluscs of South America and Africa, except for a Miocene species in the genus *Brasilennea* (van Bruggen, 1967; Parodiz, 1969). This has been placed in the subfamily Enneinae, with a close resemblance to *Ptychoon*, a subgenus of *Prychotrema* of recent tropical Africa. The present relationships between the 2 continents at the family level are demonstrated largely by taxa with a worldwide distribution; although constituent families of the superfamilies Strophocheilacea and Polygyracea are represented in both areas. Amongst groups other than terrestrial molluscs affinities are more readily demonstrated between inhabitants of the humid-tropic or aquatic environments (Fittkau, 1969) and indeed there are shared taxa amongst the freshwater molluscs (Paradiz and Bonetto, 1963). The apparent isolation of the faunas is, however, rather surprising considering that the 2 continents were not separated until 100 or 140 million years B.P. and were probably closer to one another until the Eocene, than were South and North America. It has been suggested that in South America a rich biota developed in forests and humid environments with a comparatively poor representation in the savannah and steppe habitats, and even the taxa found in these areas is frequently derived from forest forms, while in Africa the converse could be true (Fittkau, 1969). The 2 regions would appear to exhibit differences in climate, with South America having a more pluvial environment than Africa and the latter undergoing catastrophic climatic changes during the Pleistocene, the effect of which cannot be underestimated (Moreau, 1966). Nevertheless, it is difficult if not impossible to find reconstructions of climatic patterns which are consistent between all authorities. To what extent the molluscan faunas reflect these environmental differences is difficult to estimate, but the emergence of the Achatinidae as a dominant family in Africa (van Bruggen, 1977) might be associated with this pattern of fluctuating climates.

Questions arise regarding the origin and dispersal of taxa which have a marked disjunct distribution in South America or Middle America and a trans-Pacific representation. Amongst molluscan families included in this group are the families Camaenidae, Bulimulidae and the prosobranch Poteriidae and Helicinidae; the latter is represented on many Pacific Islands and for these taxa a trans-oceanic dispersal could be considered feasible. Nevertheless, for the majority of groups the evidence is against extensive trans-oceanic dispersal and the problem is to select between a later northern or earlier southern dispersal routes or a pre-drift Pangaean distribution. A possible solution arises from geological evidence suggesting a dual origin for South America—a Cretaceous to mid-Tertiary fusion between a Gondwanaland fragment and a circum-Pacific system of islands or land belts (Harrington, 1965). The attractiveness of this hypothesis has been pointed out by Fittkau (1969), but, at present, such suggestions must be considered extremely tenuous or even dubious, for the initial fragments would have been derived from an early, possibly Palaeozoic, land mass prior to the evolution of the biota. Curiously, such a hypothesis is returning very close to that proposed by Pilsbry (1900), of a Pacific continent which fragmented and foundered, but with the sophistication of a concept of continental drift superimposed! Nevertheless, the palaeogeographical evidence of Menard and Hamilton (1963) indicates that the isolation of the land masses in the Pacific Basin would not have been so marked during the Tertiary, so the distances involved in trans-oceanic dispersal need not have been so great as the present pattern of islands would indicate.

E. Quaternary faunas

The taxonomic units used in these descriptions of "large-scale" patterns have tended to be either the family or genus, but these are often too coarse for the recognition of local or fine-scale patterns and possibly too narrow for the identification of still older and broader patterns. There are examples where present species patterns can be correlated with distribution of land masses during the Pliocene as can, for example, the New Zealand general *Rhytida* and *Liarea* (Fleming, 1975; Climo, 1975), but it is not until the Quaternary, that the record is adequate for detailed analyses of faunal changes at the species level to be undertaken.

The catastrophic and widespread disturbances of the biota that resulted from the Pleistocene glaciations were not confined to the higher latitudes in each hemisphere. Major vegetational belts were compressed towards the Equator, although there is no evidence, at least in the northern hemisphere, that any were destroyed or totally admixed (Udvardy, 1969). In Africa their effect on the avifauna was profound (Moreau, 1966), but the fragmentation and disappearance of many major biotypes or changes in the environmental regime must also have had a catastrophic effect on the distribution of terrestrial molluscs (Verdcourt, 1972; van Bruggen, 1977). It is, however, in North America (Taylor, 1965) and Europe that the most detailed and thorough analyses of the faunal changes have been undertaken. A dichotomy has developed, however, in the objectives of these studies which, although not mutually exclusive, do reflect differences in the questions posed; one group has tended to concentrate on the information molluscs can provide regarding local environments (e.g. Lozek, 1962, 1965, 1967), while the other has attempted to trace the evolution of faunas, emphasizing the dynamic aspect of the distributional patterns (e.g. Ant, 1963; Kerney, 1966, 1968). Amongst studies in the latter group, those undertaken by Kerney in south-eastern England provide a very detailed picture of the changes resulting from the interplay between climate, habitat and dispersal over the last 10,000 years.

Sparks (1961, 1964a) and Taylor (1965) have drawn attention to an underlying problem encountered in many studies, there is a tendency to base interpretations of former environments or faunal associations on the assumption that individual species have changed little or not at all in their requirements. Such an approach would appear to deny the existence of natural selection. There is, however, very little evidence for the evolution of new taxa during this period (Taylor, 1960; Ant, 1963) and in all the cases that have been cited the divergence could have occurred pre-Pleistocene. Taylor (1965) went so far as to suggest that subject to a single doubtful example, "the effects of the Pleistocenes have been entirely destructive, not creative, at the specific level"; however, he also commented further "the span of Pleistocene times has been brief compared to the duration of molluscan species as identified by shells". This last quotation probably reflects the crux of the problem—the identification of molluscan species (see Davis, 1978). Gould (1969) has discussed the paradox and suggested that the reasons are largely semantic or methodological, for many of the genetical and ecological attributes of the group would suggest that pulmonate snails are excellent material for studying evolutionary responses—small populations, sedentary habits and rapid response to selective pressures. Indeed, in a study of the genus *Poecilozonites* in Bermuda, Gould (1969) demonstrated that a considerable degree of differentiation and evolution had occurred during the Pleistocene in this island microcosm.

During the last (Ipswichian) Interglacial a very rich molluscan fauna existed in the British Isles, with many species that now have a southern distribution being represented (Sparks, 1964a, b; West, 1968; Sparks and West, 1972). Yet the majority of species found in this fauna would have disappeared during the last (Würm or Weichsel) glaciation when ice sheets covered the northern two-thirds of the British Isles. A few hardy species probably survived in peri-glacial conditions in the south, but it is extremely unlikely that any of the supposed "Lusitanican" elements showing affinities with the Iberian peninsula remained; their present occurrence in the British Isles can in many instances be attributed to later introductions by man (Corbet, 1962). The present fauna rapidly colonized the region during the late-glacial and post-glacial periods when there was gradual amelioration of the climate (Kerney, 1966, 1968). Initially, during the late-glacial, the molluscan fauna of south-eastern England was composed predominantly of species with a palaearctic or holarctic distribution and a tolerance of low temperature, for example, *Cochlicopa lubrica* (Müll.), *Pupilla muscorum* (L.) and *Vitrina pellucida* (Müll.). A gradual improvement in the climate (Alleröd Interstadial) was reflected by the occurrance of species with a more southern distribution, for example, *Abida secale* (Drap.) and *Helicella itala* (L.) and many of these managed to survive a subsequent climatic deterioration. Thus the molluscs of the late-glacial period were predominantly species of open habitats, together with a few moderate thermophiles.

Although the fauna of the post-glacial (Flandrian) period exhibited continuance of the open grassland element, there was a change in the species composition associated with the spread of forests of south-eastern England. Molluscs which preferred shaded habitats increased and there was conversely a decline in open habitat forms. Temperatures throughout this period rose until the onset of the "climatic optimum", when summer temperatures were about 2 °C higher than at present; this was followed by a period of decreasing temperatures to the time in which we are now living. Additions to the molluscan fauna associated with the climatic optimum include species such as *Pomatias elegans* (Müll.), *Azeca goodalli* (Fér.) and *Acicula fusca* (Mont.). Species adapted to woodland continued to increase, while those of grassland must have survived only in isolated refuges; at least 2 grassland species, *Helicella geyeri* (Soos) and *H. striata* (Müll.), became extinct and many woodland snails which are not regarded as local were probably universally distributed.

A major perturbation was imposed on this pattern about 3000 BC, at the end of the climatic optimum, with the initiation by Neolithic Man of the forest clearances and the development of primitive agriculture. These changes continued until the middle of the sixteenth century and resulted in the devistation of the woodland environment and the associated biota. There was a corresponding "drying up" of many habitats and the spread of xerophilous species was favoured at the expense of more hygrophilous forms. Consequently many species that had been common during the late-glacial period, favouring open habitats, exhibited a resurgence. While woodland species, whose prime demand is shade, moisture and absence of disturbance, developed restricted and highly fragmented ranges, a feature which many still exhibit. Nevertheless man has not had a completely deleterious effect on the environment, for as diversity and range of habitats multiplied there has been a concomitant increase in the size of the fauna largely through the introduction of alien species; candidates for inclusion in this category are both *Helix aspersa* (Müll.) and *H. pomatia* (L.). So profound has been man's influence on the present fauna that it is now virtually impossible in many areas, including the British Isles, to distinguish or recognize the level to which man has influenced and is still influencing such phenomena as dispersal and colonization.

IV. Recent distribution

Although stylommatophoran species dominate the recent faunas of the major land masses, they coexist in many areas with representatives of both basommatophoran and prosobranch groups. The latter, in the case of prosobranch species, reaching over 50% of the fauna on some Caribbean Islands like Jamaica (Russell-Hunter, 1955). The Archaeogastropoda is the most important of the prosobranch orders in this area, but is replaced in South-east Asia and associated islands by representatives of a number of distinct superfamilies belonging to the Mesogastropoda. Amongst the Stylommatophora the 2 orders, Sigmurethra and the Orthurethra, tend to be the most numerous in nearly all regions of the world, but the distribution of these higher groups is neither uniform nor random.

A. Distribution of families

In a series of recent computerized studies Te (1972, 1976) has attempted to make a more objective analysis of the distribution patterns at the family level, using geographical units derived largely from Wallace's (1876) regions and subregions and an amalgam of the data contained in the taxonomic synthesis of Zilch (1959–1960) and the revised classification of Taylor and Sohl (1962). The latter differs from that published by Solem (1978) and consequently a few changes have been incorporated into Table I in rather critical areas; for example, the exclusion of the genus *Prestonella*

Table I. Distribution of stylommatophoran families: regional patterns. Based on analyses by Te (1972, 1976), but including taxonomic innovations suggested by Solem (1978). A family name in parentheses indicates the possibility of it being classified in a particular region, although it is actually included in another group.

A. OLD WORLD—NEW WORLD PATTERNS
- I. World-wide
  - Endodontidae[a]
  - Helicarionidae (incl. Euconulidae, Ariophantidae)
  - Pupillidae (incl. Vertiginidae, Orculidae, Chondrinidae)
  - Succineidae
  - Zonitidae (incl. Trochomorphidae, Vitrinidae)
  - (a) *Nearly world-wide*
    - Clausillidae
    - (Valloniidae)
    - Strobilopsidae
- II. Pantropical
  - Streptaxidae
  - Subulinidae
  - (a) *Extended to Western Palaearctic*
    - (Ferrussaciidae)
- III Holarctic
  - Cionellidae
  - (a) *Nearly Holarctic*
    - Arionidae
    - Valloniidae
    - (Limacidae)
- IV. South and East Asia, Australian–Pacific, Middle and South America
  - Camaenidae

[a] See Solem (1976); this recent revision has split this group into a number of families with differing distribution patterns, but the evidence is incomplete for geographical distributions to be given for all taxa.

Table I—continued

V. East Asia–America
Philomycidae
(a) *Middle America extended to Mediterranean*
(Spiraxidae)

B. OLD WORLD PATTERNS

VI. Africa–Orient–Palaearctic
Enidae

VII. Palaearctic
Pyramidulidae
(Helicidae)

VIII. Western Palaearctic
Trigonochlamydidae
Parmacellidae
Testacellidae
(a) *Extended to Siberian Palaearctic*
Helicidae
(b) *Extended to the Tropics*
Ferrussaciidae
(c) *Extended to North America*
Limacidae

IX. Africa
Urocyclidae
Achatinidae
Dorcasiidae
Aillyidae
Aperidae
? Thyrophorellidae

X. East Asia—Africa
Bradybaenidae
Corillidae

C. NEW WORLD PATTERNS

XI. North, Middle and South America
Helminthoglyptidae

XII. North America
Polygyridae
Haplotrematidae
? Ammonitellidae
Oreohelicidae
(a) *Extended to Middle America*
(?Thysanophoridae)

XIII. Middle America
Urocoptidae
Sagdidae
Oleacinidae
Ceriidae

Table I—continued

(a) *Extended to North America*
? Thysanophoridae
(b) *Extended to Mediterranean*
Spiraxidae
XIV. South America
? Systrophiidae
Strophocheilidae
Macrocyclidae
(a) *Extended to Australian–Pacific*
Bulimulidae (incl. Orthalicidae, Odontostomidae, Amphibulimidae)
(Megaspiridae)
D. AUSTRALIAN–PACIFIC PATTERNS
XV. Australian–Pacific
Amastridae
Partulidae
Athoracophoridae
Otoconchidae[a]
(a) *Extended to East Asia*
Achatinellidae
(b) *Australia*
Caryodidae
E. SOUTHERN DISTRIBUTION PATTERNS
(a) *Madagascar–Ceylon*
Acavidae
(b) *South Africa–Australia*
Rhytididae
(c) *Australia–South America*
Megaspiridae
(Bulimulidae)

from the Amphibulimidae removes the trans-Atlantic links of that family (Solem, 1959). Similarly, major changes in the classification of the Acavidae (Table II) have altered the overall Austral distribution of this family, although some of the previous constituent members still demonstrate trans-oceanic links across the southern hemisphere (Baker, 1955, 1962). Changes in the classification of the Arionacea, and in particular the "Endodontidae" *sensu* Zilch, will also necessitate revisions in the distribution patterns of the constituent families, but these await the publication of a taxonomic revision.

It is possible to superimpose another level of analysis on these data, for contemporary distribution patterns are, in some measure, the consequence of past opportunities for dispersal or dispersion. Indeed they provide another opportunity for reconstructing patterns prior to, or subsequent to,

Table II. Distribution of the family Acavidae, illustrating the effect of 3 different classifications on the distribution patterns recognized at the family level.

	Zilch (1959–1960)	Taylor and Sohl (1962)				Solem (1978)			
	Acavidae	Dorcasiidae	Stropho-cheilidae	Acavidae	Dorcasiidae	Stropho-cheilidae	Macro-cyclidae	Acavidae	Coryodidae
South America	x		x			x	x		
Africa	x	x			x				
Madagascar	x			x				x	
India									
Ceylon	x			x				x	
Australia	x			x					x

the fragmentation of land masses during the Mesozoic. Data derived from the fossil record is largely inadequate, although the extensive changes in distributions wrought as recently as the Quaternary demonstrate the need for caution in extrapolating from modern faunas. In Table III distributional data for families have been plotted in a manner which emphasizes continental patterns, and in particular the division of the Mesozoic land-masses into 2 major units; 2 minor divisions have been retained as the recent faunas are distinctive and on geological grounds the areas do not fall readily into either of the major groups. A number of generalizations emerge from the data present here and in Table I:

(i) Very few families exhibit southern disjunct distributions with representatives shared between either South America, Africa or Australia and none have representatives in all 3 continents, but it is these few that have been recognized as a Gondwanaland element (e.g. Van Bruggen, 1969). For the majority of families it can be argued that the recent distributions reflect a late Mesozoic or Tertiary arrangement of continents, subsequent to the fragmentation of the land masses, far more satisfactorily than any simple division into Gondwanaland and Laurasia. While the large group of taxa with widespread distributions demonstrate a range of more diffuse and problematical patterns possibly reflecting recent dispersal or differences in regional extinctions.

(ii) The faunas of the southern continents are characterized by families with distributions restricted to individual land masses where they have undergone, in some instances, considerable autochtonous radiation. A contrast is provided by the faunas of the 2 northern land masses where the characteristic taxa are widely distributed, frequently across both continents and in adjacent geographical regions. These 2 patterns would appear to be associated with differing degrees of isolation, both temporally and spatially, in the northern and southern continents from the Mesozoic onwards. For even though there are variations in the estimates for the opening of the major ocean barriers during this time span (Smith *et al.*, 1973; Raven and Axelrod, 1974; Smith and Briden, 1977) there is considerable agreement regarding the sequence, with the southern continents having been isolated for much longer and by far greater distances than the northern.

(iii) Many families with predominantly holarctic distributions have ranges that extend to the southern continents, while in only a single case does the converse appear to occur. There is therefore a pronounced bias in terms of wider distributions in favour of families with northern distributions. Such a pattern could reflect differing rates of dispersal and colonization or differential survival rates giving rise to strongly disjunct or relict distributions. Undoubtedly such factors play an important role in determining ranges, but there would appear to be no evidence supporting

Table III. Distribution of stylommatophoran families: continental pattern

	Southern Continents					Northern Continents				
	S. America	Africa	India	Australia	Middle America	N. America	Eurasia	S. Asia	Pacific Islands	
Megaspiridae	x			x						Southern fauna (15 families)
Bulimulidae	x			x	x	x				
Strophocheilidae	x				x					
Systrophiidae	x				x					
Macrocyclidae	x									
Rhytidae		x		x						
Dorcasiidae		x								
Achatinidae		x								
Aperidae		x								
Thyrophorellidae		x								
Urocyclidae		x								
Aillyidae		x								
Acavidae		(x)	(x)							
Caryodidae				x						
Athoracophonidae				x						
Cerionidae					x					Middle American fauna (6 families)
Urocoptidae	x				x	x				
Sagdidae					x					
Oleacimidae					x					
Helminthoglyptidae					x	x				
Spiraxidae					x	x	x		x	

Strobilopsidae	x				x	x	x	x		Widespread fauna (11 families)
Subulinidae	x	x	x		x	x	x	x	x	
Streptaxidae	x	x	x	x			x	x		
Philomycidae	x			x	x	x	x	x	x	
Helicarionidae		x	x	x	x		x	x	x	
Corillidae		x	x				x	x		
Camaenidae	x		x	x	x		x	x		
Pupillidae	x	x	x	x	x	x	x	x	x	
Succineidae	x	x	x	x	x	x	x	x	x	
Clausiliidae	x	x	x		x		x	x		
Ferussaciidae		x	x	x	x		x	x	x	
Bradybaenidae		x		x			x	x		Northern fauna (15 families)
Enidae		x					x	x		
Haplotrematidae	x				x	x				
Cionellidae			x			x	x	x		
Arionidae		x			x	x	x	x		
Vallonidae		x			x	x	x			
Limacidae		x				x	x			
Zonitidae					x	x	x	x	xx	
Polygyridae					x	x				
Ammonitellidae					x	x				
Oreohelicidae						x				
Trigochlamydidae							x			
Parmacellidae							x			
Testacellidae							x			
Pleurodiscidae							x			
Helicidae							x			
Achatinellidae	x			x				x	x	Polynesian (Pacific) fauna (3 families)
Portulidae				x				x	x	
Amastridae									x	

a distinction between northern and southern faunas in survival rates for taxa. A more rational explanation can be provided by considering dispersal. Many biologists have commented on the tendency for dispersal to run from north to south; Darwin (1859) observed that "I suspect that this preponderant migration from north to south is due to the greater extent of land in the north". Matthew (1915), Simpson (1947) and Darlington (1957, 1965) have commented on similar patterns exhibited by other taxa and on dispersal from large to smaller land mass, as between Eurasia and North America. In this situation, larger area reflects greater number of species and propagules, thereby increasing the probability of a directional bias in dispersal. Whether the extension of an alternative hypothesis that "more dominant forms" are "generated in larger area" applies is questionable; certainly this is an argument that has been developed in many different ways in relation, for example, to evolution in the tropics, and the importance of competition in evolution.

Climatic barriers probably play an important role in determining this north–south pattern. Many of the taxa found on the southern continents have a predominantly tropical or subtropical range and consequently colonization of the northern temperate and colder environments would presumably have been precarious, although not necessarily impossible. The converse presents quite different problems, for corridors or limited areas with temperate climates or microclimates exist within the mosaic of tropical ecosystems, the most obvious, for example, being found in montane or alpine regions. Southern taxa with temperate ranges are very isolated from suitable habitats in the north and frequently appear to have more restricted distributions, consequently a lower probability for successful dispersal might be anticipated. Biological attributes could also be important, but there is virtually no data on which to base any speculation. The various hypotheses which suggest broader ecological niches amongst temperate taxa compared to tropical could be relevant, for they suggest that the probability of finding a suitable environment for colonization should be higher amongst temperate taxa.

(iv) The 2 smaller faunas, the mid-American and the Polynesian, present an enigma; should they be regarded as relicts of Mesozoic distribution patterns or the product of recent dispersal? Some of the problem regarding Polynesian faunas are discussed on pp. 469–472, but for the mid-American there is no conclusive information. Pilsbry (1939–1948) did suggest a northern Eurasian origin for certain members of the latter via a West American route, but the evidence is tenuous. Some elements have an interesting extraterritorial distribution; the Spiraxidae are represented in southern Europe by both recent and fossil taxa and the prosobranch Pomatiasidae with recent African and fossil European representation.

Raven and Axelrod (1974) have suggested that islands or land masses survived in the Central American area during the Late Cretaceous and Early Tertiary. If so, these could have been isolated for long periods and have retained a distinctive fauna even though, at a much later stage, the area may have played an important role in the interchange of biotas between the two Americas. This is also a situation in which the unique elements of the region could have radiated. Indeed there is no palaeontological evidence that any of the families which exhibited such evolution ever had an extensive distribution.

(v) Table III highlights a number of anomalous features for which there are no obvious or ubiquitous explanations. For example, the absence of the Clausiliidae from North America, although present in the South, and the absence from India of the widely distributed Eurasian family, Bradybaenidae. India fails to demonstrate any clear links in the recent fauna with other southern continental masses to which it was joined in the Mesozoic; although possibly some links still occur with the Mascarene Islands (Madge, personal communication). The recent fauna would appear to be largely related to those of recently adjacent land masses, suggesting faunal interchange subsequent to the collision of the Indian Plate with Laurasia in mid-Tertiary and with the extinction of many older faunal elements (Raven and Axelrod, 1974; Holloway, 1974). In this context the Acavidae present an interesting anomaly occurring on Madagascar, Seychelles and Ceylon, but whether this can be considered as a relict of a more widespread range is equivocal.

(vi) The final group includes families united by the possession of extensive ranges covering a number of continents; in the classification of Te (1976) these taxa exhibit a number of different patterns, amongst the most obvious being those which are world- and nearly world-wide. The latter include species which are predominantly small and thus it may be inferred have a higher potential for dispersal and successful colonization than larger taxa (Peake, 1969; Vagvolgyi, 1976). These species frequently exist locally at very high densities exploiting a wide range of refuges, thereby exhibiting a potential for survival even under adverse conditions. Two or three of the families included in this group belong to the Orthurethra; an order with a long geological history (Solem, 1974) and consequently it can be argued have a wide distribution that could have been established prior to the Mesozoic fragmentation of Gondwanaland and Laurasia. Any predilection towards either of the 2 hypotheses must be for the present largely intuitive, as critical palaeontological evidence is lacking and there are no means of estimating earlier colonization rates.

The Pupillidae are a prime example of a family with a world-wide range, for in no other group are the constituent taxa so widely distributed; genera such as *Gastrocopta*, *Pupilla* and *Nesopupa*. There is an extensive fossil history during the Tertiary and this provoked Pilsbry (1927–1935) to comment that the "European Oligocene and Miocene pupillid fauna appears to be about as specialized and mature as the recent, and contains many genera and subgenera that are still widely spread". Nevertheless, to suggest that the world-wide distribution is linked to, or is evidence of, antiquity may be unnecessary, even such protagonists as Pilsbry found it difficult to reconcile the great similarity between American and Ethiopian forms with the affinities of other elements in these regions. There is evidence for considerable conservatism of shell form in pupillids, so the pattern could be ancient, but the features used in the classification of this group are limited and consequently the phylogenetic and distributional conclusions that are drawn must be tenuous. The study of Schopf *et al.* (1975) demonstrating the correlation between rates of evolution and the number of "character states" used is very pertinent here. An interesting corrollary is provided by the recent review of the "Endodontidae" *sensu* Zilch (1959–1960), a family with a supposed world-wide distribution, where the addition of new data has demonstrated that this is probably not a monophyletic group and a taxonomic revision has produced units with more restricted distribution patterns (Solem, 1976).

The majority of the families included in this group of widespread elements have either a predominantly tropical or subtropical range and an origin for many of these distribution patterns prior to, or soon after, the fragmentation of the land masses in the Mesozoic is extremely inviting. The environment, at least, in some areas of the tropics must have been comparatively stable throughout the Mesozoic and Tertiary compared to regions in higher latitudes. Consequently, it could be postulated that the extinction rate for many groups in this region would have been lower and the survival of ancient or relict faunas, as in South East Asia, might be anticipated. Moreover, the process of extinction must have been highly selective regarding taxa and environments, hence the closer affinities between the faunas of South America and Africa amongst taxa found in humid or wet (including freshwater) habitats.

All of these patterns have an overlay imposed by the presence of introduced species that reflect the cultural migrations of man and his influence on the environment. Examples are legendary, but there is every indication that the influence of man is, at present, increasing through either the adventitious dispersal of species (Elton, 1958) or the catastrophic destruction of habitats and associated faunas. Many species which have been spread by man have a European origin (Lindroth, 1957) and the ability of some forms to survive long periods of aestivation has undoubtedly been a contributory factor to the frequent representation of European helicids in exotic faunas. Tropical examples also abound, for example *Achatina fulica* (Mead, 1961, 1978), and species belonging to the families Subulinidae and Veronicellidae; the term "tropical tramps" has even been coined to cover these forms.

B. Distribution of the Helicacea

There is a wealth of information available concerning the Helicacea and it might be anticipated that this would be reflected in the taxonomic treatment of the superfamily and in a clear understanding of the distribution patterns exhibited by this taxon. Unfortunately this would be an over-optimistic viewpoint, for the superfamily clearly illustrates the limitations of the information available for many groups of Stylommatophora.

The Helicacea, *sensu* Zilch (1959–1960) and Taylor and Sohl (1962), has a world-wide distribution with representatives on all continents and in many archipelagos, but the greatest radiation is exhibited in the northern hemisphere—in North America and Eurasia—where there have been 2 largely independent developments. The major features of the group can be summarized as follows:

(i) The present classification relies heavily on the early studies of Pilsbry (1894) and subsequently those of Hesse (1931, 1934), Baker (1955) and Wurtz (1955). The unity of the group was established by Pilsbry, but he raised doubts about the wisdom of speculating too deeply on phylogeny. Recently this established position has been challenged by Solem (1978); he has suggested that a second superfamily should be recognized, the Camaenacea, and that concepts regarding the evolution of the reproductive systems should be revised (Solem, 1972). At all taxonomic levels, judgement is based extensively on anatomical features of the reproductive system and the terminal genitalia. There has been considerable convergence of shell characters and the recognition of fossil material is, therefore, problematical. The assignments of the fossil genus *Dentellocaracolus* illustrate this difficulty; it has been suggested that affinities existed with the West Indian Camaenidae or the Canary Island *Hemicycla*, but it is now included in the Sphincterochilinae, a subfamily with a circum-Mediterranean distribution (Pilsbry, 1894; Zilch, 1959–1960). If affinities with the Camaenidae were demonstrated then it would provide a most interesting extension of the range of this family.

(ii) Helicids occur in a wide range of habitats from deserts to rainforests, although the greatest diversity at the species level probably occurs in warm temperate environments. Nevertheless, in some desert regions there has been enormous autochthonous radiation that can be attributed to the fragmentation, isolation and subsequent divergence of small poulations (Bequert and Miller, 1974). The "fame" of the group is dependent on the importance of a few species that are either of economic importance or act as "weeds", but these forms are not representative of the diversity exhibited by the whole group.

A wide range of niches are occupied, from fossorial to arboreal. The

NORTH AMERICA

CAM. Camaeninae Palaeo-Mioc.
Ammonitellinae Eo.(?)-Mioc.-Rec.
Oreohelicinae Cret.-Rec.
Xanthonycinae Rec.
Humboldtianinae Rec.
HELM. Helminthoglyptinae Cret.-Rec.
Sonorellinae Rec.
Cepolininae Rec.

CENTRAL AMERICA, CARIBBEAN ISLANDS

CAM. Camaeninae Pleist.-Rec.
Xanthonycinae Rec.
HELM. Humboldtianinae Rec.
Cepolininae Rec.

EUROPE

BR. Bradybaeninae Palaeo.
Sphincterochilinae Olig.-Rec.
Helicellinae Mioc.-Rec.
Geomitrinae Eoc.-Olig.
Hygrominae Olig.-Rec.
Helicodontinae Palaeo.Rec.
Leptaxinae Olig.
Campylaeinae Eoc.(?) Olig.-Rec.
Helicinae Olig.-Rec.

NORTH AFRICA, ARABIA

HEL. Sphincterochilinae Olig.-Rec.
Helicellinae Mioc.-Rec.
Hygrominae Olig.-Rec.
Leptaxinae Mioc.
Helicinae Pleist.-Rec.

N.E. ASIA

BR. Bradybaeninae Rec.
HEL. Hygrominae

CHINA, JAPAN, S.E. ASIA, AUSTRALASIA

CAM. Camaeninae Plioc.-Rec.
Helicostylinae Rec.
BR. Bradybaeninae Plio.-Rec.

ATLANTIC ISLANDS

Canary Islands:

HEL. Helicellinae Rec.
Helicodontinae Palaeo.-Rec.
Leptaxinae Rec.
Helicinae Rec.

Madeira:

HEL. Geomitrinae Pleist.-Rec.
Leptaxinae Rec.
Helicinae Rec.

Cape Verde Islands:

HEL. Leptaxinae Rec.

SOUTH AMERICA

CAM. Camaeninae Plioc.-Rec.
HELM. Helminthoglyptinae Rec.

AFRICA

BR. Bradybaeninae Pleist.-Rec.

Fig. 4. Spatial and temporal distribution of the subfamilies of the Helicacea; for classification above the family level see pp. 527 Key to abbreviations: CAM. = Camaenidae, BR. = Bradybaenidae, HELM. = Helminthoglyptidae, HEL. = Helicidae, Cret. = Cretaceous, Palaeo. = Palaeocene, Eoc. = Eocene, Olig. = Oligocene, Mioc. = Miocene, Plioc. = Pliocene, Pleist. = Pleistocene, Rec. = Recent, ? = Record for particular geological time-span dubious.

life cycles are complex, including annual and pluriennial representatives within the same genus and species. Undoubtedly the success of the group reflects the adaptability of the life cycles and the potential for phenotypic variation depending on the prevailing environmental conditions. Survival in a wide range of climatic conditions is also linked to an ability to conserve water and escape from the extreme vicissitudes of high or low temperatures by undertaking different behavioural or physiological responses; the most obvious are aestivation and hibernation. Variations in shell form and colour are closely associated with these adaptations (see Cain, 1976, 1977b); for example, the highly calcareous and white shells of desert forms (Rensch, 1959), the highly polymorphic colour forms of many arboreal taxa and even the evolution in very humid rain-forests of a "slug-like" form, where the shell is enclosed within the mantle (Peake, 1968).

(iii) Regional patterns are clearly discernible at all taxonomic levels, but at the family and subfamily levels (Fig. 4) there is a tendency for these to be associated with major geographical boundaries. Thus the faunas of North America and Europe are distinct with no taxa shared, except for those accidentally dispersed by man; these differences have persisted since, at least, the Cretaceous, even though terrestrial connections or narrow water gaps existed across the North Atlantic until the Eocene, 49 million years B.P. There is no evidence that the Bering Straits bridge was important for the interchange of helicids between the 2 northern continents.

Although there is a characteristic fauna of the Middle American region, representatives of the helicid groups are found in adjacent areas, where the affinities are predominantly at the generic or intergeneric levels. The South American fauna is closely associated with that of Middle or North America and there are no distinctive elements at the higher taxonomic levels. Such patterns would appear to reflect the recent juxtaposition of the 2 continents which culminated in a direct link being established about 6 million years B.P.

The helicid fauna of the South-West Pacific, South-East Asia, China and Japan is dominated by 2 families, the Camaenidae and the Bradybainidae and both exhibit considerable radiation throughout this area. Both have markedly disjunct distributions. The Camaenidae also occur in Middle America and adjacent areas and the Bradybaenidae in the mountainous regions of Africa, plus a fossil record in Europe; the absence of the last family from Peninsula India is remarkable. A more extensive distribution of the Bradybaenidae in the early Tertiary would be consistent with patterns exhibited by other terrestrial molluscs. Whether it is necessary to evoke the links between Africa and Eurasia, which existed prior to the Miocene (63 million years B.P.) or more recently (post 17 million years B.P.) to explain the occurrence of the Bradybaenidae in Africa is probematical, for this extension of the range could be attributed to long range dispersal. Its absence from India may indicate that the major patterns had been established in this region prior to India abutting against Asia 45 million years B.P., but other Pulmonate taxa do not appear to show such a clear distinction between the 2 land masses. The presence of a single species of the genus *Eulota* (Bradybaenidae) in faunas of Europe since the end of the Pliocene, has been attributed to the recession of the epi-continental seas and the uniting of Europe and Asia (Likharev and Rammelmeier, 1952).

The distribution pattern for the Camaenidae is not unique; the prosobranch family Poteriidae and the subfamily Helicinae have Caribbean–Pacific distributions extending to the periphery of South-East Asia. While the Bulimulidae with a South America–Australia and Pacific distribution have many features in common with the Camaenidae. Comparable features are found in the floras of the 2 regions (Balgooy, 1971; Thorne, 1963). There are problems in analysing this type of distribution; for example whether a Southern Antarctic or a Northern Bering Straits dispersal route is favoured in preference to long distance transportation. The suggestion of an extensive southern or even northern distribution prior to the fragmentation of the southern continents appears extremely inviting for a group which has large shells and it appears a low probability for dispersal across oceanic barriers; evidence for links between the northern and southern hemispheres prior to the Cretaceous is available for many plant groups, this pattern being modified by subsequent extinction of "austral" taxa in the northern hemisphere (Raven and Axelrod, 1972). The arguments for a northern origin of these highly disjunct distributions have been forcibly expressed by Solem (1968b); he considered that the extreme southern distribution of phylogenetically more primitive taxa in the "Acavacean" groups and the Bulimulidae reflects replacement by more advanced groups in the northern parts of their range. Unfortunately such a hypothesis can readily be reversed and if a southern origin is considered then a process of centrifugal evolution along the edge of the taxon's range can be postulated!

In the Helicacea 2 notable examples of autochtonous radiation are recognized at the subfamily level; the Helicostylinae in the Philippines and the Geomietrinae on Madeira. Although these are rather exceptional examples, other patterns of radiation are found, for example, amongst the Helicellinae in circum-Mediterranean regions. While from Malaysia east to Melanesia, along the Indonesian archipelago, radiation amongst the larger forms is associated with allopatric replacement at the subspecific, specific and generic levels, both within and between island groups (Rensch, I, 1934, 1937; Rensch and Rensch, 1935, 1937); it was for such situations that Rensch (1959) coined the terms "Rassenkreise" (mosaics of races) and "Artenkreise" (mosaics of species). There are no major faunistic discontinuities in this region, similar to those recorded for mammalian and bird families. Distribution patterns would indicate that there is replacement by ecological equivalents or analogues both between and within archipelagos suggesting that in some instances some form of competitive replacement could be occurring; for example, within the Solomon Islands arboreal forms belonging to the Camaenidae ("Papuina") and the Bulimulidae appear to replace each other allopatrically along the island chain (Peake, 1969).

The fossil record provides evidence of changes in distribution and restrictions in range of taxa during the Tertiary and the Quaternary. Some areas becoming refugia for taxa which had previously enjoyed a wider distribution, examples are found in the Cape region of South Africa (van Bruggen, 1969, 1977), Caucasia (Likharev and Rammelmeier, 1952), and the Atlantic Islands.

C. Faunal elements

The authors of many recent studies of regional faunas have acknowledged the sterile nature of a process of constantly delimiting or describing faunal provinces and have followed a trend towards a more dynamic approach; references to their works are scattered throughout this review. They have recognized that faunas are composed of units or elements having different zoogeographical affinities and that the history of these elements in a region may be heterogenous. That such studies have been concerned largely with faunas in temperate regions or islands might not be entirely fortuitous, as it could be anticipated that the faunal "turn-over" in these situations is relatively high. Moreover, it is still possible to discover faunistic accounts which equate the centre of diversity or distribution with the site or centre of origin of a particular taxon, a concept which has been largely discredited for a long time. Whether the concept of a centre of origin, as frequently used, even has any validity is questionable (Croziat *et al.*, 1974).

An analysis of the molluscan faunas from South-East Asia along the island chain to Melanesia affords an opportunity to recognize faunal elements with different distribution patterns and compare these with those exhibited by other plant and animal groups. The unique features of this region, as the major transition zone between 2 vertebrate faunas, the Oriental and the Australasian, have long been recognized (Mayr, 1944). For many other groups of organisms, however, the affinities of the Melanesian biota and indeed the Polynesian must be sought in South-East Asia, these include phanerogams (Balgooy, 1971) and many insect families (Gressitt, 1961). Terrestrial molluscs follow this pattern to the extent that Solem (1968b) was forced to conclude that "on the family level, there is a single fauna of land snails that extends from South-East Asia to New Guinea and often on to the Solomon Islands and the tropical coast of Queensland"; deviations from this pattern occurring largely at the lower taxonomic levels. There is a gradual attenuation of the molluscan fauna from the core area in South-East Asia eastwards, with a secondary radiation in New Guinea (Solem, 1959, 1968b; Peake, 1969). There has been some reluctance to accept the continuity of this pattern and contrary evidence has been produced by selecting species or genera with restricted distributions (Purchon, 1977), rather than examining the complete range of patterns. Thus the biota of New Guinea and the adjacent islands to the west reflect mixed affinities, the vertebrates largely with Australia, the invertebrates and plants with South-East Asia; this phenomenon of "mixed biotas" is repeated in Hawaii, for example, where the avifauna is predominantly of North American origin and the flora and invertebrate fauna Polynesian (Mayr, 1943; Zimmermann, 1948; Cooke and Kondo, 1960; MacArthur and Wilson, 1967). There is no satisfactory explanation for these differences, even though various hypotheses have been advanced covering the geological history of the region (Raven and Axelrod, 1972), the history of the groups in the region, the ecological requirements of the taxa together with variations in the dispersal and demographic properties of the different elements of the biota (MacArthur and Wilson, 1967); more biological information is required before the relative merits of each can be distinguished. These patterns, however, must have been estab-

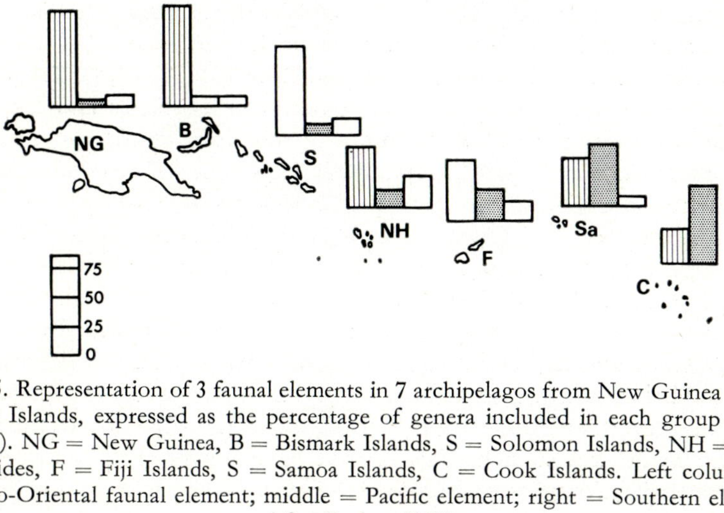

Fig. 5. Representation of 3 faunal elements in 7 archipelagos from New Guinea to the Cook Islands, expressed as the percentage of genera included in each group (scale given). NG = New Guinea, B = Bismark Islands, S = Solomon Islands, NH = New Hebrides, F = Fiji Islands, S = Samoa Islands, C = Cook Islands. Left column = Palaeo-Oriental faunal element; middle = Pacific element; right = Southern element (after Peake, 1969).

lished or radically changed subsequent to the collision of the Australasian and Asian plates during the Miocene. For plants, dispersal along high chains of mountains in the late Caenozoic was probably also important (Raven and Axelrod, 1972).

The nature of the gradual interchange or intermingling of the biota along an island chain is clearly visible in Melanesia and the adjacent archipelagos for, at least, the terrestrial molluscs (Peake, 1969). Within the 7 archipelagos stretching from New Guinea to Fiji and the Cook Islands 3 faunal elements are represented (Fig. 5). These are the Palaeo-Oriental with affinities in South-East Asia, the Polynesian (or Pacific) extending eastwards, and the southern with affinities not only in New Zealand and Australia, but also in other southern continents. The data emphasize the regular pattern for the distribution of these elements in an easterly and westerly direction and this is consistent with a process of island hopping across water gaps. Minor discontinuities in the pattern being associated with increases in the distances separating islands, for example, between the Solomon Islands and the New Hebrides.

V. Island faunas

A. Area - species curve

Within the last decade studies of island faunas have received an enormous stimulus through the development of models for analysing the structure and evolution of such biotas (Hamilton *et al.*, 1963; MacArthur and Wilson, 1967; Diamond, 1976). The concepts which they generated now occupy a crucial position in many accounts of biogeography, although they are not without their opponents (e.g. Croziat *et al.*, 1974). They have been applied extensively to the investigation of avifaunas and their influence has been seen in a few studies of molluscan faunas (e.g. Peake, 1969, 1971; Cook *et al.*, 1972; Rex, 1972; Valovirta, 1977). These models are based on the relationship that exist between faunal size, that is number of species and island area, where area is a parameter reflecting a number of biological properties. From this was developed the equilibrium model (MacArthur and Wilson, 1967), where species number was considered as a balance between the immigration rate and the extinction rate for species, with isolation and area of the island being important factors in determining the equilibrium point. The predictions of this model have been verified by observation and experimentation (Wilson, 1969) on various islands throughout the world, yet such models do not have universal application, for example they fail in studies of limited taxonomic groups where the sequence of colonization determines the species composition (Williams, 1969).

The relationship between number of species and island area may be defined by the equation $S = CA^Z$, where S = species number, A = island area, C = a parameter that depends on the taxon and the geographic region, and Z = a constant related to the slope of the regression and usually lying between 0·20–0·35. Z is a factor independent of taxonomic group or region of the world, but dependent on the degree of isolation of the islands relative to the particular taxa being considered and rising with increasing isolation. The value of approximately 0·35 obtained

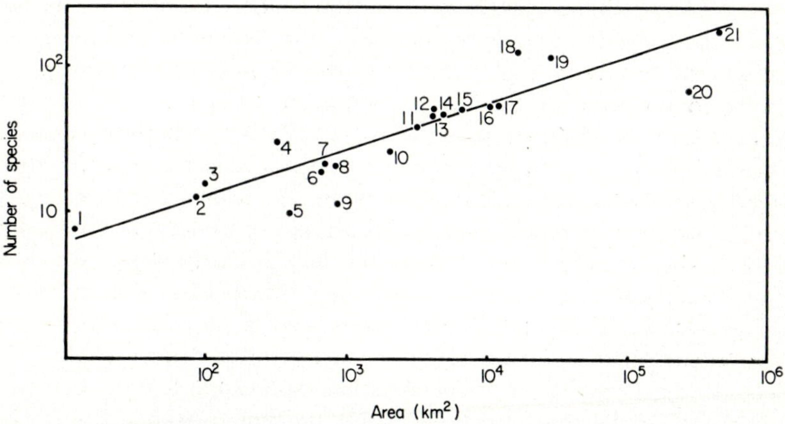

Fig. 6. Area-species curve for terrestrial molluscs on islands in the south-western region of the Pacific Ocean: the geographical limits are Sumatra in the west and Samoa in the east (after Peake, 1969). 1 = Krakatau, 2 = Nissan, 3 = Treasury, 4 = Florida, 5 = Tanna, 6 = Rennell, 7 = Tonga, 8 = Vate, 9 = Erromanga, 10 = Manus, 11 = Samoa, 12 = Malaita, 13 = Espirito Santo, 14 = San Cristobal, 15 = Guadalcanal, 16 = Viti Levu, 17 = New Ireland, 18 = New Caledonia, 19 = New Britain, 20 = Java, 21 = Sumatra. N.B. The data from Krakatau were not utilized in calculation of regression line.

during a study of the terrestrial molluscs of the Melanesian and Indonesian islands (Fig. 6) emphasizes the efficiency of oceanic barriers to the taxon (Peake, 1969); a similar relationship has also been demonstrated for islands in the Western Indian Ocean (Peake, 1971) and for the Madeiran archipelago (Cook *et al.*, 1972). An ability to apply area-species models relies initially on the availability of a faunal inventory used albeit with care, and depend to a considerable degree on the state of taxonomic knowledge and the intensiveness of collecting, for it is important to maintain internal consistency within the data.

Doubts have been expressed concerning the appropriateness of applying such models to molluscan faunas of very isolated islands in the Polynesian (Pacific) region. Solem (1973a) demonstrated that species abundance for the well sampled and studied faunas of islands in this region failed to conform to the equation relating numbers to island area. Indeed the number was much greater than might be anticipated from estimates of faunal size in less isolated situations and was the converse of that recorded for the avifauna, where increasing isolation is associated with reduction in faunal size. For molluscs elevation, if combined with area in a single analysis, provided a much closer correlation with species abundance, however, the relationship was complex. The highest species diversity being found on islands with an elevation greater than 400 m and an area between 13 and 38 km^2. There was a distinct difference between the area-species curve for the islands of Indonesia and Melanesia and the species numbers recorded for high islands of the Central Pacific Basin. It is difficult to conceive of the results being an artifact derived from inadequacies in the data, or that the close correlation between area and species diversity found in other regions does not reflect various biological properties of the taxa and the environment. Variations in the area-species curve, other than those associated with elevation, have been recorded for islands in the Western Indian Ocean, but here it has been suggested that the relatively lower diversity on limestone atolls or cays, as compared to granitic islands, is associated with differences in the water regime, the environment on the porous limestone being much drier (Peake, 1971). Solem (1973a, 1976) has provided a simple explanation for the differences in diversity on Pacific Islands; he suggests that introduced ants have decimated the native faunas of small molluscs on Polynesian islands and consequently in those areas where predatory ants occurred naturally there was always a lower level of diversity. A rich fauna could develop on those isolated islands where an ant fauna was either absent or depauperate; such an hypothesis does not explain the differences in diversity recorded between islands in the same region. Moreover rich faunas of small molluscs have also developed on other islands and in continental areas where it could be anticipated there is a rich ant fauna, witness the radiation in the family Streptaxidae in Africa, particularly the genus *Gulella* (van Bruggen, 1973), although species diversity in this group does not reach the high levels found on some islands.

The fauna of Polynesian islands, compared to those of Indonesia and Melanesia, is composed largely of smaller species (Peake, 1969; Vagvolgyi, 1976). The dispersal patterns between islands must impose a filter limiting the size of propagules, and consequently on the distribution of taxa within the area. Whether a relationship exists between body-size and island area and thus the size of viable populations is problematical and requires further analysis; a preliminary examination of data for some Melanesian islands suggests that this could be another selective factor for small size. It is these small species which have undergone enormous autochthonous radiation on virtually every high island, to the extent that at the species level endemism is over 90%. The high species diversity on these islands would involve closer packing or compression of the ecological range of individual species and here, small size could be correlated with greater compression; thus following a pattern common in many places, where more small, rather than large, animals are found coexisting (Darlington, 1965). Questions arise—does the absence of large forms increase the ecological space available to a wider spectrum of small forms or are there differences in the environments both physical and biological, on islands with low and high diversity? There is no conclusive evidence on which to base an answer, but a comparison of the faunas in the Solomon Islands and some of the Polynesian Islands demonstrates the discrete or restricted nature of the distributions of the molluscan fauna on the latter. The importance of elevation on these islands may be a reflection of increasing ecological compression, associated with the recognition of finer divisions of the environment.

High levels of diversity are found on other islands, for example, the Madeiran Archipelago, compared with the Canary Islands and the Azores (Wollaston, 1878; Cook *et al.*, 1972; Mandahl-Barth, 1950; Backhuys, 1975). However, this may not be a comparable situation with that in the Pacific, but rather a reflection of different ages of the faunas together with increased isolation of islands from colonizing propagules.

B. Dispersal

Islands provide unique opportunities for studying the consequences of the process of dispersal and colonization in a group of organisms that rely entirely on passive means for long-distance movement. The frequency of such activities are probably sufficiently rare for any comparative estimate of different methods to be impossible and "probably no single method is exclusive to any particular group of molluscs" (Peake, 1969). The interest in this problem is reflected in a number of publications and reviews, from the book of Kew (1893) to the review of Rees (1965) and more recently Peake (1969) and Vagvolgyi (1976). The success of dispersal across sea barriers is demonstrated by the data for Krakatau; 44 years after the eruption and the extinction of the original molluscan fauna the number of land snails recorded corresponded very closely to that predicted by the area-species curve (Peake, 1969). In this context dispersal is applied to the process, whether passive or active, whereby animals move from one location to another, even though the distances involved can vary enormously. Moreover, an analysis of the molluscan fauna for Aldabra in the Western Indian Ocean demonstrates the success of trans-oceanic dispersal, besides providing an indication of the rapidity of colonization after extinction of the biota by marine transgressions (Taylor *et al.*, in press). This rapidity is reminiscent of the Post Glacial colonizations in the British Isles, at a time when the molluscan fauna must have been impoverished.

The importance of long-distance aerial dispersal by wind has always been problematical, although it has been increasingly favoured by recent authors. The analysis of the fauna of Hawaii by Zimmermann (1948) was a landmark in emphasizing the importance of this method; it was suggested that a total of 22 or 24 ancestral colonizations were necessary to give rise to the 1064 indigenous molluscan taxa. More recently the studies by Gressitt and Yoshimoto (1963) have demonstrated the correlation between the presence of families of insects in the aerial plankton and their occurrence on isolated islands, but perhaps the most surprising finds in these surveys have been the presence of mineral particles up to 5 × 42·5 mm in size at altitudes of about 3000 m. (Holpzapfel and Gressitt, 1965). It has been suggested "that there is apparently no upper size limit is exceedingly interesting because this may be of importance in spreading particles of microorganisms etc., around the world". The size of many adult snails from Polynesian Islands is within the size range of these particles. Indeed the recent study by Vagvolgyi (1976) demonstrates quite convincingly the importance of aerial dispersal in determining the present nature of the Pacific fauna, yet the source of this fauna still remains problematical, whether from continental Asia (Vagvolgyi, 1976) or as a relict of a fauna that existed in this region, and perhaps others, during earlier geological periods. A Mesozoic or early Tertiary origin has been suggested (Solem, 1959; Kondo and Burch, 1971); the nature of the fauna persisting and being maintained by constant island hopping and colonization of suitable environments, thus while the fauna on an individual island is comparatively recent, the history of the constitutent taxa in the region could be much longer. The palaeogeographical evidence for the Pacific Basin would indicate that the present arrangement of islands does not reflect either the pattern or the large numbers of islands that have existed as far back as the Cretaceous (Menard and Hamilton, 1963); therefore, it might be unnecessary to postulate the presence of oceanic barriers as great as they appear today.

A consequence of aerial dispersal is that although the probability of becoming airborne must be very low, nevertheless, once airborne the probability of remaining in this state could be high or decrease very little over long distances, the limit being set by the tolerance of the animal to the physical conditions. It is possible to envisage that short-distance dispersal could be difficult; here the actual distance the snails crawl would be an important factor, but beyond that a whole range of methods could be envisaged and are indeed recorded relying extensively on the movements of other animals, water and wind.

The molluscan fauna of many islands must have been derived by more than 1 method of disperal. Witness the studies of Valovirta (1973, 1977) on islands in the Baltic where undoubtedly seaborne dispersal has been important over short distances. While the records of snails on the plumage of birds (Rees, 1965) bears witness to the potential of this method, species belonging to the Succineidae and *Vitrina pelucida* have a rather sticky mucus and would appear to use this method (Kondo, 1962; Vagvolgyi, 1976). Nor should the importance of man as a passive agent be underestimated. The overall impression is one of success, terrestrial molluscs do reach and colonize islands or isolated habitats, and the patterns produced are not random, even within the Pacific different taxa have their peculiar distribution (Cooke and Kondo, 1960), probably reflecting the ecological or dispersal properties of the individual species and the distribution of the individual land masses.

Many factors are associated with successful dispersal and colonization, and a wealth of data for analysis are found in taxonomic works, for example, a recent monograph by Cooke and Kondo (1960) of the Tornatellinidae and Achatinellidae found on Pacific Islands. A feature of the molluscan faunas of Polynesia is the high incidence of ovoviparity amongst many of the small species, and the evidence from anatomical studies would suggest that the reproductive rate in many of these species is low; only a single case of viviparity being recorded (Solem, 1972). While in some species of the Endodontidae, protection of the eggs and young is provided by the development of a brood pouch in the umbilicus (Solem, 1969b). The relationship between oviparity and ovoviparity, and the geographical distribution of genera and species, is summarized in Table IV; the widespread groups are confined to taxa that exhibit ovoviparity. The period of highest mortality in many molluscan populations that have been studied is during the egg and very young stages, the development of ovoviparity could contribute to the reduction of the

Table IV. Distribution on Pacific islands of genera belonging to the family Tornatellinidae compared with mode of reproduction (data from Cooke and Kondo, 1960). N.B. The majority of species belonging to these taxa are small (maximum altitude of adult shells less than 0·8 mm), but data in brackets refer to 4 genera found in the Hawaiian Islands where the shells are larger (height greater than 15 mm).

	Ovoviviparous	Oviparous
Single Island	1 (+ 1)	10
Archipelago	3 (+ 3)	3
Widespread	5	—

level of mortality and enhance the probability for successful colonization, particularly in species with very low reproductive rates. It has been suggested that in plants an association exists between self-fertilization or parthenogenesis and colonizing ability (see pp. 498–500); it is interesting to speculate on this relationship in molluscs, for example, the success of many basommatophoran groups in colonizing temporary ponds can be correlated with this factor. Information for other groups is usually inadequate or absent, but self-fertilization has been recorded for species belonging to the Succineidae (Patterson, 1970), *Partula* (Murray and Clarke, 1976) *Ruminia decollata* (Selander and Hudson, 1976) and some slug species (Runham and Hunter, 1970)—all of these are groups or species which exhibit an ability to successfully colonize a wide range of habitats and geographical regions.

C. "Character displacement"

The presence of large numbers of closely-related or congeneric forms on islands or "island habitats", many of which are coexisting in very small areas, presents a unique opportunity to investigate species interactions, in terms of both reproductive and ecological isolation. It is in such situations that one might anticipate the development of "character displacement" (Brown and Wilson, 1956) either to reduce competition for a particular ecological resource, or to enhance reproductive isolation. The distribution of sinistral and dextral forms of *Partula* on Moorea would appear to belong to the latter (Clarke and Murray, 1969); for in the region where the distributions of 2 "typically" sinistral species overlap, one becomes dextral, thereby it can be postulated reducing the potential for hybridization. A slightly more complex situation occurs in the genus *Mendana* on the Marquesan Islands (Baker, 1938–1941); the male reproductive systems of species occurring on the same island differ markedly in the proportions or the development of accessory organs, except for a single case on Nukahiva. Here 2 species which have similar reproductive organs coexist, but are separated ecologically—one is arboreal, the other inhabits litter on the ground (personal observations).

A striking example is provided by species of the genus *Diplommatina*, a group of prosobranch snails found in leaf litter or on rock surfaces in South-East Asia and on the islands of Indonesia and Melanesia (Peake, 1973). The shells of these snails can vary in direction of coiling, size, sculpture and shape; indeed, in some examples the final form is rather bizarre (Tweedie, 1961). In the Solomon Islands up to a maximum of 4 size groups and 5 species are found coexisting in a single locality; these size groups are always discrete and do not overlap (Fig. 7). The distribution of species within these groups results occasionally in 2 species with similar size ranges coexisting; in these instances, one is always dextral and the other sinistral. This pattern is repeated on different islands and at many localities; it would appear, moreover, to apply to coexisting populations on limestone hills in Malaysia (Peake, 1973). Indeed in this region the development of bizarre shapes always occurs in 1 species of a pair

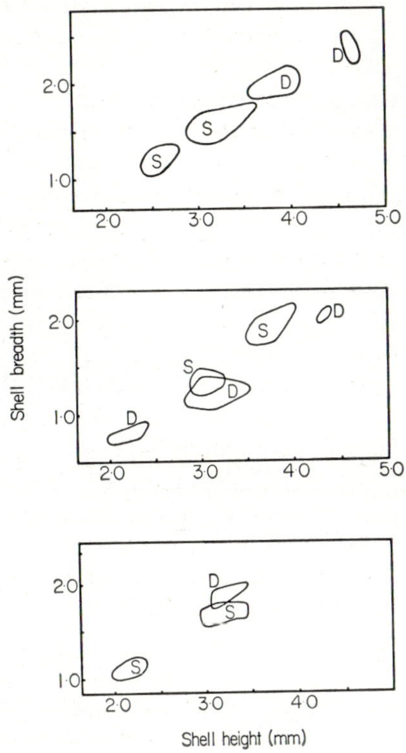

Fig. 7. Distribution of size classes in 3 samples of *Diplommatina* species from the Solomon Islands. D = dextral forms; S = sinistral forms. Upper figure, 4 species illustrating non-overlapping size classes. Central figure, 5 species with the size classes for 2 overlapping. Lower figure, 3 species with the size classes for 2 overlapping (after Peake, 1973).

with overlapping size ranges. Further, the same morphological form was found in different size groups at different localities, that is populations with the same shape, sculpture and apertural form, suggesting that there was adaptive value favouring divergence in size when 2 forms with the same direction of coiling coexisted. Thus the direction of divergence in individual species, whether to become large or small, would appear not to be predictable, but determined rather by the order of colonization and the degree of variability present in the founding population. There is no information whether these "character displacements" enhance reproductive isolation or reduce competition for limiting resources. Arguments can be advanced that both could apply, and it is interesting to note that the size differences between coexisting size groups approaches very close to the empirical ratio determined by Hutchinson (1965) for differences between structures or organs involved in feeding sufficient to allow different portions of the available food supply to be taken. Certainly the distribution of these small snails on the limestone hills in Malaysia cannot be random, as has been suggested (Purchon, 1977).

The application of the concept of "character displacement" is fraught with problems (Grant, 1972, 1975), but the suggestion of Schindel and Gould (1977) that it should be treated like any other evolutionary rule is extremely timely. Such rules apply to the "patterns of variation commonly seen, though not without counter-examples, which may or may not be explained by any one of several possible mechanisms".

VI. Distribution: concluding remarks

The development of biogeography has progressed through various stages from the descriptive and analytical to the predictive (Vuilleumier, 1975). Undoubtedly for terrestrial molluscs the descriptive phase has been paramount and will continue, of necessity, to be so for a long time. Detailed descriptions of regional patterns, both recent and past, are unavailable for many areas of the world and cannot be undertaken without further collecting and documentation. Moreover, major reassessments or modifications to the patterns at present recognized will result from taxonomic revisions and innovations; particularly where these reflect a clearer understanding of the phylogenetic relationships of the groups.

The nature of the recent distribution patterns exhibited by the group must reflect the geological age of the constituent taxa and the processes of evolution and extinction. It is hardly surprising that many families of terrestrial molluscs, including the Stylommatophora, exhibit disjunct distributions; but that many of these are not open to simple explanations or do not illustrate some common patterns might not have been anticipated. The geological evidence indicates that the facies of these patterns were recognizable in many continental regions at the onset of, or during, the Caenozoic.

Indeed, there is a close correspondence between the distributions of both fossil and recent taxa, with only a minority not being coincident, for example, the tropical taxa that had an extended range in Europe during the early Tertiary. This would suggest that differentiation of the major faunas must have occurred, at least, during the Mesozoic, possibly coincident with the major disintegration of the single continental mass during the Cretaceous. Even though the level of taxonomic diversity prior to this period could have been high, evidence regarding the level of provinciality is absent, but extrapolating from other groups it might be reasonable to suggest it would have been low. It is interesting, therefore, to speculate on the factors or features which led to the development of provinciality and on the role of extinction during the Caenozoic in determining the distinctive nature of the fauna of regions like Africa.

It is difficult to assess the importance of dispersal in determining distribution patterns in continental areas, certainly the evidence from islands and from the recolonization of post-glacial habitats indicate that this is both feasible and remarkably effective in environments depauperate in molluscs. Indeed, the changes in the fauna of Europe throughout the Caenozoic, compared to the picture of stability in North America, probably reflect differences in the isolation of the 2 regions and consequently in the probabilities of successful dispersal from adjacent areas, Europe throughout this period being closely associated with Asia, while the juxtaposition of North with South America occurring at a much later date. Closely associated with such concepts of dispersal is the notion of a centre of origin for a taxon, a concept which has been seriously challenged by Croziat *et al.* (1974); the latter emphasized the development of distribution patterns through the fragmentation of ranges (vicariance events) with dispersal playing a lesser or modifying role. For terrestrial molluscs it is feasible to consider the many examples of disjunct distributions as resulting from the fragmentation of once wider distributions; the development of the many features of the continental faunas reflecting the allopatric differentiation from once widespread and common ancestors on Mesozoic land mass. A considerable body of the evidence being obscured through subsequent extinction and dispersal; the importance of these processes varying considerably depending on the region and the taxa. Accepting such a concept would again emphasize the importance of the Mesozoic as a period when ancestral biotas subdivided in response to a changing geography, while the Caenozoic, at least in some regions, could be regarded as a time when dispersal and extinction were important in the evolution of recent faunas. Early Palaeozoic faunas could have been composed largely of small species, and in this context it is interesting to consider the habitat provided by a predominantly gymnosperm and pteridophyte vegetation. The form of the trunk or bole of the "trees" could have been more conducive to the evolution of small forms, and further the presence of a number of obnoxious plant metabolites in the leaves could have inhibited their use as food by the molluscs. The development of arboreal and indeed, larger forms was probably associated with the evolution of phanerogams, with larger leaf size, increased palatability of the plant material and a different texture to the leaf litter and humus. However, a wide range of species may not have been limited to such closed habitats, but have occurred in open environments associated with rock surfaces, where the dependence on phanerogams would not have been so strong.

VII. Ecological distribution and abundance

A. Introduction

Andrewartha and Birch (1954) conceived of the distribution and abundance of species as being different aspects of the same problem, with the physical features of the environment playing the major role in orchestrating the underlying processes. Although their controversial concepts have not received universal acclaim, many examples which appear to support their thesis can be found amongst terrestrial molluscs. In this group the relationships between distribution, abundance and edaphic factors of the environment have always been accepted, if only intuitively. The early studies of Favre (1927), Geyer (1927), Økland (1925, 1930), and many other workers recognized various aspects of this affinity, but it was Boycott (1934) in a Presidential Address to the British Ecological Society who provided an overall framework and the impetus for a considerable body of future research. He synthesized an enormous range of observations, many casual, to produce a masterly analysis of the factors affecting the distribution and abundance of terrestrial molluscs. The basis for Boycott's study had been established at the end of the previous century by the inauguration of a mapping scheme for non-marine molluscs (Roebuck and Boycott, 1921). Utilizing the data from this study and his own experience he recognized three factors as important in determining distributions, these were the history of the fauna, the climate and the nature of the habitat; with the latter usually referring to the presence or absence of shelter and calcareous material. The influence of food and competition were thought to be unimportant, although the difficulty of adequately demonstrating the action of these factors was recognized. Possibly the crux of Boycott's study was that he recognized that "human beings usually take a wider view of the environment than snails" (Sparks, 1969) and "that the ecology of a snail species may enable it to occur in a number of, what to man's eye are different, environments, but which to the snail are identical" (Evans, 1972). Both of these quotations highlight the anthropomorphic trap which has and still does bedevil many studies of terrestrial molluscs.

The phytosociological methods developed by Braun-Blanquet for describing patterns of plant associations have had an important, although largely local, impact on the analysis of habitat distributions of the molluscan fauna (Bruijns, 1947; Bruijns *et al.*, 1959). Here the qualitative and quantitative distribution of molluscan assemblages have been equated with phytocoenoses defined on the basis of the characteristic plant species. The plant communities are considered as indicators of environmental conditions, but whether such divisions provide a suitable basis for the description of molluscan habitats is questionable, certainly many other workers have laid greater emphasis on the structure of the habitat (e.g. Waldén, 1955; Peake, 1968; Cameron, 1973).

There is little evidence that in temperate regions snails and slugs form clearly defined associations or communities with interacting components. The habitat distribution of individual species being determined rather by the range of tolerance to particular environmental factors or features (Boycott, 1934; Valovirta, 1968; Ant, 1963, 1969; Wareborn, 1969). Yet it has been suggested that on the basis of quantitative, as opposed to a qualitative, analysis of species compositions that the assemblages associated with some habitats can be characterized in terms of relative densities or "constancy" rather than species composition (Ant, 1963; Kornig, 1966; Waldén, 1955; Cameron, 1973). Nevertheless, it must be recognized that many of these faunistic studies have been undertaken in areas where either major environmental discontinuities exist or where human disturbance has given rise to a mosaic of habitats with very different histories. Thus the recognition of characteristic molluscan assemblages associated with particular habitats must reflect to a considerable degree gross or artificial discontinuities in the environment. In some tropical environments the distribution of taxa provides a marked contrast, for the evidence would indicate that interactions occur between species and that some form of "biological accommodation" or competitive interaction operates in these faunas, even though the basis is not understood (see pp. 472–474; Peake, 1968, 1973).

Studies of the distributions of plant species along environmental gradients would appear to raise many problems similar to those recorded for molluscs in temperate regions. Questions, for example, regarding the replacement of species, whether these occur along sharp boundaries with those for a number of species being coincident, or whether each species is distributed in an individual manner, with no two alike (Whittaker, 1970)? It can be argued that the responses of the individual plants have evolved in relation to one another, thereby reducing competition by niche differentiation. If analogous processes have been important in the evolution of the responses of molluscan species to environmental factors, then presumably the complex histories of the faunas in mny temperate regions would have obscured the vital evidence for many taxa. It may be unwise, therefore, to speculate on the evolution of community structure in terrestrial molluscs on the basis of either a biased sample or a limited number of surveys in suitable areas. There are unfortunately few detailed investigations which provide an insight into intra- and inter-specific interactions. Indeed the extensive literature concerning the theories of community organization and problems of species diversity (for example, Hutchinson, 1959; Menge and Sutherland, 1976) appear to have made little impact on ecological studies of terrestrial molluscs.

Populations existing at the periphery of a species range frequently have a more restricted habitat distribution than those living at the centre of under optimum conditions. The northern extension of some molluscan species on isolated but south-facing slopes (Andersson, 1961) and along the coast are easily recognizable examples. In such situations many species exhibit shifts in habitat preference and these changes are frequently associated with interactions between limiting factors, where the high level of one modifies the effect of another. Compensating shifts in calcium requirements under adverse or different climatic regimes provides an example of this phenomenon. Populations at the periphery of the species distribution being obligatory calcicoles restricted to calcareous habitats, while elsewhere they are more widespread (Boycott, 1934; Boycott and Oldham, 1936; Baker, 1958; Peake, 1968; Cameron, 1973); however, this pattern is often more complex than an initial perusal of the literature might suggest (Jones and Clarke, 1969).

Many recent studies reflect the development of quantitative methods for the investigation of molluscan populations. Therefore, a brief review of the techniques employed would seem appropriate; they are basically of two types, those that rely at some stage on hand collecting and those that employ some degree of automation (Newell, 1971).

(a) Hand collecting: these are usually applied only to surface dwelling species and suffer with the disadvantage of being labour intensive. They also have the tendency to introduce a bias through underestimating small individuals (Goodhart, 1962) or species that depart from the current "search image" of the investigator. These difficulties may be partially alleviated by taking samples of soil or litter samples and using graduated sieves to sort the material into size groups. Sweep netting has been employed when the snails are living away from the ground on low vegetation.

Indirect methods rely extensively on marking-release-recapture techniques taking care that random mixing of individuals occurs; methods of marking include painting small coloured spots on the shells, insuring that the mark is not so conspicuous that predation is increased (Cain and Sheppard, 1950); boring small holes through the shell, near the lip; here the number of possible combinations of holes is so great in large helicids, that adult snails can often be individually marked (Wolda, 1963). Where the shell is unsuitable or there is no shell, the animals can be freeze-branded (Richter, 1976) or marked by feeding on a diet containing a vital dye, Neutral Red, the tissues that are stained being visible through the body wall providing the skin is sufficiently translucent (Newell, 1965; South, 1965). Radioisotopes have also been used (Webbe and Read, 1962; Newell, 1965; Francois *et al.*, 1965)—this method permits the animals to be relocated in the field using a Giger-Müller detector.

(b) Automated: these have largely been developed from the flotation methods used during investigations of soil faunas and rely on differences in the specific gravity of objects. They have been employed successfully in the study of slug populations (South, 1964, 1965; Hunter, 1966; Runham and Hunter, 1970) and in a rather ingenious manner for snails, by utilizing the bouyancy of empty air-filled shells (Vagvölgyi, 1953; Mason, 1970b) or the animals (Williamson, 1959). The use of poisoned baits (Webley, 1964, 1965) and "refuge traps" (Miles *et al.*, 1931) or a combination of these (Thomas, 1944) depends on the activity of the animals and consequently are influenced strongly by the weather. Similarly, visual methods of counting animals seen along fixed transects or walks provides only a limited amount of information on total population size, but a wealth of data concerned with the activity of the animals (Barnes and Weil, 1944, 1945; Crawford-Sidebothom, 1972). Indirect methods which rely on the amount of food eaten in a known time provide a measure of activity and population size.

Quadrating techniques have been employed in a wide range of investigations, although frequently these have taken very little account of the aggregation or non-random distribution of the animals. An overriding factor limiting the development of such methods has been the feasibility of sorting the molluscs, however, a recent study in a rather extreme habitat was able to overcome this problem of aggregation by employing an expanding quadrat technique (Yom-Tov, 1972).

The availability of automated techniques has undoubtedly enhanced the development of energetic studies which have examined the rates of assimilation and the role of molluscs in the breakdown of litter. Reviews of this work are found in Mason (1974) and Runham (1975); more recent studies include those of Richardson (1975a, 1975b), Jennings and Barkham (1975a, 1975b), and Williamson and Cameron (1976).

B. Influence of calcium and pH

Many authors have correlated the distribution of terrestrial molluscs with the presence of calcareous substrates and with variations in pH (Atkins and Lebour, 1923; Lozek, 1962; Ant, 1963; Waldén, 1965; Valovirta,

1968; Wareborn, 1969, 1970), but the relationship is so complex that some authors have questioned the importance of calcium *per se* and have emphasized other attributes associated with calcareous environments (Lozek, 1962), such as differences in texture and the generally warmer nature of chalky soils. These comments obscure the requirements of many molluscs for calcium in the construction of the shell, for successful reproduction (Kunkel, 1929; Oosterhoff, 1977) and for other physiological processes (Robertson, 1941). Indeed, in areas where calcium appears to be particularly scarce, snails are found crowded on old buildings, where the crumbling mortar provide a source of calcium (Lozek, 1962), or even rasping the shells of living snails or marine shells dropped by seabirds (Peake, 1960: Bleakney; 1966); in these extreme situations the available calcium in an ecosystem may be bound into an almost closed cycle. Under experimental conditions it has been possible to demonstrate that in the absence of calcium some species fail to copulate, have a higher mortality or produce shells with less whorls (Rensch, 1932; Voekler, 1959; Ant, 1963), but such data often have little application to natural conditions. The investigations of Oldham (1929, 1934) and Oosterhoff (1977), however, are interesting for they have demonstrated experimentally a direct relationship between shell thickness and the availability of calcium; an observation which was consistent with field observations (Boycott, 1934; Gould, 1969). Obviously other factors are involved including the genotypic response of the animal and a correlation between shell thickness and rainfall has been demonstrated (Pollard, 1975).

The relationship between egg shell thinning in birds and the presence of organochlorine insecticides has been well documented, but a similar response has been observed under experimental conditions in the shell of *Helix pomatia*. It might be anticipated that if the reaction is similar in other molluscan species than in agricultural areas, where the levels of DDT residues are high, reduction in shell growth will have occurred already (Cooke and Pollard, 1973).

The influence of pH on abundance and distribution, independent of calcium concentration, is rather difficult to assess, but some authors have claimed that within a wide range pH is unimportant (Voelker, 1959; Burch, 1955); while the field data of others indicate a strong correlation. Usually numbers of species and individuals increase with decreasing acidity, but with a slight decrease towards the very high pH values (Ant, 1963, 1969; Valovirta, 1968; Wareborn, 1969, 1970). However, Valovirta (1968) has demonstrated that in Finland there is not a uniform pattern to this response covering all species, each demonstrates an individual relationship, even though there is some degree of association into groups (Table V). This is reminiscent of the data for woodland molluscs in Southern England (Cameron, 1973), where although species numbers in-

Table V. Grouping of species by abundance and preference for certain pH classes; abundances given as number of specimens per litre of sifted litre (data from Valovitra, 1968).

Category	pH classes 4·6–5·5	5·6–6·0	6·1–6·5	6·6–7·5	Total number of specimens
1. Maximum density in pH class 5·6–6·0					
Punctum pygmaeum (Drap.)	51·3	63·1	48·7	44·1	1780
2. Numbers increase up to pH class 6·1–6·5, while in higher class numbers remain almost unchanged					
Euconulus fulvus (Müll.)	16·3	20·2	29·9	28·0	915
Discus ruderatus (Hart.)	10·8	21·6	27·7	28·7	902
Nesovitrea hammonis (Ström)	23·8	20·7	35·2	23·5	933
3. Maximum numbers at pH class 6·1–6·5 (*C. lubrica* has a slightly divergent pattern)					
Nesovitrea petronella (L. Pfeiff.)	4·6	8·7	10·1	7·5	295
Columella edentula (Drap.)	6·3	6·6	11·7	9·1	322
Vertigo substriata (Jeffreys)	3·8	10·7	14·0	5·9	329
Cochlicopa lubrica (Müll.)	1·3	6·1	13·4	5·7	276
4. Numbers mainly increase with pH; numbers of individuals at either extreme similar to category 3, but there are no peaks					
Vertigo ronnebyensis (West.)	4·6	5·9	5·1	4·2	173
Vitrina pellucida (Müll.)	6·4	3·1	4·8	7·4	182
Vertigo pusilla (Müll.)	1·7	1·4	5·5	6·5	165
Cochlicopa lubricella (Porro)	0	0·9	6·0	6·7	166
5. Numbers always small					
Clausilia cruciata (Studer)	0	0·7	2·4	3·0	73
Zoogenetes harpa (Say)	0·4	1·6	1·3	0·1	29
Vertigo alpestris (Alder)	0·4	0	1·5	1·3	36
Columella aspera (Waldén)	0	0·6	0·4	0·1	11
6. Occur mainly in areas where pH exceeds 6·0					
Acanthinula aculeata (Müll.)	0	0	3·2	4·8	102
Clausilia dubia (Drap.)	0	0	0	1·2	17
Cochlodina laminata (Mont.)	0	0	0	0·8	12
Limax marginatus (Müll.)	0	0	X	0	12
Vallonia costata (Müll.)	0	0	6·4	13·8	263
Bradybaena fruticum (Müll.)	0	0	2·0	0·6	10
Total number of specimens	301	1402	2421	2867	6991
Number of sample plots	5	15	20	25	
Number of specimens per litre	126	172	229	203	

creased with increasing pH the abundance of individual species shows a range of permutations.

Soluble calcium ions are not distributed evenly through the soil profile, but are absorbed in the humus and partly in the uppermost layers of the mineral soil (see Valovirta, 1968 for extensive bibliography; Wareborn, 1969, 1971). The source of this calcium is probably from vegetable material, and reflects the ability of some plant species to selectively accumulate certain inorganic salts (Sutcliffe, 1962; Crowell, 1973). Consequently, irrespective of other factors, the humus layer offers a particularly favourable environment for molluscs, but the organic compounds in which the calcium is bound may also influence its availability and palatability. For example, *Helix aspersa* utilizes calcium more efficiently in the carbonate combination than the phosphate (Wagge, 1952), while in calcium-poor environments the addition of calcium oxalate and citrate has positive effects on reproduction, with the latter having the more pronounced influence. This could explain the occurrence of richer molluscan faunas in litter from ash, lime, maple and elm, with high concentrations of the citrate, compared to those from oak, rich in oxalate (Wareborn, 1970). Moreover, in environments where the diet provides insufficient lime, molluscs appear to ingest the soil and actively extract calcium ions from this source (Crowell, 1973).

A study by Williams (1970a, b) of freshwater snails in Southern Africa provides a firmer basis for extrapolating between the distribution and abundance of molluscs and the pH or availability of calcium in the environment. In the field freshwater snails exhibited similar patterns to those found in terrestrial forms, in respect of their reactions to variations in concentrations of calcium. These reactions, together with those to bicarbonate, were explored in greater detail under laboratory conditions where the chemical features could be varied independently. The responses of 2 species of snails were measured; one, *Biomphalaria pfeifferi* (Krauss), was characteristic of medium and hard waters, the other, *Bulinus globosus* (Morelet), occurred over the whole range. The basis for the comparison was the biological parameter r_m; Andrewartha and Birch (1954) stated that the intrinsic rate of natural increase (r_m), when calculated under ideal conditions, would be as characteristic of a species as any distinct morphological feature. Yet as Williams points out, it might be impossible to achieve such a hypothetical resolution for r_m under experimental conditions. Nevertheless, in these experiments the species with the restricted range demonstrated a much wider range of r_m values when both calcium and bicarbonate were varied than the species with the wide range. The crucial test was the comparison of the data from laboratory experiments with that obtained in the field (Fig. 8). The total number of snails of both species collected at each station in the field was taken as a measure of abundance and compared with the r_m values obtained in the laboratory; comparisons being made at the same calcium ion concentrations—this being the common factor between the 2 sets of data. A significant positive correlation was found for the species with the restricted field distribution, while for the other species with a wide distribution there was no significant correlation. Thus, for one species the ionic composition of the water influenced abundance and thus distribution, while the other was independent of such constraints.

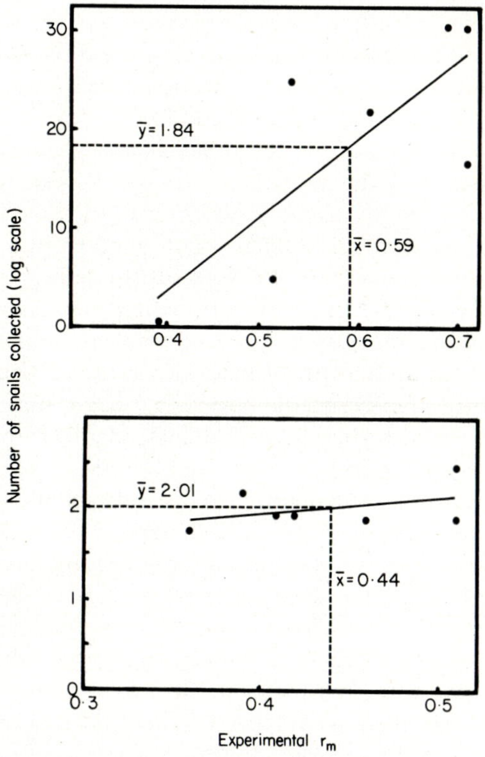

Fig. 8. Regressions of experimental values obtained for r_m against numbers of snails collected in the field, comparisons being made at comparable levels of calcium concentrations (Williams, 1970b). Upper figure data for *Biomphalaria pfeifferi*, lower figure for *Bulinus globosus*.

Although such experiments would be difficult to perform on a terrestrial species, supporting evidence is available for the influence of calcium on growth and oviposition rates in *Helix aspersa* under laboratory conditions (Crowell, 1973).

C. Climate

Information concerning the relative importance of climatic factors in the ecology of terrestrial molluscs has frequently been inferred from correlations between distribution patterns and selected parameters referring to simple temperature and rainfall patterns. Although such comparisons are usually crude and ignore the subtleties of the causal relationships they have provided a fruitful source for speculation (Kerney, 1970, 1976a, 1976d; Ant, 1963, 1969; van Bruggen, 1969). Changes in distribution patterns since the late-glacial have provided further evidence of the influence of temperature variations, and this is particularly noticeable in species at the limits of their range (Waldén, 1955; Ant, 1963; Kerney, 1968). In the British Isles these have been largely forms which appear to be limited by winter cold, for example *Lauria cylindracea* (da Costa), *Helicodonta obvoluta* and the prosobranch snail *Pomatias elegans* (Müll.), although another species which exhibits changes in distribution, *Ena montana* (Drap.), appears to tolerate cold winters but requires warm summer temperatures (Kerney, 1968). Unfortunately, speculation concerning the influence of former climatic conditions is frequently based only on variations in temperature, ignoring changes in humidity or rainfall as the recognition of these is problematical; yet the influence of these and other factors since the "climatic optimum" must not be discounted. Factors such as the destruction of habitats, changes induced through leaching of moderately calcareous soils or a failure to recolonize suitable areas after local extinctions have occurred. Undoubtedly the stimulus for ecology from this type of research will expand considerably as more detailed information on distribution patterns becomes available through the sophisticated mapping schemes that are now being developed (Kerney, 1976b).

Boycott (1934) when discussing climate, emphasized the mitigating influence of shelter in producing local microclimates where conditions were more favourable than the ambient environment. In this respect the structure of the habitat is of prime importance in providing an insulating effect, whether, for example, from exposure to the desiccating effects of wind or extreme fluctuations in temperature. Indeed the presence of localized areas of favourable conditions or refugia such as fallen branches in temperate woodlands (Lloyd, 1963) or arboreal ferns in tropical rainforests, will frequently influence the patterns of dispersion, as will localized sites for feeding, hibernation, aestivation or oviposition. Here the size of the shell or animal will undoubtedly determine the type and form of the refugia utilized. For slugs or snails that live in the soil, favourable conditions may be associated with increasing depth or particular textures of the soil (Stephenson, 1966), while the evolution of a cavernicolous fauna could be considered the ultimate in the selection of refugia with stable climatic conditions; indeed the comparatively recent recognition of such elements in a number of continents demonstrates how investigations of this habitat have been neglected (Boettger, 1940; Tesclefs, 1961; Climo, 1975).

Active snails or slugs frequently move away from resting sites or shelter when conditions are favourable, returning at the cessation of activity to the same region or even site; while slime trails provide important "clues" in this behaviour (see p. 491), orientation to physical landmarks is probably equally important in some of the larger species. Under experimental conditions the latter can distinguish dark shapes against a light background (Frankael and Gunn, 1961), while in the field under low light intensities at night they can perceive and move towards shapes silhouetted against the sky such as trees and shrubs, which are objects frequently associated with suitable resting places (personal observations).

The importance of climatic factors in determining local distribution patterns and probably the outcome of competitive interactions between ecologically analogous or closely related species has been suggested by various authors (e.g. Heller, 1975) but are highlighted in a series of detailed studies of *C. nemoralis*, *C. hortensis* and *Arianta arbustorum* (Cameron, 1969, 1970a, b, c; Harvey, 1973, 1974). Earlier Oldham (1929) and Boycott (1934) had concluded that the exclusion of *C. hortensis* from sand dunes by competition, within the range of *C. nemoralis*, was the only satisfactory, although circumstantial, evidence for competition amongst British land molluscs. Cain and Currey (1963) had explained the distributions of the 2 *Cepaea* species on chalk grasslands in southern Britain in a similar manner, although the evidence was open to alternative explanations (Cain *et al.*, 1969). The 3 species cited above are similar in shell-size and shape and are polymorphic for shell-colour and banding patterns and although they can coexist, there are differences in geographical distribution, ecology and behaviour. Under laboratory conditions there is a graded response in activity to variations in temperature and humidity, *A. arbustorum* favouring cooler conditions and higher humidities, with *C. nemoralis* at the other end of the spectrum and *C. hortensis* occupying an intermediate position. Although these responses were obtained under highly artificial conditions, they accord remarkably well with the field data for the distributions and climatic conditions under which the species live; *A. arbustorum* occurring further north and at higher altitudes than *Cepaea* species and in areas favoured by higher rainfall, the 2 species of *Cepaea* exhibiting a corresponding pattern, with *C. nemoralis* occurring in the drier and warmer environments. There are a wide range of other responses which can be correlated with climatic factors, including diurnal or nocturnal activity, survival, a range of reproductive activities and growth (Wolda, 1967; Cameron, 1970b, c). These studies have demonstrated that, although an understanding of some of the factors affecting the distribution of species in 1 area is feasible, there are no criteria available which allow these to be extrapolated uncritically to other regions. Moreover, within 1 area there are situations where the available criteria are inadequate to explain the distribution of the *Cepaea* species, and here it has been suggested the 2 species could be in competition. The evidence has been derived largely from studies of distribution patterns where there are sharp delineations between colonies of the 2 species, a significant excess of single species populations in regions where both species exist, or changes in habitat distribution of *C. hortensis* in areas where *C. nemoralis* is absent. The outcome of competition being determined possibly by climatic conditions in, at least, marginal environments where the vegetation is less diverse than those where mixed colonies occur.

The importance of the water economy and temperature regimes in the ecology of pulmonates has been emphasized by virtually every author who has considered the group, and Machin (1975) provides an excellent review of the available information. However, the interactions between these factors are complex and they are frequently not clearly understood, as demonstrated, for example, by Crawford-Sidebotham (1972) in a study of activity patterns in slugs. Although the Stylommatophora are dependent on water and there are limitations to the temperature extremes which they can tolerate, the group is remarkably successful in colonizing a wide range of habitats, including deserts where the availability of water, food and high temperatures pose major problems (Yom-Tov, 1971; Schmidt-Nielsen, 1972). It is clear that a wide range of the structural and behavioural traits exhibited by terrestrial forms can be considered as adaptations to survival in inhospitable conditions, for example, many pulmonates and their eggs can withstand high levels of water loss for at least short periods (Machin, 1975; Runham and Hunter, 1970). Another series of problems are posed at the other extreme of the climatic spectrum, where in the completely wet environments of some tropical rain-forests pulmonate snails and slugs are either absent from the litter or occur in only very low numbers. Here death from "drowning" or "over-hydration" can occur and it is, therefore, interesting to note that prosobranch species, with a permanent barrier to the shell aperture, an operculum, are common in such habitats (Peake, 1969). In temperate regions flooding can also result in the death of many snails and slugs (Boycott, 1934); moreover the eggs of *Arianta arbustorum*, a species which frequently lives in habitats with very high water tables, are laid on the surface and not buried in the soil, as are those of many of its ecological analogues found in drier environments

(Cain, in Heller 1975). The snail shell can be regarded, somewhat paradoxically, as an organ not only preventing water loss but also water entry, besides providing thermal insulation and protection.

The majority of terrestrial molluscs, however, live in fluctuating environments which have more moderate or mesic climates. Here the availability of "pallial water" could be an important factor in enabling the animal to survive short-term fluctuations, although its presence has so far been demonstrated in only a few species of snails (Blinn, 1964) and slugs (Lyth, 1973). Similarly, variations in activity cycles between individual molluscs within the same population result in only a proportion of the population being active and susceptible to sudden perturbations (Howes and Wells, 1934; Barnes and Weil, 1944, 1945). Individuals which are in an inactive phase are often concealed in refugia and insulated from extreme conditions (personal observations). It is possible that such behaviour patterns could also enable higher population densities to be maintained.

Changes in climate can result in the occurrence of a wide range of coincident, but independent, events affecting both the form and species diversity of the land snails; as in the "chronocline" of Sacchi (1965). Such patterns are clearly visible in the faunas recorded from Pleistocene glacial and interglacial deposits in continental areas of both Europe and North America, while a more detailed analysis of fossil faunas from Pleistocene Bermuda has demonstrated that variations occur not only in the frequency of taxa, but in size, shape and thickness of the shells (Gould, 1969, 1970). It has been suggested that the underlying causal agent which unified these responses was varying access to calcium carbonate. The desire to find such simple generalizations, which hold for a wide range of situations, has led authors to formulate various ecological rules. Rensch (1932, 1937, 1959) has been foremost in linking climate and variations in shell form, but even he emphasized that exceptions to the rules he proposed frequently occurred. The validity of many of his rules have since been doubted (van Bruggen, 1973; Sacchi, 1965). Greatest size was linked to optimal climatic conditions; relative weight of the shell or shell thickness being highest in forms exposed to the highest radiation or the greatest aridity. Snails with smooth glassy brown shells tending to predominate in cold climates and those with white or strongly sculptured shells in dry climates.

The correlation between white shells and aridity of the environments is undoubtedly strong (Pilsbry, 1939–1940; Cain, 1977a), for colour and shell-form are important in reflecting a very high percentage of the total incident radiant energy (Yom-Tov, 1971; Schmidt-Nielsen *et al.*, 1972). Yom-Tov (1971) presented an interesting example where the different colouration of the young and the adult was associated with different behaviour patterns; the brownish juvenile occurring in more concealed positions than the white adult. Klemm (1939) and Knipper (1939) both attempted to apply rules to studies of European molluscs. The former found that the ribs on the surface sculpture were broader at more southerly localities, reflecting variations in speed of growth with temperature (Mayr, 1940), while Knipper tried to establish a wide range of correlations, including the presence of heavy armature in the aperture of shells in drier conditions, and lighter in more moist environments. The latter applies to a limited extent to helicids and related groups in Europe although its application is not universal, e.g. amongst North American polygyrids there is a correlation between heavy armature and elevation and here elevation implies increased precipitation (Vagvolgyi, 1968). Indeed various suggestions have been made to account for the development of heavy armature in land snails, including protection from predators (Clench in Vagolgyi, 1968; Solem, 1976) and although they could afford some protection in some species and situations this does not appear to be the complete picture. Other explanations include dependence upon the abundance of calcium in the environment, reduction in water loss and orientation of the internal organs. Exceptions to correlates of size are frequently found on adjacent islands where both large and small forms can be found (Knipper, 1939; Mayr, 1942; Sacchi, 1961).

The process of searching for universal patterns is clearly illustrated in a recent study of shell shape by Cain (1977b); in many regions the shapes fell within 2 distinct groups on the basis of the height/breadth ratio, but here again the causal agents were not clearly understood. Somewhat paradoxically Rensch (1937, 1959) had pointed out that a considerable degree of convergence occurred in shell shape amongst a number of circum-Mediterranean helicid species belonging to different genera. The 2 extremes were rounded globose shells with relatively smooth surface, as in *Cepaea*, compared to shells with the whorls highly keeled, a depressed spire and some surface sculpture. These extremes were recorded either within the same species or groups of taxa, but none of the available ecological rules appeared to apply in these situations. The only common feature was that the forms with highly specialized shells, e.g. strongly keeled, had a narrower geographical range and limited ecological distribution (Sacchi, 1955, 1965).

D. Density-dependant reactions

Until recently the evidence for any form of density-dependent reaction in terrestrial molluscs was largely circumstantial and intuitive. Competition for food had been suggested as a basis for such interactions in a study by Pomeroy (1969), but even this was in an unusual situation, where the species being studied had been introduced. Detailed studies have been undertaken on the role of food in a number of species, under both field and laboratory conditions (Grimes and Blythe, 1969; Grimes *et al.*, 1970; Grimes *et al.*, 1968; Wolda *et al.*, 1971c; Richardson, 1975; Jennings and Barkham, 1975a, b; Williamson and Cameron, 1976) and while food preferences, seasonal variations and differences in food requirement between areas within the same species have all been demonstrated, no conclusive evidence has been produced that indicates quantity of food is a limiting factor, yet many of these studies have relied on subjective assessments of availability. There is some degree of agreement with Boycott's statement that "food has no influence by its quality or quantity on the recurrence of our land molluscs", however, more than an energy source is required (Wolda *et al.*, 1971c) and various authors have been forced to speculate on the quality of the food source, whether it is the presence of phosphorus (Grimes *et al.*, 1970) or the effect of unknown trace elements, vitamins, plant metabolites and other compounds (Wolda, 1973; Richardson, 1975; Yom-Tov, 1972; Thomas, 1973; Thomas *et al.*, 1975). In laboratory experiments variations in growth rates of juvenile snails of *C. nemoralis* have been associated with different diets and attributed to differences in the amounts of food consumed and assimilated rather than the nutritional values of the diets. These results apparently contradict those derived from investigations of natural populations by Williamson and Cameron (1976) who suggested that "the presence of certain foods facilitate the use of others". In field populations the same authors found it impossible to correlate adult population densities with the proportion of palatable herbs in the vegetation. If some essential component of the diet is limiting in the field it could be through some subtle mechanism affecting mortality at different stages in the life cycle, the number of eggs laid or growth and the onset of sexual maturity. The studies of the reproductive biology of *Melampus* by Russell-Hunter (1970, 1978) are relevant here in demonstrating that limitations to fecundity are imposed by variations in assimilation rates and the availability of organic nitrogenous material throughout the breeding season.

Although there are a wealth of records or studies concerned with predation, there are remarkably few which suggest it is a key factor influencing population densities (Wolda, 1963, 1973). Nevertheless, the level of thrush predation on the larger helicids can be very high in some regions of North-West Europe, even though there is a considerable degree of seasonal variation depending, it has been suggested, on the availability of other food supplies; the molluscs being utilized as emergency rations (Wolda, 1963, 1973). The snails are also an important reservoir or conservator of calcium in the environment and are a major source of that element to some bird species just prior to egg-laying (Schifferli, 1977). Moreover, a few species appear to have a diet restricted exclusively to snails, usually the larger amphibious forms and both reptiles and amphibia have been recorded as feeding on terrestrial forms (Peake, 1968; Wolda, 1972; Parker and Grandison, 1977). A wide range of arthropod predators and parasites have also been recorded feeding on slugs or snails, including carabid beetles, glow-worms sciomyzid flies (Wild and Lawson, 1937; Hyman, 1967; Wolda, 1973; Runham and Hunter, 1970; Knutson *et al.*, 1965; Stephenson and Knutson, 1966). However, the influence of these and other predators and parasites (see Wright, 1971; Brown, 1978) on the population dynamics of molluscs has frequently been ignored, even though they may have a profound effect on fecundity (Foster, 1958). A marked contrast is provided by a study of 2 helicid snails where it has been possible to demonstrate that the higher mortality in 1 area can be attributed to prey selection and differential distribution of the predatory rodents (Yom-Tov, 1970). An analogous situation has been analysed by Mordan (1973, 1977), but here both the predator and prey were both molluscs.

Recently a number of studies of larger pluriennial species have provided evidence for the operation of a density-dependent mechanism. Yom-Tov (1971a, 1971b, 1972) investigated 2 populations of a small desert snail, *Trochoidea seetzeni* Pf, living on the north and south slopes of a dry valley. Different population densities existed on the slopes largely through differential predation, but the 2 populations also varied in terms of reproductive effort. The highest density was on the north, with 4·1 individuals per m^2, compared to 1·25 individuals per m^2 on the south. Table VI

Table VI. Densities of eggs and adults of *Trochoidea seetzeni* in 2 areas of 100 m^2.

Slope facing	Number of adult snails	Total number of eggs in 1967–1968	Number of eggs/ number of adults
South	108	1794	16·60
North	495	964	1·95

shows the mean number of eggs produced per adult snail in the 2 areas; the higher number occurring on the south facing slope, and this has been ascribed to higher number of eggs per clutch and a larger percentage of snails laying eggs. These differences could be caused by either variations in density, climate, or the genetic constitution of the 2 populations. The problem was partially resolved in a series of experiments which demonstrated that the snails on both slopes responded to artificial changes in density, there was also some evidence for genetical differences between the 2 populations. Thus a density-dependent mechanism negatively affecting reproductive rate was operating, although it was impossible to detect whether it was due to the excretion of a substance (pheromone) secreted by the snails in their mucus or the influence of a limiting resource. The subtle nature of the behavioural responses to information contained in the mucous trail has recently been demonstrated in a species of stylommatophoran slug (Cook, 1977); here a dual mechanism has been suggested—one serving as an olfactory beacon, the second being perceived only on contact.

A series of extremely detailed analyses of the population dynamics of a single species, *C. nemoralis*, have been undertaken in the Netherlands by Wolda (1963–1973) and Oosterhoff (1977) using information derived from laboratory, artificial and natural populations. These studies are undoubtedly the most exhaustive available for any molluscan species and reference should be made to the original publications. Wolda followed marked individuals over a number of years and estimated egg production and survival, juvenile production and survival, and adult mortality in closed populations where it was also possible to assess the influence of genetic variation. Although there was typically a heavy mortality amongst adult snails, largely through predation by thrushes and other vertebrates, this did not appear to have been either density-dependent, or the key factor controlling the dynamics of the populations. Likewise Wolda considered that the production of juveniles and variations in growth rates were not involved, however, considerable differences occurred in the patterns of mortality between populations that were expanding and those that were decreasing (Table VII) and this led to the hypothesis that the survival of the young in the first year was of paramount importance and here the key factors were drought and predation (Wolda, 1973). The evidence for the action of density governing processes was considered unsatisfactory, and even if some density-dependent mortality was involved at the juvenile stage, the "effect on population density was probably almost negligible compared with the influence of the factors that caused the difference between the increasing and decreasing populations, factors that, considering the outcome, did by no means govern the density".

In contrast Oosterhoff (1977) analysed variations in growth rate and associated phenomena both in natural populations and in a series of laboratory studies, but always in situations where it was considered that food was in super abundance and not a limiting factor. The results led to the formulation of a hypothesis linking changes in the structure and dynamics of the population with various ecologically important factors (Fig. 9). Growth rate and adult size were affected by temperature and humidity, with growth inhibited by deviations in either direction from an optimal state, although after a prolonged drought there was a catch-up effect. Marked reductions in light intensity and low levels of calcium also reduced growth rate and, in the case of calcium, the final adult size and

Table VII. Survival tables for 2 populations of *Cepaea nemoralis* (L.): data from Wolda *et al.* (1973). Data for the population which is expanding in size covers the 1967 generation, while that for the other, which is decreasing, covers the 1958 generation.

Age in months	Numbers	Expanding population: Survival during previous period	Expanding population: Accumulative survival	Declining population: Survival during previous period	Declining population: Accumulative survival
0 (eggs)	89,260				
1 (small juveniles)	52,370	0·578	0·578	0·452	0·452
13 (large juveniles) (new adults)	22,260 1627 } 23,887	0·456	0·268	0·023	0·011
21 (large juveniles) (new adults)	11,000 546 } 11,546	0·483	0·129	0·389	0·0041
25 (new adults) (old adults)	5000 463 } 5463	0·473	0·061	0·743	0·0031
45 (adults)	760	0·139	0·0008		

thickness of the shell. The most illuminating results, however, were obtained from an analysis of the effects of population density. Under laboratory conditions, as well as in artificial populations, there was a negative correlation between density at which juvenile snails developed and both growth rate and adult shell size. There was also a positive correlation between density and the time between hatching and the formation of the adult shell, but here was no catch-up effect in growth rate if density was reduced. It was suggested that the mechanism through which density operates involved mucus produced by active snails, not direct through the inhibition of feeding but rather via a pheromone or as a toxic effect on metabolic processes. While Oosterhoff favoured the latter, the studies of Cook (1977) are relevant to the former (see p. 491).

Differences in growth rate and adult size in natural populations could be correlated with the same suite of factors; indeed Oosterhoff suggested that density and to some extent temperature were the key factors in natural populations studied over a 2 year period. While genetical factors or parental influences obviously contribute to differences in growth rate and adult size these were considered inadequate to explain the observed patterns.

The importance of variations in growth rate and adult size on population dynamics was reflected in mortality and fecundity patterns and differences in the tendency to migrate. The mortality rate was higher amongst the slow growers compared to the relatively fast. Further, the fast growers were found to have a longer life expectancy as adults and consequently a longer reproductive life, besides producing young which exhibit lower mortality. The influence on reproductive activity was equally profound, the fast grown and relatively large adults laid earlier, more frequently and produced larger clutches with larger eggs that had a better hatching success. The young produced from these eggs were also larger. Moreover, as there was a positive correlation between growth rate and the time to reach sexual maturity the fast growing forms could have produced successive generations in a shorter time.

Although initially there might appear to be some disagreement with the conclusions of Wolda (1973), the factors which he considered important were those affected by density through the growth rate and condition of the adults, juveniles or both. Certainly the data of Wolda indicated that there were higher mortality rates at higher densities.

The implication of these studies is clear, growth rate and associated features would appear to play a crucial role in the population dynamics of *C. nemoralis*. The hypothesis summarized in Fig. 9 would suggest a causal mechanism for the intra-specific regulation of population density through a number of interacting features which respond in some way to density itself. The responses would appear to be largely phenotypic with the emphasis on increased plasticity in growth rate and adult size and where environmental conditions may have an over-riding effect on gentypic slow or fast growers. Moreover, it is possible to foresee how selection under different environmental conditions could favour either fast or slow growing forms and the evolution of different life history strategies. The extent to which this hypothesis is applicable to other taxa is at present debatable, but certainly the studies of Yom-Tov (1972) indicate its aptness for, at least, related taxa. While the behavioural and physiological reactions of other species, largely basommatophoran, would suggest that the potential for such a mechanism could also exist in those taxa.

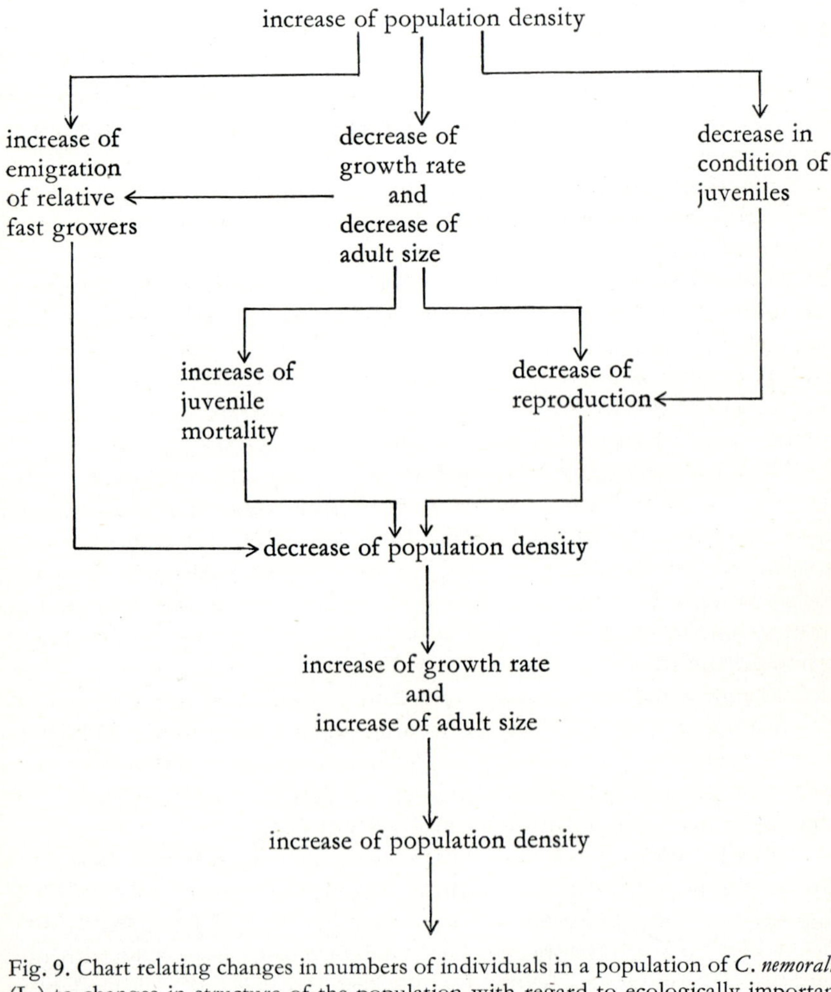

Fig. 9. Chart relating changes in numbers of individuals in a population of *C. nemoralis* (L.) to changes in structure of the population with regard to ecologically important characteristics (from Osterhoff, 1977).

Another study of *C. nemoralis*, on an area of chalk grassland in southern England (Williamson *et al.*, 1976) has shown that variations in shell-size of adult snails, in this case a 6–9% decrease, could be associated with an increase in density from 0·5–5·5 adults per m^2. Moreover, there appeared to be no simple relationship between population density and geographical features even though considerable differences in density occurred. Changes in adult size also occurred between different age classes, for the first sample measured early in the year contained adult snails surviving from previous years and the second, taken later, predominantly adults which had matured and formed a lip in the current year. Consequently it was possible to demonstrate that recruitment at lower densities increased mean adult size and *vice versa*. Such changes in size would be reflected in fecundity (Wolda, 1963, 1967) and possibly in ability to survive hibernation (Cook and O'Donald, 1971). There was no indication that food quality or quantity was a limiting factor and therefore, on the basis of largely circumstantial evidence, the authors concluded that interactions between the snails was the important component.

An attractive feature of this hypothesis (Fig. 9) is the ability to incorporate other factors or features which will affect reproductive potential, for example, shortage of food on which many workers have speculated, including Pomeroy (1967, 1968, 1969) and Butler (1976). Food could act in possibly 2 ways, through a general or specific food deficiency or through its uneven distribution, whereby it was difficult for the animal to find sufficient food in a given time or without inordinately high energy expenditure. Many of these statements regarding the importance of food have been based on subjective impressions, but the problem is difficult to quantify as has been aptly demonstrated by the studies of Williamson and Cameron (1976). There is frequently an additional problem of identifying a limitation where the food source cannot readily be identified in the field, nor its availability quantified. Both Pomeroy and Butler approached this problem by the manipulation of populations, increasing and decreasing population densities artificially in order to recognize regulatory mechanisms. They both recorded a negative correlation in growth, reproduction, activity and survival with increasing density of the snail *Cernuella virgata*, but felt that under natural population densities the data was inconsistent with an absolute shortage of "completely accessible food". Instead Butler concluded that "snails at high density interfere with each other by means other than competition for food"; a conclusion not inconsistent with the hypothesis of Oosterhoff (1977), although Pomeroy (1969) has favoured absolute shortage of food.

VIII. Life histories

A. Introduction

There is a wealth of anecdotal information concerning the life histories of terrestrial molluscs hidden in faunistic works such as those of Taylor (1894–1921) Germain (1930) and Pilsbry (1939–1948) and in the compilations of Pelseneer (1935) and Fromming (1954), yet there have been remarkably few attempts to draw any general conclusions or test the applicability of the various hypotheses that are currently available. A number of largely theoretical works have organized information from a wide variety of taxonomic groups into a series of models that enable various deductions and predictions to be made; summaries of some of these schemes are presented in Stearns (1976) and Calow (1977).

Molluscan life histories have been divided into annuals and biennials breeding only once and pluriennials with either a definite life span or forms where increased mortality is not correlated with age (Pelseneer, 1934). Another level of variation is imposed by the growth pattern of the shell or animal, which may be either determinate or indeterminate. However, such distinctions can obscure a rather complex situation for probably a continuum exists from species with very short life spans of less than 1 year to others extending for as long as 15–20 years in the wild; moreover different populations of the same species can exist at different points across this spectrum. The analogy between life history strategies of molluscs and fish has been drawn by Comfort (1957) in terms of patterns of mortality and fecundity, and indeed can be extended even further (see pp. 512–514).

A major constraint to studies of these phenomena has been imposed by the procedures employed in obtaining the data; for example, measuring a sample of a population at regular intervals throughout the year will, in a species with deterministic growth, only enable short life spans to be recognized, where there are either non-overlapping generations or only a limited degree of spread in the variation for size at each age class (Powell, 1950; Rigby, 1963; Baker, 1965, 1968; Chatfield, 1968).

For pluriennial species with determinate growth other methods must be employed, and the alternatives are either to follow marked individuals (Wolda, 1963) or age the shells. The latter has been achieved in *H. pomatia* for a colony in Cambridgeshire, England, where the adult snails lay down laminations around the lip of the shell on an annual basis, here it has been possible to demonstrate that individual snails aged 15 years or more may not be uncommon in the population (Pollard, 1973, 1975;

Pollard *et al.*, 1977). This estimate is very different from the 3 years in the wild or 6–7 years in captivity quoted by Comfort (1957, 1964) and is approaching the record of 33 years for a snail living in an artificial environment in a garden. Analogous techniques have been employed in marine groups, where daily growth lines are visible in shell sections (Panella and MacClintock, 1968; Taylor *et al.*, 1969, 1973) and in small prosobranch snails living on limestone hills in Malaysia where Berry (1962) has shown that the surface sculpture is laid down in a regular manner, 1 rib per day.

The pattern of mortality, whether it occurs predominantly at the juvenile or adult stages, is an important factor in the selection of the life history strategies (Stearns, 1976). Many studies of temperate species have demonstrated or indicated that a high mortality occurs at the egg and young or juvenile stages, even though variations can occur in the levels between adjacent populations of the same species (Yom-Tov, 1971; Wolda, 1973). The survivorship curve for such populations is probably strongly concave for at least the early stages in the life cycle and, therefore, approximates closely to type III as defined by Deevey (1947). At later stages in the life cycle mortality can be almost coincident throughout a population or show no increase in probability with age; here survivorship curves of types I and II could be more appropriate (Deevey, 1947). Very little information is available, however, for species with low annual fecundity or those that are ovoviviparous, but the success of the breeding programme undertaken on *Partula* (Murray and Clarke 1976a, b) would suggest that even though the species are ovoviviparous and the number of young produced is small, the probability for their survival is high even under laboratory conditions; this is linked to a long life span in excess of 8 years. It might be anticipated that the survivorship curves in such species would approximate closely to a straight line.

The cause of mortality in the egg and young stages is difficult to ascertain, but certainly dessication of the eggs and young, disturbance of the clutches by other snails and earthworms (Pollard, 1975) and predation by other species of snails or by soil and litter inhabiting arthropods can all contribute to the destruction of a large percentage of the eggs and young in many populations (Taylor, 1894–1921; Fromming, 1954; Wolda and Kreulen, 1973; Pollard, 1975).

Considerable effort and risk is expended or encountered by stylommatophoran molluscs during mating and, therefore, it may be anticipated that features which reduce the risk or insure successful mating will have a high selective advantage. Many species have developed complex pre-mating displays (Duncan, 1975) and these may last for several hours, but once copulation has commenced it will proceed even if conditions for normal activity disappear. The risks are also partially resolved by limiting the times of the year when mating can occur and, in pluriennial species, by storing the sperm for several years. In desert areas the complex structures associated with the terminal genitalia and with pre-mating displays are sometimes reduced in size or lost and there is presumably a concomitant reduction in water loss (Biggs, 1962; Solem, 1974); the problem of a possible increase in the error of finding or copulating with an unsuitable mate might in these situations be unimportant, as frequently the species diversity is low with only a single species occurring at any locality. The development of "aerial" copulation in some species of limacid slugs, where the animals are suspended in mid-air, could be important in reducing the attention of predators.

In temperate regions many of the larger pluriennial species hibernate during the winter, sometimes in groups in fixed sites to which they return annually. Edelstam and Palmer (1950) demonstrated in *H. pomatia* a homing response to area rather than a site and Pollard (1975) has recorded similar behaviour in the same species. A mechanism based on pheromones in the mucus could account for such behaviour (Cook, 1977), but at present, very little is known about the information content of such trails, but it appears to be a remarkably fruitful field for further research.

B. Self-fertilization

Self-fertilization has been recorded in a number of species of pulmonates (Pelseneer, 1934, 1935; Paterson, 1970; Duncan, 1975; Murray and Clarke, 1976a, b) and it obviously has a polyphyletic origin. However, many earlier claims do not bear close examination, as the data were not collected under sufficiently rigorous conditions, nor was the possibility of parthenogenesis adequately explored. Self-fertilization or parthenogenesis may also occur in aphallic or hemiphallic forms (de Larembergue, 1939; Watson, 1923, 1934; Riedel, 1953). Although some species or populations may have the propensity for self-fertilizing there is a range of variation extending to those species that are obligatory outbreeders and even individuals can exhibit both self- and cross-fertilization within the same clutch or at different times in the same breeding season (Duncan, 1975; Murray and Clarke, 1976a).

Theories regarding the adaptive significance of selfing fall into 2 groups, those that regard it as little more than a provision for colonization or an adjustment to low population densities and those that consider it as part of a strategy of adaptations to environmental heterogeneity (Selander and Hudson, 1976; Jain, 1976). The latter introduces the concept of general purpose genotype as has been developed for weedy plants, where the successful forms are often tolerant of wide variations in environmental conditions as long as they are not subject to competition from native species (Baker, 1965, 1972).

Selander and his co-workers have explored problems associated with self-fertilization in sessile or semi-sessile animals, like pulmonate molluscs, and have drawn attention to the convergence between the breeding systems in these organisms and plants. The patterns of variation in *Rumina decollata*, a snail exhibiting self-fertilization (Selander and Kauffman, 1973, 1975; Selander *et al.*, 1974; Selander, 1976), have been contrasted with that found in *H. aspersa*, an obligatory outbreeder (Selander and Kauffman, 1975). *Rumina decollata* would appear to fulfil many of the criteria for consideration as a weed; it occurs in a wide range of habitats in situations where other molluscan species are absent and exists under many different climatic types. It is also highly sedentary, compared to *H. aspersa*, and the inherent variability exhibited by colonies, already low because of the founder effect, is unlikely therefore to be augmented by migration. It is in such situations that a general purpose genotype would be important in conferring a wide range of phenotypic tolerance to short-term environmental conditions, rather than attempting to achieve the impossible and produce frequent genetic adjustments. Thus Selander and Hudson (1976) are suggesting that self-fertilization has 2 adaptive functions in *Rumina*, to enable colonies to be founded by single eggs or unfertilized individuals and to maintain highly adaptive general purpose genotypes once they have evolved. Strongly monomorphic genotypes are developed, each being adapted to particular, but wide, ranges of environmental conditions, with occasional outcrossing giving rise to new genotypes.

The development of self-fertilization must be considered amongst a suite of breeding systems exhibited by molluscs, including those of obligatory outbreeders, the significance of which will only be understood against a background of information concerning ecological and genetical factors affecting the population structure of the organism. The evolution of hermaphroditism must be viewed in a similar manner, for although a number of hypotheses are available to explain the adaptive significance of this strategy, they can only be evaluated against the totality of factors influencing the organism. Ghiselin (1969) has summarized the available data and models; the latter extending to a consideration of breeding systems and consequently self-fertilization. There are basically 3 hypotheses—the low density, size advantage and gene dispersal models. The first refers to the classical explanation where there is a reduction in the opportunities for mating; the second to the situation where the reproductive functions of the animal are correlated with size—eggs are generally larger than sperm and there is a minimum size limit, hence the majority of pulmonate species are protandrous with only a few protogynous forms being found;

the third hypothesis relates to problems of inbreeding and populations size in relation to genetic drift and limitations on the size of the genome. Murray (1964) has also suggested that simultaneous hermaphroditism increases the effective population size, with multiple matings and sperm storage having similar and reinforcing effects.

C. Variations in life cycles

There is a positive correlation between body size and generation time amongst a wide range of organisms (Bonner, 1965) and for many groups of molluscs the relationship would appear to hold. For example, a comparison of species belonging to the Helicidae demonstrates that in the British Isles many of the smaller species have typically annual life cycles, while the larger forms may take 2–3 years to reach sexual maturity and can then live for a further period which can extend for 15 years or more. However, the relationship becomes confused if the comparisons are extended to include a wider range of taxonomic groups or data from different climatic zones. Within the Helicidae members of the *Sonofella*, a desert group, can take up to 7 years to reach sexual maturity. Differences in patterns of activity and growth will make major contributions to this breakdown, for undoubtedly life span accords inversely with the rate of living (Pearl, 1928; Comfort, 1964) or with total metabolic rate (Bourlière, 1957; Southwood *et al.*, 1974).

The association between climate and life history phenomena is self evident, but the degree to which the latter can vary amongst terrestrial molluscs, both within and between populations of the same species, tends to be ignored or underestimated, yet the flexibility in the strategies adopted by some species is an important factor in determining the distribution and success of the taxon. The small helicid snail, *Cochlicella acuta*, has an extensive distribution and has been successfully introduced into a number of continents. Populations exhibit considerable differences in life cycles between northern and southern Europe (Lewis, 1968); in Britain and Northern France the life span is generally short, usually about 1 year, while in southern Europe the snails take more than a year to reach maturity and then continue to live and reproduce for a further year or two. As these differences are maintained under laboratory conditions it suggests that some level of genetical control must be involved analogous to that found in the evolution of voltinism in insects (Richards, 1961; Lewis, 1968). The considerable variation in the growth rates shown by individual snails, even those derived from the same clutch, is a feature common to this and a wide variety of pulmonate species (Wagge, 1952; Blinn, 1963; Wolda, 1970; Pollard, 1975). The observation of Wright (personal communication) that considerable variations can occur in the quantities of proteins in individual eggs is relevant, for it suggests that an important factor in determining future growth patterns could be the nutritional content of the egg. Lewis (1968) recorded that some of the slow growing forms developed at the same time as a subsequent generation after remaining small for nearly a year, although *C. nemoralis* forms with relatively slower growth have a relatively higher mortality rate (Oosterhoff, 1977).

Variations in the life cycles of 3 species belonging to the Vitrinidae have been correlated by Uminski (1975) with differences in altitude and ultimately climate. The study was undertaken in the Polish Tatra Mountains where at the lowest altitude, 980 m, 2 species, *Vitrina pellucida* (Müll.) and *Semilimax kotulai* (West.), had annual cycles with all individuals dying after 1 year, including those that had failed to reproduce or reach sexual maturity. At the higher stations, 1240 m and 1420 m, both species had biennial cycles, but in these populations there was a higher proportion of large specimens than at the lower altitude. This observation accords with a phenomenon found in a variety of species, that increased size is associated with growth extending over a longer period, as in the example above where the life span has been increased from 1–2 years (Abeloos, 1944). The data for the third species, *Eucobresia nivalis* (Dumont and Mortillet), did not present such a clear picture; at the higher altitudes the cycle appears to be biennial, although irregularities suggest that variations occur depending on the weather conditions.

There is a vast array of data concerning the life cycles of economically important species of slugs (Künkel, 1916; Carrick, 1938; Bett, 1960; Stephenson, 1968; Runham and Hunter, 1970; Hunter, 1968, 1978; Hunter and Symonds, 1971; Duncan, 1975). These confirm the very close relationship between climate and breeding activity, indeed in the opportunistic or weedy species, *Deroceras reticulatum* (Müll.), some populations are divided into sub-sets each responding to climatic factors but with the groups leap-frogging each other (Hunter and Symonds, 1971).

Breeding seasons may be curtailed or limited by increasing climatic severity or seasonality, and consequently there is a close synchronization of the whole population with favourable periods. In more open habitats, like sand dunes, a very simple correlation exists between oviposition site, breeding season and climate in helicids. The smaller species either scatter their eggs in small groups on, or near, the sand surface where they are extremely susceptible to desiccation, these breed during the colder and wetter months when the climate is more predictable. The larger species bury their eggs in holes excavated in the ground, the depth of the hole being determined by the size of the animal, here the eggs are in a comparatively protected site and these forms breed during the warmer months when the water regime is less predictable (Peake, 1960).

In temperate woodland habitats the population densities of slugs and

Table VIII. Life history strategies of 4 species belonging to Achatinacea. Data from Owen (1967) and Owiny (1972, 1974).

Species	Ovoviviparity or oviparity	Breeding season	Notes
Limicolaria martensiana	oviparous	rainy season	Consequently breeding continuous throughout year.
	facultative ovoviviparous	dry season	
Limicolaria turriformis	oviparous	eggs laid rainy season; twice during year	Young aestivate in egg (form epiphragm); hatch during following rainy season, i.e. delaying hatching. Adults aestivate; if conditions unfourable during predicted favourable period, then break aestivation to lay eggs. Also aestivate during unfavourable period—obligate diapause.
Burtoa nilotica	oviparous	eggs laid rainy season; twice during year	—
Subulina octona	ovoviviparous	continuous	—

snails appear to be stable, or demonstrate only minor fluctuations (Mason, 1970b, 1974; Jennings and Barkham, 1975). Possibly many of the species living in these habitats have extended breeding seasons or low reproductive rates and life spans covering more than 1 year (Taylor, 1894–1914; Fromming, 1954; Morton, 1954; Mason, 1970b). These patterns provide a marked contrast, however, to those exhibited by many small helicids and succinids found in open habitats; the life spans of the latter are typically annual in the British Isles, but the breeding seasons are synchronized and consequently major fluctuations occur in population densities. Analogous features are found amongst populations of small pulmonate and prosobranch snails living on limestone hills in Malaysia. Here there are fluctuations in population densities but these are rather damped, with the breeding seasons appearing to extend over the whole year in the more favourable habitats and with minor peaks coinciding with rainfall patterns (Berry, 1963, 1964, 1965, 1966). Climatic factors appear to affect adult population densities through variations in mortality rather than recruitment of young, thus although major rainfall patterns exhibit an overall influence on the populations, on a finer scale local microclimates produce a modifying effect.

The life cycles of pluriennial species probably exhibit considerable variations, but at present there is little information on which to base generalizations. Similarly data concerning taxa with tropical distributions is restricted to a few groups or taxa, for example, there is a wealth of casual information concerning the economic species *Achatina fulica* (Mead, 1961, 1978). The studies of Owen (1965, 1967) and Owiny (1972, 1974) of 3 closely related species of the Achatinacea living in the vicinity of Kampala, Uganda, however, are extremely illuminating; they demonstrate a series of multiple solutions to the problem of a fluctuating environment (Table VIII). In this case the important feature of the climate would appear to be the predictable nature of the alternating wet and dry periods and it is these that determine the reproductive strategies employed. *Limicoloria martensiana* demonstrates 2 marked periods when greater numbers of eggs are found inside the animal, these correspond with the dry periods, when there is a switch from an oviparous condition, with the eggs being laid in holes or in the litter, to facultative ovoviviparity, with the eggs being retained inside the aestivating snail and where they frequently hatch. An intermediate stage in the evolution of behaviour is seen in *A. fulica*, the eggs are accumulated in the oviduct under unfavourable conditions and are laid just prior to hatching; the latter can then extend in a single clutch from 24 hours of the eggs being laid to a week or more (Nisbet, 1974). *Subulina octona*, another member of the Achatinacea, also occurs in the vicinity of Kampala and provides another solution, it is ovoviparous and breeds throughout the year. While viviparity, the ultimate form of protection, is possibly found in *Achatina zanzibarica* (Mead, 1950).

D. Egg and clutch size

Variations in egg and clutch size have been recorded in species with both determinate and indeterminate growth patterns. For example, in *C. nemoralis*, a species with determinate growth, both clutch size and oviposition frequency are positively correlated with shell size, clutch size increasing by 4·9 eggs for every mm increase in shell diameter, but decreasing with age (Wolda, 1963, 1973). During a single breeding season clutch size diminishes due to decrease in the nutritive value of the food available during the growing season; this would appear to be analogous to the pattern in *Melampus* where the gradual depletion of food reserves within the animal cannot be balanced during the breeding season by assimilation from ingested food (Russel-Hunter, 1970). However, Oosterhoff (1977) has produced an alternative hypothesis for *C. nemoralis* (see pp. 491–494). In *Eulota fruticum* (Müll.), another species with determinate growth, an increase and subsequent decrease in annual fecundity is associated with age (Künkel, 1928); similar results have also been recorded in slug species (Künkel, 1908; Fromming, 1954). In the Achatinacea, where species exhibit indeterminate growth, the pattern is more complex; there is a positive correlation of egg-size with shell-size, both within and between species, thus within an individual snail's life span there is a continuous increase in the size of eggs produced. However, if egg-size is considered as a percentage of shell-size the correlation is negative, suggesting the major reproductive effort is towards increasing potential clutch-size with increasing age and shell-size (Figs 10 and 11). It is interesting to speculate on the selective advantages of large and small eggs; whether within a species or individual there are variations in the nutritional values of the egg contents, and corresponding changes in the developmental stages at which they hatch or in the growth rate of the young. For *A. fulica* a reduction in shell-size has been associated with an increase in colony-size and population density, although the causal relationships are unknown (see pp. 491–495), the interaction with egg and clutch-size can be anticipated. There are exceptions, however, to the general trends in egg and clutch-size within the superfamily, for example, in *Archachatina marginata* there is an enormous increase in egg-size and a corresponding decrease in clutch-size (Plummer, 1975). While in the ovoviviparous *Subulina octona* the birth rate is low, but continuous over the whole year and not restricted to favourable periods (Owiny, 1972, 1974).

Other taxonomic groups exhibit a similar relationship between egg-size and shell or animal size, but the data for species of Helicacea in Europe suggest they employ 2 distinct strategies (Fig. 12); one emphasizes increase in clutch-size, the other egg-size. This dichotomy is also reflected in data produced by Yom-Tov (1971a) for 2 species of helicids living under desert conditions. Possible correlates of the 2 strategies are length of the life span and differences in adult and juvenile mortality rates; that which favours increasing clutch-size being associated with predominantly annual or short-lived species, while the converse is true

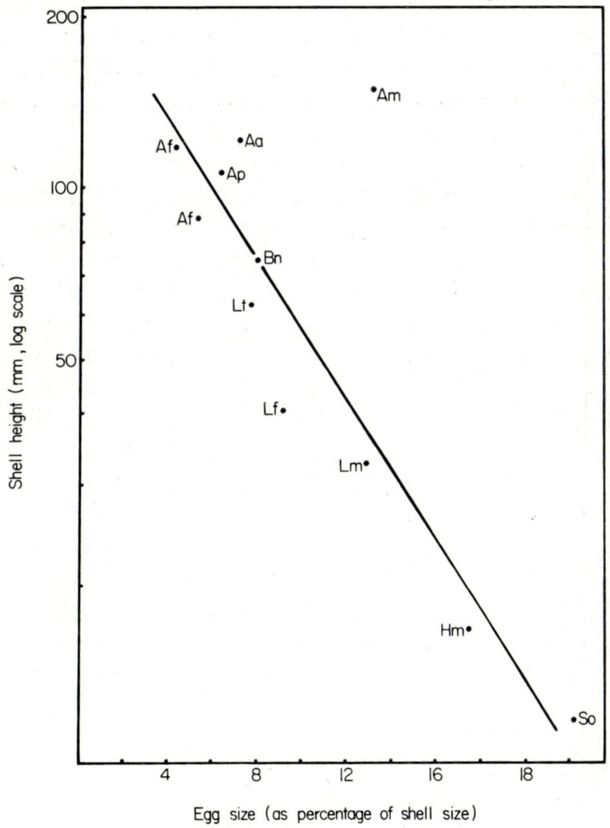

Fig. 10. Relationship between shell and egg-size in the Achatinacea: shell-size expressed as shell-height, egg-size, length expressed as a percentage of shell-length (data from many sources). Af = *Achatina fulica* Bowdich (2 localities), Aa = *Achatina achatina* (L.), Ap = *Achatina panthera* (Fér.), Am = *Archachatina marginata* (Swainson), Bn = *Burtoa nilotica* (Pfr), Lt = *Limicolaria turriformis*, Lf = *Limicolaria flammea* (Müll.), Lm = *Limicolaria martensiana* (Smith), Hm = *Homurus marginata*?, So = *Subulina octona* (Brug.)

for those with larger eggs. It is interesting that the latter group includes species which exhibit some of the highest levels of polymorphism seen amongst molluscs in Europe (Clarke, 1972; Cain, 1977a; Stearns, 1976).

There are a wide spectrum of other variables associated with eggs including calcification of the egg shell, differences in oviposition sites and the development of ovoviviparity (see pp. 471–472). Tompa (1976) has reviewed the data concerning calcification, and while it was impossible to draw any conclusions in relation to the site of deposition, he suggests that the development of heavily calcified eggs is a primitive trait associated with adaptation to terrestrial conditions. Moreover, the correlation between heavily calcified eggs and the development of large eggs up to 50 mm in maximum diameter indicates the importance of this trait in the evolution of size in terrestrial molluscs; uncalcified or partially calcified eggs are usually between 1 and 8 mm maximum diameter. Regarding

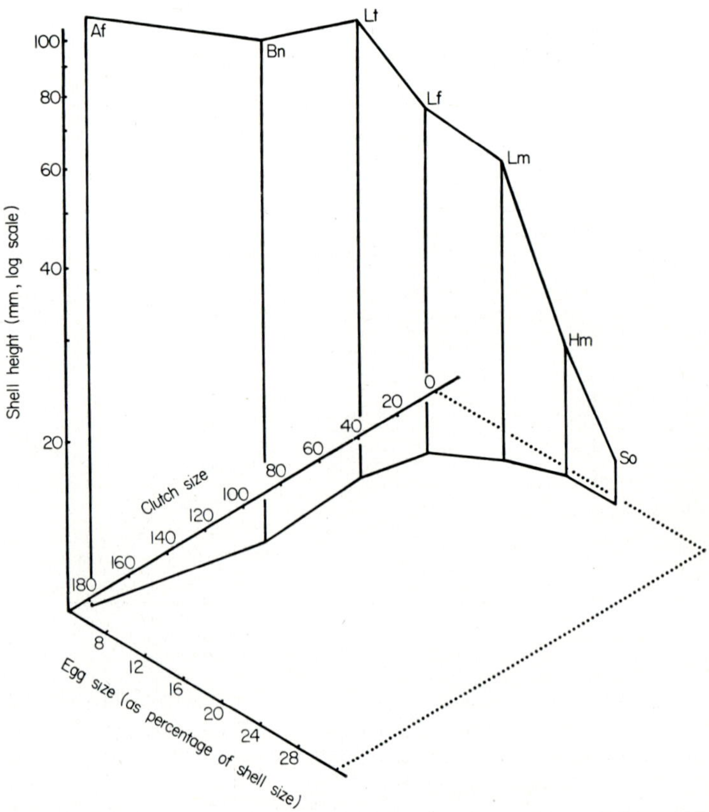

Fig. 11. Relationship between shell, egg and clutch-size in the Achatinacea. See Fig. 10 for abbreviations.

oviposition sites, these are usually in damp places near or in the ground, except for tropical arboreal forms which either lay their eggs in axils of leaves and branches, make nests or have rather gelatinous coverings. A few species, for example *Discus*, appear to scatter the eggs singly in the leaf litter or under rotten logs and some other members of the Endodontidae utilize the shell umbilicus as a brood pouch (Solem, 1969b, 1976).

It is obvious that there is a fine balance between egg and clutch-size and the size of the mature snail, the latter reflecting the size of the gonad, the oviduct and the pallial cavity. As there is possibly a minimum viable egg-size, very small shell-size places limitations on clutch-size, and here a correlation with an extended breeding season could be anticipated. Data included in many taxonomic works provides an indication of the small

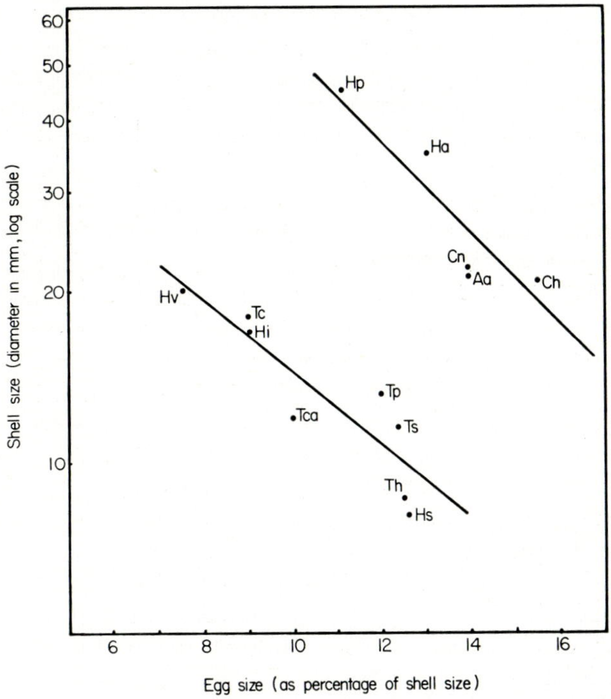

Fig. 12. Relationship between shell and egg-size in the Helicacea; shell-size expressed as shell-diameter, egg-size expressed as a percentage of adult shell-size. Data for populations in the British Isles and Northern France. Hp = *Helix pomatia* (L.), Ha = *Helix aspersa* (Müll.), Cn = *Cepaea nemoralis* (L.), Ch = *Cepaea hortensis* (Müll.), Aa = *Arianta arbustorum* (L.), Hv = *Cernuella virgata* (Da Costa), Tc = *Monacha cantiana* (Mont.), Hi = *Helicella itala* (L.), Tca = *Monacha cartusiana* (Müll.), Tp = *Theba pisana* (Müll.), TS = *Trichia striolata* (C. Pfeiff.), Th = *Trichia hispida* (L.), Hs = *Zenobiella subrufesceus* (Miller).

number of eggs or young produced in certain families, e.g. in the Pupillidae (Steenberg, 1925), Tornatellidae (Cooke and Kondo, 1960), Microcystinae (Baker, 1938–1941) and in the small species of Streptaxacea (Berry, 1963, 1965). Indeed the evidence would suggest that in some forms, particularly where ovoviparity is developed, the eggs or young are produced on a conveyor belt principle—the individuals are arranged in an age sequence in the oviduct. Production of very large eggs is restricted to a few tropical genera with large shells, these include *Archachatina*, *Acavus* and *Strophocheilus* and it is noticeable that species belonging to these genera occur predominantly in rain forests.

E. Concluding remarks

The diversity of life history patterns displayed by any group of organisms must be viewed against the background of environmental variables in which they were evolved, and also against the attributes of the organism itself, for they represent a series of compromises between many features including fecundity, age at sexual maturity, determinate or indeterminate growth, mortality patterns, age structure of the population and body-size. The recognition of the causal relationships involved in the selection of such attributes has been a major objective of many recent considerations of life history strategies; in many instances, however, the final product appears as a series of correlations associated with a body of intuitive reasoning. Emphasis has been placed on the recognition of the minimum number of selective pressures to explain many common components or strategies, for example the development of the concepts of *r* and *K*-selection (MacArthur and Wilson, 1967; Southwood *et al*., 1974; Stearns, 1976) has drawn attention to the importance of population density in relation to the availability of resources and to the stability of the environment. An examination, however, of the diversity of life history phenomena exhibited by terrestrial molluscs and the comparative lack of detailed information, can rapidly induce a feeling of sympathy towards the point of view expressed by Wilbur *et al*. (1974)—that to explain the evolution of life histories in such a simplistic manner could obscure rather than elucidate an understanding of the underlying processes. Moreover, in arguing from the particular to the general it is extremely easy to forget that the information available for terrestrial molluscs represents an extremely biased sample. It is derived largely from temperate species of open or temporary habitats and frequently those whose distribution has been strongly influenced by man through alteration of the environment or introduction.

To interpret data for pulmonates in terms of hypotheses developed for other organisms is extremely inviting, but frequently difficult as the basic template is very different. A study by Randolph (1973) has provided a lead by examining the influence of environmental stability on the population characteristics of 2 species of snails. The data have been analysed in terms of a general theory that relates environmental variability to ecosystem diversity and stability. Even though the results could have been anticipated, an interesting feature of this investigation was the attempt to manipulate populations and vary the physical features of the environment in order to test various predictions of the hypothesis. The 2 species of snails live in different habitats, one in woodlands, the other in more open conditions of grassland and to a lesser extent woodlands. The species which was subjected to the more variable environment exhibited a wider tolerance or niche, a higher *r* value or intrinsic rate of increase, shorter life span, smaller clutch size, shorter development time, more general food requirements and smaller maintained biomass than did the species which occurred in the less fluctuating environment.

A summary of some of the life history traits exhibited by some stylommatophoran molluscs are given in Table IX; although individual traits have not been assigned to particular strategies, a comparison of features correlated with *r* and *K*-selection by such authors as Pianka (1970) and Stearns (1976) will demonstrate that many have been recognized as making contributions to fitness in other organisms. In this context *r*-strategies are those that enable an organism to maximize its rate of population increase (*r*) and rapidly exploit a situation where the environment is not saturated with species, or "where the niche hypervolume is frequently empty" (Whittaker *et al.*, 1973; Southwood *et al.*, 1974). Species which fall within this groups are opportunists or generalists often in an unpredictable environment, where unpredictability is frequently associated with irregular and erratic climatic conditions (Pianka, 1970). *K*-strategies increase the efficiency of the organism in a saturated environment where there are many competing species or "maximizes its share of

Table IX. Life history traits exhibited by stylommatophoran molluscs; annotated list based on data derived from many sources.

Reproduction

a. Hermaphroditism; obligatory cross-fertilization, self-fertilization and parthenogenesis.
b. Complex pre-copulatory displays; copulation can be extended for several hours.
c. Transfer of sperm can be either reciprocal or 1 partner may act as male.
d. Sperm can be stored for longer than single breeding season.
e. If reproduction unsuccessful, resorbtion of gonadial material may occur.
f. Reproduction involves a cost in terms of growth and length of life; individuals that fail to reproduce may grow larger and live longer.

Table IX—continued

Life span

a. Life cycles can be either annual or short, pluriennial with determined growth, or pluriennial with indetermined growth.
b. Variations can occur between and within populations of the same species, and between individuals from the same clutch; development of "multiplier effect".
c. Diapause presents special problems as periods spent in it almost certainly count little, if at all, towards the physiological age of the animal (Comfort, 1964).
d. Diapause, and indeed patterns of activity, may not be synchronized throughout the whole population.

Shell and body size

a. Correlation between body size and generation time.
b. Behavioural changes with increasing age, size and shell thickness (Pollard, 1975).
c. Correlation with egg size and clutch size; but dichotomy in strategies possible either towards increasing egg size or clutch size.

Eggs

a. Development of eggs with outer layer heavily calcified.
b. Ability to withstand considerable desiccation for short periods (Duncan, 1975)
c. Oviposition sites frequently afford protection to eggs; including deposition in umbilicus of shell.
d. Ovoviviparity has polyphyletic origin; young produced regularly—"conveyor belt" system. Facultative ovoviviparity and egg retention also developed.
e. Viviparity rare.

Clutch size and annual fecundity

a. Clutch size very variable, factors influencing this largely unknown, but extreme deprivation of calcium can have an effect.
b. Clutch size and annual fecundity influenced by density-dependent mechanisms.

Mortality

a. High mortality amongst eggs and young snails frequently recorded; although considerable reduction in rate amongst adult snails particularly long-lived species. Survivorship curve Type III (Deevey, 1947).
b. Probably Types I and II (Deevey, 1947) will be recorded: existence inferred.
c. "Key factors" in mortality patterns have been recorded as climate and predation.

a fully packed niche hypervolume". These species will tend to live in predictable or stable environments and have been classed as equilibrium (MacArthur, 1960) or specialist species. These 2 strategies are at opposing ends of a continuum, and consequently the terms *r* and *K* are comparative. Their use frequently implies that a distinction is being drawn between species or groups of organisms existing at different points along this continuum, rather than at opposing extremes, therefore it can be anticipated that species will frequently exhibit a blend of correlates with both *r* and *K*-selection. A number of authors have suggested that *r*-strategists will be predominant in temperate areas where physical factors are the most limiting and *K*-strategists in tropical areas where selection favours ability to compete (Dobzhansky, 1950; Stearns, 1976).

Many species of Stylommatophora will appear to be *r*-strategists, exhibiting combinations of traits that can include high egg and young mortality, fluctuations in population size, annual or short life cycles, small body-size, large clutch-size or high annual fecundity. It could be anticipated that *K*-strategists will be found in tropical or cloud forests, which are basically non-seasonal, or on oceanic islands with very equable climates, but consideration must also be given to those species that exist in refugia or niches where the environment is comparatively stable. An example of the latter could be the litter or humus layers in some temperate woodlands, particularly those where the canopy is closed and the environment is buffered against climatic extremities; in this context the data produced by Mason (1970b, 1974) demonstrating almost stable population densities throughout the year in some species are most intriguing.

Although the *r*-strategists exhibit a wide range of solutions to the problems posed by the environment, in many instances they are responding to the availability of a single resource, water, and to a lesser extent perturbations imposed by temperature fluctuations. The most obvious response is the ability to enter and maintain a state of diapause when conditions are unfavourable. While minor environmental fluctuations and consequently short-term responses are observed in the different patterns of activity, frequently only a proportion of a population is active at any time, as the activity cycles are not synchronized between individuals, and consequently the whole population is not subjected to a potentially catastrophic mortality through sudden changes in the climate (Barnes and Weil, 1944, 1945; Crawford-Sidebotham, 1972; Pollard, 1975; Machin, 1975; personal observations). The variability in growth rate of individual snails both within and between clutches is another facet of the adaptive responses to climatic fluctuations, as is the development of leap-frogging generations (Hunter and Symonds, 1971) and the potential for developing biennial life cycles (Mordan, 1978). Indeed, many of the life history strategies exhibited by terrestrial molluscs appear to be adaptations to variations or fluctuations in the environment, particularly those related to climate.

The application of models for *r* and *K*-selection to many populations of terrestrial molluscs is open to criticism as frequently they fail to consider variations in mortality and fecundity, whether adult survival or reproductive success is the parameter principly influenced by fluctuating conditions. Schaffer (1974) has pointed out there is no *a priori* reason for assuming that increased environmental unpredictability will necessarily select for increased reproductive output, although frequently this may be the case. For these situations alternative hypotheses have been proposed employing a system of bet-hedging; nevertheless, it appears extremely difficult to discriminate between situations where different types of models apply (Schaffer, 1974; Stearns, 1976). The model for bet-hedging would suggest that where juvenile mortality is variable, and this would appear to be true for many molluscan populations, the following predictions would apply (Stearns, 1976):

Stable environments	*Fluctuating environments*
early maturity	late maturity
repeated reproduction	repeated reproduction
larger reproductive effort	smaller reproductive effort
shorter life	longer life
more young per brood	fewer young per brood
fewer broods	more broods

An examination of the data presented in Tables IX and X would suggest that at least some of these predictions are realized in the more extreme environments where favourable conditions are either distributed randomly or occur for only short periods. Under other regimes the relationship is not so clear and here a complicating factor is the increased diversity in animal size, with concomitant variation in life span and reproductive potential, but in these less extreme situations models for *r* and *K*-selection or bet-hedging would appear to produce similar predictions. A classification of different forms of environmental fluctuations or stress situations has been proposed by Stearns (1976), and while this is not applicable rigidly to terrestrial molluscs it does provide a stimulating basis for further speculation. The latter has been attempted in Table X, but this must be regarded as only provisional and necessarily treated with considerable caution. Within the continuum from stable to unpredictable environments it is obvious that even populations within the same species exhibit a range of solutions. Many exhibit a range of traits analogous with the multiplier effect described by Wourms (1972) for the variations introduced into the developmental pathways of annual fish, these lead to

Table X. Life history traits associated with fluctuating environmental conditions.

Environmental conditions		Life history traits
Stable or seasonal, but conditions constantly or predominantly favourable	Increasing unpredictability of favourable conditions → Frequency or length of favourable conditions decreasing	Extended breeding season, Life span pluriennial } stable population densities Small clutch size—"conveyor belt"
Seasonal: Tendency for conditions to be favourable but short periods may be unfavourable Conditions to some degree unpredictable, but predominantly favourable Strongly seasonal; marked favourable and unfavourable periods Conditions only favourable for short periods		Breeding season synchronized with favourable conditions; some degree of opportunism "Multiplier effect"; within clutch, population and species variability in growth and activity patterns. Dichotomy in life-span strategies; both short or annual or pluriennial. Selection probably dependent on size of mollusc Large clutch size Ovoviviparity; corresponding reduction in brood size Wide niche or tolerance Self-fertilization
Unpredictable; favourable conditions distributed randomly in time		Long periods spent in diapause; then tolerant of environmental extremes Long life span; repeated reproduction Breeding strongly seasonal or flexible Ability not to breed if conditions unfavourable Rapid growth during favourable periods Reduced clutch size

a wide range of sub-sets of the population with different growth, diapause and activity patterns—all of these sub-sets being derived from an initial single age class. Thus the multiplier effect, together with other adaptations, guarantees that some portion of the population will survive. Similar responses are observed in terrestrial molluscs and this is clearly examplified by the different patterns shown within a contiguous distribution of the same species.

The concepts of *r* and *K*-selection or evolution in fluctuating environments provide invaluable frameworks in which to examine the wide range of adaptations shown by stylommatophoran molluscs. Nevertheless, the subsequent recognition of certain associations of features or correlates does not warrant the uncritical acceptance of the causal relationships implied in the original formulations of the concepts, indeed, data on life history tactics are often consistent with more than a single hypothesis (Wilbur *et al.*, 1974). The absence of information on carrying capacities or the mechanisms of population control demands that a degree of scepticism is introduced into a discussion of causal relationships when *r* and *K*-selection are evoked, in spite of the recent studies of Wolda (1963–1973), Yom-Tov (1970, 1971, 1972), Williamson *et al.* (1976) and Oosterhoff (1977). As already indicated, such a continuum could be produced by variations in habitat or environmental stability, where an important feature is the length of time the habitat remains favourable compared to generation time (Southwood *et al.*, 1974). Certainly for terrestrial molluscs such a causal relationship is attractive.

There is an obvious need for detailed studies of populations under a wider range of environmental conditions, particularly in areas which are considered stable and where there is a high diversity of species. The opportunities to undertake such investigations in tropical areas are rapidly disappearing with the massive destruction of unique habitats like the rain forests. Already the rich endemic faunas of many Polynesian Islands have largely disappeared in the wake of human disturbance and exploitation, however, remarkably little is known about the life cycles of many common and widespread forms and consequently speculation tends to be based on a few comparatively well documented taxa.

IX. Concluding comments

A review of such a broad subject as distribution and ecology of the Stylommatophora must be highly selective in the topics and examples selected for consideration. Here scant attention is given to problems associated with diet, the influence of parasites on population dynamics or medium scale distribution patterns, yet other limitations are imposed by the lack of data or the availability of adequate syntheses or models. Here the availability or the application of new techniques will undoubtedly enhance our knowledge of the group. For example, the recent studies of Cates and Orians (1975) on the relationship between palatability of plants to herbivorous molluscs and the position of the plant species in the vegetational succession suggests new ways to explore the coevolution of the 2 groups. They have argued that plants occurring early in the succession commit less of the available energy resource to defence than those occurring later and thus it is the former which are favoured as food sources by the herbivores. Moreover, the introduction of taxonomic innovations and the addition of new fossil material will necessitate revisions in our concepts of the distribution of the Stylommatophora both in time and space. Similarly, the extension of ecological, physiological and anatomical studies to include a wider range of taxonomic groups will probably revolutionize our ideas regarding the evolution of the group.

At present, there is a tendency to emphasize the plasticity of the responses exhibited by the group to a wide range of environmental parameters. Indeed it is this plasticity which has contributed so much to the success of many taxa. Moreover, for a group which cannot fly, or whose dependence on water is constantly stressed to the extent that they can be described as an association of "neurotic hydrophils", they have a remarkably wide geographic and ecological distribution! Nevertheless our knowledge of the biology of the group tends to be restricted to a few types which are probably not typical, consequently, there is a need for integrated studies of a wider range of taxa in a more diverse range of habitats.

X. References

Abeloos, M. (1944). *Bull. biol. Fr. Belg.* **78,** 215–256.

Andersson, J. S. (1961). *Oikos* **12,** 126–156.

Andrewartha, H. G. and Birch, L. C. (1954). "The Distribution and Abundance of Animals". University of Chicago Press, Chicago.

Ant, H. (1963). *Abh. Landesmus. Naturk. Münster* **25,** 1–125.

Ant, H. (1969). *Vegetatio* **18,** 374–386.

Atkins, W. R. G. and Lebour, M. V. (1923). *Sci. Proc. R. Dubl. Soc.* **17,** 233–240.

Backhuys, W. (1975). "Land and Freshwater Molluscs of the Azores". Backhuys and Meesters, Amsterdam.

Baker, H. B. (1925). *Nautilus, Philad.* **38,** 86–89.

Baker, H. B. (1938–1941). *Bull. Bernice P. Bishop Mus.* **158,** 1–102; **165,** 1–201; **166,** 202–370.

Baker, H. B. (1955). *Nautilus, Philad.* **68,** 109–112.

Baker, H. B. (1958). *Nautilus, Philad.* **71,** 141–148.

Baker, H. B. (1962). *Nautilus, Philad.* **75,** 116–121.

Baker, H. G. (1965). *In* "The Genetics of Colonising Species" (H. G. Baker and G. L. Stebbins, Eds) pp. 147–168. Academic Press, New York and London.

Baker, H. G. (1972). *In* "Taxonomy, Phytogeography and Evolution" (D. H. Valentine, Ed.) pp. 327–347. Academic Press, New York and London.

Baker, R. E. (1965). *Proc. malac. Soc. Lond.* **36,** 269–265.

Baker, R. E. (1968). *Proc. malac. Soc. Lond.* **38,** 41–54.

Balgooy, M. M. J. van (1971). *Blumea* Suppl. **6,** 1–222.

Barnard, P. D. S. (1973). *In* "Organisms and Continents Through Time" (N. F. Hughes, Ed.). *Spec. Pap. Palaeont.* **12,** 175–187.

Barnes, N. F. and Weil, J. W. (1944). *J. anim. Ecol.* **13,** 140–175.

Barnes, N. F. and Weil, J. W. (1945). *J. anim. Ecol.* **14,** 71–105.

Bequert, J. C. and Miller, W. B. (1974). "The Mollusks of the Arid Southwest with an Arizona Check List". University of Arizona Press, Tucson.

Berry, A. J. (1962). *Proc. malac. Soc. Lond.* **35,** 46–49.

Berry, A. J. (1963a). *Proc. malac. Soc. Lond.* **35,** 139–150.

Berry, A. J. (1963b). *Proc. zool. Soc. Lond.* **141,** 361–369.

Berry, A. J. (1964). *Proc. zool. Soc. Lond.* **142,** 655–663.

Berry, A. J. (1965a). *Proc. malac. Soc. Lond.* **36,** 221–228.

Berry, A. J. (1965b). *Proc. zool. Soc. Lond.* **144,** 219–227.

Berry, A. J. (1966). *J. Zool., Lond.* **150,** 11–27.

Bett, J. A. (1960). *Proc. zool. Soc. Lond.* **135,** 559–568.

Biggs, H. E. J. (1962). *J. Conch. Lond.* **25,** 64–72.

Bleakney, J. S. (1966). *Nautilus, Philad.* **79,** 131–134.

Blinn, W. C. (1963). *Ecology* **44,** 498–505.

Blinn, W. C. (1964). *Physiol. Zool.* **37,** 329–337.

Boettger, C. R. (1936). *C. R. XIIe Congr. Intern. Zool., Lisbon* **2,** 1033–1050.

Boettger, C. R. (1940). *Bull. Mus. r. Hist. nat. Belg.* **16,** 1–42.

Bonner, J. T. (1965). "Size and Cycle: an Essay on the Structure of Biology". Princeton University Press, Princeton.

Bourlière, F. (1957). *In* "The Biology of Ageing". (W. B. Yapp and G. H. Bourne, Eds). *Symp. Inst. Biol.* **6,** 27–34.

Boycott, A. E. (1934). *J. anim. Ecol.* **22,** 1–38.

Boycott, A. E. and Oldham, C. (1936). *Scott. Nat.* **1936,** 47–52.

Brown, O. S. (1978). *In* "Pulmonates" (V. Fretter and J. F. Peake, Eds) Vol. 2a. Academic Press, London and New York.

Brown, W. L. Jr. and Wilson, E. O. (1956). *Syst. Zool.* **5,** 49–64.

Bruggen, A. C. van (1967). *J. Conch. Lond.* **26,** 181–188.

Bruggen, A. C. van (1969). *Zool. Verh. Leiden* **103,** 1–116.

Bruggen, A. C. van (1973). *Malacologia* **14,** 419–424.

Bruggen, A. C. van (1977). *Malacologia* **16,** 75–80.

Bruijns, M. F. M. (1947). *Stn int. géobot. med. alp. Commun.* **96,** 1–59.

Bruijns, M. F. M., Altena, C. O. van R. and Botot, L. J. M. (1959). *Basteria* **23** (suppl.), 132–162.

Burch, J. B. (1955). *Nautilus, Philad.* **69,** 62–69.

Butler, A. J. (1976). *Oecologia* **25,** 349–371.

Cain, A. J. (1977a). *J. Conch. Lond.* **29,** 129–136.
Cain, A. J. (1977b). *Phil. Trans. R. Soc. B* **277,** 377–428.
Cain, A. J. and Currey, J. (1963). *Phil. Trans. R. Soc. B* **253,** 499–517.
Cain, A. J. and Sheppard, P. M. (1950). *Heredity, Lond.* **4,** 275–294.
Cain, A. J., Cameron, R. A. D. and Parkin, D. T. (1969). *Proc. malac. Soc. Lond.* **38,** 269–299.
Calow, P. (1977). *Adv. Ecol. Res.* **10,** 1–62.
Cameron, R. A. D. (1970a). *Proc. R. Soc. B* **176,** 131–159.
Cameron, R. A. D. (1970b). *J. Zool., Lond.* **160,** 143–157.
Cameron, R. A. D. (1970c). *J. Zool., Lond.* **162,** 303–315.
Cameron, R. A. D. (1973). *Malacologia* **14,** 355–370.
Carrick, R. (1938). *Trans. R. Soc. Edinb.* **59,** 563–597.
Carrick, R. (1942). *Ann. appl. Biol.* **29,** 43–55.
Cates, R. G. and Orians, G. H. (1975). *Ecology* **56,** 410–418.
Chaney, R. (1940). *Bull. geol. Soc. Am.* **51,** 469–488.
Chatfield, J. E. (1068). *Proc. malac. Soc. Lond.* **38,** 233–245.
Clarke, B. (1972). *Am. Nat.* **106,** 1–13.
Clarke, B. and Murray, J. (1969). *Biol. J. Linn. Soc.* **1,** 31–42.
Climo, F. M. (1975). *In* "Biogeography and Ecology in New Zealand" (G. Kuschel, Ed.). *Monographiae biol.* **27,** 459–492.
Comfort, A. (1957). *Proc. malac. Soc. Lond.* **32,** 219–241.
Comfort, A. (1964). "Ageing: the Biology of Senescence". Routledge and Kegan Paul, London.
Cook, A. (1977). *Anim. Behav.* **25,** 774–781.
Cook, L. M. and O'Donald, P. (1971). *In* "Ecological Genetics and Evolution" (R. Creed, Ed.) pp. 93–108. Blackwell, Oxford and Edinburgh.
Cook, L. M., Jack, T. and Pettitt, C. W. A. (1972). *Bolm. Mus. munic. Funchal* **26,** 5–30.
Cooke, A. H. (1895). Molluscs. *In* "Cambridge Natural History" Vol. 3, pp. 1–459.
Cooke, A. S. and Pollard, E. (1973). *Pest. Biochem. Physiol.* **3,** 230–236.
Cooke, C. M. and Kondo, Y. (1960). *Bull. Bernice P. Bishop Mus.* **221,** 1–303.
Corbet, G. S. (1962). *Sci. Progr. Lond.* **50,** 177–191.
Cox, L. R. (1960). *In* "Treatise on Invertebrate Paleontology" (R. C. Moore, Ed.) Part I, Mollusca 1, pp. 85–169. Geol. Soc. Am. and Univ. Kansas.
Cox, L. R., Nuttall, C. P. and Trueman, E. R. (1969). *In* "Treatise on Invertebrate Paleontology" (R. C. Moore, Ed.), Part N, Mollusca 6(1), N2–N129. Geol. Soc. Am. and Univ. Kansas.
Cracraft, J. (1973). *J. Zool., Lond.* **169,** 455–545.
Crawford-Sidebotham, T. J. (1972). *Oecologia* **9,** 141–154.
Croizat, L., Nelson, G. and Rosen, D. E. (1974). *Syst. Zool.* **23,** 265–287.
Crowell, H. H. (1973). *Proc. malac. Soc. Lond.* **40,** 491–504.
Dammerman, K. W. (1948). *Verh. K. ned. Akad. Wet.* **44,** 1–594.
Darlington, P. J. Jr. (1957). "Zoogeography: the Geographical Distribution of Animals. Wiley, New York.
Darlington, P. J. Jr. (1965). "Biogeography of the Southern End of the World". Harvard University Press, Cambridge.

Darwin, C. (1859). "On the Origin of Species". Murray, London.
Davis, G. (1978). *In* "Pulmonates" (V. Fretter and J. F. Peake, Eds) Vol. 2. Academic Press, New York and London.
Deevey, E. S. Jr. (1947). *Q. Rev. Biol.* **22,** 283–314.
Diamond, J. M. (1975). *In* "Ecology and Evolution of Communities" (M. L. Cody and J. M. Diamond, Eds) pp. 342–444. Belknap Press of Harvard University Press, Cambridge (Mass.) and London.
Dobzhansky, T. H. (1950). *Am. Sci.* **38,** 209–221.
Duncan, C. J. (1975). *In* "Pulmonates" (V. Fretter and J. F. Peake, Eds) Vol. 1, pp. 309–365. Academic Press, London and New York.
Edelstam, C. and Palmer, C. (1950). *Oikos* **2,** 259–270.
Ehrlich, P. R. and Raven, P. H. (1964). *Evolution, Lancaster, Pa.* **18,** 586–608.
Ellis, A. E. (1951). *J. Conch. Lond.* **23,** 171–244.
Elton, C. S. (1958). "The Ecology of Invasions by Plants and Animals". Methuen, London.
Evans, J. G. (1972). "Land Snails in Archaeology". Seminar Press, London and New York.
Faure, J. (1927). *Mém. Soc. Phys. Hist. nat. Genève* **40,** 171–434.
Fischer, P. H. (1887). "Manuel de Conchyliologie et de Paléontologie Conchyliologique". Libriaria F. Savi, Paris.
Fittkau, E. J. (1969). *In* "Biogeography and Ecology in South America". (E. J. Fittkau, J. Illies, H. Klinge, C. H. Schwabe, H. Sioli, Eds) Vol. 2a. *Monographia biol.* **19,** 624–658.
Fleming, C. (1975). *In* "Biogeography and Ecology in New Zealand" (G. Kuschel, Ed.). *Monographiae biol.* **27,** 1–86.
Forcart, L. (1956). *J. Linn. Soc. Zool.* **43,** 113–122.
Foster, R. (1958). *Parasitology* **48,** 261–268.
Fraenkel, G. S. and Gunn, D. L. (1961). "The Orientation of Animals". Dover, New York.
Francois, E., Riga, A. and Moeus, R. (1965). *Parasitica* **21,** 139–151.
Frömming, E. (1954). "Biologie der mitteleuropäischen Landgastropoden". Dunker and Humblot, Berlin.
Germain, L. (1925). *C. r. Congr. Socs sav. Paris Sect. Sci.* **1925,** 254–269.
Germain, L. (1928). *C. r. Soc. Biogéographie Paris* **2,** 1–19.
Germain, L. (1930). "Fauna de France 21". Lechevalier, Paris.
Geyer, D. (1927). "Unsere Land- und Süsswasser-Mollusken. Einführung in die Mollusken fauna Deutschlands" 3rd edition. Lutz' Verlag, Stuttgart.
Ghiselin, M. T. (1969). *Q. Rev. Biol.* **44,** 189–208.
Goodhart, C. B. (1962). *J. anim. Ecol.* **31,** 207–237.
Gould, S. J. (1969). *Bull. Mus. comp. Zool. Harv.* **138,** 407–531.
Gould, S. J. (1970). *Science, N.Y.* **168,** 572–573.
Grant, P. R. (1972). *Biol. J. Linn. Soc.* **4,** 39–68.
Grant, P. R. (1975). *Evolut, Biol.* **8,** 237–337.
Gressit, J. L. (1961). *Pacif. Insects* **2,** 1–94.
Gressit, J. L. and Yoshimoto, C. M. (1963). *In* "Pacific Basin Biogeography" (J. L. Gressit, Ed.) pp. 283–292. Bishop Museum Press, Honolulu.
Grime, J. P. and Blythe, G. M. (1969). *J. Ecol.* **57,** 45–66.

Grime, J. P., Dearman, R. S. and McPherson-Stewart, S. F. (1968). *J. Ecol.* **56,** 405–420.

Grime, J. P., Blythe, G. M. and Thornton, J. D. (1970). *In* "Animal Populations in Relation to their Food Resources" (A. Watson, Ed.). Blackwells, Oxford.

Hamilton, T. H., Rubinoff, I., Barth, R. H. Jr. and Bush, G. L. (1963). *Science, N.Y.* **142,** 1576–1577.

Harrington, H. J. (1965). *In* "Biogeography and Ecology in Antarctica" (V. Oye and V. Mieghem, Eds). *Monographie biol.* **15,** 1–71.

Harvey, P. H. (1973). *Proc. malac. Soc. Lond.* **40,** 523–530.

Harvey, P. H. (1974). *Proc. malac. Soc. Lond.* **41,** 57–64.

Hedley, C. (1899). *Proc. Linn. Soc. N.S.W.* **24,** 391–423.

Heller, J. (1975). *J. Linn. Soc. Zool.* **57,** 1–57.

Henderson, J. (1931). *Univ. Colo. Stud.* **28,** 177–186.

Henderson, J. (1935). *Spec. Pap. geol. Soc. Am.* **3.**

Hesse, P. (1931). *Zoologica, Stuttg.* **31,** 1–118.

Hesse, P. (1934). *Zoologica, Stuttg.* **33,** 1–59.

Hesse, R., Allee, W. C. and Schmidt, K. P. (1951). "Ecological Animal Geography" 2nd edition. Wiley, New York.

Holloway, J. D. (1974). In "Ecology and Biogeography in India" (M. S. Mani, Ed.). *Monographiae biol.* **23,** 473–499.

Holzapfel, E. P. and Gressitt, J. L. (1965). *Proc. Atmos. Biol. Comp.* 151–163.

Howes, N. H. and Wells, G. P. (1934). *J. exp. Biol.* **11,** 327–343.

Hubendick, B. (1953). *Ark. Zool.* (2) **6,** 83–96.

Hughes, N. F. (1973). *In* "Organisms and Continents Through Time" (N. F. Hughes, Ed.). *Spec. Pap. Palaeont.* **12,** 189–198.

Hunter, P. J. (1966). *J. anim. Ecol.* **35,** 543–547.

Hunter, P. J. (1968). *Malacologia* **6,** 379–389.

Hunter, P. J. (1978). *In* "Pulmonates" (V. Fretter and J. F. Peake, Eds), Vol. 2a. Academic Press, New York and London.

Hunter, P. J. and Symonds, B. V. (1971). *Nature, Lond.* **229,** 349.

Hutchinson, G. E. (1965). "The Ecological Theater and the Evolutionary Play". Yale University Press, Connecticut.

Hyman, L. H. (1967). "The Invertebrates, VI. Mollusca, I". McGraw Hill, New York.

Jain, S. K. (1976). *A. Rev. Ecol. Syst.* **7,** 469–495.

Jardine, N. (1973). *In* "Organisms and Continents Through Time" (N. F. Hughes, Ed.). *Spec. Pap. Palaeont.* **12,** 59–60.

Jennings, T. J. and Barkham, J. P. (1975a). *Oecologia* (*Berl.*) **20,** 279–286.

Jennings, T. J. and Barkham, J. P. (1975b). *Oikos* **26,** 211–221.

Jones, J. S. and Clarke, B. C. (1969). *J. Conch. Lond.* **27,** 3–8.

Josse, J. (1964). *Archs néerl. Zool.* **16,** 1–103.

Keast, J. A. (1976). *In* "Major Patterns in Vertebrate Evolution" (M. K. Hecht, P. C. Goody and B. M. Hecht, Eds) pp. 249–312. Plenum Press, New York and London.

Keferstein, W. M. (1866). *In* "Die Klassen und Ordnungen des Thier-Reichs wissenschaftlich dargestellt in Wort und Bild" (G. H. Bronn, Ed.). Bd III (2), 519–1500.

Kerney, M. P. (1966). *J. Conch. Lond.* **26,** 3–14.

Kerney, M. P. (1968). *Symp. zool. Soc. Lond.* **22,** 273–291.

Kerney, M. P. (1970). *J. Conch. Lond.* **27,** 145–148.

Kerney, M. P. (1976a). *Arch. Molluskenk.* **106,** 243–249.

Kerney, M. P. (1976b). "European Invertebrate Survey. Atlas of the Non-marine Mollusca of the British Isles". Institute of Terrestrial Ecology, Cambridge.

Kew, H. W. (1893). "The Dispersal of Shells". Kegan, Paul, Trench and Trubner, London.

Klemm, W. (1939). *Arch. Naturgesch.* N. F. **8,** 198–262.

Knipper, H. (1939). *Arch Naturgesch.* N.F. **8,** 327–517.

Knutson, L. V., Stephenson, J. W. and Berg, C. O. (1965). *Proc. malac. Soc. Lond.* **36,** 213–220.

Kobelt, W. (1897–1898). 'Studie zur Zoogeographie. Bd I–II'. Kreidel, Wiesbaden.

Kondo, Y. (1962). *Bull. Bernice P. Bishop Mus.* **224,** 1–49.

Kondo, Y. and Burch, J. B. (1971). *Malac. Rev.* **5,** 17–18.

Kornig, G. (1966). *Malakol. Abh.* **2,** 1–112.

Künkel, K. (1908). *Verh. dt. zool. Ges.* **18,** 153–161.

Künkel, K. (1916). Zur Biologie der Lungenschnecken. Winter, Heidelberg.

Künkel, K. (1928). *Zool. Jb. Zool.* **45,** 317–342.

Ladd, H. S. (1957). *J. Paleont.* **32,** 183–198.

Larembergue, M. de (1939). *Bull. biol. Fr. Belg.* **73,** 19–231.

Lewis, G. (1968). Polymorphism in shell characters of certain helicid molluscs, particularly the genus *Cochlicella*. D. Phil. thesis, Oxford.

Likharev, I. M. and Rammel'meier, E. S. (1952). "Terrestrial Mollusks of the U.S.S.R." (1962, translation published by the Israel Program for Scientific Translation, Jerusalem).

Lindroth, C. H. (1957). "The Faunal Connections Between Europe and North America". John Wiley, New York.

Lloyd, M. (1963). *J. anim. Ecol.* **32,** 157–163.

Lozek, V. (1962). *In* "Progress in Soil Zoology" (P. W. Murphy, Ed.) pp. 334–342. Butterworths, London.

Lozek, V. (1965). *Spec. Pap. geol. Soc. Am.* **80,** 201–218.

Lozek, V. (1967). *In* "Quaternary Palaeoecology" (E. J. Cushing and H. E. Wright, Eds). Proc. 7th, Congr. Intern. Ass. Quatern. Res., 381–392.

Lyth, M. (1972). Aspects of the spatial distribution of slugs (*Gastropoda; Pulmonata*) with some reference to their water relations. Ph.D. thesis, University of London.

MacArthur, R. H. (1960). *Am. Nat.* **94,** 25–36.

MacArthur, R. H. (1972). "Geographical Ecology". Harper and Row, New York.

MacArthur, R. H. and Wilson, E. O. (1967). "The Theory of Island Biogeography." Princeton University Press, Princeton.

McKenna, M. C. (1973). *In* "Implications of Continental Drift to Earth Sciences" (D. H. Tarling and S. K. Runcorn, Eds) Vol. 1, pp. 295–308. Academic Press, London and New York.

McMichael, D. F. (1968). *Bull. Bur. Miner. Resour. Geol. Geophys. Aust.* **80,** 135–156.

McMichael, D. F. and Iredale, T. (1969). *In* "Biogeography and Ecology in Australia" (A. Keast, R. L. Crocker, and C. S. Christian, Eds). Monographiae biol. **8,** 224–245.

Machin, J. (1975). *In* "Pulmonates" (V. Fretter and J. F. Peake, Eds) Vol. 1, pp. 105–163. Academic Press, New York and London.

Mandhl-Barth, G. (1950). *Abh. senckenb. naturforsch. Ges.* **469,** 1–93.

Mason, C. F. (1970a). *Oecologia* **4,** 358–373.

Mason, C. F. (1970b). *Oecologia* **5,** 215–239.

Mason, C. F. (1971). *Oecologia* **7,** 80–94.

Mason, C. F. (1974). *In* "Biology of Plant Litter Decomposition" (C. H. Dickinson and G. J. F. Pugh, Eds) Vol. 2, pp. 557–591. Academic Press, London and New York.

Matthew, W. D. (1915). *Ann. N.Y. Acad. Sci.* **24,** 171–318.

Mayr, E. (1942). "Systematics and the Origin of Species". Columbia University Press, New York.

Mayr, E. (1943). *Condor* **45,** 45–48.

Mayr, E. (1944). *Q. Rev. Biol.* **19,** 1–14.

Mead, A. R. (1950). *Bull. Mus. comp. Zool. Harv.* **105,** 219–291.

Mead, A. R. (1961). "The Giant African Snail: a Problem in Economic Malacology". University of Chicago Press, Chicago.

Mead, A. R. (1978). *In* "Pulmonates" (V. Fretter and J. F. Peake, Eds) Vol. 2b. Academic Press, New York and London.

Meier-Brook, C. and Smith, B. J. (1975). *Arch. Molluskenk.* **106,** 191–198.

Menard, H. W. and Hamilton, E. L. (1963). *In* "Pacific Basin Biogeography" (J. L. Gressit, Ed.) pp. 193–217. Bishop Museum Press, Honolulu.

Menge, B. A. and Sutherland, J. P. (1976). *Am. Nat.* **110,** 351–369.

Miles, H. W., Wood, J. and Thomas, I. (1931). *Ann. appl. Biol.* **18,** 370–400.

Mordan, P. B. (1973). "Aspects of the ecology of terrestrial gastropods at Monks Wood N.N.R. and other woodlands, with special reference to the Zonitidae. Ph.D. thesis, University of London.

Mordan, P. B. (1977). *Biol. J. Linn. Soc.* **9,** 59–72.

Mordan, P. B. (1978). *J. Conch. Lond.* **29,** 247–252.

Moreau, R. E. (1966). "The Bird Faunas of Africa and its islands". Academic Press, New York and London.

Morrison, J. P. E. (1955). *J. Wash. Acad. Sci.* **45,** 149–162.

Morton, J. E. (1954). *Proc. malac. Soc. Lond.* **31,** 30–46.

Murray, J. (1966). *Proc. malac. Soc. Lond.* **37,** 167–181.

Murray, J. and Clarke, B. (1976a). *Heredity, Lond.* **37,** 253–269.

Murray, J. and Clarke, B. (1976b). *Heredity, Lond.* **37,** 271–282

Nelson, G. (1974). *Syst. Zool.* **23,** 555–558.

Nemejc, F. (1970). *Paläont. Abh. Berl. B* **3,** 553–562.

Newell, P. F. (1965). *Anim. Behav.* **13,** 579.

Newell, P. F. (1971). *In* "Methods of Study in Quantitative Soil Ecology: Population, Production and Energy Flow". (J. Phillipson, Ed.). IBP Handbook 18, pp. 128–149.

Nisbet, R. H. (1974). *Proc. malac. Soc. Lond.* **41,** 171–183.

Økland, F. (1925). *Skr. norske Vidensk-Akad. mat.-nat. Kl.* **8,** 1–168.

Økland, F. (1930). *Z. Morph. Okol. Tierre* **16,** 748–804.
Oldham, C. (1929). *Proc. malac. Soc. Lond.* **18,** 143–144.
Oldham, C. (1934). *Proc. malac. Soc. Lond.* **21,** 131–138.
Oldham, C. (1942). *Proc. malac. Soc. Lond.* **25,** 9–10.
Oosterhoff, L. M. (1977). *Neth. J. Zool.* **27,** 1–132.
Owen, D. F. (1965). *Proc. zool. Soc. Lond.* **144,** 361–382.
Owen, D. F. (1967). *Revue Zool. Bot. afr.* **75,** 1–2.
Owiny, A. M. (1972). A comparative study of the genital system and the breeding biology of selected species of tropical land snails belonging to the superfamily Achatinaceae. Ph.D. thesis, University of London.
Owiny, A. M. (1974). *J. Zool., Lond.* **172,** 191–206.
Panella, G. and MacClintock, C. (1968). *J. Paleont.* **42,** (suppl.), 64–79.
Parker, H. W. and Grandison, A. G. C. (1977). "Snakes—a Natural History". British Museum (Natural History), London.
Parodiz, J. J. (1969). *Ann. Carneg. Mus.* **40,** 1–242.
Parodiz, J. J. and Bonetto, A. A. (1963). *Malacologia* **1,** 179–214.
Patterson, C. M. (1970). *J. Conch. Paris* **8,** 61–62.
Peake, J. F. (1960). *In* "Scolt Head Island" (J. A. Steers, Ed.) pp. 246–259. Heffers, Cambridge.
Peake, J. F. (1968). *Symp. zool. Soc. Lond.* **22,** 319–346.
Peake, J. F. (1969). *Phil. Trans. R. Soc. B* **255,** 285–306.
Peake, J. F. (1971). *Phil. Trans. R. Soc. B* **260,** 581–610.
Peake, J. F. (1973). *Malacologia* **14,** 303–312.
Pearl, R. (1928). "The Rate of Living". Knopf, New York.
Pelseneer, P. (1934). *Annls Soc. r. zool. Belg.* **64,** 93–104.
Pelseneer, P. (1935). "Essai d'Ethologie Zoologique". Palais des Acadèmies, Bruxelles.
Pianka, E. R. (1970). *Am. Nat.* **104,** 592–597.
Pilsbry, H. A. (1894). *Manual Conch.* **2,** 9.
Pilsbry, H. A. (1900). *Proc. Acad. nat. Sci. Philad.* **1900,** 568–581.
Pilsbry, H. A. (1911). *Rep. Princeton Univ. Exped. Patagonia* **3,** 513–633.
Pilsbry, H. A. (1927–1935). *Manual Conch.* (2) **28.**
Pilsbry, H. A. (1939–1948). *Acad. nat. Sci. Philad. Monogr.* **3,** Parts I–II.
Pilsbry, H. A. and Olsson, A. A. (1954). *Notul. Nat.* **266,** 1–4.
Plummer, J. M. (1975). *Proc. malac. Soc. Lond.* **41,** 395–413.
Pollard, E. (1973). *Oecologia* (*Berl.*) **12,** 209–212.
Pollard, E. (1975). *J. anim. Ecol.* **44,** 305–329.
Pollard, E., Cooke, A. S. and Welch, J. M. (1977). *J. Zool., Lond.* **183,** 269–279.
Pomeroy, D. E. (1967). *Trans. R. Soc. S. Aust.* **91,** 181–186.
Pomeroy, D. E. (1968). *Aust. J. Zool.* **16,** 857–869.
Pomeroy, D. E. (1969). *Aust. J. Zool.* **17,** 495–514.
Powell, A. W. B. (1950). *Rec. Aukland Inst. Mus.* **4,** 961–981.
Purchon, R. D. (1977). "The Biology of the Mollusca". 2nd edition. Pergamon Press, Oxford and New York.
Randolph, P. A. (1973). *Ecology* **50,** 933–955.
Raven, P. and Axelrod, D. I. (1972). *Science, N.Y.* **176,** 1379–1388.
Raven, P. and Axelrod, D. I. (1974). *Ann. Mo. bot. Gdn* **61,** 539–673.

Rees, W. J. (1965). *Proc. malac. Soc. Lond.* **36,** 269–282.

Rensch, B. (1932). *Z. Morph. Okol. Tiere* **25,** 757–807.

Rensch, B. (1937). *Z. indukt. Abstamm. VererbLehre* **72,** 564–588.

Rensch, B. (1959). "Evolution Above the Species Level". Methuen, London.

Rensch, I. (1934). *Arch Naturgesch. N.F.* **3,** 445–488.

Rensch, I. (1937). *Arch. Naturgesch. N.F.* **6,** 526–644.

Rensch, I. and Rensch, B. (1935). *Rev. suisse Zool.* **42,** 51–86.

Rex, M. A. (1972). *Breviora* **391,** 1–15.

Richards, O. W. (1961). *Symp. R. ent. Soc. Lond.* **1,** 1–10.

Richardson, A. M. M. (1975a). *Oecologia* **19,** 59–70.

Richardson, A. M. M. (1975b). *Oecologia* **19,** 141–164.

Richter, K. O. (1976). *J. moll. Stud.* **42,** 146–151.

Riedal, A. (1953). *Annls Mus. zool. pol.* **15,** 83–100.

Rigby, J. E. (1963). *Proc. zool. Soc. Lond.* **141,** 311–359.

Robertson, J. D. (1941). *Biol. Rev.* **16,** 106–133.

Roebuck, A. E. and Boycott, A. E. (1921). *J. Conch. Lond.* **16,** 165–212.

Rosen, D. E. (1975). *Syst. Zool.* **24,** 431–464.

Runham, N. W. (1975). *In* "Pulmonates" (V. Fretter and J. F. Peake, Eds) Vol. 1, pp. 53–104. Academic Press, London and New York.

Runham, N. W. and Hunter, P. W. (1970). "Terrestrial Slugs". Hutchinson, London.

Russell-Hunter, W. (1955). *Glasg. Nat.* **17,** 173–183.

Russell-Hunter, W. (1964). *In* "Physiology of Mollusca" (K. M. Wilbur and C. M. Yonge, Eds) Vol. 1, pp. 83–126. Academic Press, New York and London.

Russell-Hunter, W. D. (1970). "Aquatic Productivity". Macmillan, New York.

Russell-Hunter, W. D. (1978). *In* "Pulmonates" (V. Fretter and J. F. Peake, Eds.), Vol. 2. Academic Press, New York and London.

Sacchi, C. F. (1955). *Boll. Zool.* **22,** 50–78.

Sacchi, C. F. (1961). *Colloques int. Cent. natn. Res. scient.* **94,** 321–358.

Sacchi, C. F. (1965). Proc. 1st. Europ. Malac. Congress (L. R. Cox and J. F. Peake, Eds).

Schaffer, W. M. (1974). *Am. Nat.* **108,** 783–790.

Schifferli (1977). *Orn. Beob.* **74,** 71–74.

Schindel, D. E. and Gould, S. J. (1977). *Palaeobiology* **3,** 259–269.

Schmidt-Nielsen, K., Taylor, C. R. and Shkolnik, A. (1972). *Symp. zool. Soc. Lond.* **31,** 1–13.

Schopf, T. J. M., Raup, D. M., Gould, S. J. and Simberloff, D. S. (1975). *Palaeobiology* **1,** 63–70.

Schouw, J. F. (1823). Grundzüge einer Pflanzengeographie. Gebrudt und Verlegt, Berlin.

Sclater, P. L. (1858). *J. Proc. Linn. Soc.* **2,** 130–145.

Selander, R. K. and Hudson, R. O. (1976). *Am. Nat.* **110,** 695–718.

Selander, R. K. and Kaufman, D. W. (1973). *Proc. natn. Acad. Sci. U.S.A..* **70,** 1186–1190.

Selander, R. K. and Kaufman, D. W. (1975). *Evolution, Lancaster, Pa.* **29,** 385–401.

Selander, R. K., Kaufman, D. W. and Ralin, R. S. (1974). *Veliger* **16,** 265–270.
Simpson, C. T. (1894). *Proc. U.S. natn. Mus.* **17,** 423–450.
Simpson, G. G. (1947). *Evolution, Lancaster, Pa.* **1,** 218–220.
Smith, A. G. and Briden, J. C. (1977). "Mesozoic and Caenozoic Paleocontinental Maps". Cambridge University Press, Cambridge.
Smith, A. G., Briden, J. C. and Drewry, G. E. (1973). *In* "Organisms and Continents through Time" (N. F. Hughes, Ed.). *Spec. Pap. Palaeont.* **12,** 1–39.
Solem, A. (1959). *Fieldiana Zool.* **43,** 1–359.
Solem, A. (1968a). *Proc. malac. Soc. Lond.* **38,** 251–266.
Solem, A. (1968b). *Symp. on Mollusca* **1,** 231–247. Mar. Biol. Assoc., India.
Solem, A. (1969a). *Am. Malac. Union Ann. Repts* **1968,** 10–12.
Solem, A. (1972). *Proc. malac. Soc. Lond.* **40,** 93–114.
Solem, A. (1973a). *Malacologia* **14,** 397–400.
Solem, A. (1973b). *J. malac. Soc. Aust.* **2,** 377–385.
Solem, A. (1974). "The Shell Makers: Introducing Mollusks". Wiley, New York.
Solem, A. (1976). "Endodontoid Land Snails from Pacific Island. 1. Family Endodontidae". Field Museum of Natural History, Chicago.
Solem, A. (1976). *In* "Pulmonates" (V. Fretter and J. F. Peake, Eds.), Vol. 2a. Academic Press, New York and London.
South, A. (1964). *Ann. appl. Biol.* **53,** 251–259.
South, A. (1965). *J. anim. Ecol.* **34,** 403–417.
Southwood, T. R. E., May, R. M., Hassell, M. P. and Conway, G. R. (1974). *Am. Nat.* **108,** 791–804.
Sparks, B. W. (1961). *Proc. Linn. Soc. Lond.* **172,** 71–80.
Sparks, B. W. (1964a). *J. anim. Ecol.* **33,** 87–98.
Sparks, B. W. (1964b). *Proc. malac. Soc. Lond.* **36,** 7–25.
Sparks, B. W. (1969). *In* "Sciences in Archaeology" (D. Brothwell and E. Higgs, Eds) pp. 395–406. Thames and Hudson, London.
Sparks, B. W. and West, R. G. (1972). "The Ice Age in Britain". Methuen, London.
Stearns, S. C. (1976). *Q. Rev. Biol.* **51,** 3–47.
Steenberg, C. M., (1925). "Études sur l'Anatomie et la Systématique des Maillots (Fam. Pupillidae s. lat.)". Reitzel, Copenhagen.
Stephenson, J. W. (1966). *J. Conch. Lond.* **26,** 141–145.
Stephenson, J. W. (1968). *Proc. malac. Soc. Lond.* **38,** 169–178.
Stephenson, J. W. and Knutson, L. V. (1966). *J. econ. Biol.* **59,** 356–360.
Sutcliffe, J. F. (1962). "Mineral Salts Absorption in Plants". Pergamon Press, New York.
Taylor, D. W. (1960). *Prof. Pap. U.S. geol. Surv.* **337,** 1–94.
Taylor, D. S. (1965). *In* "The Quarternary of the United States" (H. E. Wright and D. G. Fry, Eds) pp. 596–611. University Press, Princeton.
Taylor, D. W. and Sohl, N. F. (1962). *Malacologia* **1,** 7–32.
Taylor, J. D., Kennedy, W. J. and Hall, A. (1969). *Bull. Br. Mus. nat. Hist.* Suppl. **3,** 1–125.
Taylor, J. D., Kennedy, W. J. and Hall, A. (1973). *Bull. Br. Mus. nat. Hist. Zool.* **22,** 253–294.

Taylor, J. D., Braithewaite, C. J. R., Peake, J. F. and Arnold, E. N. (in press).
Taylor, J. W. (1894–1921). "Monograph of the Land and Freshwater Mollusca of the British Isles". Taylor Bros., Leeds.
Te, G. A. (1972). *Malac. Rev.* **5,** 18–20.
Te, G. A. (1976). *Malac. Rev.* **9,** 39–53.
Tesclefs, R. R. (1961). *Annls Soc. r. zool. Belg.* **41,** 85–116.
Thomas, D. C. (1944). *Ann. appl. Biol.* **31,** 163–164.
Thomas, J. D. (1973). *Adv. Parasitol.* **11,** 307–394.
Thomas, J. D., Goldsworthy, G. J. and Benjamin, M. (1975). *J. Zool., Lond.* **175,** 421–437.
Thorne, R. F. (1963). *In* "Pacific Basin Biogeography" (J. L. Gressit, Ed.), pp. 311–350. Bishop Museum Press, Honolulu.
Tompa, A. S. (1976). *J. Morph.* **150,** 861–871.
Tweedie, M. W. F. (1961). *Bull. Raffles Mus.* **26,** 49–65.
Udvardy, M. D. F. (1969). "Dynamic Zoogeography". Van Nostrand Reinhold, New York.
Uminski, T. (1975). *Annls zool. Warsz.* **33,** 17–33.
Vagvolgyi, J. (1953). *Annls hist.-nat. Mus. natn. hung. N.S.* **3,** 101–104.
Vagvolgyi, J. (1968). *Bull. Mus. comp. Zool. Harv.* **136,** 145–254.
Vagvolgyi, J. (1976). *Syst. Zool.* **25,** 465–488.
Valentine, J. W. (1973). *In* "Organisms and Continents Through Time" (N. F. Hughes, Ed.). *Spec. Pap. Palaeont.* **12,** 79–92.
Valovirta, I. (1968). *Ann. Zool. Fenn.* **5,** 245–253.
Valovirta, I. (1977). *Malacologia* **16,** 271–277.
Verdcourt, B. (1963). *Palaeontographica A* **121,** 1–37.
Verdcourt, B. (1972). *J. Conch. Lond.* **27,** 291–348.
Voekler, J. (1959). *Mitt. hamb. zool. Mus. Inst.* **57,** 37–78.
Vuilleumier, F. (1975). *In* "Avian Biology" (D. Farner and J. R. King, Eds) Vol. 2, pp. 421–496. Academic Press, London and New York.
Wagge, L. E. (1952). *J. exp. Zool.* **120,** 311–342.
Waldén, H. W. (1955). *Ark. Zool.* **7,** 391–448.
Waldén, H. W. (1963). *In* "North Atlantic Biota and their History" (A. Löve and D. Löve, Eds) pp. 153–171. Pergamon Press, Oxford, London and New York.
Waldén, H. W. (1965). Proc. 1st. Europ. Malac. Congress (L. R. Cox and J. F. Peake, Eds) pp. 95–109.
Wallace, A. R. (1876). "The Geographical Distribution of Animals". Macmillan, London.
Wallace, A. R. (1881). "Island Life". Macmillan, London.
Wäreborn, I. (1969). *Oikos* **20,** 461–479.
Wäreborn, I. (1970). *Oikos* **21,** 285–291.
Watson, H. (1920). *Proc. malac. Soc. Lond.* **14,** 6–30.
Watson, H. (1923). *Proc. malac. Soc. Lond.* **15,** 270–280.
Watson, H. (1934). *J. Conch. Lond.* **20,** 33–42.
Webbe, G. and Read, W. W. (1962). *Ann. Trop. Med. Parasit.* **56,** 206–209.
Webley, D. (1964). *Ann. appl. Biol.* **53,** 407–414.
Webley, D. (1965). *Ann. appl. Biol.* **56,** 37–45.

Wegner, A. (1929–1936). "Die Entstehung der Kontinente und Ozeane". Die Wissenschaft **66,** Vieweg, Braunschweig.

Wenz, W. (1923–1930). Fossilium Cat. **1,** 4 Bände, 3387 pp., Berlin.

West, R. G. (1968). "Pleistocene Geology and Biology with Special Reference to the British Isles". Longmans, London.

Whittaker, R. H. (1970). "Communities and ecosystems". Macmillan, New York.

Whittaker, R. H., Levin, S. A. and Root, R. B. (1973). *Am. Nat.* **107,** 321–338.

Wilbur, H. M., Tinke, D. W. and Collins, J. P. (1974). *Am. Nat.* **108,** 805–817.

Wild, S. V. and Lawson, A. K. (1937). *J. Conch. Lond.* **20,** 351–361.

Williams, E. E. (1969). *Q. Rev. Biol.* **44,** 345–389.

Williams, N. V. (1970a). *Malacologia* **10,** 153–162.

Williams, N. V. (1970b). *Malacologia* **10,** 165–180.

Williamson, M. H. (1959). *J. anim. Ecol.* **28,** 153–155.

Williamson, P. and Cameron, R. A. D. (1976). *Oikos* **27,** 493–500.

Williamson, P., Cameron, R. A. D. and Carter, M. A. (1976). *Nature, Lond.* **263,** 496–497.

Wilson, E. O. (1969). *In* "Diversity and Stability in Ecological Systems". *Brookhaven Symp. Biol.* **22,** 38–47.

Wolda, H. (1963). *Archs néerl. Zool.* **15,** 381–471.

Wolda, H. (1967). *Evolution, Lancaster, Pa.* **21,** 117–129.

Wolda, H. (1971a). *In* "Dynamics of Populations" (P. J. den Boer and G. R. Gradwell, Eds) pp. 98–108. Centre for Agric. Publish. Docum., Wageningen.

Wolda, H. (1971b). *Res. Pop. Ecol.* **12,** 185–204.

Wolda, H. (1972). *Neth. J. Zool.* **22,** 428–455.

Wolda, H. and Kreulen, D. (1973). *Neth. J. Zool.* **23,** 168–188.

Wolda, H., Zweep, A. and Schuitema, K. A. (1971c). *Oecologia (Berl.)* **7,** 361–381.

Wollaston, T. V. (1878). "Testacea Atlantica". Reeve. London.

Woodward, S. P. (1856). "A Manuel of Mollusca". Weale, London.

Wourns, J. P. (1972). *J. exp. Zool.* **182,** 389–414.

Wright, C. A. (1971). "Flukes and Snails". Allen and Unwin, London.

Wurtz, C. B. (1955). *Proc. Acad. nat. Sci. Philad.* **108,** 95–143.

Yeu, T. C. (1943). *Proc. Acad. nat. Sci. Philad.* **95,** 267–308.

Yom-Tov, Y. (1970). *Ecology* **51,** 907–911.

Yom-Tov, Y. (1971a). *Israel J. Zool.* **20,** 231–248.

Yom-Tov, Y. (1971b). *Proc. malac. Soc. Lond.* **39,** 319–326.

Yom-Tov, Y. (1972). *J. anim. Ecol.* **41,** 17–22.

Zilch, A. (1959–60). "Gastropoda: Euthyneura". *In* "Handbuch der Paläozoologie", Bd 6, Teil 2, Lief 1–4, 834 pp.

Zimmerman, E. C. (1948). *Insects Hawaii* **1,** Introduction.

Appendix

A collection of 4 schemes for the classification of pulmonates is presented. Attention has been drawn to major differences in the arrangement or status of families by numbering those taxa whose alleged affinities have varied; minor changes in the order of superfamilies or families from that proposed by the original authors have been introduced, thereby increasing the clarity of the chart, while not affecting the content of any group.

References

Hubendick, B. (1978). Chapter 1, this volume.

Solem, A. (1978). Chapter 2, this volume.

Taylor, D. W. and Sohl, N. F. (1962). *Malacologia* **1**, 7–32.

Thiele, J. (1931). "Handbuch der Systematischen Weichtierkunde". Fischer, Jena.

Zilch, A. (1959–1960). Euthyneura. *In* "Gastropoda" Vol. 2. (W. Wenz, Ed.), Handbuch der Paläozoologie 6. Borntraeger, Berlin-Nikolassee.

Abbreviations

SO = Superorder
O = Order
So = Suborder
IO = Infraorder
SF = Superfamily
* = Family known only from fossils
? = Taxonomic position or status uncertain
(incl. . . .) Taxa (family, subfamily, tribe or genus) included with nominate family
(= . . .) Alternative name applied to nominate family

Thiele (1931)	Zilch (1959–1960)	Taylor and Sohl (1962)	Solem (1978)	Examples
Subclass Pulmonata			Subclass Pulmonata	
O STYLOMMATOPHORA	O SOLEOLIFERA	O SOLEOLIFERA	SO SYSTELLOMMATOPHORA	
Stirps Onchidiacea		SF Onchidiacea	O Onchidiacea	
Onchidiidae	Onchidiidae	Onchidiidae	Onchidiidae	
Stirps Soleolifera		SF Veronicellacea	O Soleolifera	
Vaginulidae	Veronicellidae	Veronicellidae	Veronicellidae (= Vaginulidae)	
Rathouisiidae	Rathouisiidae	Rathouisiidae	Rathouisiidae	
			Hubendick (1978)	
O BASOMMATOPHORA	O BASOMMATOPHORA	O BASOMMATOPHORA	O BASOMMATOPHORA	
Stirps Actophila	SF Ellobiacea	SF Ellobiacea	SF Ellobiacea	
Ellobiidae	Ellobiidae	Ellobiidae	Ellobiidae	*Carychium*, *Melampus*, *Pythia*
Otinidae	Otinidae	Otinidae	Otinidae	*Otina*
Stirps Amphibolacea	SF Amphibolacea	SF Amphibolacea	SF Amphibolacea	
Amphibolidae	Amphibolidae	Amphibolidae	Amphibolidae	*Salinator*, *Amphibola*
Stirps Patelliformia	SF Siphonariacea	SF Siphonariacea	SF Siphonariacea	
Siphonariidae	Siphonariidae	Siphonariidae	Siphonariidae	*Siphonaria*
Gadiniidae	Trimusculidae (= Gardiniidae	Trimusculidae	Trimusculidae	*Trimusculus*
	*? Acroreiidae	*? Acroreiidae		*Acroreia*
Stirps Hygrophila	SF Lymnaeacea		SF Lymnaeacea	
		SF Unnamed		
Chilinidae	Chilinidae	Chilinidae	Chilinidae	*Chilina*
Latiidae	Latiidae	Latiidae	Latiidae	*Latia*
		*? Payettiidae		
		SF Acroloxacea		
	Acroloxidae[1]	Acroloxidae[1]	Acroloxidae[1]	*Acroloxus*
		SF Lymnaeacea		

Lymnaeidae (incl. Lancinae)	Lymnaeidae (incl. Lancinae)	Lymnaeidae	Lymnaeidae (incl. Lancinae)	*Radix*, *Stagnicola*, *Galba*, *Fossaria*
		Lancidae		*Lanx*
		SF Ancylacea		
Physidae	Physidae	Physidae	Physidae	*Physa*, *Aplexa*

Thiele (1931)	Zilch (1959–1960)	Taylor and Sohl (1962)	Hubendick (1978)	Examples
	Rhodacmeidae		Ancyloplanorbidae (incl. Rhodacmeinae, Bulininae, *Ferrissia*, Planorbinae, Ancylini	
Ancylidae (incl. Ancylinae, *Protoancylus* (*Ferrissia*), *Rhodacmea*, Acroloxinae)	Ferrissiidae	Ancylidae (incl. Ferrissiidae, Rhodacmeidae)		*Ancylus*
Planorbidae (incl. *Neoplanorbis*)	Planorbidae	Planorbidae (incl. Neoplanorbidae)		*Bulinus*, *Anisus*, *Gyraulus*, *Biomphalaria*, *Helisoma*
	Neoplanorbidae			*Neoplanorbis*

Thiele (1931)	Zilch (1959–1960)	Taylor and Sohl (1962)	Solem (1978)	Examples
O STYLOMMATOPHORA	O STYLOMMATOPHORA	O STYLOMMATOPHORA	SO STYLOMMATOPHORA	
	So ORTHURETHRA	So ORTHURETHRA	O ORTHURETHRA	
Stirps Achatinellacea	SF Achatinellacea	SF Achatinellacea	SF Achatinellacea	
Achatinellidae	Achatinellidae	Achatinellidae (incl. Tornatellinidae)	Achatinellidae (incl. Tornatellinidae)	*Elasmias*
Tornatellinidae	Tornatellinidae			
	Partulidae[2]	Partulidae[2]		*Partula*
Stirps Vertiginacea	SF Pupillacea[2]	SF Cionellacea	SF Cionellacea	
Amastridae	Amastridae	Amastridae	Amastridae	*Leptachatina*, *Carelia*
Cochlicopidae	Cochlicopidae	Cionellidae (= Cochlicopidae)	Cionellidae (= Cochlicopidae)	*Azeca*, *Cochlicopa*
		SF Pupillacea	SF Pupillacea	

Thiele (1931)	Zilch (1959–1960)	Taylor and Sohl (1962)	Solem (1978)	Examples
Vertiginidae (incl. Pupillinae, Pyramiduleae, Orculinae, Chondrininae)	Pupillidae Vertiginidae Orculidae Chondrinidae	Pupillidae Vertiginidae Orculidae Chondrinidae	Pupillidae (incl. Vertiginidae, Orculidae, Chondrinidae)	*Lauria*, *Leiostyla*, *Argua*, *Columella*, *Nesopupa*, *Sterkia*, *Orcula*, *Pagodulina* *Abida*, *Gastrocopta*, *Fauxulus*
Pleurodiscidae	Pleurodiscidae	Pleurodiscidae	Pleurodiscidae (incl. Pyramidulidae)	*Pleurodiscus*
	Pyramidulidae	Pyramidulidae		*Pyramidula*
Valloniidae (incl. *Strobilops*)	Valloniidae (incl. Strobilopsinae)	Valloniidae (incl. Strobilopsidae)	Valloniidae Strobilopsidae	*Pupisoma*, *Acanthinula* *Strobilops*
			SF Partulacea Partulidae[2]	
Enidae	Enidae	Enidae	Enidae (= Buliminidae)	*Chondrula*, *Jaminia*, *Zebrina*, *Buliminus*, *Rachis*
		O MESURETHRA SF Clausiliacea	O MESURETHRA SF Clausiliacea	
Clausiliidae[3]		Clausiliidae[3]	Clausiliidae[3]	*Phaedusa*, *Nenia*, *Acopia*, *Balea*
		Ceriidae (= Cerionidae)[4]	Cerionidae[4]	*Cerion*
		Megaspiridae[5]		
		*? Filholiidae[6]		*Filholia*
		SF Corillacea Corillidae[7]		
		SF Strophocheilacea	SF Strophocheilacea	
		Dorcasiidae[8]	Dorcasiidae[8]	*Trigonephrus*, *Dorcasia*
		Strophocheilidae[9]	Strophocheilidae[9]	*Strophocheilus*, *Gonyostomus*
	SO HETERURETHRA	SO HETERURETHRA		
Stirps Succineacea	SF Succineacea	SF Succineacea		

	Succineidae[10]		Succineidae[10]		Succineidae[10]			*Quickia*, *Catinella*,
								Oxyloma
					Aillyidae[11]			*Aillya*
Stirps	Tracheopulmonata	SF	Athoracophoracea	SF	Athoracophoracea			
	Athoracophoridae[12]		Athoracophoridae[12]		Athoracophoridae[12]			
		So	SIGMURETHRA	So	SIGMURETHRA	O	SIGMURETHRA	
		SF	Clausiliacea					
			Clausiliidae[3]					
		?	Filholiidae[6]					
				IO	HOLOPODOPES	So	HOLOPODOPES	
Stirps	Achatinacea	SF	Achatinacea	SF	Achatinacea	SF	Achatinacea	
	Ferussaciidae		Ferussaciidae		Ferussaciidae		Ferussaciidae	*Cecilioides*, *Ferussacia*
	Subulinidae		Subulinidae		Subulinidae		Subulinidae	*Homorus*, *Lamellaxis*,
								Opeas, *Ruminia*
					Spiraxidae[13] (part Oleacinidae *sensu* Zilch incl. Spiraxinae, Streptostyleae, Euglandineae)		Spiraxidae[13]	*Volutaxis*, *Spiraxis*
	Megaspiridae[5]		Megaspiridae[5]			?	Megaspiridae[5]	*Megaspira*, *Perrieria*
	Achatinidae		Achatinidae		Achatinidae		Achatinidae	*Burtoa*, *Limicolaria*,
								Archachatina
Stirps	Streptaxacea	SF	Streptaxacea	SF	Streptaxacea	SF	Streptaxacea	
	Streptaxidae		Streptaxidae		Streptaxidae		Streptaxidae	*Gonaxis*, *Gulella*, *Ennea*
				SF	Rhytidacea	SF	Rhytidacea	
	Haplotrematidae		Haplotrematidae		Haplotrematidae		Haplotrematidae	*Haplotrema*
		?	Systrophiidae[14]				Systrophiidae[14]	*Systrophia*, *Happia*
	Paryphantidae		Rhytididae		Rhytididae		Rhytididae (= Paryphantidae)	*Ougapia*, *Natalina*, *Paryphanta*
	Aperidae		Chlamydephoridae (= Aperidae)		Chlamydephoridae		Aperidae (= Chlamydephoridae)	*Chlamydephorus*
Stirps	Acavacea							
	Macrocyclidae[15]					?	Macrocyclidae[15]	*Macrocyclis*
		SF	Acavacea			SF	Acavacea	

	Thiele (1931)	Zilch (1959–1960)	Taylor and Sohl (1962)	Solem (1978)	Examples
	Acavidae	Acavidae (incl.	Acavidae (part *sensu*	Acavidae (incl.	*Helicophanta*, *Acavus*,
	Clavatoridae	Strophocheileae,[9]	Zilch; incl.	Clavatoridae)	*Clavatus*
	Dorcasiidae[8]	Dorcasiidae,[8]	Caryodidae,		
	Strophocheilidae[9]	Caryodidae,	Macrocyclidae[15])		
		Macrocyclis,[15] *Clavator*)			
	Caryodidae			Caryodidae	*Hedleyella*, *Caryodes*
Stirps	Bulimulacea	SF Bulimulacea	SF Bulimulacea	SF Bulimulacea	
	Urocoptidae	Urocoptidae	Urocoptidae	Urocoptidae	*Eucalodium*, *Holospira*
	Bulimulidae (incl.	Bulmulidae	Bulimulidae	Bulimulidae (incl.	*Drymaeus*, *Bothriembrion*,
	Orthalicinae,	Orthalicidae	Orthalicidae	Orthalicidae,	*Placostylus*, *Sultana*,
	Odontostominae,			Amphibulimidae,	
				Odontostomidae)	*Liguus*
	Amphibuliminae,	Odontostomidae	Odontostomidae		*Anostoma*
	Aillya,[11]	Amphibulimidae (incl.	Amphibulimidae		*Simpulopsis*
	Bulimulinae,	*Aillya*[11])			
	Partula[2])				
	Cerionidae[4]	Cerionidae[4]			*Cerion*
		*? Anadromidae	*? Anadromidae		
		*? Grangerellidae			
			IO AULACOPODA	So AULACOPODA	
Stirps	Zonitacea	SF Endodontacea	SF Endodontacea	SF Arionacea	
	Endodontidae (incl.	Endodontidae (incl.	Endodontidae (incl.	Endodontidae	
	Laominae, *Punctum*,	Endodontinae,	Charopidae, Punctidae)	Charopidae (incl.	
	Endodontinae,	*Charopa*,		Otoconchidae[16])	
	Charopa)	Punctinae)		Punctidae	*Punctum*, *Laoma*
				Helicodiscidae	
				Discidae	*Discus*
	Arionidae	Arionidae	Arionidae	Arionidae	*Arion*, *Geomalacus*
	Philomycidae	Philomycidae	Philomycidae	Philomycidae	*Philomycus*
		Otoconchidae[16]	Otoconchidae[16]		*Otoconcha*
	Thyrophorellidae	? Thyrophorellidae	? Thyrophorellidae		*Thyrophorella*
				SF Succineacea	
				Succineidae[10]	

				Athoracophoridae[12]	
		SF Zonitacea	SF Zonitacea	SF Limacacea Group "B"	
	Zonitidae	Zonitidae	Zonitidae	Zonitidae (incl. Trochomorphidae,[17] Vitrinidae)	*Vitrea*, *Oxychilus*, *Poecilozonites*,
	Vitrinidae (incl. *Vitrinopsis* (*Otoconcha*)[16])	Vitrinidae	Vitrinidae		*Vitrina*, *Semilimax*, *Phenacolimax*
				? Thyrophorellidae	
	Trigonochlamydidae	Trigonochlamydidae	Trigonochlamydidae	Trigonochlamydidae	*Trigonochlamys*
		Parmacellidae	Parmacellidae	Parmacellidae	*Parmacella*
	Limacidae (incl. Parmacellinae, *Milax*)	Limacidae	Limacidae	Limacidae (incl. Milacidae)	*Lehmannia*, *Deroceras*, *Limax*
		Milacidae	Milacidae		*Milax*
	Systrophiidae[14]		? Systrophiidae[14]		
			SF Testacellacea		
			Testacellidae[18]	? Testacellidae[18]	*Parmacellina*, *Testacella*
	Polygyridae[19] (incl. Ammonitellinae[20])				
	Sagdidae[21] (incl. Thysanophorinae[22])				
	Corillidae[7]				
Stirps	Ariophantacea	SF Ariophantacea	SF Ariophantacea	SF Limacacea Group "A"	
		Helicarionidae	Helicarionidae	Helicarionidae (incl. Euconulidae, Ariophantidae)	*Kaliella*, *Ryssota*, *Erepta*
		Euconulidae	Euconulidae		*Microcystis*, *Mendana*
		Urocyclidae	Urocyclidae	Urocyclidae	*Trochonania*, *Africarion*,
	Ariophantidae (incl. Helicarioninae, Kaliellinae *Euconulus*, Urocyclinae, Trochomorphinae[17])	Ariophantidae	Ariophantidae		*Dyakia*, *Nanina*, *Macrochlamys*
		Trochomorphidae[17]	Trochomorphidae[17]		*Trochomorpha*, *Bertia*
				Aillyidae[11]	

Thiele (1931)	Zilch (1959–1960)	Taylor and Sohl (1962)	Solem (1978)	Examples
		IO HOLOPODOPES	So HOLOPODOPES	
	SF Polygyracea	SF Polygyracea	SF Polygyracea	
	Polygyridae[19] (incl. Thysanophorinae[22])	Polygyridae[19]	Polygyridae[19]	*Mesodon*, *Allogona*, *Triodopsis*
	Corillidae		Corillidae (= Plectopylidae)[7]	*Sculptaria*, *Plectopylis*
		? Thysanophoridae[22]		*Microconus*, *Thysanophora*
		? Ammonitellidae[20]		
		SF Oleacinacea		
	Sagdidae[21]	Sagdidae[21]	Sagdidae[21]	*Sagda*
Stirps Oleacinacea	SF Oleacinacea		?SF Oleacinacea	
Oleacinidae (incl. *Spiraxis*[13])	Oleacinidae (incl. Spiraxinae,[13] Oleacininae [=Varicelleae + Streptostyleae[13] + Euglandinae] [13])	Oleacinidae (part Oleacininae *sensu* Zilch, incl. Varicelleae)	Oleacinidae	*Varicella*
Testacellidae[18]	Testacellidae[18]			
Stirps Helicacea	SF Helicacea	SF Helicacea		
			SF Camaenacea	
Pleurodontidae	Camaenidae (incl. Ammonitellinae[20] Oreohelicinae)	Camaenidae	Camaenidae	*Polydontes*, *Chloritis*, *Papuina*
		Oreohelicidae	Oreohelicidae	*Oreohelix*
			Ammonitellidae[20]	*Glyptostoma*, *Megomphix*
			SF Helicacea	
Fruticicolidae	Bradybaenidae	Bradybaenidae	Bradybaenidae	*Helicostyla*, *Cochlostyla*, *Euhadra*
	Helminthoglyptidae	Helminthoglyptidae	Helminthoglyptidae	*Epiphragmophora*, *Sonorelix*, *Cepolis*
Helicidae	Helicidae	Helicidae	Helicidae	*Sphincterochila*, *Otala*, *Cepaea*

Subject Index

Page numbers in italic refer to figures or tables.

Q

R

S